T0313195

Introduction to Natural Products Chemistry

Edited by

Rensheng Xu
Yang Ye
Weimin Zhao

科学出版社
Science Press

CRC Press
Taylor & Francis Group
Boca Raton   London   New York

CRC Press is an imprint of the
Taylor & Francis Group, an **informa** business

The original Chinese language work has been published by:
SCIENCE PRESS, Beijing
© 2010 by Science Press
All rights reserved.

CRC Press
Taylor & Francis Group
6000 Broken Sound Parkway NW, Suite 300
Boca Raton, FL 33487-2742

© 2012 by Taylor & Francis Group, LLC
CRC Press is an imprint of Taylor & Francis Group, an Informa business

No claim to original U.S. Government works

Printed in the United States of America on acid-free paper
Version Date: 20110608

International Standard Book Number: 978-1-4398-6076-2 (Hardback)

This book contains information obtained from authentic and highly regarded sources. Reasonable efforts have been made to publish reliable data and information, but the author and publisher cannot assume responsibility for the validity of all materials or the consequences of their use. The authors and publishers have attempted to trace the copyright holders of all material reproduced in this publication and apologize to copyright holders if permission to publish in this form has not been obtained. If any copyright material has not been acknowledged please write and let us know so we may rectify in any future reprint.

Except as permitted under U.S. Copyright Law, no part of this book may be reprinted, reproduced, transmitted, or utilized in any form by any electronic, mechanical, or other means, now known or hereafter invented, including photocopying, microfilming, and recording, or in any information storage or retrieval system, without written permission from the publishers.

For permission to photocopy or use material electronically from this work, please access www.copyright.com (http://www.copyright.com/) or contact the Copyright Clearance Center, Inc. (CCC), 222 Rosewood Drive, Danvers, MA 01923, 978-750-8400. CCC is a not-for-profit organization that provides licenses and registration for a variety of users. For organizations that have been granted a photocopy license by the CCC, a separate system of payment has been arranged.

Trademark Notice: Product or corporate names may be trademarks or registered trademarks, and are used only for identification and explanation without intent to infringe.

Visit the Taylor & Francis Web site at
http://www.taylorandfrancis.com

and the CRC Press Web site at
http://www.crcpress.com

Introduction to Natural Products Chemistry

Contents

Preface

This book is the revision of Science Press of China in 2006. It is one in the series "Basic Modern Chemistry" used to introduce modern chemistry and as a reference book for scientists and graduate students.

China is a country that is rich in natural resources because of its vast territory, diverse geography, and different climates. Traditional Chinese Medicine (TCM) is very important to the health of the Chinese and has been practiced in China for more than 5,000 years. Most TCM products come from plants. Some also come from animals, marine products, and minerals. Natural products chemistry is one of the most developed scientific fields in China and almost every Chinese university has a research or teaching unit for natural products chemistry or chemistry of natural medicines. The discovery of the remarkable antimalarial drug, artemisinin (Qinghausu) and its derivatives from TCM herbs, *Artemisia annua*, encouraged many scientists around the world to study TCM herbs. Chinese scientists have published papers in many well known scientific journals including *Nature, Science, Journal of the American Chemical Society, Tetrahedron, Journal of Natural Products, Phytochemistry*, as well as many journals in China.

This book has collected the most important research results of natural products chemistry in China. It covers the basic principles of isolation, structure, characteristics, and the current research techniques of structure elucidation with wet chemistry and spectroscopic analyses (UV, IR, MS and NMR, especially 2d-NMR, HMBC and HMQC), bioactivity, biosynthesis, and chemical synthesis. These concepts are illustrated with examples from the authors' own scientific experiments. All the authors are scientists from Shanghai Institute of Materia Medica, Chinese Academy of Sciences; Shanghai Institute of Pharmaceutical Industry; College of Pharmacy, Fudan University; and Chinese Pharmaceutical University. They are professors and experts in their fields.

Many thanks to Dr. Wenwen Ma and Mr. Michael Zahn of Washington State, USA, for reviewing the English version and providing editorial help.

Editors

2010 May

Contributors

Preface

Ren-Sheng Xu, Professor of Shanghai Institute of Materia Medica, Chinese Academy of Sciences.

E-mail: rsxu09@gmail.com

Chapter 1   Introduction

Ren-Sheng Xu, Professor of Shanghai Institute of Materia Medica, Chinese Academy of Sciences.

Chapter 2   Extraction and Isolation of Natural Products

Wei-Min Zhao, Professor of Shanghai Institute of Materia Medica, Chinese Academy of Sciences.

E-mail: wmzhao@mail.shcnc.ac.cn

Chapter 3   Chemistry of Fungal Products

Zhi-Hong Xu, Visiting professor at Duke University, Department of Chemistry, USA

E-mail: Zxu9@yahoo.com

Chapter 4   Alkaloids

Ren-Sheng Xu, Professor of Shanghai Institute of Materia Medica, Chinese Academy of Sciences.

Yang Ye, Professor of Shanghai Institute of Materia Medica, Chinese Academy of Sciences.

E-mail: yye@mail.shcnc.ac.cn

Chapter 5   Sesquiterpenoids

Zhong-Liang Chen, Professor of Shanghai Institute of Materia Medica, Chinese Academy of Sciences.

E-mail: zlchen70@hotmail.com

Chapter 6   Diterpenoids

Zhi-Da Min, Professor at Chinese Pharmaceutical University, College of Chinese Medicines.

E-mail: ZHDMIN32@hotmail.com

Chapter 7   Saponins

Wei-Min Zhao, Professor of Shanghai Institute of Materia Medica, Chinese Academy of Sciences.

Chapter 8   Amino-acids and Peptide

Chang-Qi Hu, Professor at Fudan University, College of Pharmaceutical Sciences.

E-mail: huchangqi@gmail.com

Jie-Cheng Xu, Professor of Shanghai Institute of Organic Chemistry, Chinese Academy of Sciences.

Chapter 9   Flavonoids
De-Yun Kong, Professor of Shanghai Institute of Pharmaceutical Industry.
E-mail: deyunk@yahoo

Chapter 10   Anthraquinones
Yang Lu, Professor at Jiaotong University, Department of Chemistry.
E-mail: huaxue@shsmu.edu.cn

Chapter 11   Coumarins
Ze-Nai Chen, Professor at Jiaotong University, Department of Chemistry.
E-mail: huaxue@shsmu.edu.cn

Chapter 12   Lignans
Chang-Qi Hu, Professor at Fudan University, College of Pharmaceutical Sciences.
E-mail: huchangqi@gmail.com

Chapter 13   Other Natural Bioactive Compounds
Ren-Seng Xu, Professor of Shanghai Institute of Materia Medica, Chinese Academy of Sciences.

Chapter 14   Marine Natural Products
Yue-Wei Guo, Professor of Shanghai Institute of Materia Medica, Chinese Academy of Sciences.
E-mail: ywguo@mail.shcnc.ac.cn
Wen Zhang Associate professor of School of Pharmacy, Second Military Medical University.
E-mail: zhangwen68@hotmail.com

Chapter 15   Structural Modification of Active Principles from TCM
Da-Yuan Zhu, Professor of Shanghai Institute of Materia Medica, Chinese Academy of Sciences.
E-mail: dyzhu@mail.shcnc.ac.cn

Chapter 16   Chemical Synthesis of Natural Products
Wen-Hu Duan, Professor of Shanghai Institute of Materia Medica, Chinese Academy of Sciences.
E-mail: whduan@mail.shcnc.ac.cn

CHAPTER 1

Introduction

Natural products chemistry is the chemistry of metabolite products of plants, animals, insects, marine organisms and microorganisms. The metabolic products include alkaloids, flavonoids, terpenoids, glycosides, amino acids, proteins, carbohydrates etc. The applications of natural products range from medicines, to sweeteners and pigments. The development of new technology, including information technology, promotes the mutual penetration of different scientific fields. Natural products chemistry has been involved in the investigation of many biological phenomena, such as drug mechanisms to gametophytes and receptors, drug metabolism in the human body, and protein and enzyme chemistry. Natural products chemistry is also associated with the chemistry of endogenous products and biochemistry. The book titled *Comprehensive Natural Products Chemistry* and edited by D.H. R. Barton et al., mainly expounded on the chemistry of enzyme, protein, DNA, RNA, polysaccharide and other life related compounds.

Currently, one third of clinically used drugs come from nature. They are either directly isolated from natural products, synthesized or semi-synthesized by structural modification of their natural compounds. High-throughput screening, combinatorial chemistry, computer chemistry, gene chip research and gene-drug research are promoting new drug discovery. However, these are no substitutes searching for new drugs from natural products, because a new structure with unique bioactivity can only be found from natural products. The well-known aspirin is derived from salicylic acid isolated from plants by acetylation. Local anesthetic procaine is derived from cocaine, a plant active principle. In China, our ancestors found a crystal substance from *Aconitum carmichaeli* back in the early 17th century, but the systematic study of aconitine began in the 19th century by German chemists. Well-known plant drugs such as the pain killer morphine, the antitussive codeine, the antimalarial quinine, the parasympathetic inhibitor atropine, the anticholinergic scopolamine, the antispasmodic hyosyamine, the pupil shrinking pilocarpine, the cholinergic physostigmine, the antiasthmatic ephedrine, the uterine contractor ergometrine, the antihelminthic santonin, the cardiotonic digoxin, deslanoside and others are all isolated from plants. Penicillin, streptomycin and many antibiotics come from fermentation materials of microorganisms. Most of them are still used in clinics. In the 1950s, the discovery of reserpine and vincristine, and later the anticancer taxol and camptothecine, further encouraged scientists to search for new medicines from natural products.

China is a country rich in natural resources due to its vast territory, diverse geography and disparate climate. It grows a wide range of medicinal plants including as many as 12,000 herbs. The recorded history of Chinese people using medicinal herbs for treating diseases can be traced back thousands of years. Traditional Chinese medicine (TCM) has always been an important research area for Chinese organic chemists. Organic chemists who returned home from abroad in the 1900's often started their research with Chinese medicinal herbs. Now common internationally used plant drugs, such as ergometrine, santonin, scopolamine, reserpine, vincristine, digoxin, deslanoside, and camptothecine have been successfully manufactured by using plants grown in China. Many antibiotics have been isolated and screened from domestic soils and produced for domestic and international markets. In addition, Chinese scientists have discovered many new medicines from natural products, especially from Chinese medicinal herbs; the most famous ones are the antimalarial compound artemisinin (or quinghausu) isolated from *Artemisia annua* and its derivatives artemether, artesunate and dihydroartemesinin; the acetylcholinesterase inhibitor huperzine, isolated from *Huperzia serrata* and its pro-drug shiperine, which is undergoing clinical trials for treatment of Alzheimer disease; the cholinesterase inhibitors anisodamine and anisodine, isolated from *Scopolia tangutica*; the antihepatic (lower ALT) schsandrin from *Schisandra chinensis* and its synthetic derivative bifendate (biphenyl-dimethyl-dicarboxylate); the antifungal diallyltrisulfide from *Allium sativum*; the antibacterial decanoylacetaldehyde from *Houttuynia cordata*; the sedative tetrahydropalmatine from corydalis and berberine for treatment of gastrointestinal inflammation are popularly used in clinics. Ginseng saponin Rg3 and Kanglaite, an emulsion injection, isolated from oil of coix semen are used in China for treatment of cancer patients. Both of these TCM drugs may have immnunostimulating activity and synergy effects when used with other chemotherapy drugs, and all these TCM medicines have no side effects. However, in some cases these new inventions have not been the subjects of patent applications and the discoverers have neglected the protection of their intellectual property!

In addition to the development of drugs, natural products chemistry is also involved in industry and agriculture. Stevioside isolated from *Stevia rebaudiana* and glycyrrhizin from licorice are used as flavor sweeteners. Gardennin from *Gardenis jasminoides*; tangerine from orange peel and shikonin from *Lithospermum erythrorhizon* are used as natural pigments. The pyrethrin from *Pyrethrum cinerariaefolium* and its many derivatives as well as azadirachtin separated from *Melia azadirachta* are natural insecticides. Their separation, purification, structure identification, production process and product quality control require the knowledge and technology of natural products chemistry.

Recently, people have attempted to use natural medicines to avoid chemical side effects. Scientists have found that EGCE (epigalallocatechol gallate) from green tea, resveratrol from grape seeds, genistein from soy beans and some others are useful nutritional products that may help prevent cancers. These are further contributions of natural products chemistry to human health.

The application of computers has pushed technology and instrumentation to new levels of precision, accuracy, sensitivity and automation. In natural products chemistry, separation has developed from conventional solvent separation chromatography to high speed, high efficiency and high automation. Now HPLC, CPC (centrifugal partition chromatography) and SFE (supercritical fluid extraction) have been widely used. HRMS, 2D-NMR, and x-ray diffraction have become common tools for structure elucidation. Isolation and identification with simultaneous LC-MS or LC-NMR are possible. Separation, purification and structural elucidation of natural products have gradually become routine and automatic. Tasks of isolation and structure determination required years to complete in the past, now only need a few days or even hours. Sample operation is reduced to the microgram level from milligrams. For example, the extended problem of explaining the difference between the HIF (hypothalamic inhibitory factor) and ouabain distinction, has now been solved by the K. Nakanishi group by using micrograms of sample with the ^{1}H-NMR spectroscopy method.

Crossover and coordination between different scientific fields are two of the driving forces of progress. Collaboration among natural products chemistry, medicine, biology and agriculture produce great vitality. When a new compound discovered by natural product chemists is found to have useful activity in a life science application such as the treatment of HIV, cancer, Alzheimer's or other hard-to-treat diseases, it will be a great scientific discovery. Otherwise the compound is only a record in literature. If the compound is contained in nature, but only a trace amount from a natural source, it can be developed by chemical synthesis, tissue culture or biosynthesis and developed in mass volume for manufacturing production and then used in clinics.

When we review the history of natural products chemistry and a summary of its status, it is clear to see that it has produced enormous results and made great contributions to human health, industry and agriculture. However, compared with the abundance of plants, animals and microbial resources only a very small part of the materials have been studied. Chinese traditional medicine is a "gold mine" and further exploration with modern scientific technology will lead to discovery and development of even more useful medicines to treat diseases in the future and many subjects can be further studied by future-generations. As long as we persevere, there will be a continuing series of major scientific discoveries that will make a significant contribution for the benefit of mankind.

Ren-Sheng XU

CHAPTER 2

Extraction and Isolation of Natural Products

In order to determine the structure and evaluate the physical property and biological activity of a natural product, it needs to be purified. However, the purification of natural products is a relatively tedious and time-consuming work, even with the unceasing development of separation technology today. Once an interesting natural product is identified, it is critical to find a low cost and effective separation method to isolate the compound. The extraction and separation of natural products are important in natural products chemistry.

It is common to adopt techniques with high capacity and low resolving power in early separation steps, such as solvent extraction, precipitation and filtration, and simple open column chromatography. A judicious choice of extraction solvent or solvents is the first step of isolation and purification procedures. Extraction with low-polarity solvents yields the more lipophilic components, while alcoholic solvents may give extracts containing both polar and non-polar components. If a polar solvent is used for the initial extraction, subsequent solvent partition allows it to be divided into different polarity fractions.

There are many approaches to extract and isolate the constituents. In this chapter, some frequently used extraction and separation techniques will be introduced briefly from a practical point of view.

Section 1 Extraction of Natural Products

Extraction is the first step to investigate the chemical constituents of a natural material. The application of an adequate extraction method not only guarantees the target ingredients to be extracted, but also avoids the interference of other unnecessary components, and therefore, will also simplify the subsequent separation work. In some cases, one extraction step may yield a pure compound.

1.1 Traditional solvent extraction methods

Traditional solvent extraction methods include soaking, percolation, decoction, reflux, and continuous reflux extraction.

Solvents with gradually increasing polarity, for example, dichloromethane, methanol, and water in turn, may be used to soak a plant material.

The percolation method is widely adopted for its relatively high efficiency and it is often used to extract large numbers of plant samples either with a single solvent or several solvents. Soaking and percolation methods are generally performed at room temperature, therefore, the yielded extracts normally contain fewer impurities.

Compared with the above two methods, decoction, reflux, and continuous reflux extraction methods are operated at higher temperature with higher extraction efficiency, but also produce more impurities. The continuous reflux extraction method is characteristic for its simple operation and lower solvent consumption. When the stability of natural products is unknown, extraction at high temperature should generally be avoided to prevent decomposition.

Among all the solvents that can be used for extraction, water is the cheapest and safest. Several commercially available natural products, such as berberine, rutin, and glycyrrhizin, are extracted with water. However, extraction with water may also yield more salt, protein, sugar and starch in the extract, and thus bring difficulty to the purification process. Ethanol, a solvent with low toxicity and cost, adequate boiling point, and strong ability to penetrate plant cells, is widely used for extraction. Except for protein, phlegm, pectin, starch, and polysaccharide, most organic compounds can be dissolved in ethanol. When a plant material contains only a few major components, an appropriate solvent can be selected to extract the required compound(s) according to the polarity or solubility, and leave other compounds remaining in the plant residue.

1.2   Water steam distillation

This method can be applied to natural products which can be extracted with steam distillation but not degraded. These compounds should be water immiscible or only slightly soluble in water, and have a vapor pressure of around 100°C. When water is heated to boiling, the substances will be extracted with steam. For example, volatile plant oils, several small molecule alkaloids such as ephedrine and nicotine, and certain small molecule acidic substances such as paeonol, can be extracted with the water steam distillation method. Some volatile components with large solubility in water should be further extracted with low polar and low boiling point solvents such as petroleum ether and ethyl ether.

1.3   Supercritical fluid extraction

Supercritical fluid extraction (SFE) is used to extract samples using fluid's special properties under supercritical conditions. This method has been developed rapidly since the 1980s. Many supercritical fluids have better diffusivity and lower viscosity compared with a liquid, which makes them more suitable for a faster extraction of plant components. The solvent strength can be modified by varying the pressure and addition of other solvents, and the extraction selectivity can be achieved and cleaner products obtained by the adjustment of temperature and pressure conditions. Carbon dioxide is the most commonly used supercritical fluid; and therefore, constitutes a safer extraction method than using bulk organic solvent. SFE is especially suitable for thermally or chemically unstable compounds and is mainly employed for the extraction of lower polarity components from a matrix[1,2].

The list of natural products extracted by SFE is getting longer. For example, extraction of nicotine from tobacco, caffeine from coffee and tea, and tanshinone IIA from *Salvia miltiorrhiza, etc.* Liu reported that anethole (~90% purity) could be obtained from star anise with supercritical carbon dioxide, a procedure considerably more efficient than the classical solvent extraction method, which is tedious, time-consuming, and less economic[3,4].

1.4   Solid phase extraction

Solid phase extraction can be used in the following two ways: 1. The interfering matrix elements of a sample are retained on the cartridge while the components of interest are eluted. 2. The required compounds are retained in the column while interfering matrix elements are eluted. In the second case, a concentration effect can be achieved. The required compounds can then be eluted from the cartridge by changing the solvent.

A petroleum ether extract of tomato puree was free of carotenoids after passing through a silica cartridge eluted with petroleum ether. Lycopene retained on the silica cartridge was then eluted with chloroform and purified further by semi-preparative HPLC[5].

Section 2   Separation of Natural Products

2.1   Classical separation method

The classical separation methods mentioned herein have been used for a long time, and their operations are relatively simple without requiring complex and expensive equipment.

2.1.1   Solvent partition

The structures of different natural products may vary considerably, and the number and position of polar functions in the molecules determine their solubility in different solvents. Polar compounds are easy to dissolve in polar solvents, and non-polar compounds are easy to dissolve in non-polar solvents.

Once a biologically active extract has been identified, the use of the solvent partition method may rapidly yield the fraction containing the interested compound, and simultaneously remove a large proportion of extraneous material. This method has been used to search for anti-tumor[6] and anti-HIV[7] agents from plant sources and for the detection and isolation of acetogenins from the Annonaceae plants.

During an investigation of saponin components from plant resources, crude materials can be extracted with industrial ethanol. After vacuum evaporation of ethanol, the aqueous residue can then be extracted with chloroform, ethyl acetate, and n-butanol, successively. The saponins are concentrated in the n-butanol fraction, with the low polarity components removed. As to the isolation of alkaloids, adjustment of the aqueous solution to different pH values followed by extraction with organic solvent can yield the fraction rich in alkaloid components, and may also provide a preliminary separation between strong and weak alkaline alkaloids.

The organic solvents used for separation are normally inert, that is, they cannot react with the compounds to be purified. But in some cases, some alcohols, such as methanol, ethanol or n-butanol, can react with natural products containing free carboxylic groups to yield the corresponding carboxyl esters. Additionally, extraction with ethyl acetate may reduce the acetylation of natural products containing free hydroxyl functions and artificial ketal derivatives may be yielded when natural products possessing adjacent hydroxyl groups are treated with acetone.

2.1.2   Fractional distillation

Based on the boiling point differences of the components in an extract, fractional distillation can be used for isolation. For example, coniine and hydroxyconiin from the total alkaloid fraction of *Cicuta maculatum*, and pseudo-punicine, isopelletierine, and methyl isopelletierine from the peel of *Punica granatum* can be separated initially by using distillation under normal pressure or vacuum before further purification.

2.1.3   Precipitation

The solubility of organic compounds and their characteristic reaction with some reagents to yield precipitation can be used as an initial separation method. The precipitation reaction should be reversible for the compounds to be isolated.

Neutral lead acetate or alkaline lead acetate can react with a lot of compounds and generate an insoluble lead salt or other type of salt precipitation in water or diluted alcohol. This property enables the required components to be separated from unwanted impurities. Lead may be removed from the precipitates by filtration and suspension in clean water or diluted alcohol, and then treated with hydrogen sulfide gas to change the lead salt into insoluble PbS. Sulfuric acid, phosphoric acid, sodium sulfate, and sodium phosphate are also used to remove lead due to simplicity, but lead sulfate and lead phosphate are slightly soluble in water, and thus, reduce the effectiveness. Potassium acetate, barium hydroxide, phosphotungstic acid, and silicotungstic acid are also used as precipitants. Furthermore, polysaccharides and proteins can be precipitated by acetone, ethanol or ethyl ether.

2.1.4   Membrane separation

A mixture can be separated by using the property that small molecules in solution can pass through a membrane with a certain pore size while macromolecules cannot. The membrane separation method is frequently used in the separation of macromolecules, such as proteins, peptides, and polysaccharides from small molecular compounds such as inorganic salts, monosaccharides, and disaccharides. For example, the membrane separation technique is used in the industrial production of soybean protein. According to pore size, the membrane separation technique can be classified into ultrafiltration and nanofiltration. Membrane separation has a great advantage in that it avoids the use of large amounts of organic solvents. With the improvement of technology, the membrane separation method will be used more widely in the research and production of natural products.

2.1.5   Sublimation

Sublimation of a compound is a transition from the solid phase directly into gas phase with no intermediate liquid stage. Natural products with such a property, for example, the camphor in camphorwood, the caffeine in tea, and the benzoic acid existing in some plants, can be purified by the sublimation method. Although the method is simple, the yields are usually low and decomposition of compounds may happen in some cases.

2.1.6   Crystallization

Most natural products are solid compounds, and some of them can be purified by crystallization. Crude crystals usually still contain some impurities; repeated recrystallization may remove the impurities to yield pure compounds.

In some plants in which the content of certain constituents is particularly high, crystals can be obtained after extraction with suitable solvent by cooling or slight concentration. Purification of natural products with the crystallization method is cheaper compared with preparative chromatographic separation methods because it can be performed without the use of complicated equipment and is suitable for mass production. The solvent selected for crystallization should have different solubility for the interested component at different temperatures. The impurities in the solvent should be insoluble or hardly dissolvable. However, a solvent in which impurities have great solubility but the interested substance cannot be dissolved or hardly dissolved can also be used. After removing impurities by washing, another suitable solvent can be selected for crystallization.

2.1.7   Removal of impurities

2.1.7.1   Removal of chlorophyll

Chlorophyll is soluble in common organic solvents, particularly in chloroform, ethyl ether, and other low-polarity solvents, and also in alkaline solutions, such as sodium hydroxide. In the case of removing chlorophyll from digoxin extract from the leaves of *Digitalis lanatae*, the fermented leaf powder was first extracted with 70% ethanol, and the concentrated extract was washed with chloroform and diluted alkali solution. The residue was then extracted with acetone to obtain digoxin[8]. As long as there are no solubility problems, a convenient method of removing chlorophyll from crude extract is to pass the extract through a C_{18} silica column. For example, prior to the separation of flavonoid glycosides from *Dryas octopetala* (Rosaceae), a silica gel cartridge was used to eliminate tannins from an ethanol extract and then chlorophyll was removed by elution on a C_{18} cartridge[9].

2.1.7.2   Removal of wax

Wax is soluble in petroleum ether and ether, which can be used to remove the wax in plant materials before further extraction with other solvents. Acetonitrile can also be used to treat plant extracts to remove the wax. Elliger et al. reported the suspension of the chloroform extract of *Petunia integrifolia* (Solanaceae) in boiling acetonitrile and kept stirring for 1 h. A waxy, solid material was formed after cooling to 5°C[10]. Zheng et al. reported the isolation of cytotoxic flavonoids and terpenoids from *Artemisia annua* (Asteraceae). The aerial parts of the plant were extracted with hexane, the extract was dissolved in chloroform (20 ml), and wax was precipitated with acetonitrile (180 ml)[11].

2.1.7.3   Removal of tannin

It is often necessary to remove tannins from plant extracts or fractions before submission to biological testing. Tannins can be effectively removed by chromatography over a polyamide column. This method was employed by Tan et al. during the evaluation of plant extracts for inhibition of HIV-1 reverse transcriptase and the various methods for tannin removal have also been compared by these same authors[12]. For the bioassay of plant extracts in small quantity, 3 mg of sample can be dissolved in a minimum volume of water and applied to a glass column (10×0.6 cm) packed with polyamide powder (400 mg), and then eluted with water (2 ml) followed by 50% methanol (2 ml) and finally absolute methanol (5 ml). Elution with methanol may give non-tannin compounds with two or three phenolic hydroxyl groups, i.e., most flavonoids can be recovered. The problem is that this method can also remove non-tannin compounds with phenolic hydroxyl groups.

2.2   Chromatographic separation methods

As early as the beginning of the 20th century, Tswett first successfully separated pigments from plants by application of liquid-solid adsorption chromatography. After Kuhn and Lederer separated α and β-carotene on preparative alumina and calcium carbonate columns in the 1930s, chromatographic separation technology attracted chemists' attention. The development of HPLC was begun at the end of the 1960s and has great advantages in separation speed and resolution compared to classical liquid chromatography. HPLC can match with different types of detectors such as UV, refractive index, evaporative light scattering and fluorescence for the analysis and separation of a variety of compounds. In recent years, HPLC has been coupled with the mass spectrometer and nuclear magnetic resonance spectrometer, which greatly improved the application scope and effectiveness of HPLC. In this section, various chromatographic methods will be briefly introduced.

2.2.1   Basic principles

Chromatography can achieve separation by using the different equilibrium distribution coefficient of different materials in the stationary phase and the mobile phase. In chromatographic separation, the sample mixture continuously distributes to balance between the stationary phase and mobile phase. Because of the differences of their physical and chemical properties, the quantities of different compounds are not the same in the two phases.

2.2.2   Classification of chromatography

The mobile and stationary phases used in chromatographic separation cannot dissolve each other. The mobile phase can include both gas and liquid phases, while the stationary phase consists of a liquid or solid phase. When the mobile phase is liquid, the method is called liquid chromatography; when the mobile phase is gas, it is called gas chromatography.

Another classification method is based on the mechanism of chromatographic process. Separation based on the different adsorption performances of components on absorbent surfaces is defined as adsorption chromatography. Separation based on the different partition coefficients of different components between the mobile phase and stationary phase belongs to partition chromatography. Exclusion chromatography refers to the separation of components with different molecular sizes using different blocking effects. Ion-exchange chromatography separation is based on the different affinity of different components to an ion exchange agent. In addition, according to the relative polarity of the stationary phase and the mobile phase, chromatography can be categorized as normal phase or reversed phase.

2.2.3   Liquid-solid chromatographic separation

Liquid-solid chromatography means the stationary phase is solid, while the mobile phase is liquid. Solid stationary phase may be spread in the form of a thin layer on a glass plate or other carrier, and may also be packed into a column. The former is called thin layer chromatography (TLC), while the latter is known as column chromatography.

2.2.3.1   Preparative thin layer chromatography (Preparative TLC)

Samples can be separated by a preparative TLC method in which adsorbent is spread uniformly on a glass plate, sample is applied and then eluted with suitable solvent to achieve separation. The advantages of preparative TLC are that it is simple, rapid, and sensitive. Silica gel and alumina are commercially available absorbents and can be used for the separation of hydrophilic or lipophilic substances. In traditional preparative TLC, the mobile phase flows through the stationary phase by capillary force. In addition, the mobile phase can flow through the stationary phase by external force, such as in centrifugal TLC and compression TLC[13,14].

1. Traditional preparative TLC (PTLC)

Samples on the gram scale can be purified by traditional preparative TLC, but, in most cases, this method is used to isolate compounds in milligram quantity. Traditional preparative TLC method is the most simple preparative separation method and can be used in combination with column chromatography in the purification of natural products.

The plates used for preparative TLC should be pre-washed to remove impurities on the adsorbent. The sample should be dissolved in a small quantity of solvent and applied on the TLC as a narrow band to guarantee a better resolution. For a band which is too broad, concentration can be achieved by allowing the migration of a polar solvent to about 2 cm above the applied band. Volatile solvents are preferred as developing agents since band broadening problems occur with less volatile solvents. There are many variables in PTLC, but as a general guideline, 10 to 100 mg of sample can be separated on a 1 mm thick 20 × 20 cm silica gel or aluminum oxide layer. The size of the TLC plate depends on the amounts of sample and the size of the developing tank. When the sample size is large, several preparative plates can be eluted in the developing tank at same time. To reduce the edge effect, a sheet of filter paper dipped with the mobile phase can be put inside the tank along the wall, which keeps the tank saturated with the mobile phase. Choice of mobile phase is determined by a preliminary investigation with analytical TLC. Since the particles of the adsorbent are approximately the same, the analytical TLC mobile phase is directly transferable to PTLC.

2. Centrifugal thin layer chromatography (CTLC)

Traditional preparative TLC separation requires the scratching of adsorbent bands containing purified substances from the plate and then eluting the substances from the adsorbent.

The time required for solvent development through the plate is long, and adsorbent impurities may exist in the sample. In order to overcome some of these problems, an approach called centrifugal TLC (CTLC) has been attempted, in which the flow rate of the mobile phase is accelerated by the action of a centrifugal force. The centrifugal TLC separation process is illustrated in Figure 2-1. After introduction of sample, solvent elution gives concentric bands of components on the TLC plate. The rotor unit is housed in a chamber covered with a quartz glass, which enables the observation of colorless but UV active substances with the aid of a UV lamp. At the periphery, bands are spun off and collected through an exit tube in the chamber. Fractions of eluents thus obtained can be further analyzed by TLC.

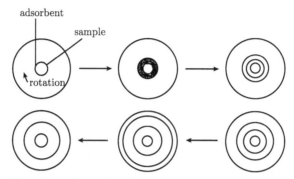

Figure 2-1   Separation of mixture using centrifugal TLC

Preparative centrifugal TLC can be used to separate mixtures of around 100 mg depending on the thickness of the plate. Resolution of preparative centrifugal TLC is inferior to that of preparative HPLC but operating conditions are simple and separations are rapid. Its major advantage over PTLC is compound elution without having to scrape it from the adsorbent. It is also possible to wash in gradient eluents. After separation, the coated plate can be regenerated and used again. Besides silica gel and alumina which are commonly used in TLC, other adsorbents, such as ion exchangers, and polydextran gels, can also be used in CTLC.

2.2.3.2   Column chromatography

Compared to preparative TLC, larger diameter columns and more stationary phases are used in column chromatography for the separation of larger numbers of samples.

1. Open column chromatography

Open column chromatography is a separation method that allows a mobile phase to flow through a stationary phase by gravity. Because of the simplicity of the operation, its application is universal; however, the speed is slow. It is necessary to use a large particle stationary phase to guarantee that the flow rate of the mobile phase will be fast enough. The sample can be dissolved in a small amount of initial eluting solvent, and then added to the top of the stationary phase. When the solubility of the samples to be separated is poor in the eluting agent, a solid sampling method can be performed by dissolving the sample in a suitable solvent with low boiling point, and then adding a small amount of stationary phase or diatomite into the solution. Then, after evaporation of the solvent at low temperature, the powder can be added to the top of the column. In order to prevent destruction of the sample interface, a layer of sand or glass beads can be covered on the top of the sample before elution. Open column chromatography is usually used in the initial separation of crude extracts. Gradient elution may increase the resolution of the open column chromatography.

To overcome the shortcoming of open column chromatography, a series of improvements have been made. Several commonly used preparative column chromatographic separation methods are introduced below.

2. Preparative pressure liquid chromatography

Preparative pressure liquid chromatography includes any method involving the use of devices to put pressure on the liquid chromatography. The application of pressure can accommodate finer granulometry packing material in liquid chromatography and thereby give better resolution. It could also accelerate the flow rate of eluents, leading to a shortened separation time.

According to the pressure employed for the separation, different preparative techniques can be classified into flash chromatography (about 2 bar), low pressure liquid chromatography (<5 bar), medium pressure liquid chromatography (5-20 bar), and high pressure liquid chromatography (>20 bar). This is only for classification purposes, as there are considerable overlaps between low pressure, medium pressure and high pressure liquid chromatography. The size of the columns and absorbent particles are dependent upon the individual separation problem. For samples difficult to separate, longer columns with small particles should be used, but higher pressure will also be required.

A. Flash chromatography

Flash chromatography was first published in 1978 as an attempt to reduce the operation time of open column chromatography[15]. The setup of a typical flash chromatography column is shown in Figure 2-2[16]. A glass column of suitable length, fitted with an exit tap, is either dry-filled or slurry-filled with suitable packing material. Dry-filling gives better packing but requires the passage of a large amount of solvent before the support is fully moistened. It is advisable to introduce a layer of sand at the top of the absorbent. Enough space should be left above the absorbent to allow repeated filing of solvent. Alternatively, a dropping funnel can be attached to the top of the column to act as a solvent reservoir. After the sample is loaded, about 1 bar of pressure can be applied in order to elute the

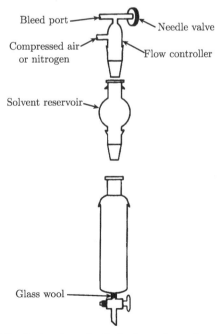

Figure 2-2   Basic setup of a flash chromatography apparatus

sample. The air inlet is fitted with a needle valve to control the compressed air supply. Depending on the size of the column, samples in the range 0.01 to 10.0 g can be separated within 15 minutes[16].

The most widely used stationary phase in flash chromatography is silica gel. Some spherical silica gels with narrow particle size are specially designed for this particular technique. In addition, non-spherical silica gels with certain sizes can also be used[15].

Flash chromatography can be used as the final purification procedure of a natural product or as a pre-purification method of crude extracts or mixtures before other techniques with higher resolution are employed. Nakatani, et al. reported the isolation of antifeedant limonoids by a combination of flash chromatography and semi-preparative HPLC. These compounds are unstable and can gradually decompose on open silica gel columns[17].

B. Low pressure liquid chromatography

The most widely used low-pressure liquid chromatography system is the Lobar range (E. Merck). Ready-filled columns are made of glass, and the support can be silica gel, RP-18, RP-8, NH$_2$-, CN-, or diol bonded phase silica gel. The packing material is sealed into the column with a glass frit and the connection to the pump is provided by a metal cannula attached to a PTFE ring which is held in place by a screw cap (Figure 2-3)[18].

Figure 2-3   Lobar pre-packed column

The resolution of Lobar columns sometimes can approach those of HPLC. In any case, for greater resolution or larger quantity of sample, several columns can be coupled in series. The simplicity of Lobar separation is the reason why these columns are so widely used. Lobar columns can be reused, and the lifespan of the reversed-phase column is generally much longer than that of the normal phase column. Its particle size is large (40 to 60 μm) which enables high flow rates at relatively low pressures. The sample, dissolved in proper solvents, can be added into the column by use of an injector or infusion pump. Because of the pumping arrangement, isocratic separations are normally adopted. In our case, triterpenoid saponins were isolated from the herb *Mussaenda pubescens* (Rubiaceae) by using a Lobar RP-18 column, in which ethanol–water (1.1/1) and acetonitrile/water (1/1) was used as solvent system[19].

C. Medium pressure liquid chromatography

The advantages of Lobar chromatography systems are that they are easy to use and have high resolution. However, separations of more than 5 g of sample are rarely possible with this type of column. In order to separate more sample at one time, medium pressure liquid chromatography with longer columns and larger internal diameters can be used. The particle size of the support used in this system is smaller and its resolution is higher than that of low pressure LC, which requires higher pressures than low pressure LC to enable sufficiently high flow rate. The required pressure can be supplied by compressed air or the reciprocating pumps. Appropriate solvent systems can be selected efficiently with analytical HPLC, and the transformation of the HPLC conditions to MPLC is straightforward and direct (Schaufelberger and Hostettmann)[20].

Büchi markets a complete MPLC system with a wide range of column combinations, providing a very flexible separation capacity from 100 mg to 100 g sample sizes. A piston pump and exchangeable pump heads allow flow rates from 3 to 160 ml/min at a maximum pressure of 40 bar. A gradient former can also be added to the system. Different UV detectors are also available (Figure 2-4)[21]. The chromatography columns are of strengthened glass with a plastic protective coating, giving a visual control of the separation. They can be coupled together simply by means of flanges to increase the resolving power.

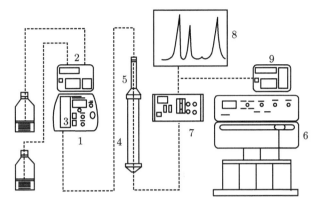

Figure 2-4   Büchi B-688A medium-pressure liquid chromatography system:
1. B-688 pump; 2. Gradient former; 3. Sample introduction; 4. Glass column; 5. Pre-column; 6. Fraction collector; 7. UV detector; 8. Recorder; 9. Peak detector

In the case of MPLC with silica gel, the columns may be filled with dry media or with a slurry of the stationary phase in a suitable solvent. The dry-filling method gives a 20% higher packing density than the slurry technique. Although the pressure limit on the glass column is only 40 bar, the use of 15 μm particles gives efficiencies and separations similar to those obtained on HPLC columns[22,23]. For bonded stationary phases, the slurry method should be used for packing. Samples can be injected through a septum directly onto the column or via a sample loop. It is also possible to perform solid introduction of the sample with the aid of a small Prep-Elut column connected just before the main separation column. If a pre-column (attached directly onto the preparative column) is used during chromatography, the contaminated packing material at the top of the column can be removed after each separation. Silica gel supports can be regenerated by washing in the sequence methanol-ethyl acetate-hexane, but after a certain time the support should be thrown away. Bonded-phase columns are easier to clean and have a longer working life. Sometimes polyamide or cellulose can also be used as the stationary phase[24].

D. High pressure liquid chromatography

Preparative high pressure liquid chromatography apparatus is used more and more frequently in the research of natural products chemistry. In preparative HPLC, with the adoption of microparticulate stationary phases (5 to 30 μm) of a narrow size range, high pressure is necessary to enable mobile phase flow. Although the complexity and costs of the systems are greater, there is a large gain in separation efficiency.

Semi-preparative chromatography normally refers to those separations of 1 to 100 mg mixtures with columns of 8 to 10 mm i.d., often packed with 10 μm particles[25]. Isocratic conditions are often used in preparative HPLC, which makes it convenient to inject a sample repeatedly. However, gradient elutions are also favorable in many cases to reduce separation time[26,27].

A certain amount of pure natural product can be obtained either by a single injection of the sample onto a large dimension column or by repetitive injections onto a column of more modest dimension. A glycosidic lupin alkaloid has been isolated from *Lupinus hirsutus* (Leguminosae) using HPLC as the final purification step with an eluent of 25% MeOH in Et$_2$O-5% NH$_4$OH (50:1) (Suzukion et al.)[28].

3. Vacuum liquid chromatography

Vacuum liquid chromatography (VLC) is a chromatographic separation method which uses a vacuum to speed up eluent flow rates. Vacuum liquid chromatography has several advantages, such as simple equipment, shorter separation time, better resolution, and large separation capacity.

Vacuum liquid chromatography can be performed by packing a short column or a Büchner filter funnel with a glass frit using dry adsorbent. Then, the vacuum is applied to make the adsorbent into a hard layer. After the vacuum is released, a solvent of low polarity is poured quickly onto the surface of the adsorbent and then vacuum is re-applied. When the eluent is removed, the column is ready for loading. The sample, in a suitable solvent, is applied directly to the top of the column and is drawn gently into the adsorbent under vacuum. Alternatively, the sample can be preabsorbed on silica gel, aluminum oxide or Celite. The column is developed with appropriate solvent mixtures, starting with a solvent of low polarity and gradually increasing the polarity, pulling the column dry between each fraction collected. When pungent constituents were separated from ginger (*Zingiber officinale*) by VLC, there was a virtually complete separation of the gingerols from the shogaols[29].

4. Dry column chromatography

Dry column chromatography requires the filling of a chromatography column with dry packing material. The sample is added as a concentrated solution or absorbed onto a small amount of adsorbent before introduction. The solvent is allowed to move down the column by capillary action until the solvent front nearly reaches the bottom. The solvent flow is stopped and the bands on the column removed by extrusion, slicing or physical removal. They are then extracted by a suitable solvent. There is no liquid flow down the column and sample bands are sharp. Dry column chromatography is also time consuming and very little solvent is needed.

Separation effects can be extrapolated directly from analytical TLC plates by choosing the same adsorbent in the column. In fact, dry column chromatography is just a variant of preparative TLC, with the same resolution. Elution of a dry column with solvent mixtures may not always give the resolution of analytical TLC. In this case, it is recommended to presaturate the dry column adsorbent with about 10% mobile phase before packing the column.

The easiest way of removing a chromatography column support after development is to use a nylon column. The column or tube can be cut with a sharp knife into sections corresponding to the migrated bands and the separated compounds can then be extracted and filtered. Another advantage of using a nylon column is that the colorless bands can be observed with a UV lamp to guide sectioning.

The support is the main factor to determine the column efficiency; the smaller particle diameter of the support and the more narrow the range of the particle size, the higher the column efficiency. If the support is filled uniformly, the separation zone will be orderly and the resolution will also be high. But if the particles are too small, the linear velocity of mobile phase and mass transfer will be slow, and resolution may be affected by the longitudinal diffusion of samples.

A 95% ethanol extract of *Salvia miltiorrhiza* roots was partitioned with several solvents and the resulting fraction (45 g) was separated by dry column chromatography on 2.3 kg of silica gel ($CHCl_3$-MeOH-HCOOH 85:15:1). The column was cut into 16 sections which were eluted with warm methanol. Sections 13 and 14 were purified on a Sephadex LH-20 column to give 2.06 g of salvianolic acid B[30].

2.2.3.3   Stationary phases commonly used in liquid-solid chromatographic separation

There are a variety of stationary phases commonly used in liquid-solid chromatographic separation, such as silica gel, bonded phase silica gel, alumina, polyamide, polydextran gel, and ion-exchange resin. Each of these stationary phases has its own characteristics, and can be selected for the separation of different types of compounds. Multiple chromatographic separations with different stationary phases are often adopted to obtain pure compounds. The following is a brief review of several commonly used stationary phases.

1. Silica gel

Silica gel is a polyporous material with porous siloxane and crosslinking structure of -Si-O-Si-, which can be represented by the formula $SiO_2 \cdot xH_2O$. The silanol groups on the surface of silica gel can interact with polar or unsaturated molecules through hydrogen bonds. The absorbability of silica gel depends on the number of silanol groups and water content in it. With the increase of moisture, adsorption capacity will decrease. If the water content in silica gel exceeds 12%, the adsorption capacity will be too low to be used for adsorption chromatography, and only suitable as a carrier of partition chromatography. The surface area and surface structure of silica gel, as well as the volume and radius of the micropores of the silica gel, have a direct impact on chromatographic resolution.

Certain chemical reagents can be mixed with silica gel to improve adsorption performance and increase resolution. For example, silica gel treated with silver nitrate is typically used to separate unsaturated hydrocarbons with similar structures. Modified adsorbent can be prepared by addition of 1 to 10% of the chemical reagents in water or acetone to silica gel, and can be used after blending and drying at 110°C.

More varieties of mobile phases can be chosen for normal phase chromatography than for reversed phase chromatography, and mixtures of solvents in different proportion are often used as eluents for better resolution. The selection of mobile phase is usually guided by TLC, and gradient elution is achieved by gradually increasing the solvent polarity from low to high for the actual chromatography process.

Deactivation of silica gel is necessary in some cases to avoid decomposition of samples on columns[31]. About 10 to 30 mg of sample can normally be loaded on 1g of 50 to 200 μm silica gel. Silica gel columns can also be overloaded for filtration purposes. Verzele et al. reported the filtration of 1g of sample through 10 g of silica gel during the separation of hydrocarbon terpenes and oxidized terpenes from essential oils[32].

Currently, most preparative liquid chromatography separations are still carried out on silica gel, mainly due to its low cost, high speed of separation, application of a broad range of eluent solvents, and the low boiling points of most solvents easily removed after fractionation.

2. Bonded phase silica gel

Octadecyl silane, dihydroxypropyl, amino, and cyano groups can be chemically bonded to porous silica gels. to form bonded phase silica gels. These bonded phase silica gels have different selectivity for the separation of compounds, occupy an important position and are widely used in liquid chromatography. Bonded phase silica gel has many advantages, such as repeated use, less risk of sample decomposition, and irreversible adsorption.

Methanol-water, ethanol-water, and acetonitrile-water systems are commonly used eluents in reversed phase chromatography with bonded phase silica gel as adsorbent. Preparative column chromatography eluents can be determined according to analytical TLC or analytical HPLC results which use the same bonded phase silica gel.

3. Alumina

Alumina is made by dehydration of aluminum hydroxide under high temperature (about

600°C), and is often slightly basic due to manufacturing. This makes it an ideal adsorbent for the separation of alkaline ingredients in plants, but not suitable for the purification of aldehyde, ketones, esters, and lactones because basic alumina may react with the above constituents during reactions such as isomerization and oxidation.. Alumina can be washed with water to remove the basic impurities, and then activated to obtain neutral alumina. Apart from the separation of basic substances, such as alkaloids, alumina is rarely used as stationary phase in liquid chromatography.

The activity of alumina is also related to its content of water, and heating of alumina at around 200°C for 4 to 6 h may remove the water and increase its activity. On the contrary, addition of a certain amount of water to reduce the activity is also necessary in some cases.

As in the case of silica gel chromatography, the elution power of a polar solvent is larger than that of a non-polar solvent in alumina chromatography. By gradually increasing the polarity of eluents, compounds adsorbed on the alumina column can be eluted one by one according to their polarity to achieve the intended separation. Eluents used in alumina column chromatography can be selected by alumina thin layer chromatography analysis. The alkaloid (-)-argyrolobine was separated from *Lupinus ergenteus* (Leguminosae) by using basic alumina chromatography with a mixture of ether and methanol as the elution agent[33]. Xu et al. reported the isolation of licorice chalcone from the plant *Glycyrrhiza inflata* Batal, in which the chloroform extract was first separated over a polyamide column, and then over an alumina column with chloroform as eluents. Crystalline products were obtained successfully[34].

4. Activated carbon

Activated carbon chromatography is one of the major methods used for the separation of water-soluble substances. It is effective in separating certain glycosides, carbohydrates and amino acids in plant materials. Because of its easy availability and low price, activated carbon chromatography is suitable for large-scale preparative separation.

The adsorption of activated carbon is strongest in water, and weak in organic solvent. When ethanol-water is used as an eluent in activated carbon chromatography, elution power increases with the increase of ethanol concentration. Diluted methanol, acetone, and acetic acid solutions are also used as eluents in some cases.

The adsorption of activated carbon to aromatic compounds is greater than those of aliphatic compounds. Additionally, the adsorption of activated carbon to a macromolecule is greater than those of smaller molecular compounds, and adsorption of compounds with more polar groups (COOH, NH_2, OH, etc.) is greater than those compounds with fewer polar groups. These differences in adsorption can be used for the separation of water-soluble aromatic compounds from aliphatic compounds, amino acids from peptides, and polysaccharides from monosaccharides.

Prior to the use of activated carbon, it should be heated at 120°C for 4 to 5 h to remove adsorbed gas. Used activated carbon can be re-activated by treatment with dilute acid and dilute alkali solution alternatively, and then washed with water and activated by heating. Powder activated carbon can also be converted to granular nylon-activated carbon (1/2) or mixed with diatomaceous earth (1/1) before packing into a chromatographic column in order to increase the flow rate. However, the adsorption of the granular activated carbon is lower than that of powder activated carbon. During the purification of water-soluble sarmentosin from *Sedum sarmentosum*, granular activated carbon was used to remove impurities before repeated silica gel column chromatography[35].

5. Ion exchanger

An ion exchanger is a high molecular weight compound with dissociated ion exchange groups, which can exchange with other cations or anions in aqueous solution. This exchange is reversible. When two or more components are 'absorbed' on the ion exchanger and eluted, the elution power is determined by the equilibrium constants of eluting reaction of each material. Most ion exchangers are synthetic agents. The exchange groups can be monomers before polymerization or are introduced after polymerization. The most widely used exchanger is ion exchange resin. An ion exchanger can also be made by introduction of ion exchange groups into saccharide or cellulose, and most of them are used in the separation and purification of macromolecular proteins, nucleic acids, enzymes, and polysaccharides in plant materials.

Ion exchange resins can be divided into two broad categories, namely cation exchange resin and anion exchange resin:

Cation exchange resin	strong acid type	$-SO_3H$
	weak acid type	$-COOH$
		$-PO_3H$
Anion exchange resin	strong base type	$-N-(CH_3)_3X$
		$-N-(CH_3)_2(C_2H_4OH)X$
	weak base type	$-NR_2$
		$-NHR$
		$-NH_2$

According to the principle above, different types of ion exchange resins can be used to separate water soluble acid, alkali, and amphoteric components in plants (Figure 2-5).

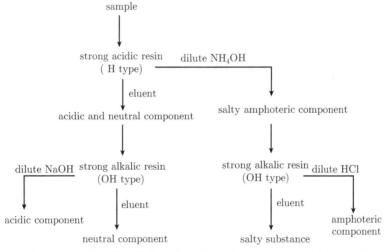

Figure 2-5   Separation of acidic, alkali, and amphoteric components with ion exchange resins

6. Macroporous resin

Macroporous resin is a lipophilic polymer adsorbent with large porous structure but without ion exchange groups. Macroporous resins have a variety of surface properties. For example, hydrophobic polystyrene can adsorb compounds with different chemical properties effectively, and desorption is also easy to perform. The surface area of macroporous resin is usually 100 to 600 m^2/g. The larger the surface area of adsorbents, the higher the adsorption capacity. But for some organic molecules with large stereochemical structure, the porous size of the resin should be considered in order to allow the penetration of the molecules.

Macroporous resins have the characteristics of good selectivity, high mechanical strength, and fast adsorption speed, and are also convenient to regenerate. Macroporous resins are therefore suitable to be used in the separation of low polar or non-polar compounds from water solutions. The greater difference between the polarities of constituents, the better separation effect. After a sample is adsorbed in a macroporous resin, the column is generally eluted with a gradient of water and methanol, ethanol or acetone in 10%, 20% ... stepwise (v/v), and finally eluted with alcohol or acetone. A macroporous resin can be regenerated by soaking and washing with methanol or ethanol, when necessary, soaking with l mol/L hydrochloric acid and sodium hydroxide and then washed with distilled water to neutral pH. A macroporous resin may be preserved in methanol or ethanol, and exchanged with alcohol with distilled water before use.

At the beginning stage of purification, hydrophilic impurities (such as amino acids, carbohydrates, etc.) can be separated by passing a polar sample through a polymer column. A macroporous resin is normally the first adopted adsorbent before silica gel chromatographys reversed phase adsorbent, and Sephadex LH-20[36]. Amberlite XAD-2 model macroporous resin can adsorb polyphenolic compounds, and has been used for the separation of flavonoids[37]. The obtained flavonoid fraction can be further purified by using other adsorbents to yield pure components.

Since the end of 1970s, macroporous resin has been applied in the extraction and separation of chemical constituents in Chinese traditional and folk medicines. Commonly used resins include D-101, DA-201, MD-05271, GDX-105, CAD-40, XAD-4, and D-type resin, etc.

Jin et al. reported the extraction of total glycosides from *Paeonia lactiflora* by using DA-201 type macroporous resin with a yield of 1.5%. The operation is simple and the yield and quality of products are consistent[38].

7. Gel

Gel chromatography is a separation and analytical technique developed in the 1960s. The gel used as a stationary phase is solid with a pourous network structure, and has the nature of a molecular sieve. When the molecular size of materials to be separated is different, their ability to enter the the gel is also different. Pore size of gel has a similar magnitude to the molecular size. When a mixture is passing through the gel phase, molecules that are smaller than the pore size may enter the gel freely, and molecules that are larger than the pore size are kept outside. Therefore, differences in their mobile speeds are observed. Larger molecules cannot access some of the pores and exit the column more rapidly. Small molecules are retained due to diffusion or movement into the pore, and fall behind larger molecules to be separated (Figure 2-6).

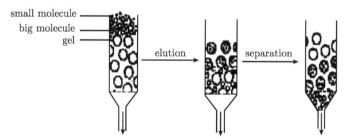

Figure 2-6   Gel chromatography

In theory, the separation efficiency of gel filtration columns is based on the molecular exclusion effect. When gel is used for the separation of small molecules, the interactions among solvent, solute, and stationary phase become important.

Sephadex LH-20 gel filtration can not only be used as an effective means of the initial separation, but also for the final purification to remove the last traces of solid impurities, salts or other foreign substances. When the quantity of sample is very small, it is best to use Sephadex LH-20 gel filtration in the final stages of separation. because it leads to very little sample loss.

Generally speaking, the used gel does not need any treatment and can be used repeatedly. When the sample loading area becomes dark in color, this part can be removed. If the color of an entire gel cartridge becomes dark, it can be treated with 0.2 N NaOH (containing 0.5 N NaCl), and then washed with water.

8. Polyamide chromatography

A polyamide is a class of high molecular weight compounds formed by polymerization of amide groups. Phenols, acids, quinines, and nitro compounds can be absorbed to it through hydrogen bonding, and can be separated from other compounds that cannot form hydrogen bonds.

The more phenolic hydroxyl groups present in a molecule, the stronger the adsorption effect will be. The adsorption effect is also related to the number of aromatic fragments and conjugated double bonds. The existence of intramolecular hydrogen bonding in a molecule may decrease its absorption to polyamide. The elution of adsorbed compounds is achieved through the formation of new hydrogen bonds between the polyamide and solvent molecules. For example, flavonoid glycosides are eluted earlier from the polyamide column than its aglycones with diluted alcohol as eluents; while flavonoid aglycone is eluted out earlier than its glycoside eluted with non-polar solvent. This suggests that the polyamides possess different chromatographic properties due to the existence of both the non-polar lipid bond and the polar amide groups. When polar solvent containing water is used as the mobile phase and with polyamide as the non-polar stationary phase, the chromatographic performance is similar to that of reversed-phase partition chromatography, and the flavonoid glycoside is desorted more easily than its aglycone. When non-polar chloroform-methanol is used as the mobile phase, a polyamide acts as a polar stationary phase. The chromatographic behavior is similar to that of normal phase partition chromatography, and flavonoid aglycone is desorped easier than its glycoside. Besides phenolic compounds, polyamide chromatography can also be used for the separation of terpenoids, steroids, alkaloids, and carbohydrates.

Polyamide film chromatography is an important means to detect the compounds indicated above. It can be performed by dissolving a polyamide in formic acid and then spreading on polyester film. The film can be used after formic acid is volatilized and the film dried. Polyamide film chromatography can be used to explore the separation conditions and can check the composition and purity of each column fraction. Addition of a small amount of acid or base to the solvent systems may overcome chromatographic tailing effects and also makes the spots clear.

2.2.4   Countercurrent chromatography[39]

Countercurrent chromatography (CCC) is a separation which relies on the partition of a sample between two immiscible solvents. The relative proportion of a certain compound passing into each of the two phases is determined by the respective partition coefficient. It is an all-liquid method which is characterized by the absence of a solid support. Countercurrent chromatography is basically an outgrowth of countercurrent distribution, a method developed for the batchwise or continuous fractionation of mixtures. In practice, the CCC apparatus consists of several hundreds of elements, in each of which a partition of the solute between two liquid layers is performed before transferring one of the layers to the next element. In this process, equilibration is complete before phase transfer. CCC is a continuous, non-equilibrium process comparable to LC. Similar to LC, CCC can avoid irreversible adsorption of the sample, and can also avoid sample damage because of the interaction between some solid stationary phase and the sample.

2.2.4.1 Droplet countercurrent chromatography

Droplet counter-current chromatography (DCCC) is a technique which uses a group of vertical separation tubes filled with liquid stationary phase. Mobile phase passes continuously through the stationary phase in a liquid droplet form, prompting the solute to consecutive partition between the two phases (Figure 2-7)[40]. Before separation, the mobile phase and the stationary phase should be shaken well to make a balance of the heavier phase (lower) and the lighter phase (upper). Any of the two phases can be used as the stationary phase. If the heavier phase is selected as the stationary phase, it is called an ascending method. The reverse is called a descending method. When the droplets are passing through the separation tube, the stationary phase between the droplets and the wall forms a film which contacts the droplets and continuously creates a new surface, which prompts all the components in the mixture to participate repeatedly between the two phases. This method is much less cumbersome and complex than conventional countercurrent distribution machines and can avoid the problems of emulsion or foam formation. In addition, the separation time is shorter and solvent consumption is considerably reduced. Sample loads up to 6.4 g of crude extract on columns of 3.4 mm i.d. have been performed,[41] and even sample sizes of 16 g have become possible with 10 mm i.d. columns[42].

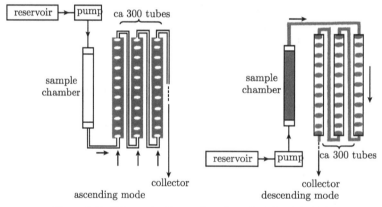

Figure 2-7   Schematic illustrating the principle of DCCC

Binary solvent systems are impractical for the formation of suitable droplets because of the large difference in polarity between the two components. Ternary or quaternary systems are required for the preparation of the two phases, such that the addition of a third or fourth component, miscible with the other components, diminishes the difference in polarity between the two phases. One of the most commonly used solvent systems is a mixture composed of $CHCl_3$-MeOH-H_2O.

DCCC is suitable for the separation of polar natural products, especially phenols and glycosides. The acidic hydroxyl groups in polyphenols often cause irreversible adsorption on solid stationary phases during conventional column chromatographic procedures. Mahran et al. reported the isolation of a triterpenoid saponin from the leaves of *Zizyphus spinachristi* (Rhamnaceae) using DCCC. The solvent was $CHCl_3$-MeOH-*n*-PrOH-H_2O 9:12:1:8 and the mode was descending[43].

Compared with HPLC, solvent consumption is generally less in DCCC, but separation time is longer and resolution is lower. Before separation, consideration should be given to choosing a suitable solvent system. With the application of centrifugal partition chromatography, the use of DCCC is dwindling.

2.2.4.2   Centrifuge partition chromatography (CPC)

There are two basic types of countercurrent chromatography: the hydrostatic equilibrium system (HSES) and the hydrodynamic equilibrium system (HDES).

The DCCC system as mentioned before is an extension of this gravity method. Its disadvantage is that the elution speed cannot be elevated, with separations taking two days or more. This basic CCC system uses only 50% of the efficient column space for actual mixing of the two phases. A more effective way of using the column space is to rotate the coil while eluting the mobile phase. A hydrodynamic equilibrium is rapidly established between the two phases and almost 100% of the column space can be used for their mixing. Thus, the interfacial area of the phases is dramatically increased. Solutes, which may be injected as solution in the mobile phase, stationary phase or a mixture of both, are partitioned between the two phases and separated according to their partition coefficients. In fact, as there is no solid support, solute retention depends only on the partition coefficient. Because these systems utilize centrifugal force on the two phase solvent systems, this technology is called centrifugal partition chromatography with more accuracy than countercurrent chromatography.

Hydrodynamic equilibrium systems (HDES): In HDES, the column revolves around the central axis of the centrifuge and simultaneously rotates about its own axis at the same angular velocity (Figure 2-8)[44]. The motion of the coil causes vigorous agitation of the two solvent phases and a repetitive mixing and settling process, ideal for solute partitioning, occurs at over 13 times per second. Therefore, this technique can achieve efficient separations with small volumes of solvent[45]. This technique is also known as high-speed countercurrent chromatography (HSCCC), and is very useful for the separation of grams of samples within a few hours.

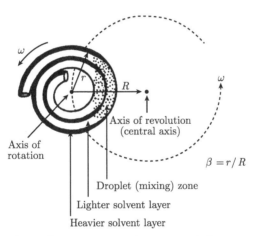

Figure 2-8   High speed countercurrent chromatograph (ω = rotational speed)

Hydrostatic equilibrium systems (HSES): This centrifugal countercurrent chromatography is different from the HDES systems in that column units are fixed in a centrifuge and do not rotate, i.e. they only have a single axis.

Cartridge instruments: Murayama and collaborators at Sanki Engineering, Kyoto, Japan, described the operation of their chromatograph in a paper dating from 1982[46]. In the original instruments, the separation columns, in the form of cartridges, are arranged around the rotor of a centrifuge with their longitudinal axes parallel to the direction of the centrifugal force and connected by narrow-bore tubes. The columns are first filled with stationary phase, while the rotor starts spinning, mobile phase is pumped into the separation columns, forming a stream of droplets which crosses the stationary phase. The different parameters of the cartridge instrument have been investigated by Berthod and Armstrong, who found that a high rotational speed gives better peak resolution[47]. Suitable separations are only achieved when no more than 20% to 30% of the stationary phase is eluted with the mobile phase. Judicial choice of rotor speed and pumping pressure is necessary to fulfill this condition. In addition, efficiency increases with increasing flow rates.

The right choice of solvent system is crucial for successful CPC separation. The most efficient combinations yield partition coefficients of 0.2 to 5, with values approaching 1 are the best. It is vital that the setting time for the two phases is short. For HSCCC, this should be shorter than 30 seconds[48]. It is also an advantage if the chosen solvent system provides roughly equal volumes of upper and lower phases.

CPC has already been used widely as a routine preparative separation technique in the field of natural products for crude and semi-purified fractions with sample sizes ranging from several milligrams to grams. CPC can also be exploited as an extraction technique for the concentration of trace component(s) from a large volume of solvent. In this case, the solvent in which the compound is dissolved is employed as the mobile phase, while a stationary phase which has a high affinity for the product in question is chosen.

Extracts of the leaves of *Gingko biloba* are widely used in the treatment of circulatory disorders of the brain. Flavonoid glycosides from these leaves have been successfully purified by a combination of CPC and semi-preparative HPLC methods. The liquid-liquid step involved gradient elution, starting with water as stationary phase and eluting with ethyl acetate. Increasing amounts of isobutanol were then gradually added to the ethyl acetate until the proportion of ethyl acetate:isobutanol (3:2) was reached at the end of the elution. Seven flavonoid glycosides were obtained from 500 mg of leaf extract[49].

CPC is characteristic for its fast separation speed, high resolution, large sample-loading capacity, and low solvent consumption; however, it has not become a competitor of preparative HPLC. When faced with a problem that cannot be solved by other types of chromatography, there is no harm in choosing CPC.

Section 3   Concluding Remarks

It is time-consuming, although exciting, work to isolate a pure bioactive component or components from plant material which contain complex ingredients and may involve many separation steps. In many cases, a combinational use of different separation techniques is the best approach. It is essential to pick steps which differ as much as possible in selectivity to determine a separation strategy. In the very early stage, solvent participation and precipitation followed by filtration are the simplest methods which allow dealing with large numbers of samples. Macroporous resin, ion exchange resin, and silica gel are relatively inexpensive stationary phases, and are often used in column chromatography. Subsequent chromatographic steps for smaller quantity components can be performed with more expensive adsorbent and equipment, such as HPLC. If only one stationary phase is available throughout the purification steps, selectivity should be maximized by varying the eluents. The rational choice of separation method can effectively improve the efficiency and avoid sample loss.

Bibliography

[1]   Xu R S, Chen Z L. Extraction and Separation of Bioactive Components from Chinese Medicinal Herbs. 2nd Edition, Shanghai: Shanghai Science and Technology Publishing House, 1983.

[2]   Institute of Materia Medica, Chinese Academy of Medical Sciences. Study on the Chemical Constituents of Chinese Traditional and Folk Medicines. Beijing: People's Medical Publishing House, 1980.

[3] Hostettmann K, Marston A, Hostettmann M. Preparative Chromatography Techniques—Application in Natural Products Isolation. Berlin: Springer-Verlag, 1998.

[4] Yao X S. Chemistry of Natural Medicines. Beijing: People's Medical Publishing House, 1998.

[5] Lin Q S. Chemistry of the Components of Chinese Traditional and Folk Medicines. Beijing: Science Press, 1977.

[6] Xu R S. Natural Products Chemistry. Beijing: Science Press, 1993.

References

[1] McHugh M, et al. Supercritical Extraction: Principles and Practice. Boston: Butterworth, MA, 1986.

[2] Westwood S A. Supercritical Fluid Extraction and Its Use in Chromatographic Sample Preparation. New York: Chapman and Hall, 1993.

[3] Li P Y, et al. Natural Products Research and Development, 1992, 4: 30.

[4] Tang Z J, et al. Science in China, 1974, 6: 15.

[5] Hakala S H, et al. J Agric Food Chem, 1994, 42: 1314.

[6] Pettit G R, et al. J Med Chem, 1995, 38: 1666.

[7] Gustafson K R, et al. J Med Chem, 1992, 35: 1978.

[8] Xu R S, Chen Z L. Extraction and Separation of Bioactive Components from Chinese Traditional and Folk Medicines. 2nd Edition, Shanghai: Shanghai Science and Technology Publishing House, 1983: 305.

[9] De Bernardi M, et al. J Chromatogr, 1984, 284: 269.

[10] Elliger C A, et al. J Chem Soc Perkin Trans I, 1990, 525.

[11] Zheng G Q. Planta Med, 1994, 60: 54.

[12] Tan G T, et al. J Nat Prod, 1991, 54: 143.

[13] Nyiredy S Z. Anal Chim Acta, 1990, 236: 83.

[14] Erdelmeier C A J, et al. Phytochem Anal, 1991, 2: 3.

[15] Still W C, et al. J Org Chem, 1978, 43: 2923.

[16] Hostettmann K, et al. Preparative Chromatography Techniques—Application in Natural Products Isolation. Berlin: Springer-Verlag, 1998: 73.

[17] Nakatani M, et al. Phytochemistry, 1994, 36: 39.

[18] Hostettmann K, et al. Preparative Chromatography Techniques—Application in Natural Products Isolation. Berlin: Springer-Verlag, 1998: 83.

[19] Zhao W M, et al. J Nat Prod, 1994, 57: 1613.

[20] Schaufelberger D, et al. J Chromatogr, 1985, 346: 396.

[21] Hostettmann K, et al. Preparative Chromatography Techniques—Application in Natural Products Isolation. Berlin: Springer-Verlag, 1998: 89.

[22] Zogg G C, et al. Chromatographia, 1989, 27: 591.

[23] Zogg G C, et al. J Liq Chromatogr. 1989, 12: 2049.

[24] Hostettmann K, et al. Preparative Chromatography Techniques—Application in Natural Products Isolation. Berlin: Springer-Verlag, 1998: 90.

[25] Snyder L R, et al. Introduction to Modern Liquid Chromatography, 2nd ed. New York: John Wiley, 1979.

[26] Slimestad R, et al. J Chromatogr A, 1996, 719: 438.

[27] Govindachari T R, et al. J Liq Chromatogr, 1996, 19: 1729.

[28] Suzuki H, et al. Phytochemistry, 1994, 37: 591.

[29] Zarate R, et al. J Chromatogr, 1992, 609: 407.

[30] Ai C, et al. J Nat Prod, 1988, 51: 145.

[31] Bruno M, et al. Phytochemistry, 1996, 42: 1059.

[32] Verzele M, et al. J Chromatogr Sci, 1980, 18: 559.

[33] Arslanian R L, et al. J Org Chem, 1990, 55: 1204.

[34] Xu R S, et al. Acta Chim Sinica, 1979, 37: 289.

[35] Fang S D et al. Acta Chim Sinica, 1962, 28: 244.

[36] Mizui F, et al. Chem Pharm Bull, 1990, 38: 375.

[37] Rosler K H, et al. J Nat Prod, 1984, 47: 188.

[38] Jin J S, et al. China Journal of Chinese Materia Medica, 1994, 1: 31.

[39] Hostettmann K, et al. Preparative Chromatography Techniques—Application in Natural Products Isolation. Berlin: Springer-Verlag, 1998: 135.

[40] Hostettmann K, et al. Preparative Chromatography Techniques—Application in Natural Products Isolation. Berlin: Springer-Verlag, 1998: 137.

[41] Hostettmann K, et al. Nat Prod Rep, 1984, 1: 471.

[42] Komori T, et al. Liebigs Ann, 1983, 2092.

[43] Mahran G E H, et al. Planta Med, 1996, 62: 163.

[44] Hostettmann K, et al. Preparative Chromatography Techniques—Application in Natural Products Isolation. Berlin: Springer-Verlag, 1998: 164.

[45] Ito Y. J Chromatogr, 1984, 301: 377.

[46] Murayama W, et al. J Chromatogr, 1982, 239: 643.

[47] Berthod A, et al. J Liq Chromatogr, 1988, 11: 547.

[48] Oka F, et al. J Chromatogr, 1991, 538: 99.

[49] Vanhaelen M, et al. J Liq Chromatogr, 1988, 11: 2969.

Wei-Min ZHAO

CHAPTER 3

Chemistry of Fungal Products

Section 1   Introduction

Nature has proven to be the richest source of biological and chemical diversity. Natural products still provide the majority of the world with a wide range of drugs and pharmaceuticals. Historically, fungi and actinomycetes have been found to be the most prolific producers of secondary metabolites among all microorganisms studied[1]. Fungi are a large group of ubiquitous microorganisms with more than 200,000 species distributed widely in soil, air, water, and found residing inside or on the skins of plants and animals. According to Whittaker in 1969[2], Kingdom Fungi includes slime molds and true fungi, and can be further classified into three subkingdoms: Gymnomycota (slime molds), Dimastigomycota (oosphere fungi), and Eumycota. Subkingdom Eumycota includes four phyla: Chytridiomycota (true chytrids and related fungi), Zygomycota (conjugation fungi), Ascomycota (sac fungi) and the Basidiomycota (club fungi). As a result of updates in nomenclature and systematic information over the last three decades (intensified even more by DNA sequence analyses), numerous changes have been made in the fungi names to reflect the phylogenetic situation[3]. Currently, Kingdom Fungi (true fungi) is divided into four major phyla: Chytridiomycota, Zygomycota, Ascomycota and Basidiomycota. The deutoromycetes are an additional true fungi group but are not recognized as a formal phylum[4]. The fungal body (thallus) is either a single cell or a threadlike multi-cell structure (hyphae), which constitutes the mycelium, and may give rise to multiform spore-producing cells. In most cases, these spore-producing cells form part of a special structure called the fruit body (sporocarp). "Mushroom" is a popular term applied to fungi that have this structure as seen with the naked eye[5]. Higher fungi (macrofungi or macromycetes), which produce these characteristic macroscopic fruiting bodies used to disperse their spores, include basidiomycetes, known as mushrooms, and members of the Clavicipitaceae, the Hypocreaceae, and the Xylariaceae in ascomycetes. There are approximately 10,000 species of higher fungi known in China, including edible and inedible mushrooms representing about 600 and 500 species, respectively, and some 100 strongly toxic and often lethal species[6].

Fungi have played an important role in the history of human life. According to the relationship between humans and fungi, some fungi are considered beneficial to humans while some others are harmful and pathogenic. The beneficial aspects can be summarized in the following major areas: i). Food. For example, *Lentinus edodes*, *Auricularia auricula-judae*, *Tremella fuciformis*, *Flammulina velutipes*, and some other fruiting bodies of fungi, namely mushrooms, have been valued as delicious food since ancient times. Some nonvolatile amino acids and volatile compounds containing eight carbons are believed to be the origin of delicious tastes and fragrant odors[7]. These fungi are nutritious and found to contain vitamins B_1, B_2, PP, C, and D, among others. Most edible mushrooms contain vitamins at levels 2 to 8 times higher than those in most vegetables. ii). Food processing. *Sacharomyces cerevisiae*, *Aspergillus oryzae*, and other fungi from the genus *Penicillium* have been used for making foodstuffs such as cheese, bread dough, preserved bean curd, and soy sauce. They are also used in alcoholic fermentation processes. The fermentation products of *Monascus*, especially those produced by solid-state fermentation of rice, have been used in China as food and health remedies for over 1000 years. iii). Chemical and biochemical product synthesis. The common products synthesized by various fungi include active enzymes such as amylase, albumenase and fibrinase, organic acids such as gallic acid and citric acid, and chitin and water-soluble chitosan (polysaccharides composed from N-acetyl-D-glucosamine units). iv). Environmental recycling. Fungi are heterotrophic eukaryotes that play one of the major roles in the decomposition of dead plant tissues (cellulose and lignan) and to a lesser extent animal tissues such as keratin and chitin[4]. They encode a variety of peptidases that are likely used to degrade proteins in their environment[8]. In particular, soil fungi are the principal degraders of biomass in terrestrial ecosystems and are responsible for much of the organic re-cycling in the environment[9]. v). Production of antibiotics. The most prolific producers of clinically used compounds are the ascomycetes. Sir Alexander Fleming, in his landmark discovery in 1929, noted the effects of *Penicillium notatum* (now known as *Penicillium chrysogenum*) and penicillin (β-lactam) on bacteria. Later, penicillin was developed as the first highly active antibiotic. It is well known that the β-lactam antibiotics exert their effects by interfering with the cross-linking structure of peptidoglycans in bacterial cell walls. Ever since penicillin was isolated from *P. notatum*, chemists have been involved in the discovery of other novel bioactives from microbial metabolites. In addition to penicillin derivatives, cephalosporins, β-lactam antibiotics originally derived from *Acremonium*, are used widely for the treatment of bacterial infections. Another important fungal compound often used in clinics is cyclosporin, an immunosuppressant isolated from *Cordyceps subsessilis* (asexual state: *Tolypocladium inflatum*). Other noted fungal metabolites in successful clinical use include ergotamine and the antihypercholestemic drug lovastatin and its derivatives[10].

Currently, many antibiotics are manufactured in large quantities through the use of fermentation processes. For example, *Penicillium chrysogenum* is used to manufacture penicillin; *Penicillium griseofulvum* is used to manufacture griseofulvin; and *Cephalosporium acremonium* is used to manufacture cephalosporin. vi). Pesticides, treatment of plant diseases, and herbicides. Some fungi produce a wide variety of toxic metabolites. These fungi are important natural regulators of weed and insect populations. *Beauveria bassiana* and *Puccinia xanthii* are two good examples. *B. bassiana* is a known natural enemy of a number of insect pests of crop plants, and *P. xanthii* is often used to prevent the intruding of the giant ragweed *Ambrosia trifida*. vii). Fungal remedies. Medicinal fungi used for the prevention and treatment of various diseases occupy a very small fraction of the total fungal resource. Most belong to the basidiomycetes and ascomycetes. According to Mao's investigation,[11] over 1,000 edible wild fungi species were identified in China, of which approximately 500 species were classified as medicinal fungi. Basidiomycetes, especially polypores, have a long history of folk medicinal use. The use of hot water extracts of medicinally important macrofungi as remedies for various diseases is known in the folk medicines of many eastern countries.

The first written reports (*Compendium of Materia Medica*) associated with the medicinal value of mushrooms date back to the Eastern Han Dynasty (25 to 220 AD) in China. Several polysaccharide preparations from basidiomycetes, including polypores such as *Grifola frondosa*, *Ganoderma lucidum*, and *Trametes versicolor*, are being used in clinical trials in China. Other noted medicinal fungi include *Poria cocos*, *Cordyceps sinensis*, *Coriolus versicolor*, *Omphalia lapidescens*, *Lasiosphaera fenzlii*, *Polyporus umbellatus*, *Lentinus edodes*, *Fomes fomentarius*, *Fomitopsis officinalis*, *Laetiporus sulphureus*, *Inonotus obliquus*, and *Wolfiporia cocos*.

The harmful impact that certain fungi have on humans as well as other animals can be categorized as: i). Human and livestock infections by fungi. Genera *Trichophyton*, *Microsporum* and *Epidermophyton* are the common pathogens that cause scald head, ringworm of the body, and athlete's foot. *Aspergillus* species are ubiquitous molds that cause life-threatening infections in immunocompromised patients[12]. *Aspergillus fumigatus* is well known to cause aspergillosis in chickens. ii). Food poisoning. Food poisoning deaths of fungal origin are reported every year. Many toxins have been identified in wild fungi, including at least thirty types that were determined to be lethal, such as muscarine produced by *Clitocybe cerussata*[13]. Although a few species of *Amanita* are edible, fatal poisoning is almost exclusively attributable to members of the genus *Amanita*, with Phalloideae responsible for 90% of the fatal human mushroom poisonings[14]. iii). <u>Food decay and goods rot</u>. Some *Penicillium* spp. and *Aspergillus* spp. can cause food decay, and clothes, leather and fibre mildew and rot. iv). <u>Wood degradation and plant diseases</u>. Wood degradation by *Trichoderma harzianum* occurs frequently. In China, more than 800 fungal species were recorded to have caused wood degradation[11]. Around 80% of known crop diseases are generated by pathogenic fungi. *Verticillium* spp. are soil-borne plant pathogens responsible for Verticillium wilt diseases in over 200 hosts, including many economically important crops. Disease symptoms may comprise of wilting, chlorosis, stunting, necrosis and vein clearing[15].

Whether fungi are beneficial or harmful to humans, animals and plants is mainly determined by the metabolites contained within, or generated by, the specific fungi. Medicinal fungi can be used to treat various diseases because they produce beneficial bioactive compounds; while some fungi are lethal because they produce potent toxins.

Macrofungi are distinguished as important natural resources of immunomodulating and anticancer agents. These agents belong mainly to polysaccharides (especially β-D-glucan derivatives), glycopeptide/protein complexes (polysaccharide-peptide/protein complexes), proteins and triterpenoids. These compounds have been found to act upon the immune effecter cells involved in the innate and adaptive immunity, resulting in the production of biologic response modifiers[5]. One of the well-studied antitumor polysaccharides present in polypores is PSK, also known as Krestin, a protein-bound polysaccharide or glycoprotein, isolated from *Trametes versicolor* (as *Coriolus versicolor*). Currently, it is used in Japan as an immunotherapeutic agent to treat colorectal, gastric, and lung cancers. It was also reported to have antiviral effect, including *in vitro* anti-HIV (human immunodeficiency virus) activity[16,17].

Some fungi generate both bioactive compounds and toxins. For example, *Monascus* spp. were found to produce valuable metabolites including food colorants, cholesterol-lowering agents, and antibiotics. However, several species of the genus *Munascus* also produce citrinin, a mycotoxin harmful to the hepatic and renal systems[18]. The poisonous metabolites produced by fungi are often called mycotoxins. In order to better understand mycotoxins, phytotoxins need to be introduced here. As reviewed by Berestetskiy,[19] phytotoxins (PTs) are largely represented by low molecular weight secondary metabolites capable of deranging the vital activity of plant cells or causing their death at concentrations below 10 mM. Although PTs are formed by diverse organisms including bacteria, plants, and even phytophagous insects, fungi, particularly those that are phytopathogenic, are best known as PT producers. A considerable number of PTs are toxic to animals and/or microorganisms. For example, *Amanita phalloides*, a poisonous basidiomycete fungus, is known to produce the fatal amatoxins and phallotoxins that damage the liver and kidneys[13]. Amatoxins are mainly produced by four fungal genera: *Amanita, Galerina, Lepiota* and *Conocybe*. Amatoxins are known to inhibit eukaryotic DNA-dependent RNA polymerase II, resulting in the cessation of transcription that leads to target cell death. Phallotoxins are produced by two fungal genera: *Conocybe and Amanita*, and have been found to directly act on actin[12]. Regarding the economic importance, phytotoxic metabolites are arbitrarily classified as mycotoxins or antibiotics.

Bioactive secondary metabolites of fungal origin cover a broad range of the function spectrum in addition to antibiotic activity. Plant endophytic fungi and marine-derived fungi have proven to be rich sources of biologically active and structurally novel secondary metabolites. The term 'endophyte' covers all organisms that, during a variable period of their lives, colonize the living internal tissues of their hosts. Almost all plants in nature are thought to symbiose with mycorrhizal and/or endophytic fungi that express different symbiotic lifestyles ranging from parasitism to mutualism. Endophyte-free plants are few. This is especially true of shrubs and trees. Endophytes are found in a wide variety of tissue types of all plant divisions from mosses and ferns to monocotyledons. The most common endophytic microorganisms found are bacteria and fungi, and endophytic fungi are the ones frequently isolated. It is generally believed that endophytes play an important beneficial role in the physiology of host plants, and together with mycorrhizal fungi, which generate various metabolites and can directly or indirectly have an influence on plant secondary metabolism, play an important role in ecological systems through shaping plant communities and mediating ecological interactions. Plants infected with endophytes are often found to be healthier than endophyte-free ones. This may be partly due to the endophytes' production of phytohormones and/or promotion of the hosts' absorption of nutrition. Endophyte-infected plants also gain protection from herbivores, pests and pathogens due to the bioactive secondary metabolites that are generated by endophytes in plant tissue. A plant's stress (such as drought, heat, salinity, and disease) tolerance can also be enhanced by fungal symbiosis[20]. As a result of gene recombination during the evolutionary process, endophytic microorganisms have developed the ability to produce compounds similar or identical to those produced by their host plants[21]. Camptothecin (CPT) and taxol are two good examples[22]. The terpenoid indole alkaloid anticancer agent CPT was first isolated from the Chinese tree *Camptotheca acuminata* in 1966, and later was found in many other unrelated orders of angiosperms. Recently, CPT was reportedly produced by the endophytic fungus isolated from the inner bark of *Nothapodytes foetida* found in the Western coast of India. Taxol, an anticancer diterpene found in *Taxus* species, was also produced by the endophytic fungus *Taxomyces andreanae* from the plant *Taxus brevifolia*. In contrast to other endophytes, fungal endophytes are seldom reported to produce polysaccharides, enzymes or proteins. Major natural products from fungal endophytes can be grouped into several categories including alkaloids, steroids, terpenoids, isocoumarins, quinones, phenylpropanoids, lignans, phenol and phenolic acids, aliphatic metabolites, and lactones[21].

Marine-derived fungi have proven to be a promising source of novel bioactive agents. Most marine fungi are capable of generating unique and unusual secondary metabolites, probably promoted by their unique and extreme growth habitat. Various fungal strains have been isolated from the inner tissues of marine invertebrates. It was evident that many facultative marine fungi belong to genera common to terrestrial habitats[23]. In general, compounds produced by marine-derived fungi bear the typical structural features expected for fungal metabolites and are structurally not related to the invertebrate hosts[24].

In this chapter, the introduction of the chemistry of fungal products, the focus is on the major groups of fungal secondary metabolites, including the potent biologically active compounds present in medicinal fungi, as well as some mycotoxins found in pathogenic species.

Section 2   Bioactive Fungal Metabolites

It is not possible to classify fungal metabolites within a particular group of chemical compounds. The majority of bioactive compounds of fungal origin are represented by terpenoids, steroids, alkaloids and other heterocyclic compounds, organic acids, acetylene derivatives, polyketides, macrolides, pyranones, lactams, peptides, proteins, polysaccharides, glycosides, sphingolipids, perylenequinones, *p*-terphenyls and other phenolic compounds, and derivatives thereof. These compounds exhibit antibiotic, antineoplastic, antitumor, antiviral, anti-inflammatory, immunomodulatory, enzyme inhibitory, cardiovascular, neuritogenic, analgesic, antidiabetic, antioxidant, phytotoxic, insecticidal, nematocidal, and other activities.

It is interesting that fungal secondary metabolites may differ for some species at different stages in the fungal life cycle. A good example is the polypore *Merulius tremellosus*. Its fruiting bodies produce antimicrobial polyketide metabolites, merulinic acids A, B and C (**1**-**3**); while its mycelial cultures do not produce merulinic acids, but instead a highly antifungal sesquiterpenoid, merulidial (**4**)[3].

Some fungi may produce different secondary metabolites when they grow in a different environment or are cultured under different conditions. For example, an induced diterpene libertellenone D (**5**), with cytotoxicity against the HCT-116 human adenocarcinoma cell line (inhibitory concentration 50%: $IC_{50} = 0.76$ µM), was obtained from the culture of marine-derived fungus *Libertella* sp. after a marine α-proteobacterium was added into the culture medium. However, when the fungus and the bacterium were cultured alone, diterpene metabolites were not observed[25].

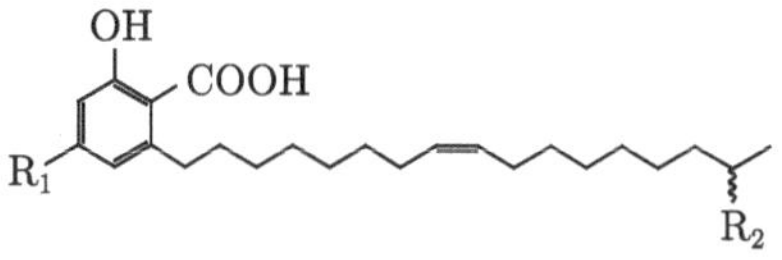

1 R$_1$=OH, R$_2$=H; **2** R$_1$=H, R$_2$=OH; **3** R$_1$=H, R$_2$=H   **4** 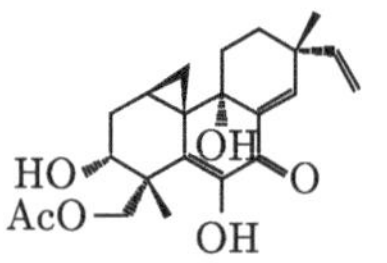   **5**

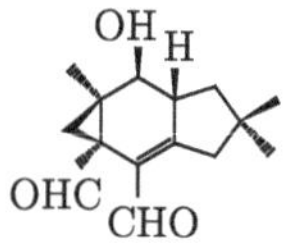

2.1   Terpenes

Terpenes are a large class of hydrocarbons derived biosynthetically from units of isoprene, which has the molecular formula C_5H_8.

Within the terpenoid group, the prominent toxins are sesquiterpenes that mainly belong to trichothecene and aristolochene derivatives. Sesquiterpene stachyflin (**6**), which contains a pentacyclic moiety including a novel *cis*-fused decalin, was isolated from the fungus *Stachybotrys* sp. RF-7260. It demonstrated potent *in vitro* antiviral activity against influenza A virus (H1N1) ($IC_{50} = 3$ nM) through a unique mechanism of inhibiting fusion between the viral envelope and the host cell membrane[26,27]. Sesquiterpene xenovulene (**7**) was isolated from *Acremonium strictum* by Ainsworth et al.[28]. It inhibited flunitrazepam binding to the GABA-benzodiazepine receptor with an IC_{50} of 40 nM in an *in vitro* assay.

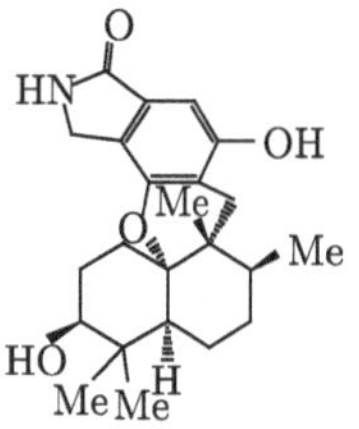

6 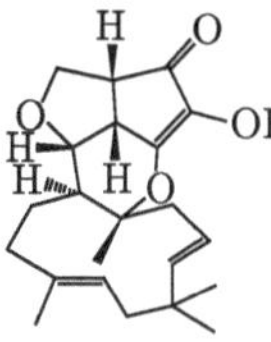   **7** 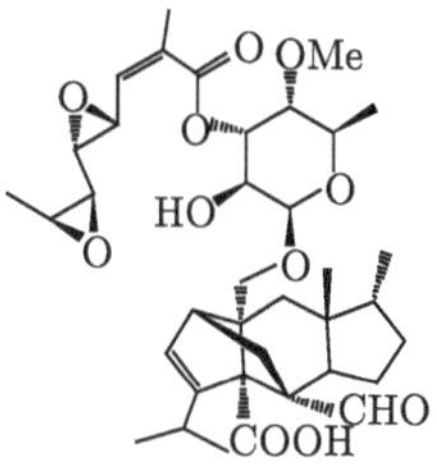   **8** 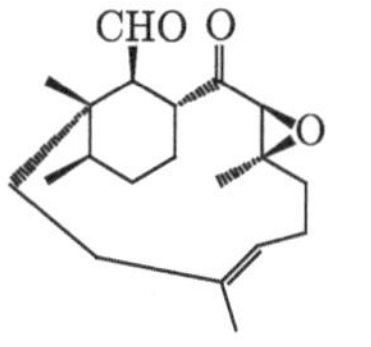   **9**

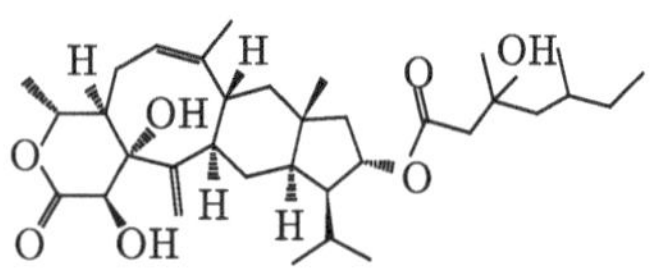

10 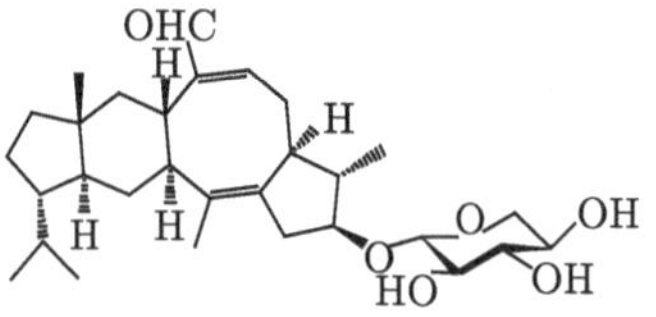   **11** 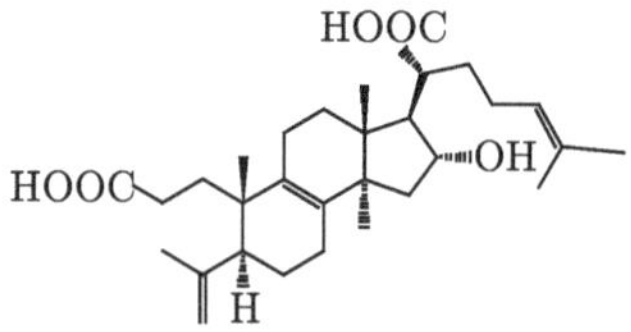   **12**

Daferner and co-workers obtained an antifungal sordarin derivative, hypoxysordarin (**8**), from a facultative marine fungus *Hypoxylon croceum* M97-25 isolated from driftwood in a mangrove estuary in the Everglades, Florida[29]. This diterpene presented selective inhibition effect toward several yeast and filamentous fungal species with a range of MIC (minimum inhibitory concentration) at 0.5 to 20 µg/mL. Sordarin derivatives have proven to selectively inhibit fungal protein synthesis and have a broad spectrum of activities. Another structurally interesting diterpene is phomactin D (**9**) reported by Sugano et al.[30]. This PAF (platelet activating factor) antagonist is produced by *Phoma* sp. SANK 11486, which was isolated from the shell of the crab *Chinoecetes opilio* found in Japan, and exhibited the inhibition of PAF-iduced platelet aggregation (IC_{50} = 0.8 µM) and binding of PAF to its receptor (IC_{50} = 0.12 µM).

Compound **10**, an unusual tricarbocyclic δ-lactone sesterterpene (a hexadecahydroindeno [5′,6′:4,5]cycloocta[1,2-c]pyranyl heptanoate derivative) from the fungus *Paecilomyces inflatus*,[31] strongly inhibited glycosylphosphatidylinositol synthesis *in vitro* of yeast microsomes (MIC = 3.4 nM). Aleurodiscal (**11**), a potent antifungal β-D-xyloside isolated from *Aleurodiscus mirabilis*, a wood-rotting polypore,[32] was selectively active against zygomycetes, especially *Mucor miehei* (MIC = 0.3∼1 µg/mL).

Numerous cytotoxic triterpenoids have been isolated from various species of the Polyporaceae. A good example is poricoic acid G (**12**), a new 3,4-secolanostane-type triterpene, isolated from the sclerotium of *Poria cocos*[33]. This compound has shown potent cytotoxicity against leukemia HL-60 cells (GI_{50} = 39.3 nM, concentration resulted in a 50% growth inhibition of cancer cells).

2.2   Steroids

Zhankuic acid A (**13**), a steroid isolated from the polypore *Antrodia cinnamomea*, a parasite of the heartwood of the Taiwan evergreen tree *Cinnamomum micranthum*, has shown cytotoxic activity against P-388 murine leukemia cells (IC_{50} = 1.8 µg/mL)[34]. *A. cinnamomea* has been utilized in Chinese medicine for the treatment of various disorders including liver cancer.

Favolon (**14**), an unusual ergosterone with a B/C-*cis* ring junction, was obtained from the culture of an Ethiopian *Favolaschia* species. This compound displayed strong antifungal activity against numerous pathogens, especially for *Mucor miehei*, *Paecilomyces varioti*, and *Penicillium islandicum* in the agar diffusion assay[35].

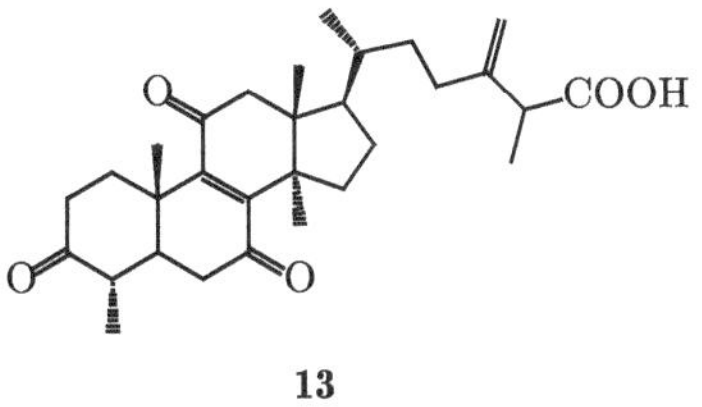

13 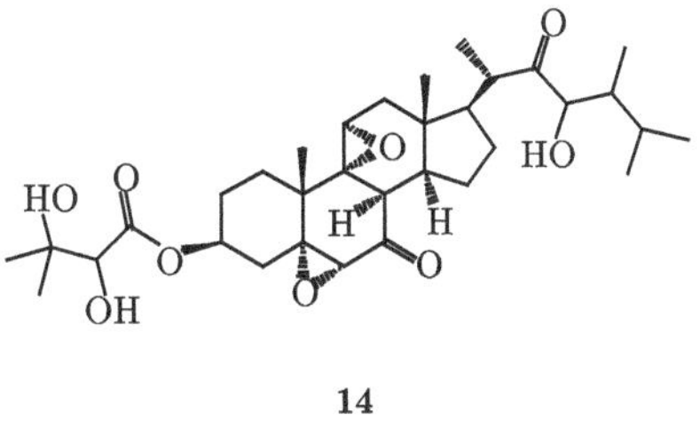   14

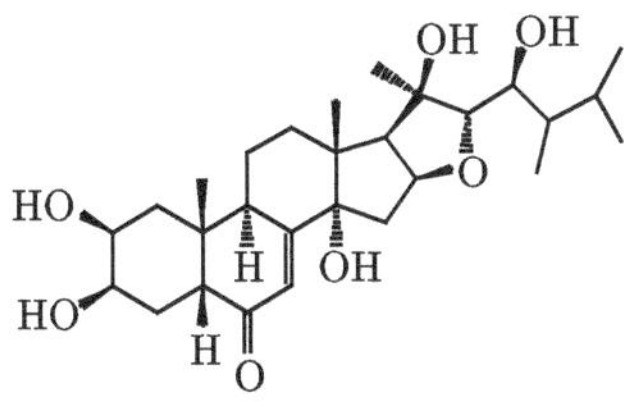

15 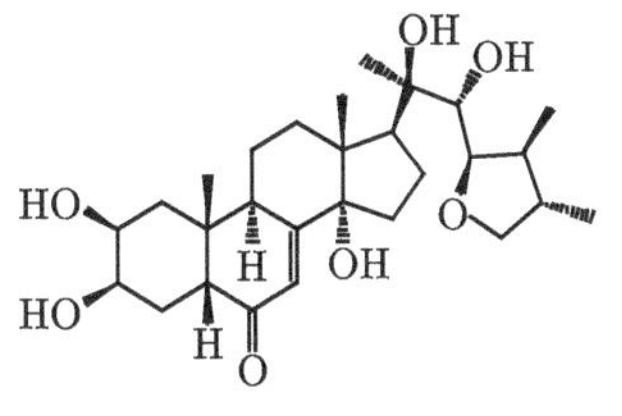   16 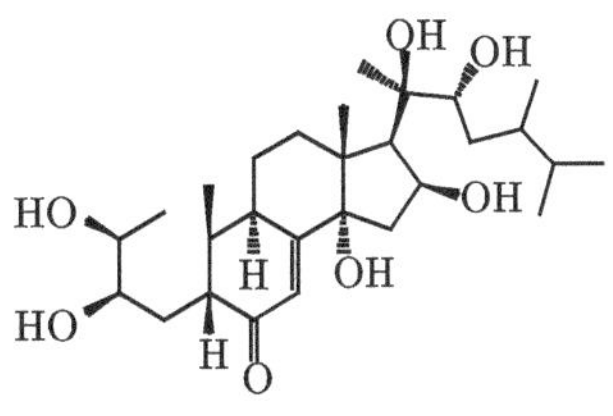   17

Recently, bioassay-guided fractionation of the sclerotia of *Polyporus umbellatus* (Zhu-Ling) resulted in three new ergostane-type ecdysteroids polyporoid A (**15**), B (**16**), and C

(**17**), together with several other known ecdysteroids. Ecdysteroids A and B, containing a furan ring on the side chain, are rare in nature. All these compounds exhibited strong anti-inflammatory activity in the test of TPA-induced inflammation (1 µg/ear) in mice (inhibitory dose 50%: ID_{50} of 0.117–0.682 µM/ear)[36].

2.3   Polyketides

Polyketides are a group of compounds containing alternating carbonyl and methylene groups. Among phytotoxic polyketides, the best known are curvulin (**18**) and cercosporin (**19**)[19]. Curvulin is a host specific phytotoxin from *Drechslera indica* that causes necroses on purslane and spiny amaranth, while cercosporin is a non-host specific phytotoxin from *Cercospora* spp. that is pathogenic to plants, causing damaging leaf spot and blight diseases.

A novel γ-lactone, acremolactone A (**20**), was obtained from *Acremonium roseum*. This compound exhibited potent herbicidal activity, especially against the harmful weeds crabgrass (*Digitaria adscendens* Henr.) and smartweed (*Polygonum blumei* Meisn.)[37].

PS-990 (**21**), from *Acremonium* sp. KY 12702, has exhibited an inhibition effect on brain calcium/calmodulin-dependent cyclic neurotide phosphodiesterase (IC_{50} = 3 µg/mL)[38].

ES-242-1 (**22**), a bioxanthracene isolated from *Verticillium* sp. SPC-15898, is an antagonist of N-methyl-D-aspartate receptor, inhibiting [³H]thienyl cyclohexylpiperidine binding to rat crude synaptic membranes at IC_{50} of 0.116 µM[39].

Saintopin (**23**), a reddish purple pigment isolated from *Paecilomyces* sp., is an antitumor agent with topoisomerase II-dependent DNA cleavage activity. It exhibited high *in vitro* cytotoxicity against human tumor cell line HeLa S3 (IC_{50} = 0.35 µg/mL)[40].

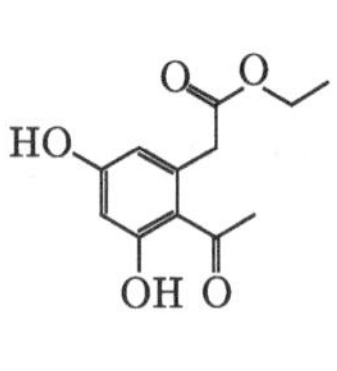

18 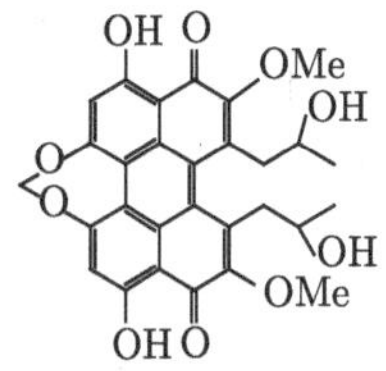   19 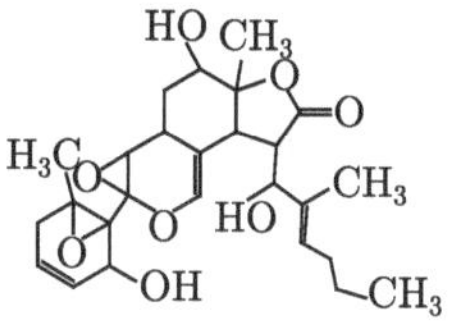   20 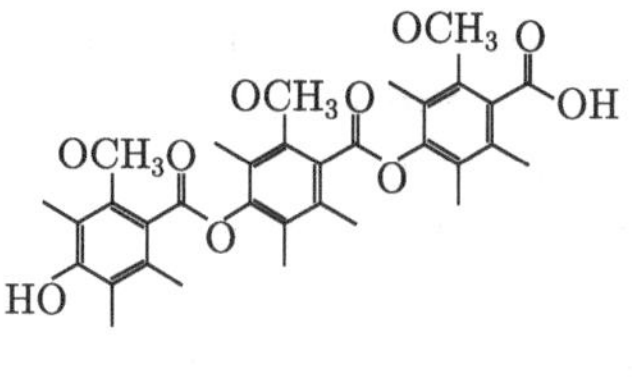   21

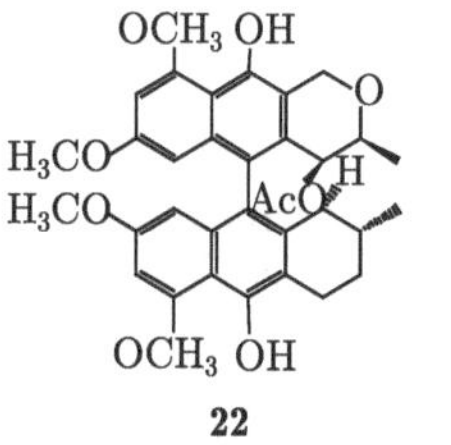

22 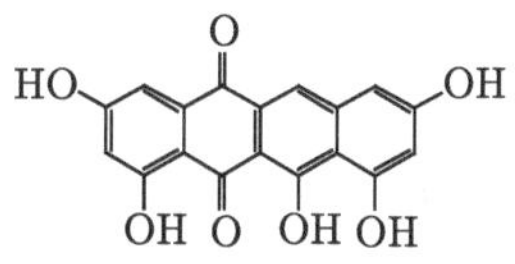   23

2.4   Phenols, quinones, and other aromatic compounds

The structures of phenol and quinone derivatives are similar to polyketides. In an effort to search for free radical scavengers, several leucomentins (**24**, **25**, **26**, **27**) were isolated from fruiting bodies of *Paxillus panuoides*[41]. These compounds showed better inhibition against lipid peroxidation in rat liver microsomes (IC_{50}s: 0.7 µM for **24**, 0.5 µM for **25** and **26**, and 0.4 µM for **27**) than the control, vitamin E (IC_{50} = 2.6 µM). Another leucomentin analog, kynapcin-12 (**28**) isolated from the fruiting bodies of the Korean mushroom *Polyozellus multiplex*,[42] was found to be an inhibitor (IC_{50} = 1.25 µM) of prolyl endopeptidase (PEP), a serine protease that plays an important role in the degradation of proline-containing neuropeptides. PEP inhibitors might be useful as lead compounds in the development of anti-dementia drugs. Bl-III (**29**) was isolated from the methanol extract of the fruiting bodies of *Boletopsis leucomelas*. This terphenol inhibited the enzyme 5-lipoxygenase which is involved in many inflammations and allergies (IC$_{50}$ = 0.35 μM)[43]. An unusual terphenol, kynapcin-9 (**30**) isolated from fruiting bodies of the Korean toadstool *P. multiplex*, was found to inhibit prolyl endopeptidase with an IC$_{50}$ at 0.212 μM[44].

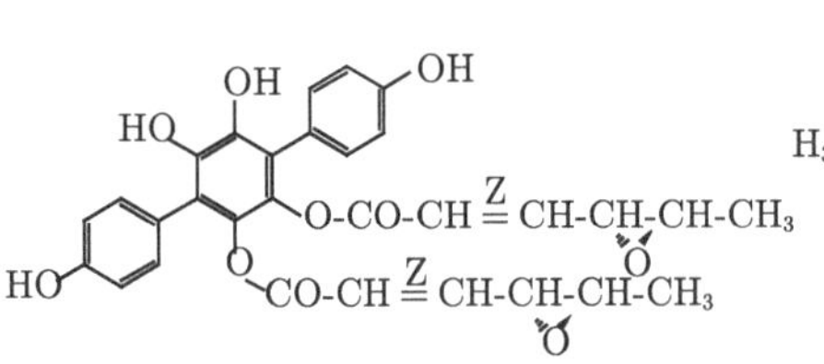

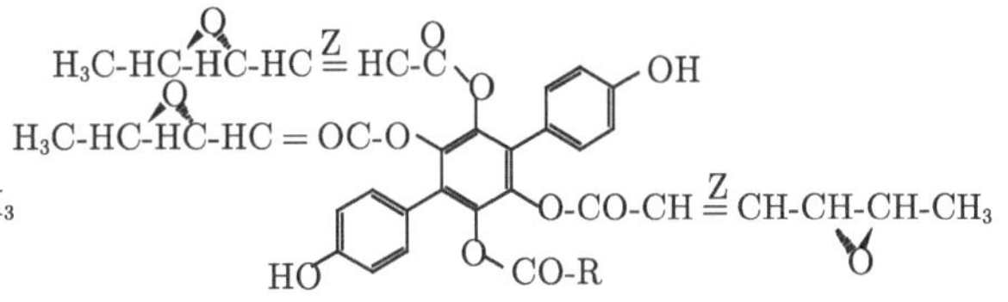

24 25 R = CH $\overset{Z}{=}$ CH-CH-CH-CH₃;   **26 R** = CH₃; **27 R** = CO-CH $\overset{Z}{=}$ CH-CH $\overset{E}{=}$ CH-CH₃

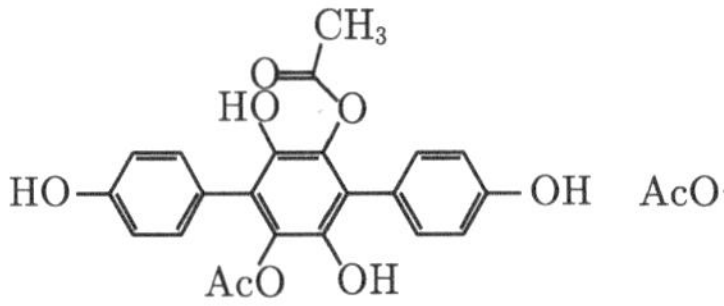

28 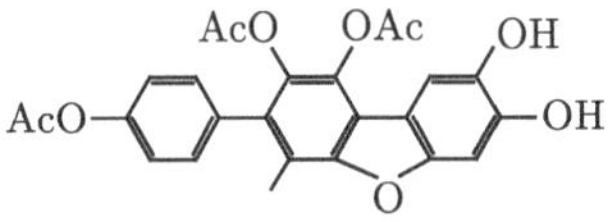   **29** 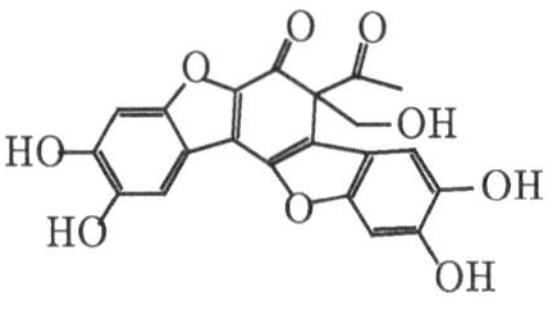   **30**

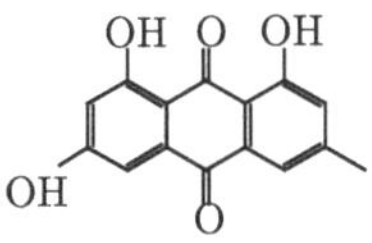

31 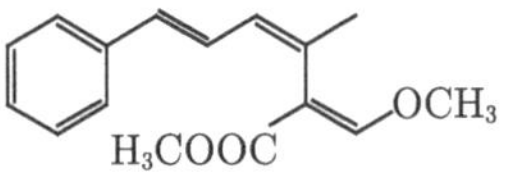   **32** 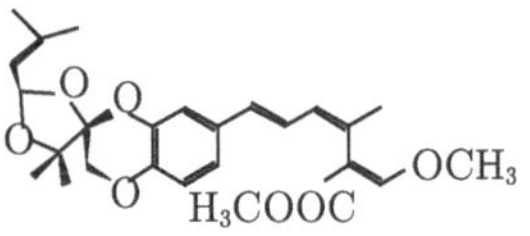   **33** 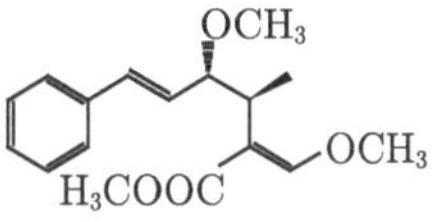   **34**

Anthraquinones are well-known metabolites present in *Trichoderma* species, free-living fungi in root, soil and foliar environments that have been used successfully against many crop pathogens. Emodin (**31**), found in *Trichoderma viride*, inhibits both monoamine oxidase and tyrosine kinase, and acts as an antimicrobial, antineoplasic and cathartic agent[45].

Aromatic antifungal compounds strobilurins A (**32**), E (**33**), and oudemansin A (**34**), non-mammalian toxic derivatives of β-methoxyacrylic acids, were isolated from basidiocarps of the genera *Agaricus*, *Favolaschia*, and *Filoboletus*. These compounds exhibited potent antifungal activity (MIC range: 0.1-1 μg/mL). They might be valuable lead compounds for the development of commercial agricultural fungicides[3].

2.5   Polyacetylenes

Acetylenic metabolites are a group of molecules containing one or more triple bonds. Some of these possess antitumor, antimicrobial, antifouling, pesticidal, phototoxic, HIV-inhibitory, and immunosuppressive properties. Mycomycin (**35**) is used as a therapeutic agent for tuberculosis and late-stage inoperable primary hepatocellular carcinoma. This compound, first isolated in 1950 by Jenkins, was later isolated from fermentations of fungal culture LL-07F275[46]. The antibiotic 10-hydroxyundeca-2,4,6,8-tetraynamide (**36**), isolated by Baeuerle and co-workers[47] from cultures of the fungus *Mycena viridimarginata*, showed encouraging activities against bacteria, yeasts, filamentous fungi, and Ehrlich ascites carcinoma. Two tricholomenyns A (**37**) and B (**38**) from the fruiting bodies of *Tricholoma acerbum*, isolated by Garlaschelli and co-workers,[48,49] exhibited efficient inhibition of T-lymphocyte mitosis.

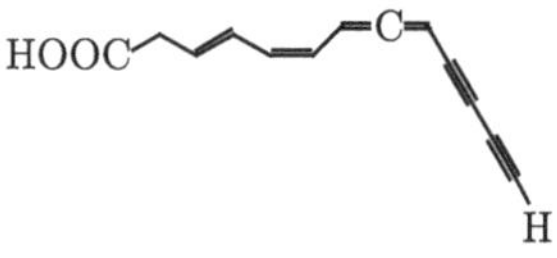

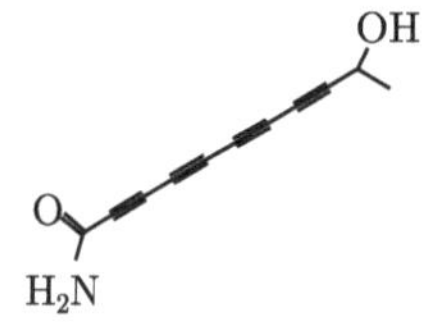

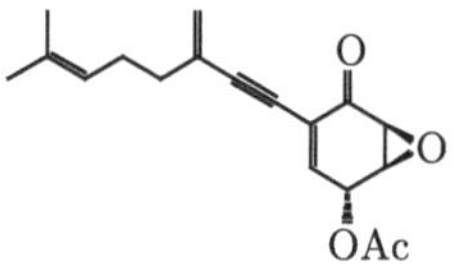

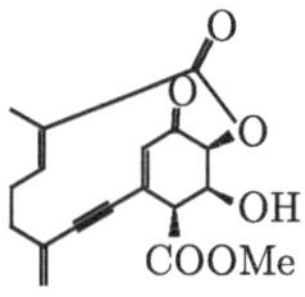

35 36 37 38

2.6 Alkaloids

Alkaloids are basic organic compounds containing at least one nitrogen atom in a heterocyclic ring. The majority of the alkaloids isolated from fungi were found in cultures of grass-associated endophytic fungi. These metabolites play an important role in inhibiting herbivores and insects.

Two tetramic acids cryptocin (**39**) and ascosalipyrrolidinone A (**40**), with major differences in the orientations of specific substituents, exhibited different bioactivities. Cryptocin was obtained from the fungus *Cryptosporiopsis cf. quercina* isolated from the bark of *Tripterygium wilfordii*. It presented potent antimycotic activities against certain plant pathogenic fungi (MIC = 1.0 μg/mL) from oomycetes (*Pythium* and *Phytophthora*) and ascomycetes (*Sclerotinia* and *Pyricularia*), especially *Pyricularia oryzae*[50]. Compound **40** was isolated from the marine fungus *Ascochyta salicorniae* found in the marine green alga *Ulva* sp. It showed potent antiplasmodial activities against two strains of malaria-causing protozoan of *Plasmodium falciparum* K1 (Thailand; resistant to chloroquine and pyrimethamine, IC_{50} = 736 ng/ml) and NF54 (an airport strain of unknown origin, IC_{50} = 378 ng/ml)[51].

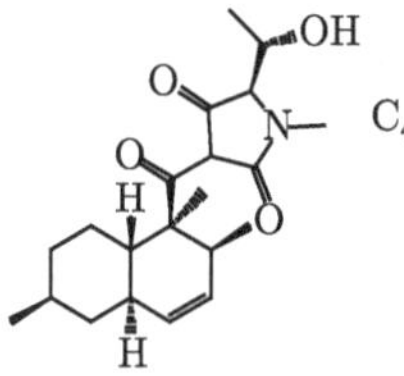

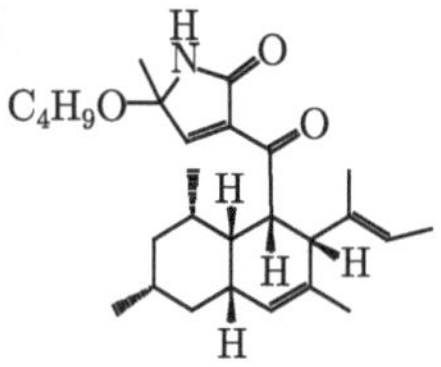

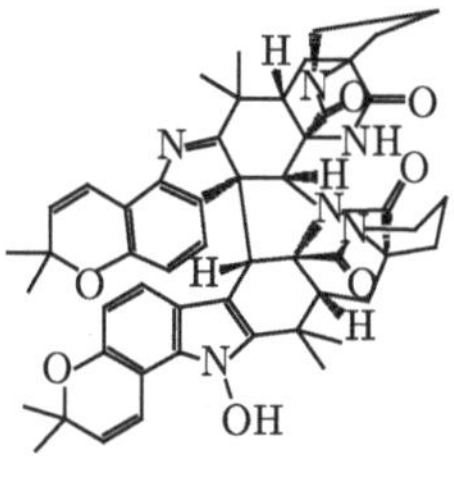

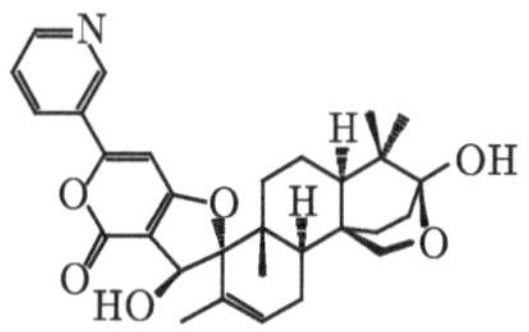

39 40 41 42

Stephacidin B (**41**), a structurally interesting antitumor agent, was obtained by fermentation of *Aspergillus ochraceus* WC76466[52]. Compound **41** showed *in vitro* cytotoxic activity against human prostate, ovarian, colon, breast, and lung cancer cell lines, and was especially selective against testosterone-dependent prostate LNCaP (IC_{50} = 0.06 μM).

The pyridine alkaloid decaturin B (**42**) from fungi *Penicillium decaturense* and *P. thiersii*, reported by J. B. Gloer and co-workers, exhibited potent anti-insect activity against the fall armyworm *Spodoptera frugiperda*[53].

2.7 Macrolides

There are more than 200 macrolides that have been isolated from fungi, including up to 16-membered ring monolactone macrolides and macrodiolides, up to 18-membered ring macrotriolides, and up to 25-membered ring macrotetrolides[54].

Two phenochalasins A (**43**) and B (**44**), novel inhibitors of lipid droplet formation, were isolated from marine sponge-derived fungus *Phomopsis* sp. FT-0211 by Tomoda and co-workers[55−57]. Compound **43** inhibited cholesteryl ester synthesis with an IC_{50} of 0.61 μM, while compound **44** exhibited stronger inhibitions in both cholesteryl ester ($IC_{50} = 0.20$ μM) and triacylglycerol syntheses ($IC_{50} = 0.38$ μM).

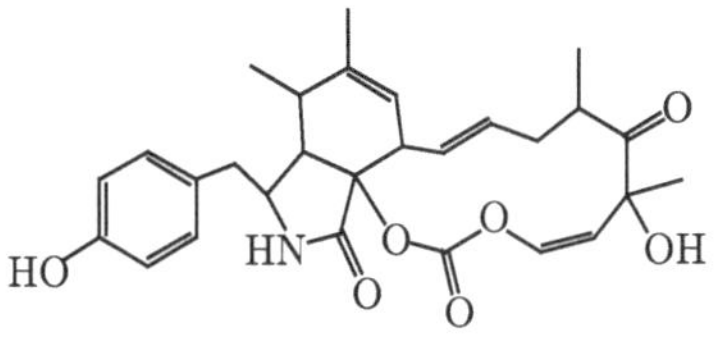

43 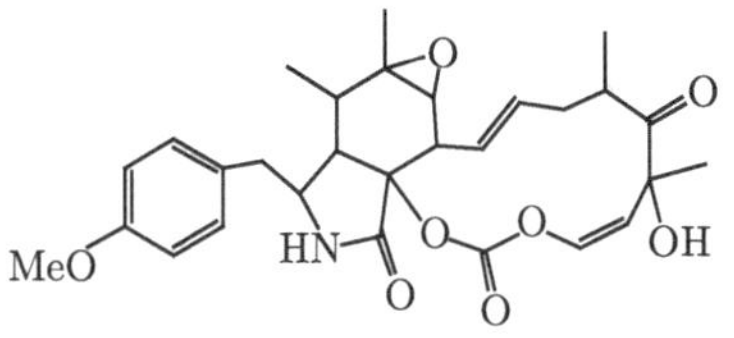   **44** 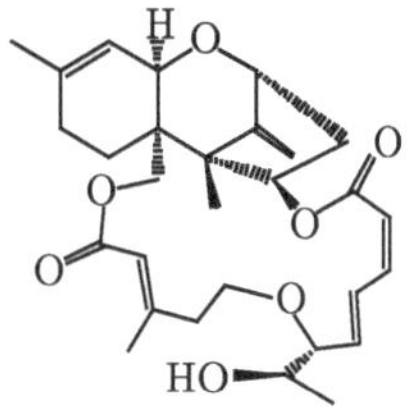   **45**

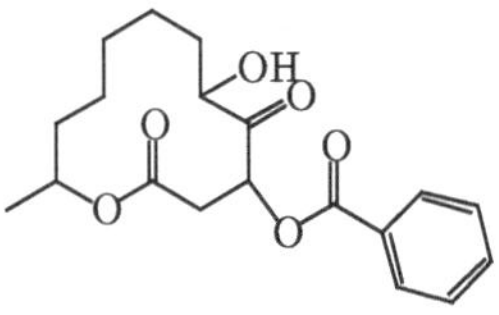

46 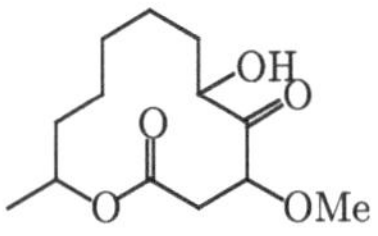   **47**

A macrocyclic trichothecene, 12,13-deoxyroridin E (**45**),[58] generated by *Myrothecium roridum* TUF 98F42 isolated from a submerged woody material in Palau, exhibited strong cytotoxicity against HL-60 ($IC_{50} = 25$ ng/mL) and L1210 cell lines ($IC_{50} = 15$ ng/mL). It is worth mentioning that, a derivative of **45**, which has a 12,13-epoxy at the exo-methylene position, showed about an 80-fold stronger activity than did compound **45**.

Sporiolides A (**46**) and B (**47**) (two twelve-membered macrolides) were obtained from the cultured broth of a fungus, *Cladosporium* sp., isolated from an Okinawan marine brown alga *Actinotrichia fragilis*[59]. Compound **46** showed cytotoxicity against L1210 cell line ($IC_{50} = 0.13$ μg/mL), antifungal activity against *Candida albicans* (MIC = 16.7 μg/mL), *Cryptococcus neoformans* (8.4 μg/mL), *Aspergillus niger* (16.7 μg/mL), and *Neurospora crassa* (8.4 μg/mL), and antibacterial activity against *Micrococcus luteus* (16.7 μg/mL). Compound **47** exhibited cytotoxicity against L1210 cells ($IC_{50} = 0.81$ μg/mL) and antibacterial activity against *Micrococcus luteus* (16.7 μg/mL).

2.8   Peptides, diketopiperazines, depsipeptides

Peptide phytotoxins include cyclic and noncyclic compounds that frequently include residues of unusual amino or hydroxy acids. The cyclic forms are more frequently reported.

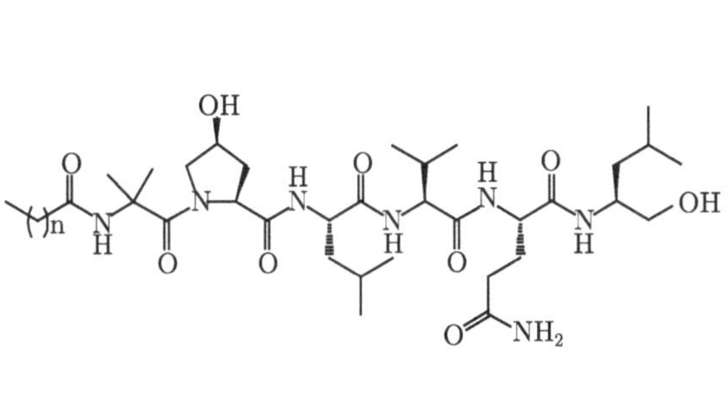

48 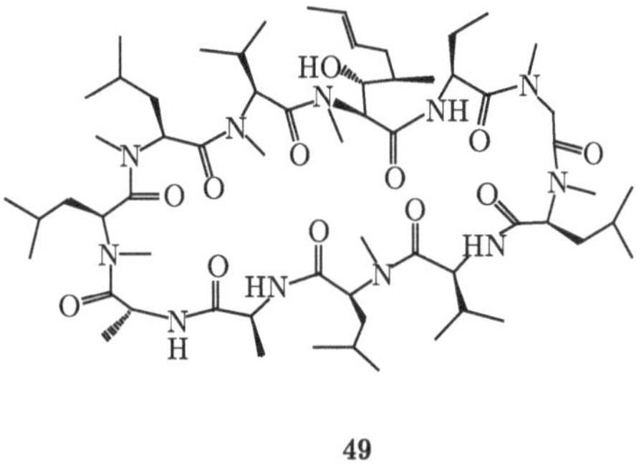   **49**

Halovirs A (**48**), a linear peptide from the marine fungus *Scytidium* sp., has shown potent antiviral activity against herpes simplex virus (HSVs) 1 (effective dose 50%: $ED_{50} = 1.1$ μM) and 2 ($ED_{50} = 280$ nM) in a whole cell assay against infectious viruses[60].

Basidiomycetes (such as *Amanita, Conocybe, Lepiota, Galerina,* and *Omphalotus* species)
are rich in cyclopeptides. In most cases, these compounds are toxic to mammals, and have
associated cytotoxic, insecticidal, antimalarial, estrogenic, sedative, nematicidal, antimicrobial, immunosuppressive, and enzyme-inhibitory activities[61]. HC-toxin (an inhibitor
of histone deacetylases (HDACs) in many plants, insects, and mammals[62]) produced by
Cochliobolus carbonum is a cyclic tetrapeptide of structure cyclo(D-Pro-L-Ala-D-Ala-L-Aeo),
where Aeo stands for 2-amino-9,10-epoxi-8-oxodecanoic acid. Cyclosporin A (**49**), a cyclic
oligopeptide with 11 amino acids originally isolated from *Tolypocladium polysporum,* has
been used in post-transplantation therapy and treatment of some autoimmune diseases, as
well as in the treatment of nephritic syndrome, refractory Crohn's disease, biliary cirrhosis,
ulcerative colitis, rheumatoid arthritis, anemia, myasthenia gravis, and dermatomyositis.
Cyclosporins are common metabolites from the genus *Tolypocladium.* The side effects of
these compounds include toxicity to kidneys and increased blood pressure and tremors[63].

50	**51** R=H	**53** R1=R2=R3=R4=R5=R6=*i*-Pr	**57**
	52 R=OH	**54** R1=*i*-Bu, R2=R3=R4=R5=R6=*i*-Pr	
		55 R1=R2=R3=*i*-Bu, R4=R5=R6=*i*-Pr	
		56 R1=R2=R3=R6=*i*-Pr, R4=R5=*s*-Bu	

Diketopiperazines are a class of lactams that result from the formation of peptide bonds
between two amino acids. They are the smallest possible cyclic peptides. Diketopiper-
azine (**50**), isolated from laver (*Porphyra yezoensis*) derived fungus M-3, displayed strong
anti-fungal activity against *Pyricularia oryzae* (MIC = 0.36 µM)[64]. Sulfur-containing diketopiperazines are generated by many unrelated fungi. Two dimeric diketopiperazines, 11,11′-
dideoxyverticillin A (**51**) and 11′-deoxyverticillin A (**52**), were obtained from *Penicillium*
sp. CNC-350 isolated from alga *Avrainvillea longicaulis* in the Bahamas. Both compounds
showed potent cytotoxicity against the HCT-116 human colon carcinoma cell line (IC$_{50}$ =
30 ng/mL)[29,65]. According to the definition of alkaloid, these three compounds could also
be classified as such.

When one or more amide (-CONHR-) bonds are replaced by ester (COOR) bonds in a peptide, the compound is called a depsipeptide. Antimalarial agents enniatins B (**53**), B$_4$ (**54**),
G (**55**), and I (**56**), produced by various *Fusarium* species, belong to a group of well-known
cyclohexadepsipeptide antibiotics. They exhibited strong properties against the K1 strain of
Plasmodium falciparum (IC$_{50}$s: 0.42 µM for B, 0.31 µM for B$_4$, 0.67 µM for G, and 0.36 µM
for I)[66]. The genus *Beauveria* is a source of various cyclodepsipeptides. An antiathroscle-
rotic agent beauveriolide III, a 13-membered cyclodepsipeptide (cyclo-[(3*S*,4*S*)-3-hydroxy-
4-methyloctanoyl-L-phenylalanyl-L-alanyl-D-allo-isoleucyl]), was isolated from the culture
broth of a soil isolate *Beauveria* sp. FO-6979. This compound inhibited the cholesteryl ester
synthesis with an IC$_{50}$ value at 0.41 µM[67]. Cyclic hexadepsipeptides destruxins are secondary metabolites first reported in 1961 and have names derived from the entomopathogenic
fungus *Oospora destructor* (Later on, *O. destructor* was re-named *Metarrhizium anisopliae*
(Metchnikoff) Sorokin). These compounds present a broad range of biological activities,
such as insecticidal, phytotoxic, cytotoxic and anti-viral effects. Destruxin B (**57**),[68] isolated from *O. destructor, Trichotecium roseum* TT103 and OS-F68576, was reported to have

caused chlorosis and necrosis on thirty species of host and non-host plants.

2.9   Helagen containing compounds

Some fungal products contain organic helagen. Those containing chlorine are common, while those with bromine are quite rare.

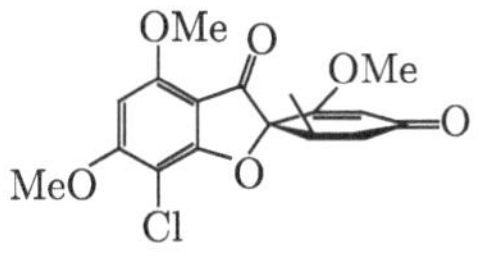

58 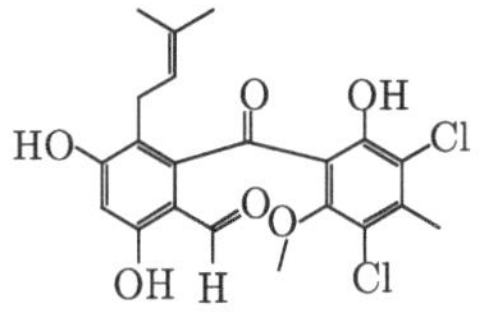   **59** 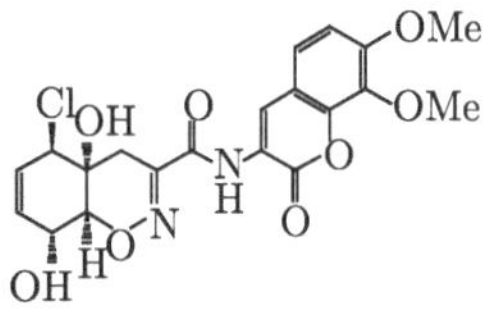   **60** 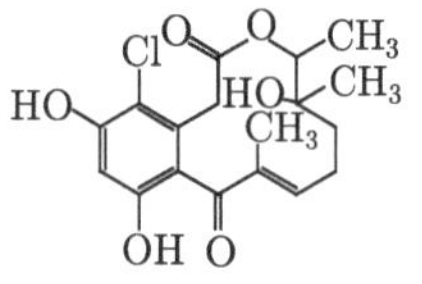   **61**

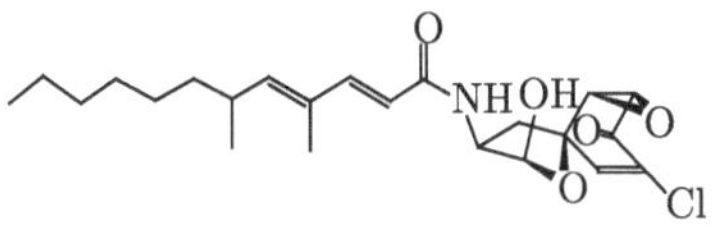

62 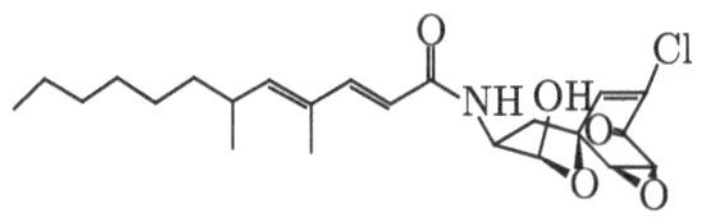   **63**

Griseofulvin (**58**), isolated from mycelia of *Penicillium griseofulvum*, was the first available oral agent to treat dermatophytoses. Currently, its main use is to treat dermatophyte infection by *Tinea capitis*[69].

Pestalone (**59**), a chlorinated benzophenone compound present in the marine fungus *Pestalotia* sp., showed potent antibiotic activity against methicillin-resistant *Staphylococcus aureus* (MIC = 37 ng/ml) and vancomycin-resistant *Enterococcus faecium* (MIC = 78 ng/ml)[70].

Trichodermamide B (**60**), a modified dipeptide isolated from cultures of the marine-derived fungus *Trichoderma virens*,[71] exhibited potent *in vitro* cytotoxicity against HCT-116 human colon carcinoma (IC_{50} = 0.32 μg/mL).

Oxacyclododecindione (**61**), a chlorinated macrocyclic lactone obtained from fermentations of the imperfect fungus *Exserohilum rostratum*,[72] displayed potent inhibition against the IL-4 induced expression of the reporter gene secreted alkaline phosphatase in transiently transfected HepG2 cells (IC_{50} = 54~67.5 nM).

Arnochlors A (**62**) and B (**63**), isolated from the fermentation broth of *Pseudoarachniotus roseus*, displayed strong inhibition against the Gram-positive bacterium *Micrococcus luteus* (MIC = 0.39 μg/mL)[73].

2.10   Miscellaneous secondary metabolites

Cordycepin (**64**), an antifungal, antiviral and antitumor nucleoside derivative, was isolated from *Cordyceps militaris* by Bentley and co-workers[74]. It was also found to be an immunoregulatory active component of *Cordyceps sinensis*[75].

Guignardic acid (**65**), the first member of a novel class of natural products, was obtained from the culture broth of *Guignardia* sp. isolated from *Spondias mombin*. The antimicrobial activity of this compound was detected using an agar overlay method. The encouraging results suggest the possibility of using this compound for the treatment of infections caused by filamentous fungi, yeasts, bacteria and actinomycetes[76].

A sphingoid derivative ISI-I (also called myriocin or thermozymocidin, **66**), an antifungal agent isolated from *Isaria sinclairii*, exhibited immunosuppressive activity 10 to 100 times more potent than cyclosporin A (**49**)[63].

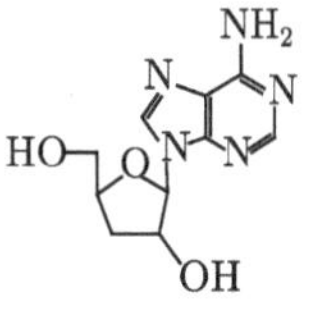

64 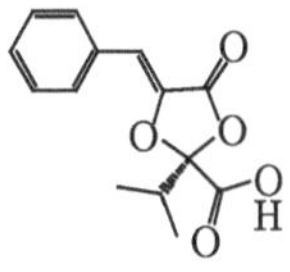   **65** 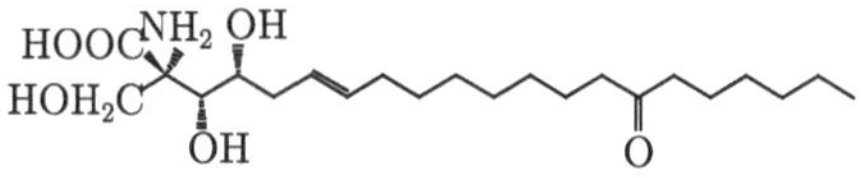   **66** 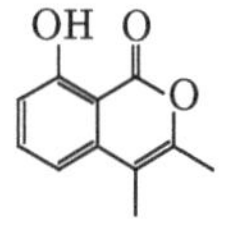   **67**

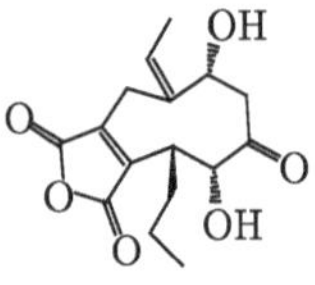

68 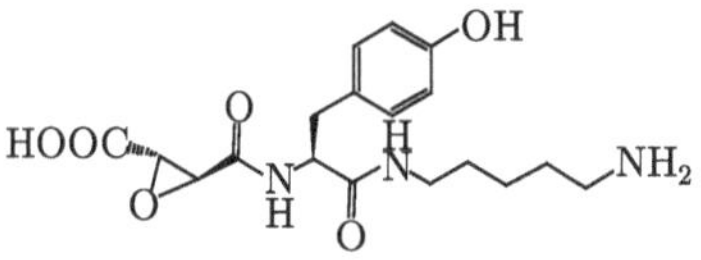   **69** 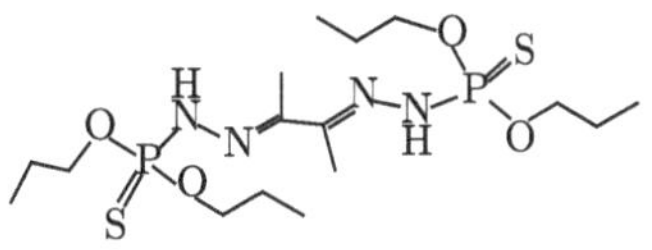   **70**

Isocoumarin oospolactone (**67**), isolated from *Gleophyllum sepiarium*,[3] was shown to be active against strains of the asexual ascomycete *Alternaria* (MIC = 12.5-25 µg/mL).

Comexistin (**68**), a nonadride from *Paecilomyces variotti* SANK 21086, displayed herbicidal activity against annual weeds[63].

Cathestatin C (**69**), an epoxysuccinyl-tyrosyl-cadaverine obtained from a sponge-derived fungus *Microascus longirostris* SF-73, displayed a potent and irreversible *in vitro* inhibition against papain and cathepsins B and L (IC$_{50}$s: 2.5, 5.0 and 15.0 nM)[77].

A phosphate derivative, compound **70** isolated from marine fungus *Lignincola laevis* from a marsh grass, showed cytotoxic activity against L1210 cells[78].

Section 3   Mycotoxins from Fungi

Phytotoxins of fungal origin are mainly produced by typical phytopathogens, such as *Septoria* and *Sclerotinia* spp., and by soil fungi[19]. Certain fungi produce very toxic phytotoxins known as mycotoxins that spoil food, cause plant diseases and poison or even kill humans and animals. For example, *Colletotrichum* species are known to produce toxins that induce symptoms similar to those of the pathogens themselves[79]. *Colletotrichum gloeosporioides* is a grape pathogen; *C. gloeosporioides*, *C. truncatum* and *C. fragaria* are found on strawberries; *C. nicotianae* causes tobacco anthracnose disease; *C. lagenarium* and *C. capsici* are pathogens on peanuts, soybeans, and cowpeas, among others; *C. dematium* is a pathogen on *Phaseolus vulgaris*. Other major mycotoxin-producing fungi are species of *Aspergillus*, *Fusarium*, and *Penicillium*. Toxic compounds produced by pathogenic fungi range from low molecular weight secondary metabolites to complex cyclic peptides and proteolytic enzymes[80]. Important mycotoxins include aflatoxins, fumonisins, trichothecenes, ochratoxins, cyclopiazonic acid, patulin, deoxynivalenol, zearalenone, citrinin, gliotoxin, phallotoxins, amatoxins, and sterigmatocystin. Aflatoxin and sterigmatocystin (**71**) probably are the most studied fungal secondary metabolites including extensive investigations of their gene cluster identification and of the factors governing their biosynthesis[81].

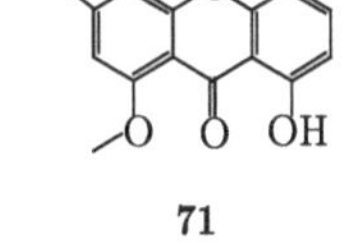

71 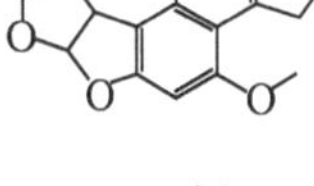   **72** 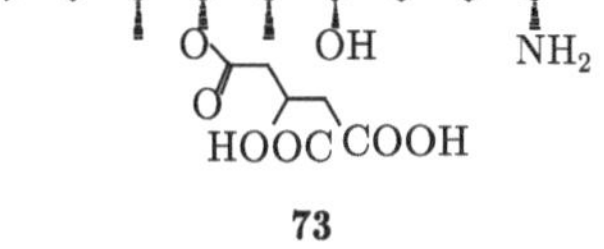   **73** 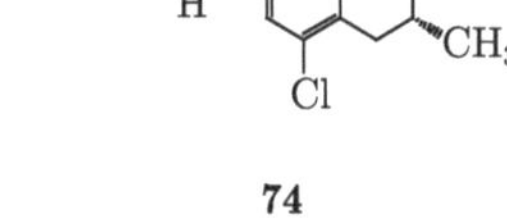   **74**

75   **76**   **77**

It is noteworthy that aflatoxin B_1 (**72**), fumonisin B_1 (**73**) and ochratoxin A (**74**) are the most toxic to mammals and have hepatotoxic, teratogenic, and mutagenic activities[82]. Aflatoxins, produced by *Aspergillus flavus* and *A. parasiticus* on various nuts and grains,[83] have been found to cause cancer in all species of animals tested and are among the most potent carcinogenic compounds yet identified[84]. Fumonisins are synthesized mainly by different species of the genus *Fusarium*. Compound **73** is the predominating fumonisin isolated from foods[85]. The fumonisins have been reported to cause human esophageal cancer, a fatal neurological disease in horses, and a fatal respiratory condition in pigs. The ochratoxins, produced on cereal grains by *Aspergillus* and *Penicillium* species,[14] have been implicated in a type of human renal atrophy[12]. Compound **74**, from *Aspergillus carbonarius*, was also found in Australian grape products[86].

Interestingly, some chlorine and sulfur containing compounds are quite toxic. Sporidesmin A (**75**), penitrems (e.g. penitrem A, **76**), and phomopsin A (**77**) are good examples. Ingestion of spores of the saprophytic fungus *Pithomyces chartarum* by sheep, cattle, deer, and goats causes the disease known as facial eczema in livestock[87]. The spores contain compound **75**, a potent mycotoxin which was also found in the fungus *Sporidesmium bakeri*[69]. This compound causes extensive liver damage, particularly to the biliary system[88]. Penitrems are primarily found in the *Penicillium* and *Aspergillus* species. The tremorgenic mycotoxin **76**, an insecticide isolated from *Penicillium crustosum* and other species, showed convulsive and insecticidal activities against *Bombyx mori*, *Spodoptera frugiperda* and *Heliothis zea*[69]. Compound **77**, the main mycotoxin isolated from *Phomopsis leptostromiformis*, causes lupinosis disease[89]. The consumption of lupins infected with *P. leptostromiformis* has been found to be the cause of lupinosis and mycotoxicosis in livestock[69].

Section 4   Isolation and Structure Studies of Fungal Products

Isolation and structure studies of fungal products are similar to those performed on medicinal plants. Fungal cultures are complex mixtures of mycelia and substrate components. Submerged liquid fermentations can either be whole-broth extracted or mycelia and culture broth extracted separately. For those fermentations performed on grains, agar, or other solid substrates, the extraction by solvent should be on whole solids. Mycelia need to be disrupted in suitable solvent, a water-miscible solvent such as alcohol or acetone, and then all solids and most of the organic solvent should be removed before the further partition between water and an immiscible organic solvent is performed. Usually either centrifugation or filtration is applied to clarify the crude extract solution, and the organic solvent is removed by vacuum evaporators. It is worth mentioning that, although the pH of the aqueous phase can be adjusted to different values to reach extraction of some ionizable compounds efficiently, some components might decompose due to varied stabilities at different pH levels. The clarified crude extract solution (with a small amount of organic solvent) could also be loaded onto polymeric resins, such as neutral polystyrenedivinylbenzene polymers, to perform solid-phase extraction.

Various chromatography techniques are used to isolate compounds of interest from the crude extract. Commonly used techniques include normal phase column chromatography, reverse-phase HPLC, Sephadex LH-20, ion-exchange chromatography, countercurrent chromatography, etc. The policy for choosing solid-phase adsorbents is to reach maximum substance recovery. For example, normal silica gel is slightly acidic because of the silanol groups, which can cause decomposition of some acid-labile metabolites and strongly adsorb some basic compounds such as alkaloids. The solution to this problem is to use diol-silica or polymeric reverse-phase columns. When the mobile phase is an aqueous solution, buffers are often used to stabilize the metabolites, but desalting of the final product is necessary in most cases. Crystallization is another separation method that allows the possibility of structure elucidation by x-ray diffraction. However, in most cases it is very difficult to obtain suitable crystals from minor components isolated in 0.5 to 1 mg pure form.

Structure identification can be achieved by the analysis of physical and chemical data of the pure compound. Currently, structure elucidation relies heavily on mass spectrometry (MS) and nuclear magnetic resonance spectroscopy (NMR), especially various 2D NMRs; while infrared (IR) and ultraviolet spectroscopy (UV) are mainly useful in the identification of functional groups in a molecule. HPLC mass spectrometry (LCMS) is typically useful for the correlation of major compounds in peaks shown on the chromatograph with individual mass spectra. The ionization techniques for LCMS are usually chemically gentle and strong molecular ion peaks are observed for most compounds tested. In addition to spectral data, organic synthesis is often used to confirm or disprove structures.

Structure modifications have proven to be fruitful approaches to obtain more potent and less toxic bioactive derivatives. Total synthesis of some bioactive compounds has been successful and feasible. However, approaches to obtain structurally more complex compounds, such as those found in fungal drugs, total synthesis has proven to be very challenging, expensive and rarely possible. Although conventional fermentation process is the major approach for commercial production of some metabolites, this method remains challenging for marine-derived fungi (especially for symbiotic species) due to their poor culturability. An alternative and very promising solution to these problems appears to be recombinant technology. This technology is a relatively new approach to drug discovery. Many specific genes encoding metabolites are highly conserved and could be cloned by hybridization probing or amplified from the fungal genome by use of degenerate PCR primers[90]. Recombinant technology has been applied to identify gene clusters encoding for bioactive natural products, which might allow assembly of structurally complicated secondary metabolites or intermediates through enzymatic catalyses. This may also provide a long term supply of compounds where total or partial synthesis fails. A good example is the vasoconstrictor ergotamine (**78**),[91] an ergot alkaloid produced by the ergot fungus *Claviceps purpurea* and related fungi in the family Clavicipitaceae. It was considered that two non-ribosomal peptide synthetases, LPS1 (gene *cpps1* encoded) and LPS2 (gene *cpps2* encoded), are responsible for the biosynthesis of **78** from D-lysergic acid (LSA), as shown in Scheme 3-1.

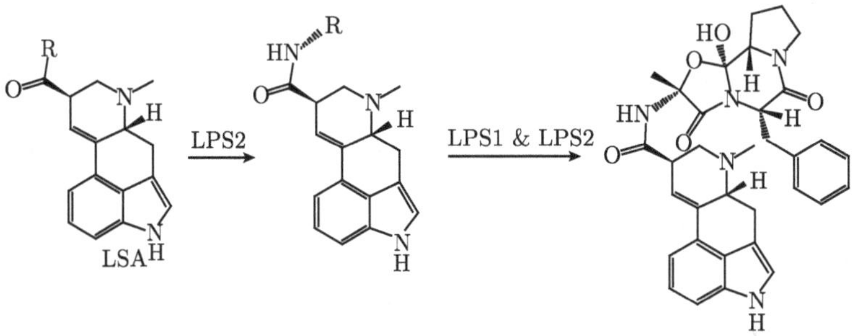

78

Scheme 3-1   Biosynthesis pathway of ergotamine

Several extraction, isolation and structural study examples of fungal products are illustrated in sections 4.1 to 4.5.

4.1   Polysaccharide preparations of Chinese traditional medicine *Poria cocos* (Fu-Ling) and *Polyporus umbellatus* (Zhu-Ling)

Polypores, the terrestrial fungi of the phylum Basdiomycota, as well as certain Ascomycota, are major sources of active secondary metabolites and polysaccharides. The majority of these polysaccharides can be classified as β-D-glucans. The classical extraction processes of polysaccharides from some fungal remedies are well-established in China.

4.1.1   Polysaccharide extract of *Poria cocos*

Recently, a neutral polysaccharide fraction from *Poria cocos* (Schw.) Wolf (Polyporaceae) was found to exhibit antiproliferative and differentiating effects on human leukemia cells[92]. *P. cocos*, an oriental fungus, has been used as a traditional medicine and also as a health-promoting food in China for centuries. The total crude polysaccharide extract may be obtained as shown in Scheme 3-2[93]. The dried sclerotia of *P. cocos* are stirred with 0.5 M NaOH for 10 min, then soaked in this basic solution at 3°C for 3 hr before centrifugation is applied to clarify the extract solution. The solution is then neutralized with 10% acetic acid and filtered. The resulting precipitate is washed with solvent in the order of water, acetone and ethyl ether, and then dried. The dried solid is then dissolved in dimethylsulfoxide followed by the addition of water to generate a precipitate. Centrifugation is then applied to obtain the precipitate. This step is repeated once, and the resulting precipitate is washed again with water, acetone and ethyl ether, and then dried to afford the crude extract of *P. cocos* polysaccharides.

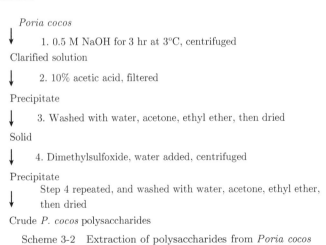

Scheme 3-2   Extraction of polysaccharides from *Poria cocos*

4.1.2   Polysaccharide extract of *Polyporus umbellatus*

The sclerotium of *Polyporus umbellatus* (Pers.) Fries (Polyporaceae), known as the traditional Chinese drug Zhu-Ling, is used for the promotion of diuresis. Reported pharmacological constituents from this fungus mainly include polysaccharides and steroids[87]. The crude polysaccharides from *P. umbellatus* may be obtained through the procedure described in Scheme 3-3[93]. The sclerotia of *P. umbellatus* are extracted with hot water for 1.5 hr and filtered. The clarified solution is concentrated to a small volume, and then three times this volume of ethanol is added to precipitate the polysaccharides. After filtration, the precipitate is redissolved in a small amount of water and filtered. The resulting aqueous solution is then subjected to MG-1 or AB-8 type macroporous resin and eluted with water. The water eluents are combined and dried to afford the crude *P. umbellatus* polysaccharide extract.

Polyporus umbellatus
↓   1. Boiling water, 1.5 hr, filtered
Clarified solution

↓   2. Concentrated, 3 times ethanol added, filtered
Precipitate
↓   3. Water added, filtered, cooling

Aqueous solution

↓   4. Subject to MG-1 or AB-8 macroporous resin, elute with water
Eluents dried

↓

Crude *P. umbellatus* polysaccharides

Scheme 3-3   Extraction of Polysaccharides from Polyporus umbellatus

4.2   Bioactive triterpenes from polypore *Ganoderma lucidum* Ling-Zhi

Ganoderma lucidum (Fr.) Krast is well known in China and other eastern countries, where it is used as a folk remedy for the treatment of cancer, hepatitis, chronic bronchitis, asthma, hemorrhoids, nephritis, hypertension, hyperlipidemia, arthritis, neurasthenia, insomnia, gastric ulcer, arteriosclerosis, diabetes, and fatigue symptoms[3,94]. Triterpenoids have been shown to be mainly responsible for some of these observed biological activities. In addition to more than 130 pharmacologically active triterpenoids isolated from this polypore, several triterpenoids, including lucidenic acid N (**79**) and methyl lucidenate F (**80**), were recently isolated from the dried fruiting bodies of this fungus. Compound **79** showed significant cytotoxic effects on the Hep G2, Hep G2,2,15, and P388 cell lines (IC$_{50}$ = 2.06 x 10^{-4}, 1.66 x 10^{-3}, and 1.20 x 10^{-2} μM, respectively)[95].

4.2.1   Isolation of lucidenic acid N and methyl lucidenate F[95]

The air-dried fruiting bodies of *G. lucidum* (603 g) were cut into small pieces and soaked in EtOH (2.5 L) at room temperature overnight. They were then extracted twice with hot water (5 L). The H$_2$O extract was partitioned with CHCl$_3$. The CHCl$_3$ layer was concentrated to a brown syrup and chromatographed repeatedly on silica gel eluted with CHCl$_3$-MeOH to give compound **79** (23 mg) and compound **80** (1 mg, after rechromatography on silica gel using Me$_2$CO-hexane as an eluent).

4.2.2   Structure elucidation of lucidenic acid N[95]

The HRMS of lucidenic acid N (**79**) revealed the molecular formula C$_{27}$H$_{40}$O$_6$. The UV absorbance at 251 nm indicated the presence of an α, β-unsaturated ketone moiety. The IR spectrum of **79** suggested the presence of hydroxyl, carbonyl, and carboxyl groups. The ^{1}H NMR, ^{13}C NMR, combined with NOE, HMQC, COSY, and HMBC data suggested that compound **79** was a lanostane-type triterpene close in structure to a known compound, lucidenic acid A, with the major differences of the H-3 signal at δ 3.22, and the upfield shifts of C-2, C-3, and C-4 to δ 27.6, 78.3, and 38.6 from δ 34.1, 216.8, and 46.6, respectively. The NMR data showed that **79** had six methyls, seven methylenes, four methines (including two oxymethines), five quaternary carbons, two sp^2 carbons, and three carbonyls. To confirm the elucidated structure, compound **79** was oxidized with K$_2$CrO$_7$ and concentrated H$_2$SO$_4$ in Me$_2$CO, and then was methylated with MeOH and H$_2$SO$_4$ to afford compound **80**, a methylation product of lucidenic acid F reported earlier by Kikuchi et al.[96].

79   **80**

Lucidenic acid N (79). Colorless powder (CHCl$_3$), mp 202 to 204°C, $[\alpha]_D$ +119.5° (c0.23, CHCl$_3$); UV (MeOH) λ_{max} (log ε) 251(3.90) nm; IR (KBr) ν_{max} 3449, 2927, 1724, 1656, 1458, 1382, 1174, 1031, 750 cm^{-1}; ^{1}H NMR (400 MHz, CDCl$_3$, ppm) δ 4.80 (1H, dd, J = 9.2, 8.2 Hz, H-7), 4.37 (1H, br s, D$_2$O exchangeable, OH), 3.22 (1H, dd, J = 10.6, 5.6 Hz, H-3), 2.83 (1H, dt, J = 13.6, 3.7 Hz, H-1β), 2.78 (1H, dd, J = 19.7, 8.4 Hz, H-16α), 2.75 (1H, d, J = 16.6 Hz, H-12β), 2.71 (1H, d, J = 16.6 Hz, H-12β), 2.43 (1H, ddd, J = 16.3, 8.6, 5.5 Hz, H-23), 2.32 (1H, ddd, J = 16.3, 8.0, 8.0 Hz, H-23), 2.19 (1H, ddd, J = 12.0, 7.4, 1.2 Hz, H-6α), 2.13 (1H, dd, J = 19.7, 9.4 Hz, H-16β), 1.99 (1H, dt, J = 18.5, 9.4 Hz, H-17), 1.79 (1H, m, H-22), 1.60 (4H, m, H-2, 6β, 20), 1.35 (3H, s, H-28), 1.03 (3H, s, H-19), 0.98 (3H, d, J = 6.2 Hz, H-21), 0.96 (1H, m, H-1α), 0.96 (3H, s, H-18), 0.88 (1H, dd, J = 8.1, 1.2 Hz, H-5α), 0.85 (3H, s, H-30); ^{13}C NMR (100 MHz, CDCl$_3$, ppm) δ 217.5 (s, C-15), 198.0 (s, C-11), 178.2 (s, C-24), 156.8 (s, C-8), 142.7 (s, C-9), 78.3 (d, C-3), 66.9 (d, C-7), 59.4 (s, C-14), 50.3 (t, C-12), 49.1 (d, C-5), 46.1 (d, C-7), 45.3 (s, C-13), 41.0 (t, C-16), 38.8 (s, C-10), 38.6 (s, C-4), 35.1 (d, C-20), 34.8 (t, C-1), 30.7 (t, C-23), 30.4 (t, C-22), 28.1 (q, C-26), 27.6 (t, C-2), 26.6 (t, C-6), 24.4 (q, C-25), 18.4 (q, C-19), 18.0 (q, C-21), 17.4 (q, C-18), 15.4 (q, C-27); EIMS m/z 460 [M$^+$] (58), 432 (29), 331 (77), 320 (37), 55 (31); HREIMS m/z460.2824 (calcd. for C$_{27}$H$_{40}$O$_6$, 460.2824).

4.3   Undecylresorcinol dimer from *Coleophoma* sp.[97]

In the search for inhibitors of protein kinases from fungal extracts, a novel undecylresorcinol dimer (**81**) was isolated from *Coleophoma* sp., a mitosporic fungus belonging to the coelomycete class isolated from a plant sample found in Malaysia. This compound inhibited cFMS receptor tyrosine kinase (IC$_{50}$ of 0.4 μM), with greater than 10-fold selectivity versus nine other protein kinases. The cFMS receptor kinase, also known as CSF-1 or m-CSF receptor tyrosine kinase, has been implicated in both neoplastic and bone diseases.

4.3.1   Isolation of undecylresorcinol dimer (81)

The fungal culture was maintained on malt extract agar at 25°C. After the solid state fermentation (50 ml) was incubated for 23 days, methanol (100 ml) was added. After 2 h, the solid substrate was broken into small pieces and left to extract overnight. The mixture was then filtered and the extract was dried by rotary evaporation. Then 5 ml of dry methanol was added. After mixing for 30 min, this methanol extract was filtered through filter paper and dried under a N$_2$ stream. The material obtained (440 mg) was dissolved in 4.4 ml of DMSO, and 2 ml was purified by preparative HPLC at room temperature on a Hypersil HyPURITY Elite C18 column (5 mm particle size, 15 cm × 21.2 mm i.d.) with the acetonitrile gradient in 20 mM NH$_4$OAc buffer: a linear gradient of 0 to 100% in 30 min, and the flow rate was 15 ml/min. The fractions with retention times of 26 to 33 min showed inhibitory activity and were further purified on the same C18 column. By eluting the material with 40 to 100% gradient of acetonitrile in 20 mM NH$_4$OAc buffer, an active compound (**81**) was isolated from the fraction with retention times of 12.5 to 14.5 min.

4.3.2   Structure elucidation of undecylresorcinol dimer (81)

Compound **81** was isolated as a colorless oil. UV absorption bands were present at 263 and 300 nm. The IR spectrum showed absorbance for hydroxyl (3599 cm^{-1}) and ester carbonyl (1644 cm^{-1}) groups. MS spectra were acquired on a Micromass LCT time-of-flight mass spectrometer. Samples were ionized via electrospray ionization (ESI). The LCT was equipped with the LockSpray apparatus to assist with mass accuracy, with reserpine used as the lock mass (m/z 609.2812 for the protonated species). Compound **81** gave an [M+H]$^+$ ion at m/z 571.4034, corresponding to the formula $C_{35}H_{55}O_6$ (calcd. 571.3999) in the HRESIMS. NMR spectra were acquired on a 500 MHz Varian Inova spectrometer equipped with a 3-mm Nalorac Indirect Detection probe (H/C/N). Compound **81** was dissolved in DMSO-d_6. The spectra were referenced by DMSO-d_5 at 2.50 ppm for ^{1}H, and by DMSO-d_6 at 39.5 ppm for ^{13}C. The spectra were acquired at ambient temperature. The carbon spectrum was acquired using a 3-mm Nalorac Carbon Probe. The ^{1}H NMR spectrum showed three aromatic protons in an A_2B spin system at δ 6.71 (2H, d, $J = 2.0$ Hz) and 6.75 (1H, d, $J = 2.0$ Hz), two *meta*-coupled protons at δ 6.11 and 6.13 (each d, $J = 2.2$ Hz), one oxymethine proton at δ 5.04, and two terminal methyl protons at δ 0.84 and 0.87. Nineteen methylene groups were identified from the ^{13}C NMR and HMQC spectra (Table 3-1).

81

(Carbons are renumbered from the original paper.)

Table 3-1   Chemical Shifts and Assignments for Undecylresorcinol Dimer

Assignment	^{1}H Chemical Shift (δ ppm)	^{1}H Integral, Multiplicity and Coupling Constant (J)	^{13}C Chemical Shift (δ ppm)
1	—	—	109.0
2	—	—	159.7
3	6.14	1H, d, $J = 2.2$ Hz	100.4
4	—	—	160.2
5	6.12	1H, d, $J = 2.2$ Hz	108.7
6	—	—	144.0
1′	2.56	2H, ∼t, $J \sim 7.8$ Hz	34.4
2′	1.45	2H, m	31.4
3′–8′	1.24–1.29	14H, m	28.6–29.2
9′	1.24–1.29	2H, m	31.2
10′	1.24–1.29	2H, m	22.0
11′	0.85	3H, t, $J = 7.4$ Hz	13.8
1″	5.05	1H, m	74.2
2″	1.32	2H, m	24.9
3″–6″	1.24–1.29	8H, m	28.6–29.2
7″	1.50	2H, m	30.9
8″	2.46	2H, ∼t, $J \sim 7.8$ Hz	35.2
1‴	—	—	142.6
2‴, 6‴	6.72	2H, d, $J = 2.0$ Hz	115.0
3‴, 5‴	—	—	153.5
4‴	6.76	1H, t, $J = 2.0$ Hz	110.3
1⁗′	1.56	2H, m	35.8
2⁗′	1.36	2H, m	18.1
3⁗′	0.88	3H, t, $J = 7.4$ Hz	13.9
C=O	—	—	169.3

Fragment ions at m/z 263.2039 and 309.2075 in the positive ESIMS spectrum were assignable to undecylresorcinol ($C_{17}H_{27}O_2$, calcd. 263.2011) and undecylresorcylic acid ($C_{18}H_{29}O_4$, calcd. 309.2066) as shown in Figure 3-1. Therefore, the structure of this cFMS inhibitor was determined to be a novel undecylresorcinol dimer, 8-(3,5-dihydroxyphenyl)-1-propyloctyl 2,4-dihydroxy-6-undecylbenzoate.

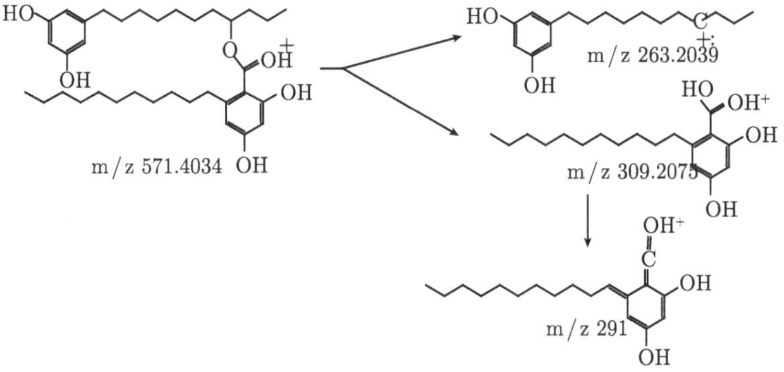

Figure 3-1   Mass fragmentation of undecylresorcinol dimer

4.4   Balanol from the fungi *Verticillium balanoides*[98] and *Acremonium* sp.[97]

Balanol (**82**) was earlier reported as a potent inhibitor of protein kinase C (PKC, $IC_{50} = 4$ to 9 nM) from *Verticillium balanoides* by Kulanthaivel et al. in 1993.[98] Extensive research exploring the inhibition mechanism and structure modifications has been conducted since the discovery. Recently, in our efforts to search for natural inhibitors of cFMS receptor tyrosine kinase from a fungal source, balanol was isolated from an extract of an *Acremonium* species from the *Quercus* roots from an English garden. Balanol inhibited cFMS receptor kinase with an IC_{50} of 1 nM and selectivities of 14 to 75-fold versus pp60c-Src and VEGF receptor kinases and greater than 10,000-fold versus seven other kinases.[97] This is a good example of a known metabolite that presented striking novel activity.

4.4.1   Isolation of balanol from *Verticillium balanoides*[98] and *Acremonium* sp.[97]

V. balanoides, collected from *Pinus palustris* needle litter near Hoffman, North Carolina, U.S. was cultured in a yeast extract, peptone, dextrose, malt, and cornmeal media. The freeze-dried culture was repeatedly extracted with MeOH, and the extract (IC_{50} 10 to 20 µg/mL against PKC) was partitioned between *n*-BuOH and H_2O. Around 12 mg of balanol was obtained from 40 L of culture through bioassay-guided fractionation of the *n*-BuOH-soluble material ($IC_{50} = 1$ µg/mL) by gel permeation on Sephadex LH-20 with a CH_2Cl_2-MeOH gradient, followed by reverse-phase (ODs) HPLC.[98] Balanol was also isolated from *Acremonium* sp. by the similar procedure in our work described in 4.3.1.[97]

4.4.2   Structure elucidation of balanol[98]

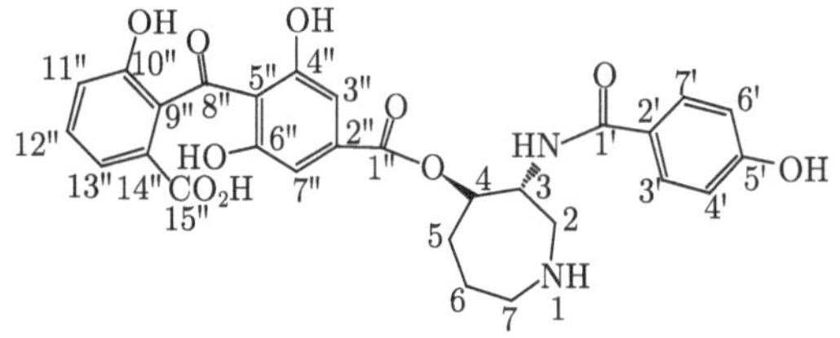

82 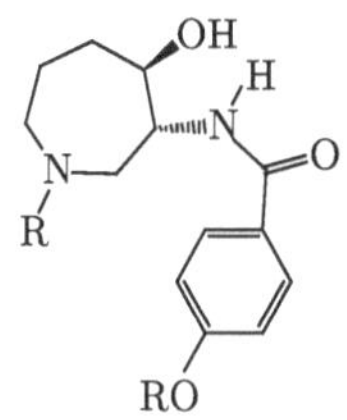   **83** R＝H   **84** R＝*p*-bromobenzoyl

Balanol (**82**) is a yellow amorphous solid, $[\alpha]_D$-129° (c 0.25, MeOH). The molecular formula was established as $C_{28}H_{26}N_2O_{10}$ by HR-FABMS ([M+H]$^+$, m/z 551.1685). The IR spectrum indicated the presence of hydroxyl, ester, and amide functionalities. The structure **82** for balanol was determined based on analysis of ^{1}H, ^{13}C (Table 3-2), and 2D NMR (COSY, NOESY, TOCSY, HMQC, and HMBC) data. The ^{1}H NMR spectrum supported the presence of 1,4-disubstituted (δ 7.62, 6.74, each 2H, d, $J = 8.7$ Hz), 1,2,6-trisubstituted (δ 6.64, br, 7.07, t, 6.96, br, each 1H, $J = 7.6$ Hz), and 1,3,4,5-tetrasubstituted (δ6.66, 2H, s) benzene rings. The main core of balanol was identified by ^{1}H-^{1}H decoupling and TOCSY experiments to be a hexahydroazepine ring with amide and ester substituents at C-3 and C-4, respectively. In the HMBC experiment, H-3 (δ_H 4.17), the amide proton (δ_H 8.08), and H-3',7' (δ_H 7.62) all showed HMBC cross peaks to the amide carbonyl (δ_C 165.9), which indicated that the 1,4-disubstituted benzene ring was attached to the nitrogen adjoining the C-3 of the hexahydroazepine ring. Significant interactions between the amide proton and H-3', H-2b, H-4, and between H-4 and H-2b were observed in the NOE experiment. Proton H-2b showed large coupling ($J_{2b,3} = 7$ Hz) with H-3 and hence should be assigned *trans* to H-3. The determination of this structural moiety was supported by the hydrolysis of balanol by base (K_2CO_3-H_2O-MeOH) to generate the alcohol **83**, followed by an x-ray analysis of **84**, a *p*-bromobenzoyl derivative of **83**. The tetrasubstituted benzene ring attached to C-4 of the hexahydroazepine ring through an ester linkage was evident in the HMBC spectrum from the cross peaks shown by H-4, H-3'', and H-7'' to the ester carbonyl (δ_C 165.4). The tri- and tetrasubstituted benzene rings linked via a carbonyl (δ_C 201.4) were also established in the HMBC experiment.

Table 3-2   Chemical Shifts and Assignments for Balanol

Assignment	δ^1H (ppm)	^{1}H Multiplicity and Coupling Constant (J)	δ^{13}C (ppm)
2a	2.90	dd; 14.6, 3.8 Hz	50.14
2b	2.78	dd, 14.6, 7.0 Hz	
3	4.17	dddd, 8.7, 7.6, 6.0, 3.8 Hz	55.36
4	5.11	ddd, 8.1, 7.6, 3.8 Hz	78.00
5a, 5b	1.89	m	29.05
6a	1.74	m	24.77
6b	1.60	m	
7a	2.81	ddd, 13.0, 6.0, 6.0 Hz	48.22
7b	2.73	ddd, 13.0, 8.1, 6.0 Hz	
1'			165.87
2'			125.66
3', 7'	7.62	d, 8.7 Hz	129.44
4', 6'	6.74	d, 8.7 Hz	115.09
5'			160.43
1''			165.40
2''			134.77
3'', 7''	6.66	s	107.99
4'', 6''			159.83
5''			120.12
8''			201.44
9''			129.90
10''			153.43
11''		br	116.59
12''	7.07	t, 7.6 Hz	129.03
13''	6.96	br	118.70
14''			141.41
15''			171.78
CONH	8.08	d, 8.7 Hz	

4.5   Pericosine A from *Periconia byssoides* OUPS-N133

Pericosines are unique C_7 carbasugar metabolites from *Periconia byssoides* OUPS-N133 fungus isolated from the sea hare, *Aplysia kurodai*. Pericosine A ((+)-**86**) was reported to exhibit significant growth inhibition against the murine P388 cell line (ED_{50} of 0.1 μg/mL), to have selective growth inhibition against human cancer cell lines HBC-5 and SNB-75 (log concentration of compound for inhibition of cell growth at 50% compared to control: log GI_{50} of -5.22 and -7.27, respectively), and to demonstrate inhibitory activity against protein kinase EGFR (100 μg/mL by 40 to 70%) and topoisomerase II (IC_{50} = 100 to 300 mM), in addition to its significant *in vivo* antitumor activity against P388 leukemia cells.[99]

4.5.1   Isolation of pericosine A[99]

The fungal strain was cultured in artificial seawater medium (containing 1% malt extract, 1% glucose, and 0.05% peptone adjusted to pH 7.5 at 27°C) for 4 weeks. Mycelia obtained by filtration of the broth were extracted with AcOEt. The AcOEt extract was separated by Sephadex LH-20 using MeOH-CH_2Cl_2 (1:1) as eluent, followed by silica gel column chromatography with a gradient CH_2Cl_2-MeOH solvent system. This separation was directed by an *in vitro* P388 cytotoxicity assay. The active fraction that was eluted with 10% MeOH-CH_2Cl_2 was further purified by reverse-phase preparative HPLC using the MeOH-H_2O solvent system to yield pericosine A.

4.5.2   Structure elucidation of pericosine A[99−102]

Prior to the accomplishment of the total syntheses of pericosines[100−102] (Scheme 3-4), the structural elucidation of pericosine A was extremely difficult. Due to its multifunctionalized cyclohexenoid structures with torsional strain, the stereo structures could not be determined by spectral analyses.[102] Spectroscopic analyses, including NMR, MS, and IR, initially led to the elucidation of pericosine A as **85**,[103] which was incorrect. After the total syntheses of **85**, pericosine A (+)-**86** and its enantiomer (−)-**86** were achieved, the structure of pericosine A was revised as (+)-**86** with a 3*S*,4*S*,5*S*,6*S* configuration. All spectral data of synthesized (+)-**86**, including specific rotation and HPLC retention time, agreed with the data of natural pericosine A.[101] This example reminds one of the importance of applying alternative methods to confirm elucidated structures.

Table 3-3 ^{1}H and ^{13}C NMR Spectral Data of Pericosine A in Acetone-d_6

Position	δ^1H, ppm		J, Hz	NOE	δ^{13}C, ppm	
1	—	—	—	—	130.28	(q)^a
2	6.93	d	4.0 (3)	3	141.89	(t)
3	4.41	br t	4.0 (2, 4)	2, 4	66.96	(t)
4	4.10	dd	4.0 (3), 2.0 (5)	3, 5	68.64	(t)
5	4.13	dd	4.5 (6), 2.0 (4)	4	75.42	(t)
6	4.90	dd	4.5 (5), 0.9 (3)	—	57.69	(t)
7	—	—	—	—	168.18	(q)
8	3.79	s	—	—	52.88	(p)

a Assigned by DEPT

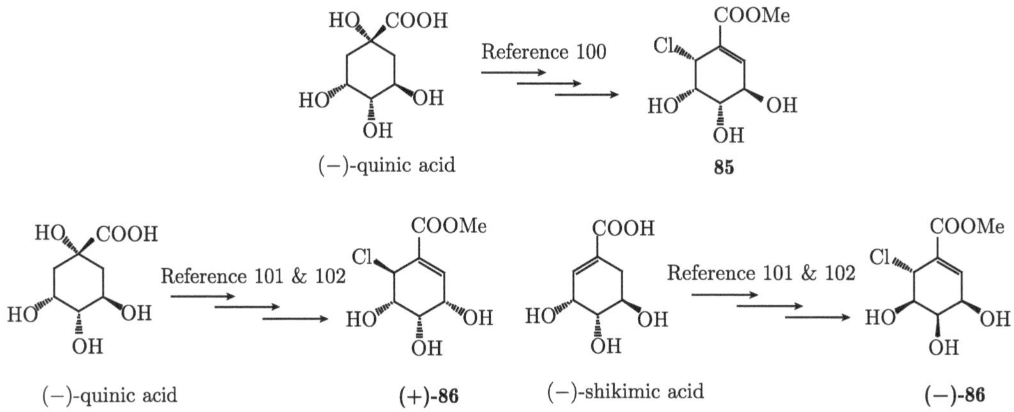

Scheme 3-4   Outline of syntheses of pericosines

Pericosine A ((+)-86). Obtained as plates, mp 95 to 97°C, $[a]_D + 57.0$ ($c3.16$ in EtOH); UV λ_{max} (EtOH) 217 nm (log ε 3.90); IR ν_{max} (KBr) 3353 (OH), 1720 (ester) and 1651 (C=C) cm^{-1}; EIMS m/z223 ([M + H]$^+$, 1.8%), 187 (MH$^+$ − HCl, 3.6) and 126 ([C$_7$H$_9$O$_4$]$^+$, 100) [m/z(HREI) found: [M + H]$^+$, 223.0364. C$_8$H$_{12}$ClO$_5$ requires 223.0363]; CD λ ($c1.40 \times 10^{-3}$ mol dm^{-3} in EtOH) 289 ($\Delta\varepsilon$ 0), 248 (−0.59), 238 (0) and 227 (+1.75) nm. ^{1}H and ^{13}C NMR data are listed in Table 3-3.

Section 5   Perspectives

The study of fungal metabolites is a relatively new area in natural product chemistry, especially of those found in the lower fungi. In recent years, there has been a significant amount of progress made in the investigations of fungal bioactive compounds, exemplified by the evaluation and development in edible fungi nutrition, the fermentation industry, the studies of secondary metabolite isolation, structure identification and modification, and the total syntheses of bioactive fungal metabolites. Currently, many fungal drugs are used regularly; however, there is still much room for further investigation and development, including:

1. Biological studies. This includes the identification of molecular targets and gene clusters that encode bioactive metabolites. New targets might lead to the development of a series of new drugs to treat diseases that are resistant to currently available medications.

Although morphology still plays a major role in taxonomy, mycologists now employ a number of techniques, including ribosomal DNA sequences, to aid in fungal identification and to organize phylogenetics. In addition, recombinant DNA techniques have been used to study the metabolite formation in fungi for nearly two decades. Currently, only a small part of the biosynthetic gene clusters encoding bioactive compounds have been isolated and identified. Further studies are urgent. Once the target gene clusters are available, the biosynthesis pathways of related compounds can be confirmed or disproved. Another advantage is that some trace components in fungi or fungal drugs could be produced on a large scale through means of recombinant technology.

2. Isolation and identification of fungal metabolites. To date, only a small portion of fungal origin natural products have been reported. Further investigation in this area is needed. Recent progress in modern analytical techniques, such as GC-MS, LC-MS, NMR, increases the importance of secondary metabolites as markers in chemotaxonomic studies of fungi. Incorporating chemotaxonomy into fungal metabolite research could prove to be very useful for comparing strains and dereplication of known metabolites.

3. Application of special metabolites. This area includes the further development of some delicious or fragrant components such as special amino acids, natural pigments, and gel

products, along with more common biochemical products such as enzymes, which could be
further developed to produce other products such as food additives and medical merchandise.

4. Structure modification and total synthesis of fungal natural products. Derivatives of
natural products are being made all the time in efforts to decrease side effects, such as toxicity, and increase the bioactive potency of lead compounds obtained from natural resources.
This area also provides the opportunity to generate new compounds with an even broader
bioactive spectrum than their parent leads. New, efficient, economical or practical synthetic
approaches are strongly encouraged.

References

[1]   Gunatilaka AAL. J. Nat. Prod., 2006, 69: 509.

[2]   Whittaker RH. Science, 1969, 163: 150.

[3]   Zjawiony JK. J. Nat. Prod., 2004, 67: 300.

[4]   Bugni TS, Ireland CM. Nat. Prod. Rep., 2004, 21: 143.

[5]   Moradali M-F, Mostafavi H, Ghods S, Hedjaroude G-A. Int. Immunopharmacol., 2007, 7:
      701.

[6]   Liu J-K. Heterocycles, 2002, 57: 157.

[7]   Wu TN. Plant Resource, Nutrition and Health Care. Shanghai: Shanghai Jiao Tong Uni-
      versity Press, 2008:129.

[8]   James MNG. Biol. Chem., 2006, 387: 1023.

[9]   Thornton CR. Eur. J. Plant Pathol., 2008, 121:347.

[10] Jimenez-Teja D, Hernandez-Galan R, Collado IG. Nat. Prod. Rep., 2006, 23: 108.

[11] Mao XL. Wandering in the Kingdom of Fungi. Zhengzhou: Haiyan Press, 2005: 3.

[12] Morris MI, Villmann M. Am. J. Health-Syst. Pharm., 2006, 63:1813.

[13] Liu DY, Yu CL, Liu QH. Development and Application of Mycotoxin. Beijing: Chemical
      Industry Press, 2007: 500.

[14] Vetter J. Toxicon, 1998, 36: 13.

[15] Fradin EF, Thomma BPHJ. Molecular Plant Pathology, 2006, 7: 71.

[16] Tochikura TS, Nakashima H, Hirose K, Yamamoto N. Biochem. Biophys. Res. Commun.,
      1987, 148: 726.

[17] Tochikura TS, Nakashima H, Yamamoto N. J. Acquired Immun. Defic. Syndr., 1989, 2:
      441.

[18] Wang TH, Lin TF. Adv. Food Nutr. Res., 2007, 53: 123.

[19] Berestetskiy AO. Appl. Biochem. Microbiol., 2008, 44: 453.

[20] Rodriguez R, Redman R. J. Exp. Bot., 2008, 59: 1109.

[21] Zhang HW, Song YC, Tan RX. Nat. Prod. Rep., 2006, 23: 753.

[22] [Sirikantaramas S, Asano T, Sudo H, Yamazaki M, Saito K. Curr. Pharm. Biotechnol., 2007,
      8: 196.

[23] Holler U, Wright AD, Matthee GF, Konig GM, Draeger S, Aust H-J, Schulz B. Mycol. Res.,
      2000, 104: 1354.

[24] Konig GM, Kehraus S, Seibert SF, Abdel-Lateff A, Muller D. Chem Bio Chem, 2006, 7:229.

[25] Oh DC, Jensen PR, Kauffman CA, Fenical W. Bioorg. Med. Chem., 2005, 13: 5267.

[26] Minagawa K, Kouzuki S, Yoshimoto J, Kawamura Y, Tani H, Iwata T, Terui Y, Nakai H,
      Yagi S, Hattori N, Fujiwara T, Kamigauchi T. J. Antibiot., 2002, 55: 155.

[27] Minagawa K, Kouzuki S, Kamigauchi T. J. Antibiot., 2002, 55: 165.

[28] Ainsworth AM, Chicarelli-Robinson MI, Copp BR, Fauth U, Hylands PJ, Holloway JA, Latif M, O'Beirne GB, Porter N, Renno DV, Richards M, Robinson N, Xenovulene A. J. Antibiot., 1995, 48: 568.

[29] Fingerman M, Nagabhushanam R. Biomaterials form Aquatic and Terrestrial Organisms. Enfield: Science Publishers, 2006: 285.

[30] Sugano M, Sato A, Iijima Y, Furuya K, Haruyama H, Yoda K, Hata T. J. Org. Chem., 1994, 59: 564.

[31] Wang Y, Oberer L, Dreyfuss M, Suetterlin C, Riezman H. Helv. Chim. Acta, 1998, 81: 2031.

[32] Lauer U, Anke T, Sheldrick WS, Scherer A, Steglich W. J. Antibiot., 1989, 42: 875.

[33] Ukiya M, Akihisa T, Tokuda H, Hirano M, Oshikubo M, Nobokuni Y, Kimura Y, Tai T, Kondo S, Nishino H. J. Nat. Prod., 2002, 65: 462.

[34] Chen C-H, Yang S-W, Shen Y-C. J. Nat. Prod., 1995, 58: 1655.

[35] Anke T, Werle A, Zapf S, Velten R, Steglich W. J. Antibiot., 1995, 48: 725.

[36] Sun Y, Yasukawa K. Bioorg. Med. Chem. Lett., 2008, 18: 3417.

[37] Sassa T, Kinoshita H, Nukina M, Sugiyama T. J. Antibiot., 1998, 51: 967.

[38] Toki S, Ando K, Yoshida M, Matsuda Y. J. Antibiot., 1994, 47: 1175.

[39] Toki S, Ando K, Yoshida M, Kawamoto I, Sano H, Matsuda Y. J. Antibiot., 1992, 45: 88.

[40] Yamashita Y, Saitoh Y, Ando K, Takahashi K, Ohno H, Nakano H. J. Antibiot., 1990, 43: 1344.

[41] Lee I-K, Yun B-S, Kim J-P, Ryoo I-J, Kim Y-H, Yoo I-D. Biosci. Biotechnol. Biochem., 2003,67:1813.

[42] Lee H-J, Rhee I-K, Lee K-B, Yoo I-D, Song K-B. J. Antibiot., 2000, 53: 714.

[43] Takahashi A, Kudo R, Kusano G, Nozoe S. Chem. Pharm. Bull., 1992, 40: 3194.

[44] Kwak J-Y, Rhee I-K, Lee K-B, Hwang J-S, Yoo I-D, Song, K-S. J. Microbiol. Biotechnol., 1999,9: 798.

[45] Reino JL, Guerrero RF, Hernandez-Galan R, Collado IG. Phytochem. Rev., 2008, 7: 89.

[46] Schlingmann G, Milne L, Pearce CJ, Borders DB, Greenstein M, Maiese WM, Carter GT. J. Antibiot., 1995, 48: 375.

[47] Baeuerle J, Anke T, Jente R, Bosold F. Arkiv. Mikrobiol., 1982, 32: 194.

[48] Garlaschelli L, Magistrali E, Vidari G, Zuffardi O. Tetrahedron Lett., 1995, 36: 5633.

[49] Garlaschelli L, Vidari G, Vita-Finzi P. Tetrahedron Lett., 1996, 37: 6223.

[50] Li JY, Strobel G, Harper J, Lobkovsky E, Clardy J. Org. Lett., 2000, 2: 767.

[51] Osterhage C, Kaminsky R, Knig GM, Wright AD. J. Org. Chem., 2000, 65: 6412.

[52] Qian-Cutrone J, Huang S, Shu Y-Z, Vyas D, Fairchild C, Menendez A, Krampitz K, Dalterio R, Klohr SE, Gao Q. J. Am. Chem. Soc., 2002, 124: 14556.

[53] Zhang Y, Li C, Swenson DC, Gloer JB, Wicklow DT, Dowd PF. Org. Lett., 2003, 5: 773.

[54] Omura S. Macrolide Antibiotics (Second edition). San Diego: Academic Press, 2002:1.

[55] Tomoda H, Namatame S, Si S, Kawaguchi K, Masuma R, Namikoshi M, Omura S. J. Antibiot., 1999, 52: 851.

[56] Tomoda H, Namatame I, Tabata N, Kawaguchi K, Si S, Omura S. J. Antibiot., 1999, 52: 857.

[57] Namatame I, Tomoda H, Arai M, Omura S. J. Antibiot., 2000, 53: 19.

[58] Namikoshi M, Akano K, Meguro S, Kasuga I, Mine Y, Takahashi T, Kobayashi H. J. Nat. Prod., 2001, 64: 396.

[59] Kobayashi J, Tsuda M. Phytochem. Rev., 2004, 3: 267.

[60] Rowley DC, Kelly S, Kauffman CA, Jensen PR, Fenical W. Bioorg. Med. Chem., 2003, 11: 4263.

[61] Pomilio AB, Battista ME, Vitale AA. Curr. Org. Chem., 2006, 10: 2075.

[62] Walton JD. Phytochemistry, 2006, 67: 1406.

[63] Isaka M, Kittakoop P, Thebtaranonth Y. Mycology Series, 2003, 19: 355.

[64] Liu X, Xu F, Shao C, She Z, Lin Y, Chan WL. Stud. Nat. Prod. Chem., 2008, 35: 197.

[65] Son BW, Jensen PR, Kauffman CA, Fenical W. Nat. Prod. Lett., 1999, 13: 213.

[66] Tirunarayanan MO, Sirsi M. J. Ind. Sci., 1957, 39: 185.

[67] Tomoda H, Doi T. Acc. Chem. Res., 2008, 41: 32.

[68] Pedras MSC, Zaharia LI, Ward DE. Phytochemistry, 2002, 59: 579.

[69] Rezanka T, Spizek J. Stud. Nat. Prod. Chem., 2005, 32: 471.

[70] Cueto M, Jensen PR, Kauffman C, Fenical W, Lobkovsky E, Clardy J. J. Nat. Prod., 2001, 64: 1444.

[71] Seephonkai P, Isaka M, Kittakop P, Boonudomlap U, Thebtaranonth Y. J. Antibiot., 2004, 57: 10.

[72] Erkel G, Belahmer H, Serwe A, Anke T, Kunz H, Kolshorn H, Liermann J, Opatz T.   J. Antibiot., 2008, 61: 285.

[73] Mukhopadhyay T, Bhat RG, Roy K, Vijayakumar EKS, Ganguli BN. J. Antibiot., 1998, 51: 439.

[74] Bentley HR, Cunningham KG, Spring FS. J. Chem. Soc., 1951, 2301.

[75] Zhou X, Luo L, Dressel W, Shadier G, Krumbiegel D, Schmidtke P, Zepp F, Meyer CU. Am. J. Chinese Med., 2008, 36: 967.

[76] Ferreira RHK, Heerklotz JA, Werner C. PCT Int. Appl., 2002, 30 pp.

[77] Yu CM, Curtis JM, Walter JA, Wright JLC, Ayer SW, Kaleta J, Querengesser L, Fathi-Afshar ZR. J. Antibiot., 1996, 49: 395.

[78] Abraham SP, Hoang TD, Alam M, Jones EBG. Pure Appl. Chem., 1994, 66: 2391.

[79] Garcia-Pajon CM, Collado IG. Nat. Prod. Rep., 2003, 20: 426.

[80] Charnley AK. Adv. Bot. Res., 2003, 40: 241.

[81] Cary JW, Ehrlich KC. Mycopathologia, 2006, 162: 167.

[82] Reddy KRN, Reddy CS, Abbas HK, Abel CA, Muralidharan K. Toxin Rev., 2008, 27: 287.

[83] He YC. Mycology. Beijing: Chinese Forestry Publishing House, 2008:3.

[84] Ames BN, Magaw R, Gold LS. Science, 1987, 230: 271.

[85] Soriano JM, Dragacci S. Food Res. Int., 2004, 37: 985.

[86] Leong SL, Hocking AD, Pitt JI, Kazi BA, Emmett RW, Scott ES. Int. J. Food Microbiol., 2006, 111: S10.

[87] Rezanka T, Sobotka M, Spizek J, Sigler K. Anti-Infect. Agents Med. Chem., 2006, 5: 187.

[88] Pedras MSC, Abrams SR, Seguinswartz G, Quail JW, Jia ZC. J. Amer. Chem. Soc., 1989, 111: 1904.

[89] Culvenor CCJ, Cockrum PA, Edgar JA, Frahn JL, Gorstallman CP. J. Chem. Soc. Chem. Commun., 1983, 21: 1259.

[90] Schuemann J, Hertweck C. J. Biotechnol., 2006, 124: 690.

[91] Misiek M, Hoffmeister D. Planta Med., 2007, 73: 103.

[92] Chen Y-Y, Chang H-M. Food Chem. Toxicol., 2004, 42: 759.

[93] Liu B, Ni J. Extraction and Examination of Traditional Chinese Medicine. Beijing: China Press of Traditional Chinese Medicine, 2007:343, 471.

[94] Shim MJ, Kim HW, Kim BK. Huazhong Nongye Daxue Xuebao, 2004, 23: 89.

[95] Wu TS, Shi LS, Kuo SC. J. Nat. Prod., 2001, 64: 1121.

[96] Kikuchi T, Kanomi (n'ee Matsuda) S, Kadpta S, Muyui Y, Tsubono K, Ogita Z. Chem. Pharm. Bull., 1986, 34: 3695.

[97] Oyama M, Xu Z, Lee K-H, Spitzer TD, Kitrinos P, McDonald OB, Jones RRJ, Garvey EP. Lett. Drug Des. Discovery, 2004, 1: 24.

[98] Kulanthaivel P, Hallock YF, Boros C, Hamilton SM, Janzen WP, Ballas LM, Loomis CR, Jiang JB, Katz B, Steiner JR, Clardy J. J. Am. Chem. Soc., 1993, 115: 6452.

[99] Yamada T, Iritani M, Ohishi H, Tanaka K, Doi M, Minoura K, Numata A. Org. Bioorg. Chem., 2007, 5: 3979.

[100] Usami Y, Ueda Y. Chem. Lett., 2005, 34: 1062.

[101] Usami Y, Takaoka I, Ichikawa H, Horibe Y, Tomiyama S, Ohtsuka M, Imanishi Y, Arimoto M. J. Org. Chem., 2007, 72: 6127.

[102] Usami Y, Ichikawa H, Arimoto M. Int. J. Mol. Sci., 2008, 9: 401.

[103] Numata A, Iritani M, Yamada T, Minoura K, Matsumura E, Yamori T, Tsuruo T. Tetrahedron Lett., 1997, 38: 8215.

Zhi-Hong XU

CHAPTER 4

Alkaloids

Section 1   General

Alkaloids are generally defined as a group of organic compounds of plant origin that contain nitrogen (except proteins, peptides, amino acids and vitamin B). However, many nitrogen containing compounds not only discovered from plants but also from marine products, microorganisms, fungi and insects, are considered alkaloids. Therefore, generally speaking, all organic compounds containing nitrogen from natural sources can be regarded as alkaloids.

Alkaloids were the first class of bioactive natural products studied by scientists. In China, Zhao Xuemin noted in his book *Bencao Gangmu Shiyi (Supplements of Compendium of Materia Medica)*, that *Baiyuan Jin,* a book written in the 17th century, kept a record of a crystalline toxin extracted from *Aconitum carmichaeli* and used as curare. This toxin is determined to be an aconitine by modern analysis. In Europe, morphine was initially isolated from *Papaver somniferum* L. in 1804 by the German pharmacist Friedrich Wilhelm Adam Sertürner. It was called "the vegetable alkali" for its basic property. Cinchonino crystal was isolated from the bark of the South American cinchona tree by the Spanish dentist Bernardino Antonio Gomes in 1810. Cinchinino was later proved to be a mixture of quinine and cinchonine. In 1819, pharmacist W. Meissner used the word "alkali-like" or "alkaloid" referring to any basic nitrogenous compound of plant origin. The term *alkaloid* has been used since then.

Most alkaloids have biological activity and often are the active constituents of various medicinal plants including many traditional Chinese medicines (TCM). Examples include the analgesic agent morphine and the antitussive agent codeine of *Papaver somniferum* L., the anti-asthma agent ephedrine from *Ephedra sinica*, the anti-spasm agent atropine of *Atropa belladonna* L., the anti-cancer agent vincristine from *Catharanthus roseus*, and the anti-inflammatory agent berberine of *Coptis chinensis*. When treated with acids, alkaloids can be converted into water-soluble salts, which can be easily absorbed *in vivo*. Moreover, most alkaloids contain complex chemical structures. Due to these characteristics, the alkaloids have always been an interesting research topic which attracts many organic chemists. So far, more than 4000 alkaloids have been identified and the number is still increasing every year. Trace amounts of alkaloids, such as the anti-cancer agent maytansine, can be discovered from plants. This nitrogen-containing macrocyclic compound has a complex molecular structure and its total synthesis has been reported. The isolation, structural elucidation and total synthesis of alkaloids are still important research areas for natural product chemists.

Most alkaloids possess biological activity and are the active components of many medicinal plants and TCM herbs. However, there are few exceptions. For example, aconitine is the major constituent of *Aconitum carmichaeli*. The cardiotonic and analgesic ingredient is not aconitine, but the minor component *dl*-demethyl-coclaurine (or higenamine). Vincristine and maytansine are trace constituents of *Catharanthus roseus* and *Maytennus hookerii*, respectively. The relationship between content and activity is worthy of further investigation.

Alkaloids are found in many plants and are commonly found in the Leguminosae, Solanaceae, Manispermaceae, Papaveraceae, Ranurculaceae and Berbereaceae families. Alkaloids are accumulated in the rootbarks, roots and stems of some plants but rich in the seeds or fruits of others. Alkaloid content varies greatly in different plants. For example, the total alkaloid content in cinchona tree bark is more than 3% while the content of vincristine in *Catharanthus roseus* is only 0.0001%. Much less is found in the content of maytansine from *Maytenus hookerii* Loes, at only 0.00002%. In general, alkaloid content is considered high if it is more than 0.1%.

Alkaloids in one plant are usually derived from the same precursor and they often exhibit similar structures. In many cases, alkaloids from the same genus and the same family have similar structural skeletons. Therefore, identification of the plant name, its species and understanding of the known constituents from the same species and genus, as well as their biogenetic relationship, prior to start the study is helpful for the study of alkaloids. Otherwise, the research may head in the wrong direction. It is more important for studying Chinese herbs, since many Chinese herbs have several different folk names, and sometimes one folk name is duplicated with another one. However, the presence of alkaloids cannot be predicted soely by the plant family or genus. This is because taxonomy is largely based on morphology of the plants and there is no definite relationship between the plant taxonomy and chemical constituents. For example, *Cephaelis ipecacuanha* (Rubiaceae) contains about 2% emetine, while *Gardenia jasminoides* and *Rubia cordifolia* from the same family do not contain alkaloids.

Alkaloids comprise the largest family of natural organic products. Their structures vary in many ways and include different subtypes. While determining the structures, scientists always study structure-activity relationships and modify the molecules for enhancing efficacy and mass production. For example, the study of morphine (**1**) resulted in the development of isoquinoline alkaloids and the discovery of the analgesic agent dolantine (**2**). The study of cocaine (**3**) resulted in the development of hyoscyamine alkaloids and the discovery of the local anesthetic agent procaine (**4**).

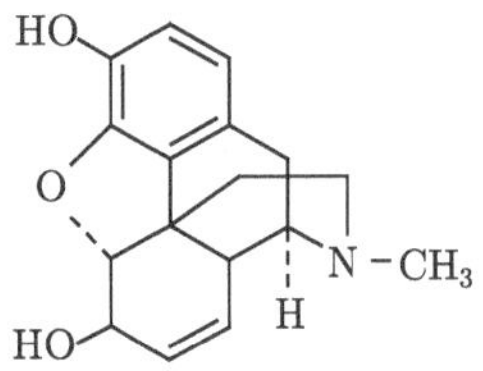

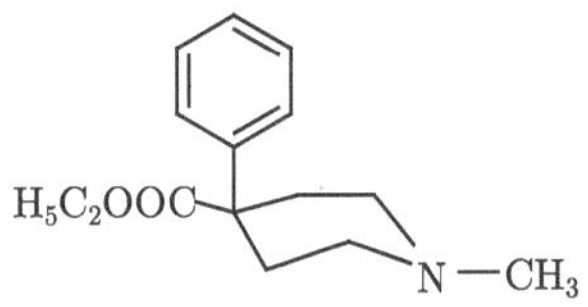

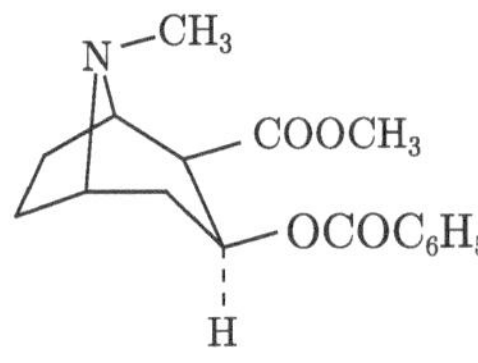

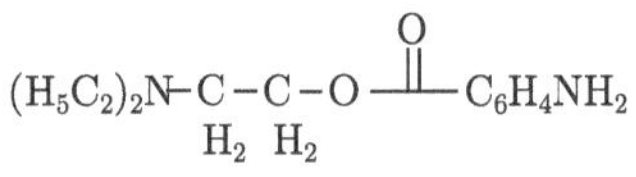

Alkaloids are recognized commonly as secondary plant metabolites, which play a protective or positive role in plant growth and metabolism. However, their real significance is controversial, because plants which do not contain or contain trace amount of alkaloids still grow and develop well. Therefore, the real function of plant alkaloids needs to be further investigated.

Section 2   Characterization, Identification and Isolation

Alkaloids are colorless crystals (a few are liquid) with specific melting points and optical rotation. Many of them taste bitter and their free forms dissolve easily in organic solvents but not in water. Most alkaloids are basic but few are not, such as the amide alkaloids. Generally, alkaloids form salts with organic or inorganic acids according to their different basic properties. Alkaloid salts dissolve in water except for salts derived from some special inorganic acids (such as silicotungstic acid and phosphotungstic acid) or organic acids (such as picric acid). This characteristic can be used for the identification and isolation of alkaloids. Phenolic alkaloids are amphiprotic compounds, which can dissolve in basic solvents. For example, morphine can make salts not only with acid but also caustic bases. Alkaloids of quaternary ammonium salts are commonly water-soluble and require special attention in their extraction and isolation procedures.

Alkaloids exhibit different colors when treated with some reagents, and this quality can be used to identify specific alkaloids[1].

Dragendorff's reagent, a solution of potassium bismuth iodide, is used to determine the presence of alkaloids. The preparation procedure is as follows: Prepare 0.85 g basic bismuth nitrate and 10 ml of glacial acetic acid in 40 ml of water (solution I); and 8 g potassium iodide in 20 ml of water (Solution II). Mix 1 ml of equal volumes of solutions I and II together with 2 ml glacial acetic acid and 10 ml water. When treated with this reagent, a solution or spot on a thin layer chromartography (TLC) plate, containing alkaloids turns brownish-yellow.

Based on their chemical properties, alkaloids can be extracted from plants using different methods. Alkaloids usually exist in plants as salt forms of organic acids like caffeic acid, citric acid, oxalic acid or papaveric acid. Ethanol, dilute ethanol (60 to 80%), water or acidic water (0.5 to 1% sulfuric acid or acetate acid) with soaking, percolation or heat treatment is used to extract the alkaloid from the plants. After removal of the water or ethanol, the concentrated syrup is then treated with 2% dilute acid to give the crude alkaloid extract. If the material is extracted with dilute alcohol, the thick solution is acidified with dilute acid and then washed with ethyl ether or chloroform to remove fats and oils. After filtration, the acidic aqueous solution is basified with ammonia, sodium carbonate or limewater to convert the alkaloids to free base forms from their salts. The basic aqueous solution is then extracted with ethyl ether or chloroform to obtain the crude alkaloids. The water-soluble alkaloids, if present, are in the water layer. Although the crude alkaloids obtained by the above method may contain lots of impurities, the method is relatively economic and safer to use for industrial scale production. The ethanol extract may be subjected to column chromatography over macroporous resin eluted with water and dilute acid. The crude alkaloids are yielded from the dilute acid fractions. If the column is washed with alcohol in different concentrations and each fraction is collected respectively, it can provide a preliminary isolation of the crude alkaloids.

Chloroform, dichloromethane, benzene or other water-immiscible solvents may also be used to extract crude alkaloids. Before extraction, mix or grind the powdery material with a basic aqueous solution, such as 1% ammonia, sodium carbonate or limewater to convert the alkaloids into free bases. The material is then soaked or percolated with organic solvents. The solution is then extracted with dilute acidic aqueous solution such as 1 to 2% hydrochloric acid three to five times until there are no alkaloids present in the organic layer by test of Dragendorff's reagent as customary. At that time, alkaloids have turned into salts and will dissolve into the acidic water layer. After filtration, ether or chloroform is used to remove the fat-soluble impurities from the acidic water. However, some alkaloids in this case may dissolve in chloroform and therefore, the chloroform layer should be treated again. The acidic water layer is basified with a base. If there are sediments or precipitates, it should be filtered and recrystallized. The basic aqueous solution is then extracted with ethyl ether and chloroform.

Weak basic alkaloids usually exist in free base states in plants and can be directly extracted with organic solvents such as benzene, chloroform, etc. Strong basic alkaloids may be present in plants as salt forms. In order to increase the extraction efficiency, the raw material powder is macerated with water to swell the tissue cells, or macerated with dilute organic acid, such as aqueous citric acid, so that some moderate basic alkaloids can remain in the plants as salts. They are then extracted with organic solvent to obtain the crude weak basic alkaloids. If the material is extracted directly with water or dilute alcohol, the condensed aqueous solution always exhibits weak acidic properties due to the presence of some organic acids. In such a case, the free weak basic alkaloids can be extracted from the aqueous solution by chloroform or other organic solvents. For example, the pH of condensed percolate from *Camptotheca acuminate* is 5.4; camptothecin can then be extracted directly with chloroform.

Water-soluble alkaloids and alkaloids of quaternary ammonium salts cannot be extracted from aqueous solutions with common organic solvents. To obtain these types of alkaloids, the aqueous solution should be acidified first and then subjected to column chromatography with cation exchange resin, or the aqueous solution directly extracted with butanol or pentanol. For example, water-soluble leonurine A was extracted directly by pentanol from *Leonuri artemisae*.

Volatile alkaloids such as ephedrine can be conveniently isolated using steam distillation. Sublimable alkaloids can be directly obtained from the dried plant material powder by sublimation. For example, when tea leaves are heated, caffeine will sublimate and be collected after condensation. Nowadays supercritical fluid (e.g., carbon dioxide) extraction (SFE) has been used due to its high speed and efficiency. For example, isolation of tetrahydropalmatine from *Corydalis yanhusuo*, and scopolamine from *Datura metel* can be performed using this method. However, SFE requires specific equipment that handles the high pressure involved.

Crude alkaloids extracted with the methods above should be checked with TLC to determine the content of the alkaloids in different fractions. Chromatographic methods will be selected for further isolation and purification[2].

Section 3   Classification

The classification of alkaloids is generally based on their chemical structures or their common molecular biosynthetic pathway precursors. However, traditional classification methodology has often been challenged by newly discovered compounds with novel structural skeletons. The classification of alkaloids became more and more specific. For example, isoquinoline alkaloids comprise a large group of compounds, which can be further divided into morphines, benzylisoquinolines, aporphines, protopines, dimeric aporphines, and bis-benzylisoquinolines, and so on.

The goal of this book is to present the basic chemistry of various compounds, rather than a comprehensive encyclopedia. It will not list all existing alkaloids. This chapter illustrates some common structural types of alkaloids including isoquinolines, quinolines, pyrrolidines, indoles, etc. The examples come from our own or others' research work on natural products from China.

3.1   Isoquinolines

Isoquinoline alkaloids represent the largest group of alkaloids whose basic nucleus is isoquinoline or tetrohydroisoquinoline. According to the different functional groups attached to the nucleus, they can be divided into nine subtypes.

3.1.1   Simple isoquinoline alkaloids

The hypotensive components salsoline (**5**) and salsolidine (**6**) from *Salsola richteri* karelin are two of the simplest isoquinoline alkaloids.

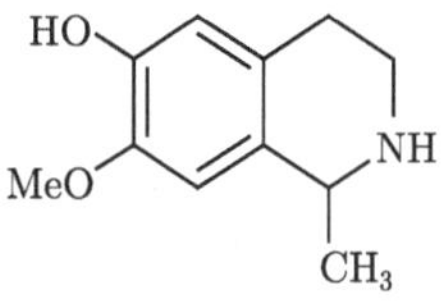

5

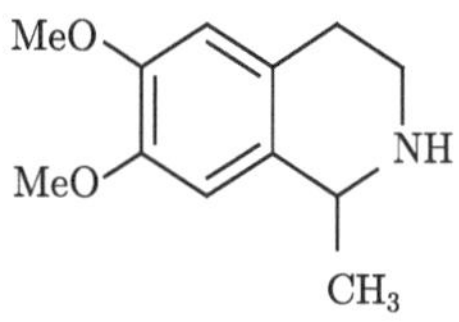

6

3.1.2   Benzylisoquinoline alkaloids

Normally, benzylisoquinoline alkaloids contain a benzyl group located at the C-1 position, such as spasmolitic papaverine (**7**) from opium; demethylcoclaurine (**8**), the cardiotonic component from *Aconitum carmichaeli*; as well as narcotine (**9**), which has an additional lactone ring in its structure.

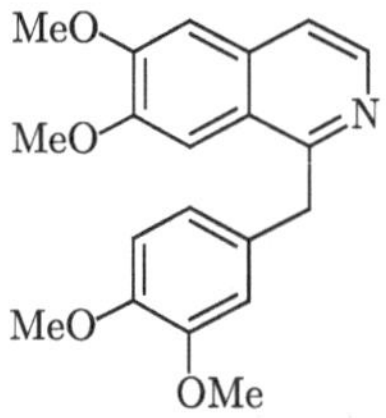

7

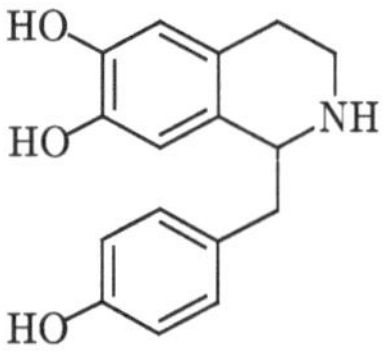

8

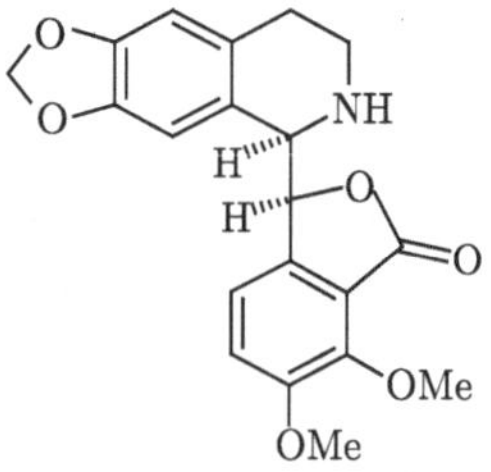

9

3.1.3   Bisbenzylisoquinoline alkaloids

Alkaloids of this type feature two benzylisoquinoline moieties, which are connected by etherification of phenolic hydroxyls. Based on the ether bonds, they can be divided into monoether, diether, triether, and so on. If the quinoline ring is the head and the benzyl ring is the tail, they can be further classified into tail-to-tail, head-to-tail and head-to-head binding types. For example, liensinine (**10**) from *Lotus Plumule* belongs to a head-to-tail monoether alkaloid.

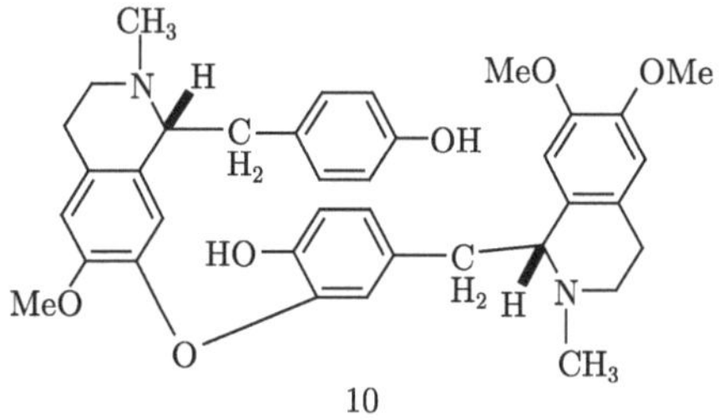

10

3.1.4   Aporphine alkaloids

Aporphine alkaloids are tetracyclic compounds derived from the benzylquinoline alkaloids, in which two benzyl ring structures are connected to form an additional ring. Such examples are stephanine (**11**) and magnoflorine (**12**), the quaternary ammonium salt alkaloid from *Magnolia grandiflora*.

3.1.5   Protoberberine alkaloids

Protoberberine alkaloids are characterized by two fused isoquinoline rings. For example, the antibiotic alkaloid berberine (**13**) from *Coptis chinensis* and *Berberis pruinosa*, and the sedation-analgesia constituent tetrahydropalmatine (**14**) from *Corydalis yanhusuo* are both protoberberine alkaloids.

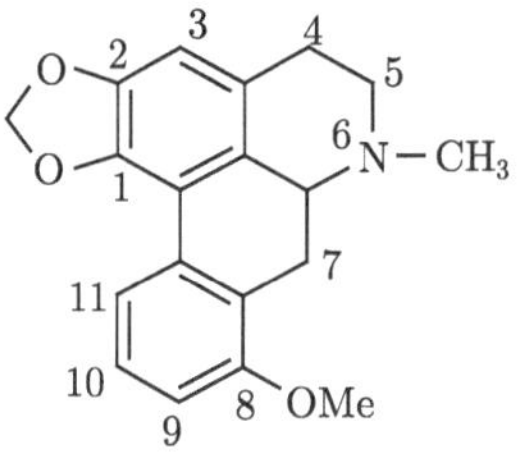

11

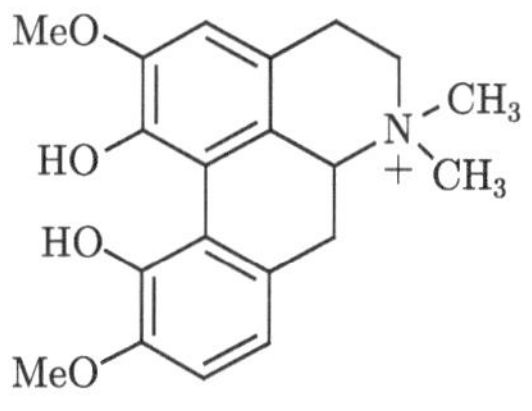

12

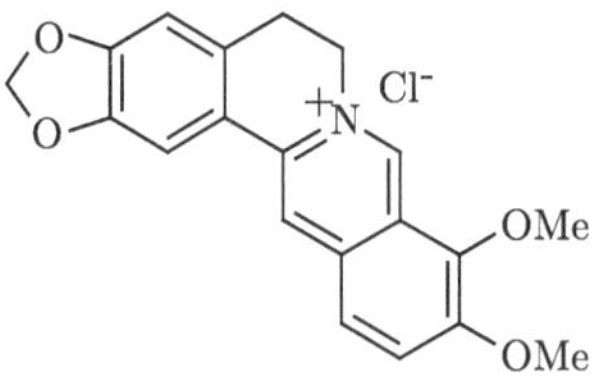

13

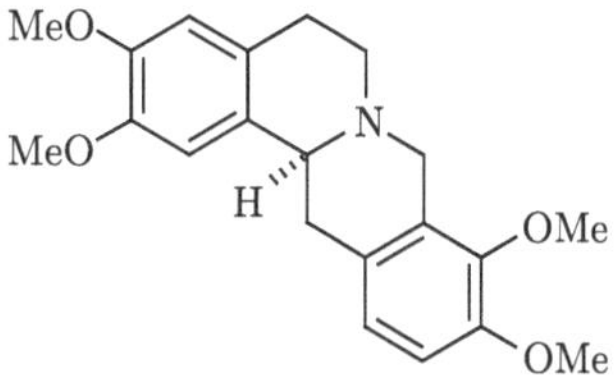

14

3.1.6   Protopine alkaloids

Protopine alkaloids, such as the protopine (**15**) and cryptopine (**16**) in *Corydalis yanhusuo*, are actually the carbonyl *seco*-compounds of berberine.

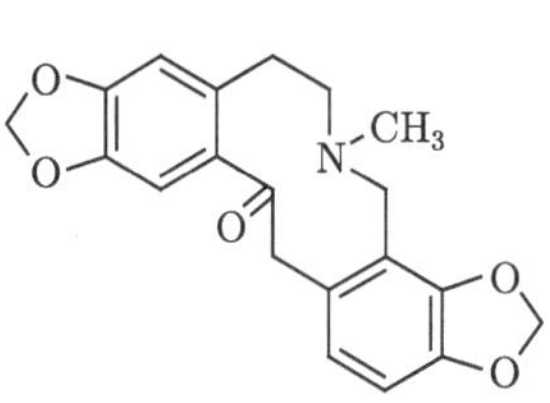

15

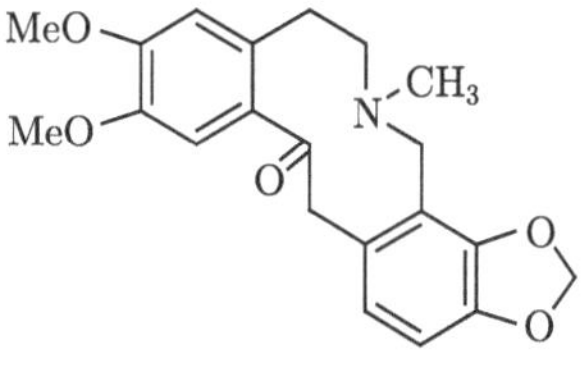

16

3.1.7   Emetine alkaloids

Emetine alkaloids are comprised of isoquinoline and benzoquinoline rings. Emetine (**17**) is the active constituent against amebic dysentery from *Uragoga ipecacuanha*.

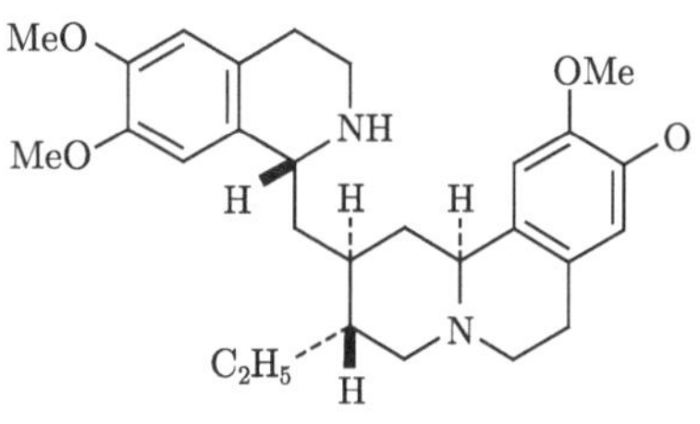

17

3.1.8   α-Naphthaphenanthridine alkaloids

Examples of theses types of alkaloids include the antibacterial sanguinarine (**18**) from *Macleaya cordata* and chelerythrine (**19**) from *Chelidonium majus*.

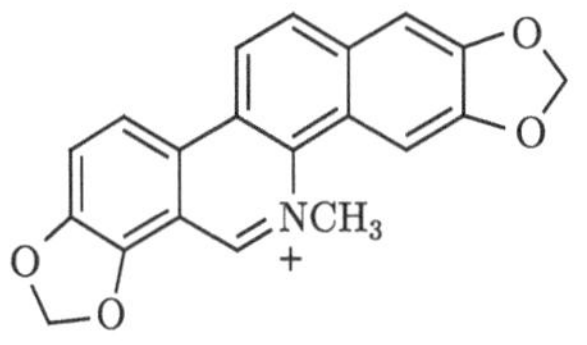

18 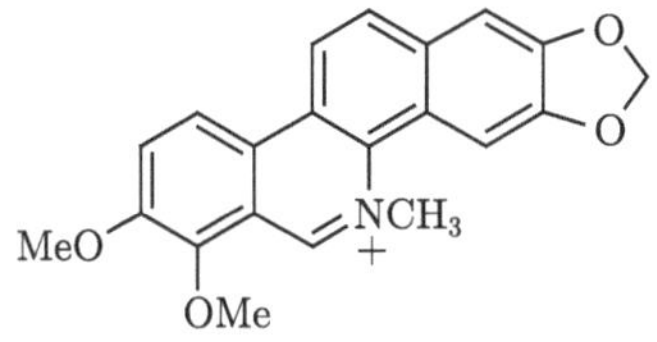   19

3.1.9   Morphine alkaloids

Morphine (**20**) is a representative compound of this alkaloid type.

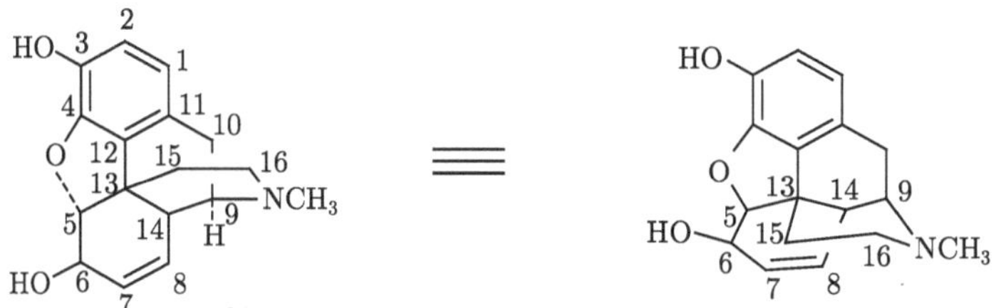

20

3.2   Quinolines

A quinoline ring is the basic structure of quinoline alkaloids. The simplest compounds are furan quinoline derivatives such as skimmianine (**21**) (mp. 194°), which exhibits sedation-analgesia activity and is extracted from the leaves of *Skimmia reevesiana*, and dictamnine (**22**) (mp. 133°C) found in the roots of *Dictamnus dasycarpus*. Others like quinine (**23**) and camptothecin (**66**) also belong to the quinoline alkaloid group.

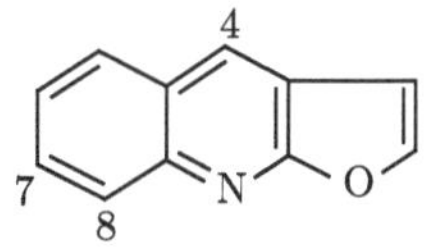

21   4=7=8=OCH$_3$
22   4=OCH$_3$,7=8=H

UV absorptions, λ_{max} (ε), of quinoline rings are 230 (37000), 270 (3600), and 314 (2750) nm. Their IR absorption bands of 1400 to 1600 cm^{-1}are similar to those of the benzyl ring.

Cinchona alkaloid is the most important of the quinoline alkaloids. Cinchona is a general term for all *Cinchona* woody plants of Rubiaceae. It is valued in South America for its antimalarial properties. Later, its effective component was found to be quinine (**23**). Quinine was also discovered from two other genera, *Remijia* and *Cuprea*, of the same family at the same time. Its content was found as high as 3%. Therefore, all three genera were widely introduced into South America and Southeast Asia.

Quinine is one of the earliest studied alkaloids following morphine. In 1792 it was isolated as a crude substance by Fourcroy. Later, in 1810 and for the first time, Gomes extracted cinchona bark with an alcoholic caustic potash solution and obtained a crystalline substance named cinchonino. In 1820, two French chemists, Pelletier and Caventou, isolated cinchonino from the bark of *Cinchona condaminea*, then they did further purification and obtained two pure compounds, quinine (**23**) and cinchonine (**27**). From then on, many chemists carried out systematic studies on alkaloids from the *Cinchona*, *Remijia* and *Cuprea* plants. More than 30 alkaloids were isolated and most of them are quinine derivatives. The major compounds are listed in Table 4-1.

Table 4-1   Major Cinchona Alkaloids

Names	M.F.	M.P. (°C)	$[\alpha]_D$ /(°)
quinine (**23**)	$C_{20}H_{24}O_2N_2$	175	-158(EtOH)
quinidine			
conchinine (**24**)			
β–quinine	$C_{20}H_{24}O_2N_2$	173.0	+255(EtOH)
epiquinine (**25**)	$C_{20}H_{24}O_2N_2$	oil	43
epiquinidine (**26**)	$C_{20}H_{24}O_2N_2$	113	+102(EtOH)
cinchonine (**27**)	$C_{19}H_{22}ON_2$	264	+224(EtOH)
epicinchonine (**28**)	$C_{19}H_{22}ON_2$	83	+120.3(EtOH)
cinchonidine (**29**)	$C_{19}H_{22}ON_2$	205	-110(EtOH)
epicinchonidine (**30**)	$C_{19}H_{22}ON_2$	104	62.8(EtOH)

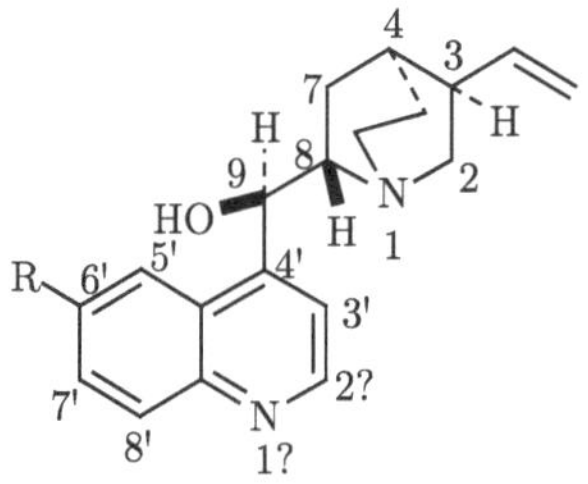

23   quinine   R=OCH$_3$
29   cinchonidine R=H

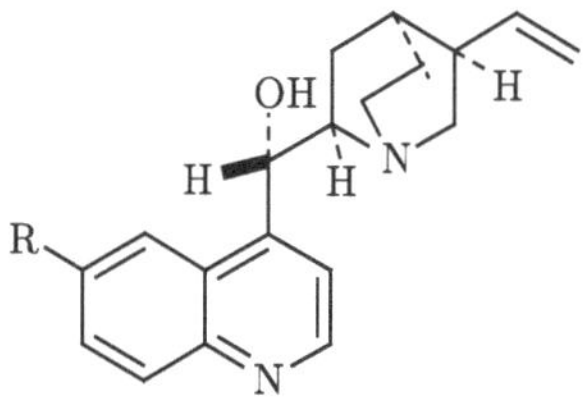

24   quinidine   R=OCH$_3$
27   cinchonine R=H

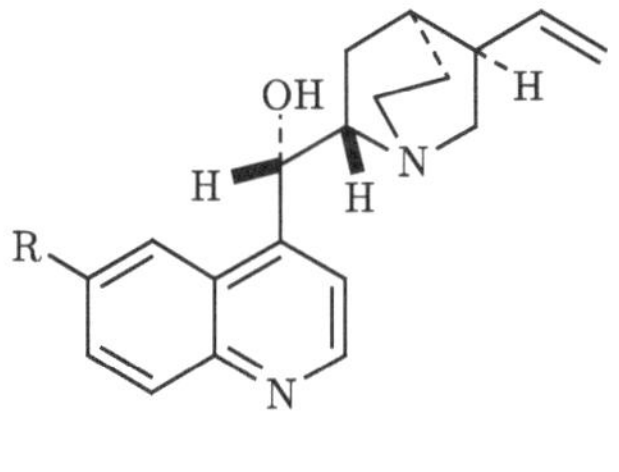

25   epiquinine   R=OCH$_3$
30   epicinchonidine R=H

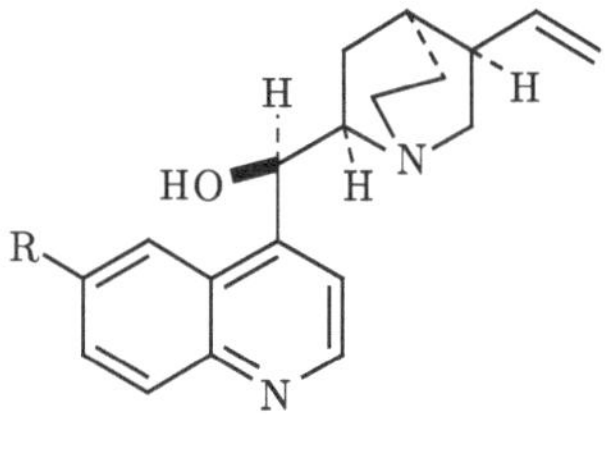

26   epiquinidine R=OCH
28   epicinchonine R=H

3.3   Pyrrolidines

3.3.1   Simple pyrrolidine alkaloids

Simple pyrrolidine alkaloids, such as hygrine (**31**; bp. 195°C; $[\alpha]_D$ 0°) and cuscohygrine (**32**; bp. 88 to 170°C; $[\alpha]_D$ 0°). Both liquid alkaloids were isolated from the leaves of *Erythroxylon coca* Lam.

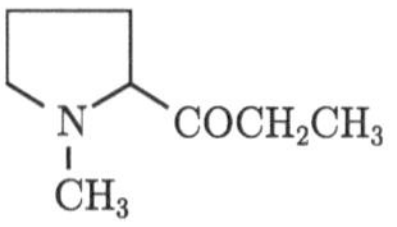

Another example is codonopsine (**33**; mp. 150°C; $[\alpha]_D$ −16° in MeOH) from *Codonopsis clematidea* (Schrenk) Clarke.

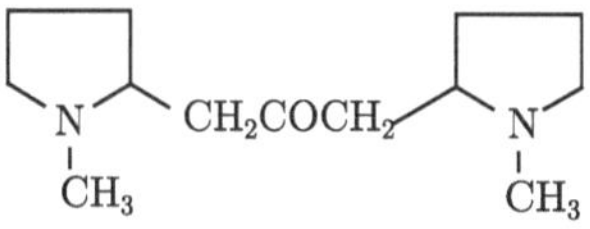

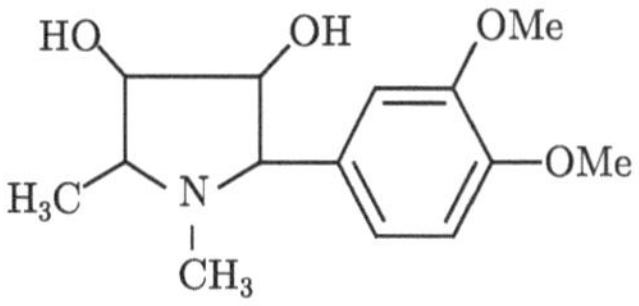

3.3.2 Pyrrolizidine alkaloids

Pyrrolizidine alkaloids comprise two pyrrolidines fused by a tertiary nitrogen. Examples include macrophylline (**34**; $C_{13}H_{21}O_3N$; mp. 44°C; $[\alpha]_D$+35° in EtOH), monocrotaline (**35**; $C_{16}H_{23}O_6N$; mp. 219°C; $[\alpha]_D$-55° in CHCl$_3$) from *Crotalaria sessiflora* L., and platyphylline (**36**; $C_{18}H_{27}O_5N$; mp. 129°C; $[\alpha]_D$-56° in CHCl$_3$) from *Senecio platyphyllus* D. C. All these compounds exist as esters formed by an organic acid and an amino-alcohol derived from pyrrolizidine. Most of these compounds are diesters as **35** and **36**, and a few are monoesters as **34**. The amino-alcohol moiety of these alkaloids is similar to platynecine (**37**), or macronecine (**38**), both are epimers at position C-1.

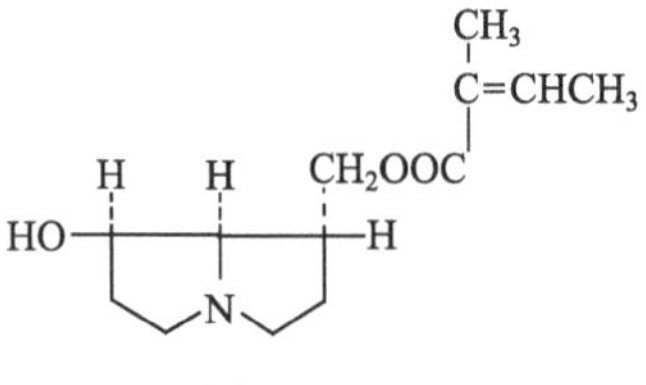

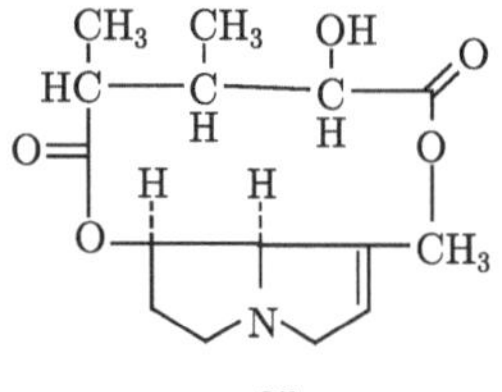

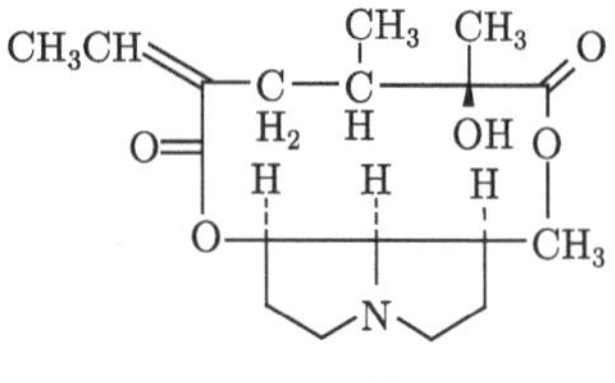

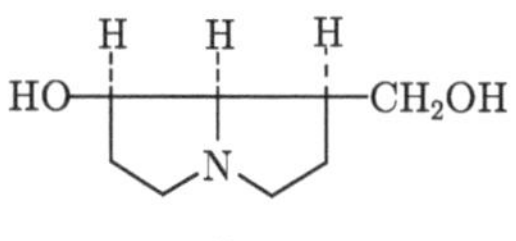

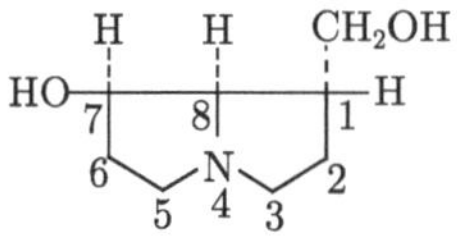

3.3.3 Indolizidine alkaloids

Indolizidine alkaloids are characterized by an indolizidine ring, consisting of a pyrrolidine fused to a piperidine ring. Examples include ipalbidine (**39**) and securinine (**40**).

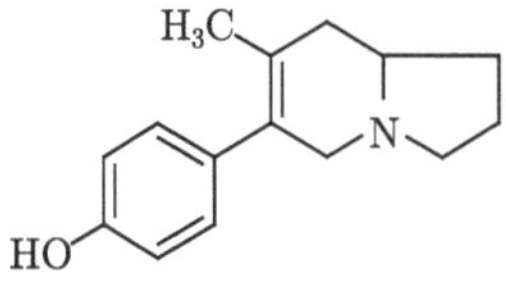

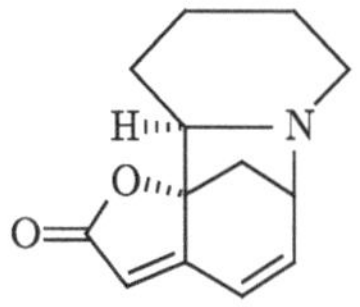

3.3.4   Tropane alkaloids

Tropane is a heterocyclic ring comprised of a pyrrolidine and a piperidine. Atropine (**41**) and its related compounds, a class of tropanes esterified with organic acids, are the most common alkaloids of this group.

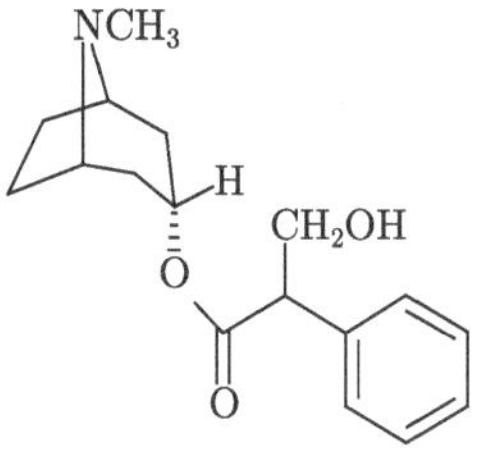

41

3.3.5   *Stemona* alkaloids

Alkaloids isolated from the roots of *Stemona* species contain a pyrrolidine skeleton and are thus classified as pyrrolidine alkaloids. Examples include protostemonine (**42**) and tuberostemonine (**43**).

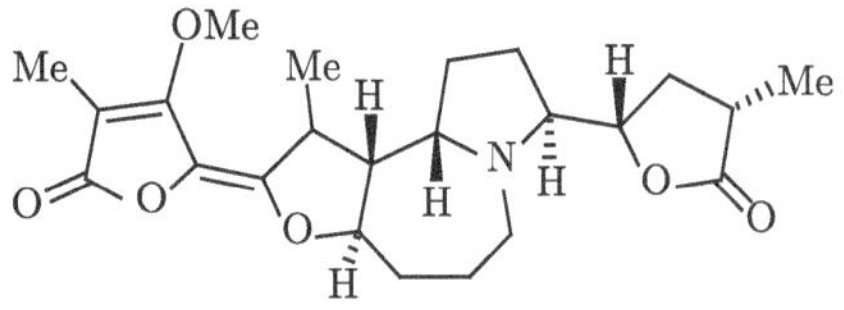

42 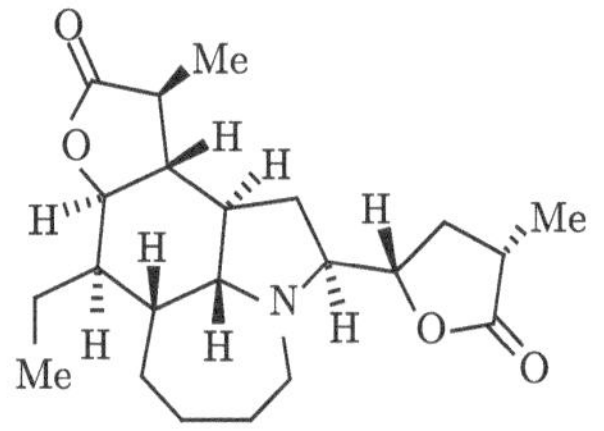   **43**

3.4   Indoles

Since the discovery of the antihypertensive reserpine (**44**) and the antineoplastic drugs vinblastine (**45**, VLB) and vincristine (**46**, VCR) in the 1950s, indole alkaloids have attracted organic chemists from many countries. The isolation, structural elucidation and synthesis of more than 400 indole alkaloids make this group of alkaloids a predominant research area for natural products.

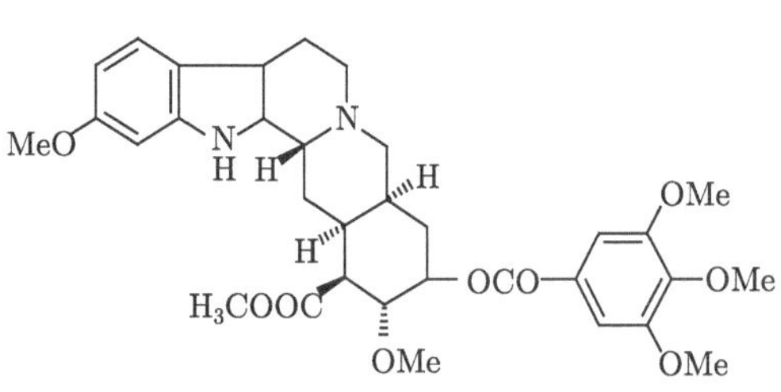

44

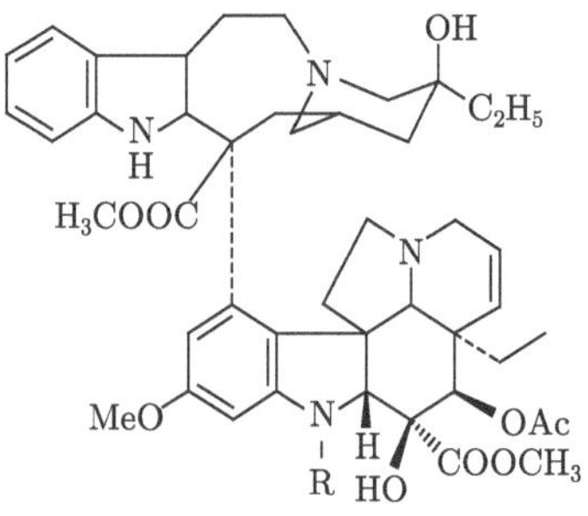

45   R = CH$_3$
46   R = CHO

Generally, the UV spectrum is an important diagnostic method for the identification of indole alkaloids. The maximal UV absorptions of indole alkaloids are at: (a) 200 nm, 280 to 290 nm for the basic indole ring; (b) 250 nm, the characteristic strong absorption for the dihydroindole due to the N, N-dimethylaniline; (c) 240 nm for the hydroxyl indole ring, similar to acetylaniline; (d) 280 nm, the strong absorption for the N-acylindole, and 257 and 290 nm due to the conjugation of the N-acyl and benzene rings.

Other tools IR spectra of this group shows sharp absorptions at 1450, 1590, 1620 and 340 cm^{-1} for NH and the ^{1}H-NMR spectrum mainly displays signals of different aromatic protons.

Examples of the simplest indole alkaloids include abrine (**47**, mp. 297°C) from *Abrus precatorius* L, and harmaline (**48**, mp. 231°C) and harmine (**49**, mp. 261 °C) from the seeds of *Peganum harmala* L.

47   48   49

Section 4   Structural Investigations of Some Alkaloids

Since chemical structures of alkaloids are greatly diverse, it is necessary to know what constituents have been previously reported in the same species or genus prior to the start of a structural investigation for a specific compound. The molecular formula of a compound can be determined by high resolution mass spectrum, and then compared with those of the known alkaloids. If this compound is new, a detailed analysis of the ^{1}H- and ^{13}C-NMR and 2D-NMR spectra, especially HMBC and HMQC spectra should be subsequently carried out. Some examples are given as below based on their classification.

4.1   *Stemona* alkaloids

Radix Stemonae is a traditional Chinese medicine and has been used as an antitussive and insecticide agent for long time. Plant species of this Radix Stemonae used in China include *Stemona tuberosa* Lour, *S. sessilifolia* (Miq.) Miq., and *S. japonica* Miq. These plants are distributed in South Yangzi River area. *Stemona parviflora* C.H. Wright grows only in Hainan island.

In the early 1930s, the German scientist Schild at al. determined the functional groups of tuberostemonine using Hofmann degradation, bromo-cyaniding, oxidation and reduction.[2] In the 1950s, the two Canadian scientists Götz and Boyri[3,4] performed extensive research work on this compound. They applied newly developed spectroscopic technologies, such as UV, IR, MS, and NMR, etc. for the structural elucidation, and succeeded in determining its planar structure for the first time. In the 1960s, Edward finally confirmed the relative configuration of tuberostemonine by single crystal x-ray diffraction. The Chinese chemist J.H.Chu started investigation of the chemical constituents of *S. sessilifolia* and *S. tuberose* in 1949 and successively identified protostemonine and tuberostemonine. In the 1980s, great progress was made in China for the structural elucidation of *Stemona* alkaloids, given the help of the rapid development of spectroscopic technology. Jia Guo et al.[7,8] obtained stemoninine from *S. tuberosa* collected from the Sichuan province. They were able to deduce its structure by chemical derivation and^{1}H- and ^{13}C-NMR, 2D-NMR spectral interpretation. For the first time Ren-Sheng Xu et al.[9,10]depended wholly on 2D-NMR technology to determine the structures of two complicated alkaloids, stemotinine and isostemotinine, that were isolated from *S. tuberosa* collected from the Yunnan province. Their structures were consistent with the result of the later x-ray diffraction experiments[11]. This work greatly promoted the application of modern spectroscopic technology for the structural elucidation of *Stemona* alkaloids. Wen-Han Lin[12−19] and Yang Ye[20−23] later carried out the systematic investigation of six Chinese Stemonaceae species for alkaloids. Several new alkaloids were separated and structurally elucidated using spectroscopic methods. Some contained novel skeletons and were confirmed by x-ray diffraction (Table 4-2). *Stemona* alkaloids commonly contain an azaazulene basic ring and one or two α-methy-γ-lactone rings (if two, one is always attached to the α-position of the pyrrole ring). These alkaloids can be further classified into seven subtypes according to their structural features. Examples include protostemonine (**42**), tuberostemonine (**43**), maistemonine (**50**), parvistemonine (**51**), stemoninine (**52**), stemofoline (**53**) and croomine (**54**) types (details see related reviews[24−26]). The major *Stemona* alkaloids are shown in Table 4-2.

Table 4-2   The Major Stemona Alkaloids[25]

Names	M.F.	M.P.(°)	$[\alpha]_D$ (°)
Protostemonine, **48**	$C_{22}H_{31}O_6N$	172	+170 (EtOH)
Tuberostemonine, **43**	$C_{22}H_{33}O_4N$	88	-47 (CHCl$_3$)
Maistemonine, **50**	$C_{23}H_{29}O_6N$	205∼207	
Parvistemonine, **51**	$C_{22}H_{33}O_5N$	296	
Stemoninine, **52**	$C_{22}H_{31}O_5N$	113∼115	-110 (EtOH)
Stemofoline, **53**	$C_{22}H_{29}O_5N$	87∼89	+273 (MeOH)
Croomine, **54**	$C_{18}H_{23}O_4N$	B.p. 210-215 $(2\times10^{-3}\text{mmHg}^{①})$	+7.8 (CHCl$_3$)
Stemonine, **55**	$C_{17}H_{25}O_4N$	151	-114 (EtOH)
Stenine, **56**	$C_{17}H_{27}O_2N$	65∼67	-30 (MeOH)
Oxotuberostemonine, **57**	$C_{22}H_{31}O_5N$	222	
Stemonamine, **58**	$C_{18}H_{23}O_4N$	172∼174	
Isostemonamine, **59**	$C_{18}H_{23}O_4N$	165∼190	
Stemonidine, **60**	$C_{18}H_{29}O_5N$	119	-5.4 (Me$_2$CO)
Stemotinine, **61**	$C_{18}H_{25}O_5N$	207∼208	+91.7 (MeOH)
Isostemotinine, **62**	$C_{18}H_{25}O_5N$	245∼246	+47.5 (MeOH)

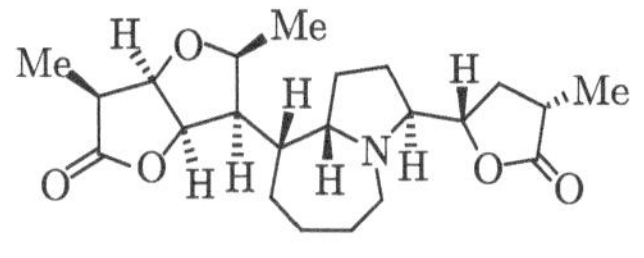

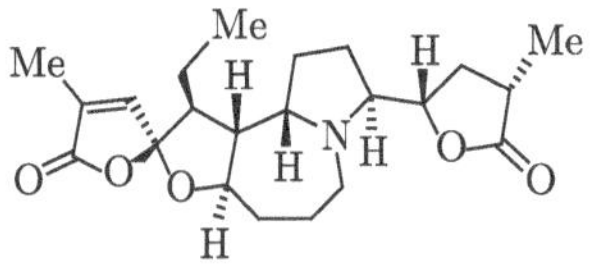

① 1 mmHg = 1.33322×10^2 Pa.

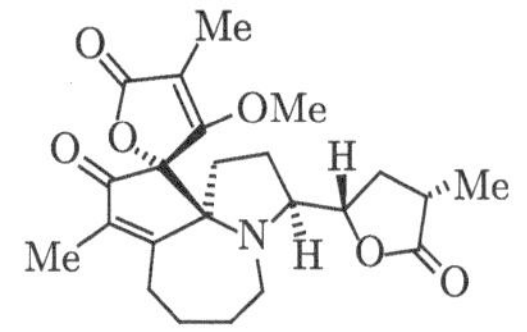

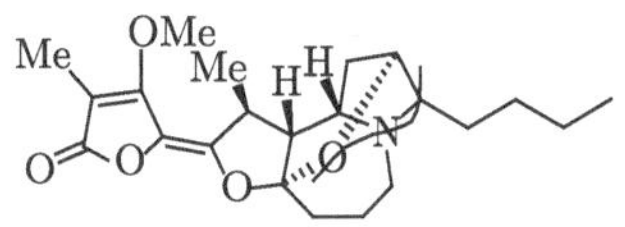

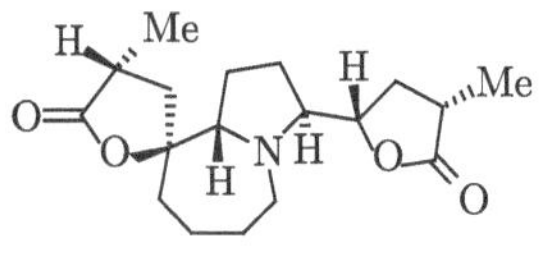

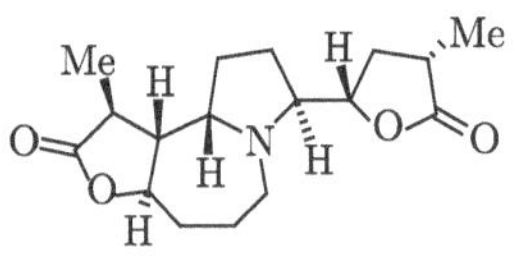

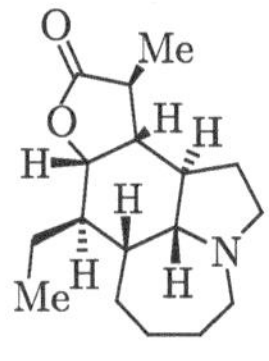

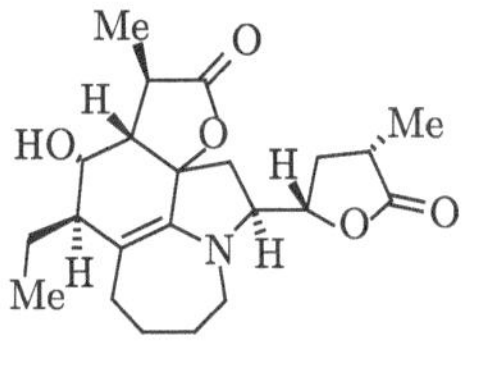

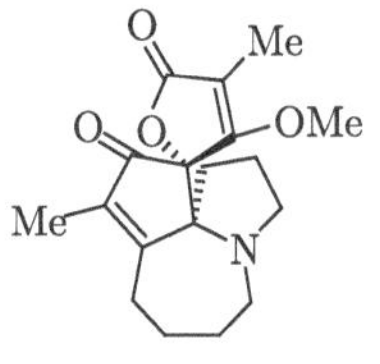

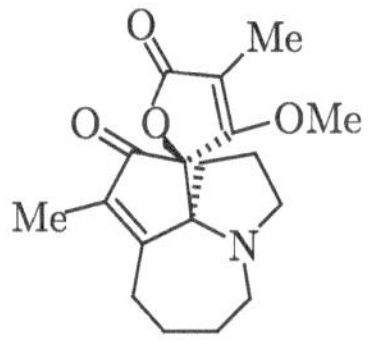

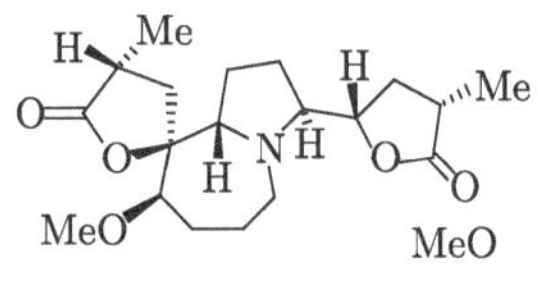

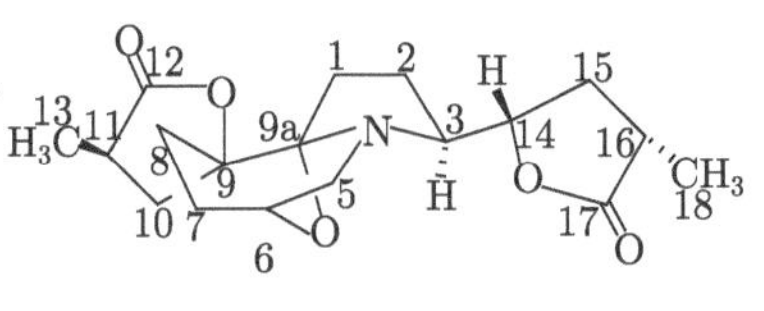

4.1.1   **Stemotinine**

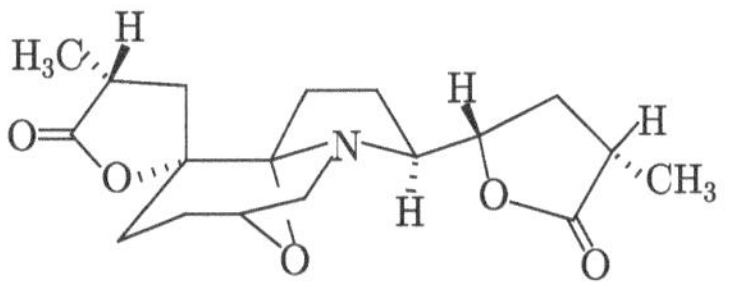

The molecular formula of stemotinine was determined to be $C_{18}H_{25}O_5N$ with the molecular ion peak M$^+$(m/z) 335.1751 by HREIMS. The UV spectrum indicated no absorption bands and IR spectrum showed absorption bands at 1764 and 1190 cm^{-1} for γ-lactones, and no absorptions for NH, OH and double bonds. Its ^{13}C-NMR spectrum displayed two carbonyl signals at δ 179.2, and 179.7, indicative of two γ-lactone rings. Considering its seven degrees of unsaturations, it was converted to a pentacyclic compound. The characteristic cleavage peak 236 (M^{+}-C$_5$H$_7$O$_2$) in its EI-MS suggested the presence of the α-methyγ-lactone annexed at the pyrrolidine ring. Its structure was finally established as **61** by ^{1}H-NMR, 2D-NMR and NOE spectra. The chemical shift of H-10α (δ 2.61) was much lower than that of H-10β (δ 1.70) due to the anisotropic effect of the tetrahydropyran ring, which indicated the relative configuration of C-9 in the spiro-γ 1 actone was S^*. The long-range W-plan coupling ($J = 1.4$ Hz) were observed between H-5α and H-7α, H-6α and H-8α which suggested that the tetrahydropyran ring was in a stable chair-conformation. H-14 in the other α-methy-γ-lactone was β-oriented. The NOE enhancements between H-5α and H-5β, H-3α and H-6 further confirmed the relative configuration of stemotinine.

4.1.2   Isostemotinine

The molecular formula of isostemotinine was the same as stemotinine, and its ^{1}H-NMR and IR spectra were also similar to those of stemotinine. But the H-10α proton resonated at δ 2.10, an up-fielded chemical shift of 0.51 ppm versus that of the corresponding proton in stemotinine (δ 2.61). This suggests that the spiraled γ-lactone in isostemontinine was different, that is, the configuration of C-9 was R^* and the tetrahydropyran ring retained its boat conformation. Such elucidation was consistent with the up-fielded chemical shift of H-10α, which is adjacent to H-10β (δ 1.71) and caused by the absence of any oxygen-ring effect (shown in **62**).

The ^{1}H-NMR data of stemotinine and isostemotinine are shown in Table 4-3.

Table 4-3 ^{1}H-NMR Data of Stemotinine and Isostemotinine

No.	Stemotinine (**61**)	Isostemotinine (**62**)
1α, β	1.86 & 1.91 m	1.88 & 1.92 m
2α, β	1.72 & 1.98 m	1.60 & 2.15 m
3α	2.86 ddd 10.8 (2β), 8.8 (14β), 5.8 (2α)	2.93 ddd 10.8 (2β), 7.8 (14β), 6.1 (2α)
5α	3.00 ddd 10.7 (5β), 6.3 (6), 1.4 (7α)	3.04 dd 10.4 (5β), 6.3 (6)
5β	3.22 d 10.7 (5α)	3.20 d 10.4 (5α)
6	4.59 m 6.3 (5α), 2.0 (7α), 2.0 (7β),1.4 (8α)	4.68 ddd 6.3 (5α), 2.0 (7α), 2.0 (7β)
7α	1.81 m 13.5 (8β), 12.6 (7β), 5.9 (8α)	
7β	1.62 bdd 12.6 (7α), 5.4 (8β), 1.8 (8α)	
8α	1.55 ddt 13.5 (8β), 5.9 (7α), 1.8 (7β), 1.8 (6)	
8β	2.34 dt 13.5 (8α), 13.5 (7α), 5.4 (7β)	
10α	2.61 dd 14.6 (10β), 11.6 (11)	2.10 dd 13.1 (10β), 10.0 (11)
10β	1.70 dd 14.6 (10α), 6.3 (11)	1.71 dd 13.1 (10α), 12.6 (11)
11	2.81 ddq 11.6 (10α), 7.7 (11-Me), 6.3 (10β)	2.80 ddq 12.6 (10β), 10.0 (10α), 7.7 (11-Me)
11-Me	1.34 d 7.7 (11)	1.28 d 7.7 (11)
14β	4.26 ddd 11.3 (15α), 8.8 (3α), 5.4 (15β)	4.14 ddd 11.3 (15α), 7.8 (3α), 5.4 (15β)
15α	1.48 ddd 12.6 (16β), 12.6 (15β), 11.3 (14β)	1.58 ddd*
15β	2.36 ddd 12.6 (15α), 9.0 (16β), 5.4 (14β)	2.36 ddd*
16β	2.67 ddq 12.6 (15α), 9.0 (15β), 7.5 (16-Me)	2.66 ddd*
16-Me	1.26 d 7.5 (16β)	1.28 d*

* J vaules are the same as those of **61**.

In 1982, these types of relative complicated structures were elucidated first by NMR and later confirmed by single crystal x-ray diffraction[11]. Nowadays, 2D-NMR technologies like HMBC, HMQC, ROESY, etc. Have been widely used in the structural elucidation of these types of alkaloid structures.

Besides stemotinine, similar alkaloids such as croomine (**54**) and stemonidine (**60**) were isolated from *Croomia japonica* and *S. japonica* and *S. sessilifolia*, respectively. Ren-Sheng Xu[9] speculated about the structure of stemonidine, which was later identified as stemospironine by x-ray diffraction[27,28].

4.1.3   Parvistemonine

Up-to-now four alkaloids with the same specific basic skeleton were isolated exclusively from *S. parviflora*, an endemic plant on Hainan Island. Parvistemonine (**51**, $C_{22}H_{33}O_5N$; mp. 296°; IR νmax 1765, 1775, 1172 and 1120 cm^{-1}) was the representative alkaloid of this type. Its structure was elucidated by decoupling, ^{1}H-^{1}H COSY, NOE and ^{1}H-^{13}C correlation spectra. The single crystal x-ray diffraction experiment of its hydrobromic acid salt further confirmed its absolute configuration as 3S, 9R, 9aS, 10S, 11R, 12R, 13R, 16S, 18S, 20S[12]. Bisdehydroparvistemonine (**63**) and parvistemoline (**64**) were isolated from the roots of the same plant. Their EIMS displayed a characteristic [M-99]$^+$fragment peak due to the loss of the α-methyl-γ-lactone ring annexed at C-3. In addition, there is a common side chain of an α-methyl-γ-lactone ring fused to an α-methyl furan ring attached to C-9 in these types of alkaloids. These structural features were indicated by the IR spectrum (νmax 1775 to 1785, 1350 to 1450, 1150 to 1200 cm^{-1}) as well as the ^{1}H-NMR spectrum (three tertiary methyl groups at δ 0.8 to 2.0 ppm, d, J= 7.0 Hz, Me-15, Me-17, Me-22).

63   64

12α-methoxylparvistemonine (**65**; $[\alpha]_D$ +80.6) was obtained from the stems and leaves of *S. parviflora*. The maximal UV absorption was at 234 nm. The HREIMS established the molecular formula as $C_{23}H_{35}NO_5$with seven unsaturations according to the molecular ion at m/z 421.2488. The EIMS fragment peak at m/z 135 suggested the presence of the basic azaazulene skeleton of *Stemona* alkaloids and its base peak at m/z322 [M-99]$^+$ suggested the presence of an α-methyl-γ-lactone ring attached to the C-3 of ring A. The IR absorption bands at 1778, 1775, 1189, 1050 cm^{-1}, together with two quaternary signals at δ_C179.7 and 176.3 ppm in the ^{13}C-NMR spectrum, indicated the existence of two γ-lactone rings. The ^{1}H-, ^{13}C-NMR and DEPT spectrum suggested the presence of three tertiary methyl groups (δ 1.34, d, Me-15; 1.19, d, Me-17; 1.25, d, Me-22), one methoxy (δ 3.35, s, MeO-23), seven CH$_2$, nine CH and three quaternary carbons. In addition, the characteristic geminate protons of CH$_2$-5(δ 3.37, m, H-5α; 2.58, m, H-5β) were indicative of the basic skeleton of azaazulene. The ^{1}H-NMR signals of **65** were very similar to those of parvistemofoline (**51**), except for the presence of an additional methoxyl signal and the absence of the H-12 proton signal. On the basis of the biogenetic relationship, this compound was deduced to be a 12-methoxyl derivative of parvistemonine. Such elucidation was further confirmed by ^{1}H-^{1}H COSY, HMQC and HMBC spectra. All ^{1}H- and ^{13}C-NMR signal assignments of this compound are shown in Table 4-4. The α configuration of the 12-methoxyl in **65** was determined by the NOESY spectrum. The ^{1}H-^{1}HNMR, NOESY, HMBC, and HMQC spectra of 12-methoxyl-parvistemonine are shown in Figures 4-1 to 4-4.

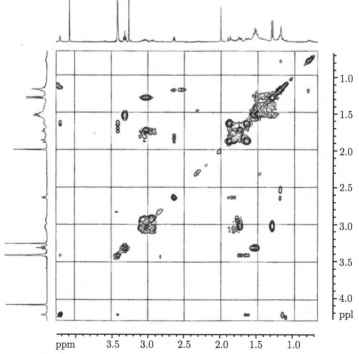

65

Table 4-4 ^{1}H- and ^{13}C-NMR Data and HMBC Correlations of 65 (in CDCl$_3$)

No.	^{1}H-NMR δ (ppm)	J(Hz)	^{13}C-NMR δ (ppm)	HMBC (C→H)
1	1.40, m; 1.64, m		27.1 (t)	H-3, H-9
2	1.59, m; 1.75, m		26.9 (t)	
3	3.40, m		63.7 (d)	H-1, H-19
5	α 2.85,m; β 3.37, m		46.6 (t)	H-3
6	1.42, m; 1.70, m		24.8 (t)	H-8
7	1.38, m; 1.77, m		28.7 (t)	H-5, H-9
8	1.45, m; 1.91, m		26.9 (t)	H-6
9	2.15, ddd		39.0 (d)	H-1, H-7
9a	3.43, ddd		62.9 (d)	H-2, H-8, H-10
10	2.30, ddd		47.4 (d)	H-8, Me-17
11	4.60	4.3	86.4 (d)	H-16
12	——		110.3 (s)	H-16, MeO-23, Me-15
13	2.89, q	7.2	44.7 (d)	
14	——		176.3 (t)	H-11, Me-15
15-Me	1.34, d	7.2	10.0 (q)	
16	4.48, dq	6.7, 7.2	79.4 (d)	H-10, H-11
17-Me	1.19, d	6.7	19.4 (q)	H-10
18	4.18, m		83.4 (d)	H-2
19	1.55, m; 2.35, m		34.2 (t)	Me-22, H-20
20	2.59, m		35.0 (d)	
21	——		179.7 (s)	H-19, Me-22
22-Me	1.25, d	7.0	15.0 (q)	H-19, H-20
23-OMe	3.35, s		51.2 (q)	

Figure 4-1 ^{1}H-^{1}H COSY(CDCl$_3$, 400 MHz)of 12-methoxyl-parvistemonine

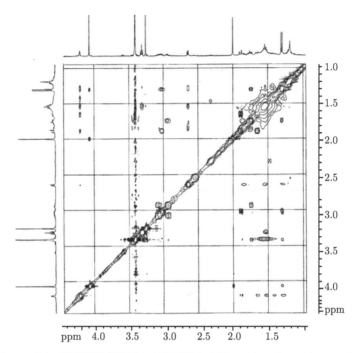

Figure 4-2   NOESY(CDCl$_3$, 400 MHz)of 12-methoxyl-parvistemonine

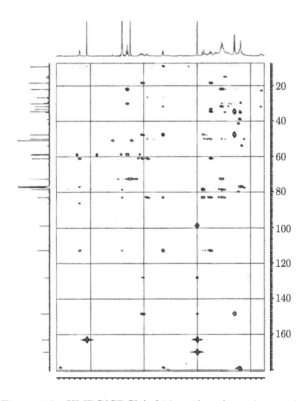

Figure 4-3   HMBC(CDCl$_3$)of 12-methoxyl-parvistemonine

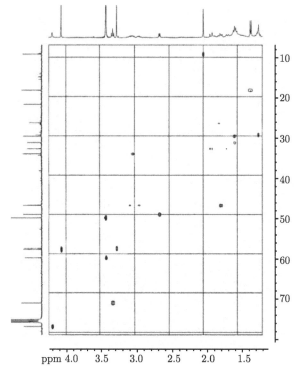

Figure 4-4   HMQC(CDCl$_3$)of 12-methoxyl-parvistemonine

4.2 Camptothecin and its analogues

The anti-cancer component camptothecin was discovered by Wall et al. in 1966 from *Camptotheca acuminata* Decne, which is a Chinese endemic plant. Camptothecin, $C_{20}H_{16}N_2O_4$, mp. 264 to 366°C, $[\alpha]_D$ +40° (CHCl$_3$:MeOH 4:1), and its absolute configuration were confirmed by x-ray crystallography by its iodide acetyl derivative. The only chiral carbon (C-20) was determined to be S[29]. Camptothecin was regarded as a potential anti-cancer drug due to its strong anti-cancer activity in animal experiments. It attracted us to study the domestic tree in 1966. We also isolated the second active compound, 10-hydroxycamptothecin (**67**, $C_{20}H_{16}N_2O_5 \cdot H_2O$; mp. 266 to 267°C; $[\alpha]_D$ −147° in pyridine) from the bark[30,31]. The study was further developed and camptothecin was first used for clinical trials in China on stomach cancer, carcinoma of urinary bladder and leukemia. However, due to side effects such as haematuria and frequent urination, its use was restricted. 10-Hydroxycamptothecin was used for the treatment of liver cancer and head and neck tumors. Its side effects were much less severe than those of camptothecin. Later, camptothecin was found to be selectively oxygenated into 10-hydroxycamptothecin by *Aspergillus flavus* T-37, which is an additional source for mass production[32].

The content of camptothecin is about 0.004%, 0.01%, 0.01%, 0.02% and 0.03 % in the branches, barks, roots, root barks and fruits of *C. acuminata*, respectively. The fruits were selected as the raw material for production because of their rich abundance and high content of camptochecin and 10-hydroxycamptothecin (0.0003%). Minor constituents, 10-methoxycamptothecin (**68**), 11-hydroxycamptothecin (**69**) and 11-methoxycamptothecin (**70**), were also isolated from the fruits. Their structures were determined by comparing ¹H-NMR data with those of methoxypyrrole[3,4] quinoline (**71**) analogues. These compounds showed strong inhibition against animal tumors. In addition, the non-active compound deoxycamptothecine (**72**) was found in the same source[30].

66 R1＝R2＝H, R3＝OH

67 R1＝H, R2＝OH, R3＝OH

68 R1＝H, R2＝OMe, R3＝OH

69 R1＝OH, R2＝H, R3＝OH

70 R1＝OMe, R2＝H, R3＝OH

72 R1＝R2＝R3＝H

Camptothecin and its analogues are a special class of pentacyclic compounds that contain quinoline, δ-lactone and δ-lactam rings. They are near neutral compounds and show negative reaction with the Dragendorff's reagent. They do not form salts with acids, nor do they dissolve in common organic solvents or water. Rather, they can dissolve in dilute alkali with an open lactone ring. These properties dictate that their extraction method is different from that of the common alkaloids. Camptothecin and its analogues can be extracted directly with chloroform from the concentrated aqueous solution (pH 5.4) after removing the ethanol. Once the oily material is removed, the chloroform extraction yields a pale yellow powder, from which pure camptothecin can be obtained by recrystallization with chloroform:methanol (1:1).

Due to the consecutive conjugation system between the quinoline ring and the unsaturated lactam ring, camptothecin is a yellow crystal and its chloroform solution shows very strong fluorescence. The UV absorptions λ_{max} (lgε) are at 218 (4.63), 254 (4.53), 290 (3.78), and 370 (4.35) nm. The IR spectrum displays absorption bands at 3440 cm^{-1} for hydroxyls, 1760 to 1750 cm^{-1} for lactones, and 1673 cm^{-1} for unsaturated lactams. The ^{1}H-NMR spectrum (in trifluoroacetic acid-d) displays signals at δ 0.69 (3H, t, $J = 7$ Hz, H-18), 1.72 (2H, q, $J = 7$ Hz, H-19), 5.40 (2H, s, H-5), 5.18, 5.54 (2H, q, $J = 16$ Hz, H-17), 7.66-8.14 (5H, m, H-9 to 12 and H-14), and 9.00 (1H, s, H-7).

In recent years, camptothecin has attracted more attention since it was found to be a DNA topoisomerase I inhibitor and different from most other anticancer drugs, which are DNA topoisomerase II inhibitors. The synthesis of camptothecin analogues was developed with the aim to increase its water solubility and bioactivity. The structural modification was commonly carried out on 10-hydroxycamptothecin due to its higher anti-cancer activity than camptothecin. The new drug topotecan (**73**), a 9-dimethylaminomethyl derivative of 10-hydroxycamptothecin, produced by SmithKline Beecham Pharmaceuticals (now GSK) has been approved for use by the United States Food and Drug Administration (FDA) in 1996. Treated with the Adams' catalyst in acetic acid, camptothecin (**66**) was first converted to 1,2,6,7-tetrahydrocamptothecin, which was then treated with lead tetraacetate to afford a mixture of 10-acetoxycamptothecin and camptothecin. This mixture was hydrolyzed in aqueous acetic acid and then subjected to preparative HPLC separation, yielding **67**. Compound **67** was further treated with 37% aqueous formaldehyde and 40% dimethylamine to give topotecan[33]. Topotecan hydrochloride has excellent solubility in water and its acidity helps avoid the possibility of a decrease in activity due to the opening of the lactone ring. Such structure modification is simple and can produce products with higher efficacy and lower toxicity. Topotecan has now been widely used to treat small cell lung cancer and ovary cancers.

Camptothecin -11 (CPT-11) or irinotecan, namely 7-ethyl-10-[4-(1-peperidino)-1-peperidino]-carbonyloxy camptothecin, is another camptothecin analogue approved for use by the FDA in 1996 to treat oophoroma and metastatic colorectal cancer. Camptosar, the trade name for CPT-11, is produced by Daiichi Seiyaku Co., Ltd. (Japan) and Pharmacia & Upjohn Inc. (USA). Camptothecin (**66**) was dissolved in a mixture of propaldehyde, ferrous sulfate and concentrated sulfuric acid, and 30% H_2O_2 was added to the mixture to produce 7-ethylcamptothecin (**75**). **75** was then heated in the solution of 30% H_2O_2 and acetic acid and converted into a N-oxidative derivative (**76**). **76** was transferred to 7-ethyl-10-hydroxycamptothecin (**77**) by photoreaction (450W mercury-vapor lamp) in a solution of dioxane and diluted sulfuric acid. **77** was then converted into **74** by reacting with 1,4'-bipiperidine-1-carboxylic chloride in pyridine. **77** could also be dissolved in a solution of dioxane and Et_3N and reacted with phosgene to produce acyloxide (**78**). **78** was then reacted with 4-piperidinopiperidine to give **74**[34]. Researchers synthesized a series of **74** analogues. CPT-11 exhibited the strongest activity in animal experiments among all the analogues and it was used by intravenous injection or oral administration. It was found that **74** decomposed into the metabolite 7-ethyl-10-hydroxycamptothecin (SN-38) in vivo, which explained why CPT-11 showed very low inhibitory activity against topo-I in vitro, but high in vivo since CPT-11 is actually a prodrug of SN-38.

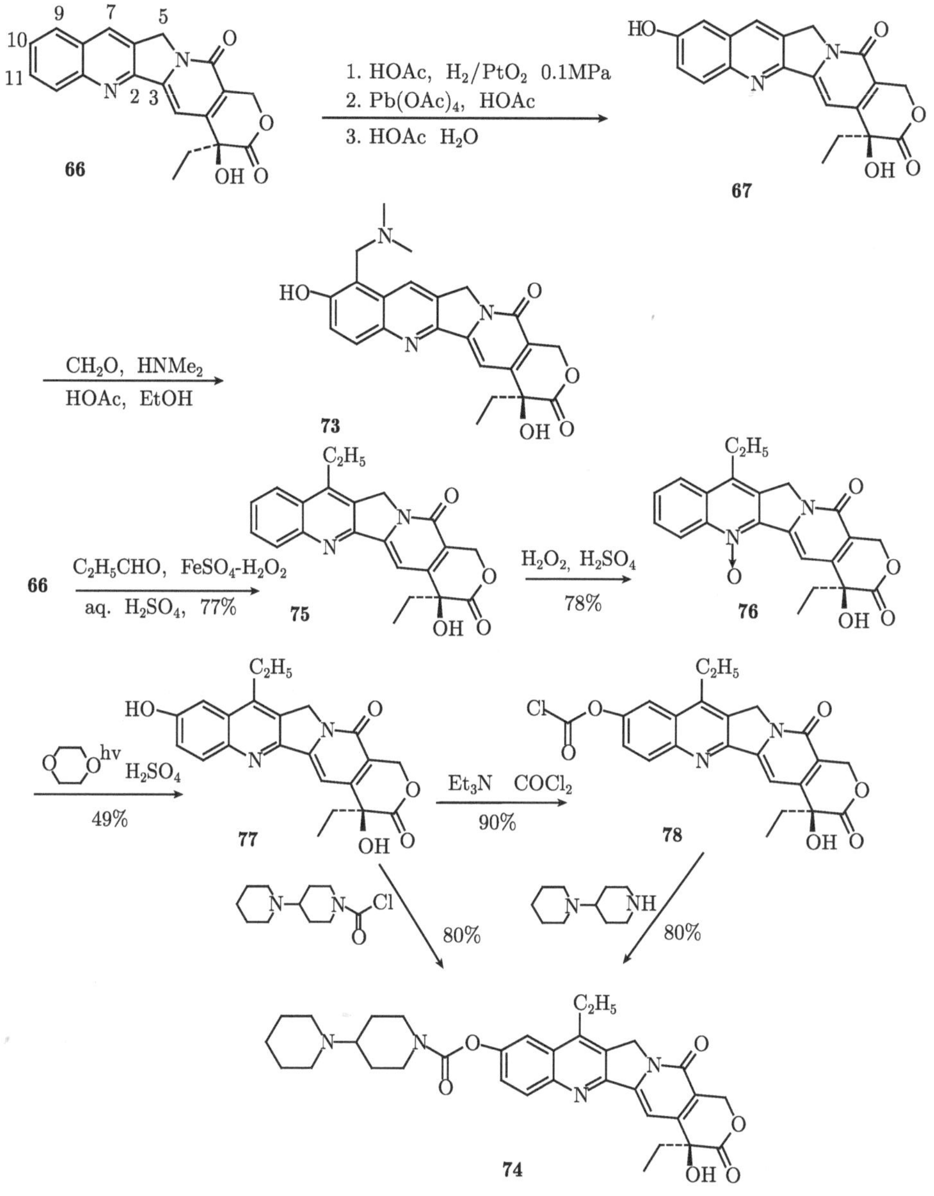

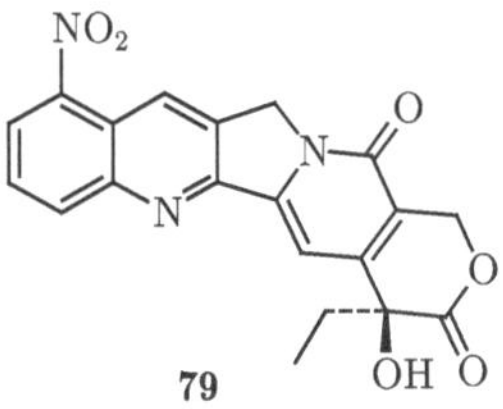

79 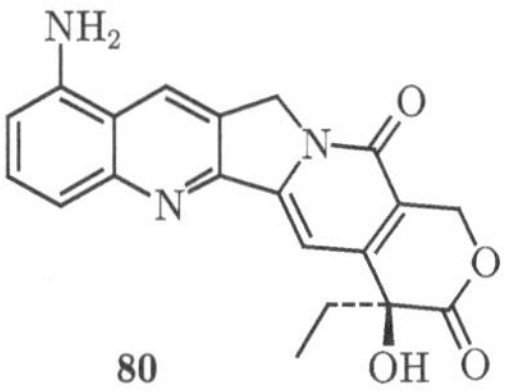   **80** 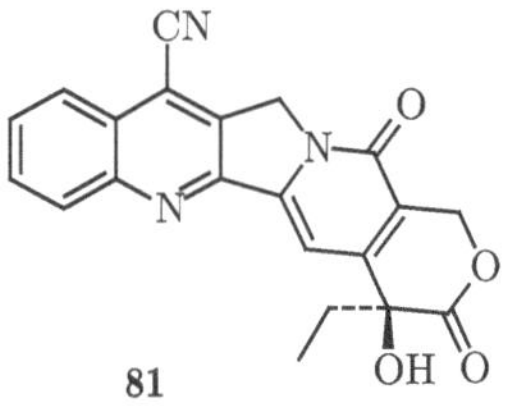   **81**

Another camptothechin analogue is 9-nitrocamptothecin (**79**, 9-NC) or rubitecan, from which oral tablets are made by Supergen (USA) for the treatment of oophoroma and colorectal cancer, and especially pancreas cancer[35]. Later, the results were denied. Other compounds such as 9-aminocamptothecin (**80**) and 7-cyanocamptothecin (**81**) are under clinical study.

4.3   Sinomenine and sinoacutine

Fangji (*Sinomenium acutum* Rehder et Wilson) is a plant from the Menispermaceae family used to ease pain and inflammation. The commercial product of Fangji is often confused as Qingfengteng, however, it is not a plant from the Sabiaceae (Qingfengteng) family. The rhizome of Fangji mainly contains sinomenine (**82**) and sinoacutine (**83**). Sinomenine is a phenolic alkaloid ($C_{19}H_{23}O_4N$; mp. 161°C, $[\alpha]_D 70°$ in EtOH). Its UV spectrum absorption at 263 nm (lgε 3.79) indicates that it belongs to a tetrahydroisoquinoline structure. The IR spectrum shows absorptions of phenol groups at 3575, 1590, 1296 cm^{-1} and the α,β-unsaturated ketone at 1700 cm^{-1}. The ^{1}H N MR (CDCl$_3$) spectrum shows the proton signals of one NCH$_3$ at δ 2.34 (3H, s), one OCH$_3$ at δ 3.40 (3H, s), one ArOCH$_3$at δ 3.70 (3H, s), two aromatic protons of an AB system at δ 6.56, 6.44 (each 1H, d, J= 7 Hz), one olefinic proton at δ 5.40 (1H, s) and one CH$_2$ adjoining to a ketone at δ2.30 and 2.40 (each 1H, d, J = 14 Hz).

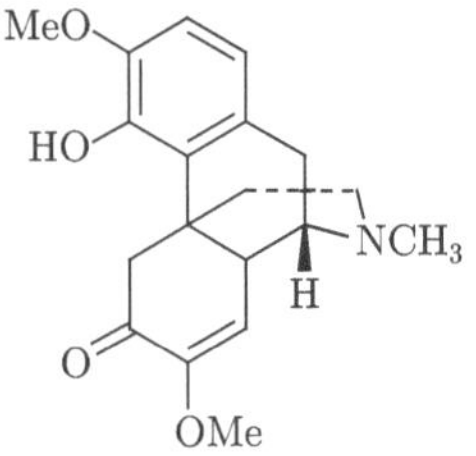

82 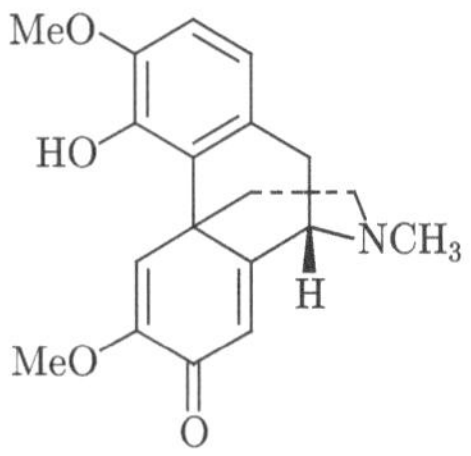   **83**

Sinoacutine (**83**; $C_{19}H_{21}O_4N$; m.p. 198°; $[\alpha]_D$–81.6° in CDCl$_3$) contains two protons less than sinomenine. The structure was elucidated by UV and chemical reaction. Its expanded formula is $C_{15}H_{11}(OCH_3)_2(OH)(>CO)(NCH_3)$. The UV absorptions λmax (lgε) were at 242 (4.32) and 281 (3.82) nm, suggesting that it possesses one more double bond than sinomenine. Both compounds possess the same tetrahydroisoquinoline moiety, but the other double bond conjugate system of sinoacutine is different from that of sinomenine. The IR spectrum of sinoacutine (νmax) shows absorptions of phenolic hydroxyl (3525, 1296 cm^{-1}), unsaturated ketone cabonyl (1882 cm^{-1}) and conjugated double bond (1646, 1626 cm^{-1}) functional groups. It is converted into an alcohol **84** with KBH$_4$, resulting in the change of the UV absorption λmax (lgε) to 242 nm (4.32). These results suggest that both of the two double bonds of **83** are conjugated with the carbonyl group and the conjugations are removed after reduction with KBH$_4$. The tetrahydroisoquinoline and $\alpha\beta - \alpha'\beta'$ conjugated ketone moieties can only be arranged as shown in **83**. However, the positions of OMes were still unknown based on the UV spectra. After dehydration with acid, **84** is converted into the known *l*-thebaine (**85**), which confirmed the position of the OMe group[36].

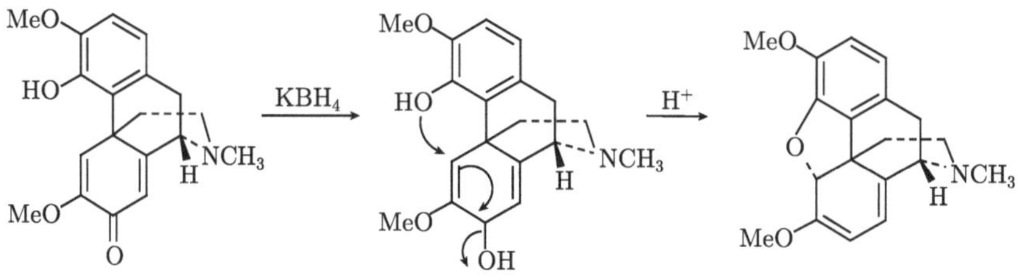

83   **84**   **85**

4.4 Pyridone alkaloids—huperzine

4.4.1 Huperzines A and B

Huperzines A and B belong to a group of pyridone alkaloids isolated from *Huperzia serrata* (Thunb) Trev. (Lycopodiaceae) collected from Zhejiang province[37,38]. Both of them showed potential cholinesterase inhibition in pharmacological studies.

Huperzine A (**86**), $C_{15}H_{18}ON_2$, mp. 230°C, $[\alpha]_D$ -150.4° (c 0.5, MeOH); UV λ_{max} (log ε): 231 (4.0), 313 (3.89) nm; IR ν_{max} 3180, 1650, 1550 cm^{-1} indicated a pyridone skeleton in the structure. Its structure was determined by NMR spectroscopic data and analysis of NOE difference spectrum. Huperzine B (**87**), $C_{16}H_{20}ON_2$, mp. 270 to 271°C, $[\alpha]_D$ −54.2° (c 0.2, MeOH); UV λ_{max}(log ε) 231 (3.95), 312 (3.85) nm; IR ν_{max} 3100, 1670, 1620, 1610 and 1560 cm^{-1}; NMR spectral analyses showed its sturcture is different from that of huperzine A at C-12 and C-13, which in **87** formed another piperidine ring. The NMR spectral data of the two alkaloids are shown in Table 4-5.

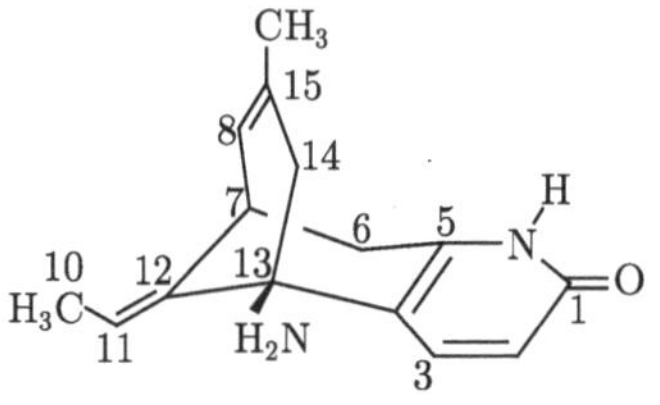

86 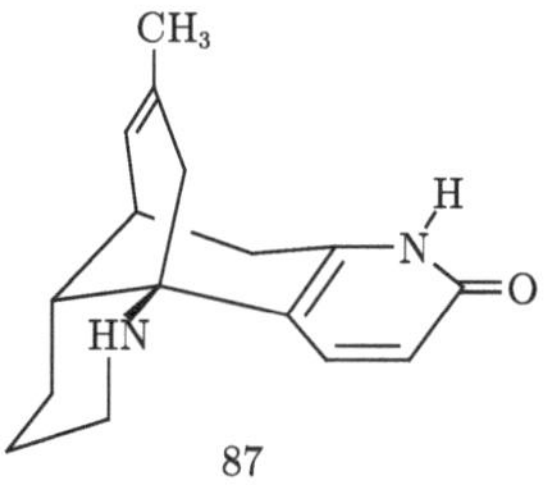   **87**

Table 4-5 ^{1}H- and ^{13}C-NMR Spectral Data of Huperzines A and B (CDCl$_3$)

No	^{1}H-NMR δ (ppm)				^{13}C-NMR δ (ppm)	
	86 (100 MHz)		**87** (400 MHz)		**86** (22.63MHz)	**87** (25.18MHz)
	δ	J (Hz)	δ	J (Hz)		
1					165.52 (s)	165.39 (s)
2	6.38, d	9	6.43, d	9	116.97 (d)	117.90 (d)
3	7.84, d	9	7.68, d	9	140.25 (d)	140.37 (d)
4					122.95 (s)	117.90 (s)
5					142.59 (s)	143.27 (s)
6	2.72, 2H,	16, 3, 0	α, 2.85, dd; β, 2.43, d	18, 5 18, 0	35.24 (t)	29.45 (t)
7	3.56, m		2.34, ddd	4, 5, 5	32.95 (d)	34.59 (d)
8	5.38, d	5	5.43, bd	5	124.36 (d)	126.12 (d)
9			α, 2.29, ddd β, 2.74, ddd	14, 13, 2 14, 3, 2		48.03 (t)

continued

No	86 (100 MHz) δ	J (Hz)	87 (400 MHz) δ	J (Hz)	86 (22.63MHz)	87 (25.18MHz)
	¹H-NMR δ (ppm)				¹³C-NMR δ (ppm)	
10	1.62, d, 3H	7	α, 1.54[1)] β, 1.43, ddddd	13,13,13,3,3	12.31 (q)	25.34[2)] (t)
11	5.46, q	7	α, 1.22, dddd β, 1.54[1)]	4, 13, 13, 13	111.23 (d)	28.13[2)] (t)
12			1.67, ddd	13, 4, 4	143.30 (s)	40.70 (d)
13					54.35 (s)	53.22 (s)
14	2.12, s, 2H		H_{endo}, 1.83, d H_{exo}, 2.02, d	17 17	49.25 (t)	41.67 (t)
15					134.09 (s)	132.23 (s)
16	1.46, s, 3H		1.59, s, 3H		22.57 (q)	22.68 (q)
NH (*inpridonecycle*)	13.20, bs		13.20, bs			

1) Overlapped signals.

2) Exchangeable.

4.4.2   Total synthesis of huperzine A

Because of its significant physiological activities, research focused on the total synthesis of huperzine A has been reported[39]. Kozikovski[40] has completed the total synthesis with a relatively shorter route, starting from the mono-protected cyclohexanedione 88 through a key intermediate β-keto ester 89 to huperzine A. At present, the e.e. (enantiomeric excess) of huperzine A has reached 90% by enantioselective synthesis[41].

4.4.3   Physiological activity of huperzine

Huperzine A is presently the most potent cholinesterase inhibitor derived from natural products. It shows nosotropic activity and improved multiplicate dysmnesia in animals and clinical trials. At the same time, pharmacological studies have shown that huperzine A has a protective neurocyte activity against toxicity and apoptosis induced by β-amyloid proteins. It even has a curative effect on nosotropic activity and improves memory and behavior of older people. The troche of huperzine A gained a Certificate of Chemical Drug in China in 1996. Its capsule and other pharmceutical forms were sent to clinical trials. It also shows great potential for the treatment of Alzheimer's disease.

Table 4-6 shows that huperzine A has better selective inhibition on acetylcholinesterase (AchE) and butyrylcholinesterase (BuChE) than tacrine and donepezilon, two new drugs currently used to treat Alzheimer's disease. The inhibitory effect of huperzine A is much better than huperzine B to acetylcholinesterase.

Recently structure modification was made to increase the activity, among them huperzine A Schiff base derivative shiprine (ZT-1, 90) is the most successive one. It showed much more effective than huperzine A[42].

Table 4-6   Comparison of Huperzine A Selectivity and Other Cholinesterase Inhibitors

	Enzyme activity inhibition concentration, 50%(nM)		The rate of IC_{50} (/AChE)
	BuChE (rat serum)	AChE (rat cerebral synaptosomes)	
Shiprine ZT-1(90)	116333.3 ± 6984.1	63.6 ± 0.5	1829
Huperzine A(86)	54500.0 ± 838.6	56.2 ± 2.6	970
Donepezil	7663.3 ± 346.4	13.8 ± 0.7	555
Tacrine	58.0 ± 1.2	82.4 ± 3.5	0.7

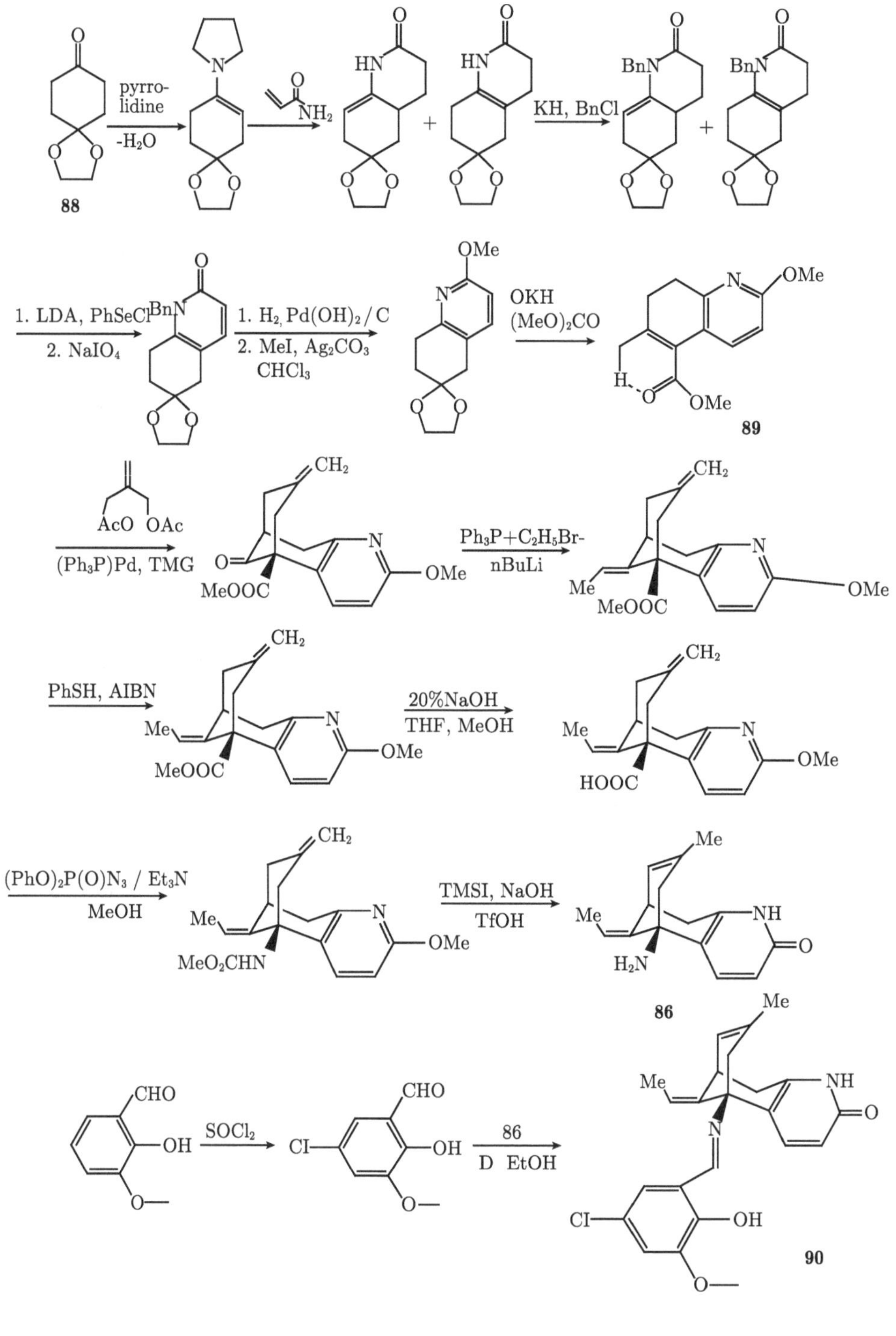

Bibliography

[1] Xu RS, Ye Y, Zhao WM. Natural Products Chemistry (Version 2), Beijing: Science Press, 2004.

[2] Lin QS. Chemical Constituents of Herbal Medicines. Beijing: Science Press, 1977.

[3] Xu RS and Chen Z L. Extraction and Isolation of Bioactive Components from of Medicinal Herbs (Version 2). Shanghai: Shanghai Science and Technology Press, 1983.

[4] Manske R H F, Brossi A, Cordell G A. The Alkaloids, New York: Academic Press, 1950 to 2007, Vol I-64.

[5] Pelletier S W. Alkaloids (Chemical & Biological Perspectives) New York: John Wiley & Sons, now Pergamon, 1983 to 2001, Vol 1 to 15.

[6] Yao XS. Natural Products Chemistry, Beijing: People's Healthcare Press, 2001.

[7] K Hostettmann, A Maston, M Hostettmann. Preparative Chromatography Technology (Translated to Chinese by Zhao W-M). Beijing: Science Press, 2000.

References

[1]   Boit N G. Ergebnisse der Alkaloid-Chemie. Berlin:Akademie Verlag, 1961, 574.

[2]   Schild et al. Ber Dtsch Chem Ges, 1936, 69:74.

[3]   Götz M et al. Tetrahedron Letters, 1961, 20:707.

[4]   Götz M et al. Tetrahedron, 1968, 24:2631.

[5]   Edwards O E. Can J Chem, 1962, 40:455.

[6]   Chu J H. Science Record, 1949, 2 :310.

[7]   Guo J et al.. Huaxue Xuebao, 1981, 36(4):865.

[8]   Cheng DL et al. Journal of Natural Products, 1988, 51(2):202.

[9]   Xu RS et al. Tetrahedron, 1982, 38:2667.

[10] Xu RS et al. Huaxue Tongbao, 1985, 12:1.

[11] Xu RS et al. Structural Chemistry, 1986, 5:97.

[12] Lin WH et al. Huaxue Xuebao, 1990, 48:811.

[13] Lin WH et al. Huaxue Xuebao, 1991, 49:927.

[14] Lin WH et al. Huaxue Xuebao, 1991, 49:1034.

[15] Lin WH et al. Youji Huaxue, 1991, 11:500.

[16] Lin WH et al. Journal of Crystallographic and Spectroscopic Research, 1991, 21(2):189.

[17] Lin WH et al. Chin Chem Lett, 1991, 2:369.

[18] Lin WH et al. J Nat Prod, 1992, 55(5):571.

[19] Lin WH et al. Phytochemistry, 1994, 36(5):1333.

[20] Ye Y et al. Chin Chem Lett, 1992, 3(7):511.

[21] Ye Y et al. J Nat Prod, 1994, 57:665.

[22] Ye Y et al. Phytochemistry, 1994, 37(4):1201.

[23] Ye Y et al. Phytochemistry, 1994, 37(4):1205.

[24] Qin GW et al. Med Res Rev, 1998, 18:375.

[25] Xu RS. In Studies in Natural Products Chemistry, Atta-ur-Rahman, Ed. Elsevier Sciences B V, 1989, 21:729.

[26] Pilli R A et al. Nat Prod Rep, 2000, 17:117 .

[27] Sakata K et al. Agri Bio Chem, 1978, 42:457.

[28] Lin WH et al. Youji Huaxue, 1991, 11:500.

[29] Wall M E et al. J Am Chem Soc. 1966, 88:3888.

[30] Xu RS et al. Huaxue Xuebao, 1977, 35:194.

[31] Shanghai Institute of Materia Medica.J. Chin. Med., 1978, 58:598.

[32] Zhu GP et al. Kexue Tongbao, 1978, 23:761.

[33] Kinsbury W D et al. J Med Chem, 1991, 34:98.

[34] Sawada S et al. chemy Pharm Bull, 1991, 39:1446.

[35] Stehlin et al. Int J Oncol, 1999, 14:821.

[36] Hsu jS et al. Scientia Sinica, 1964, 13:2016.

[37] Liu JS et al. Can J Chem, 1986, 64:837.

[38] Ayer W A et al. Can J Chem, 1989, 67:1538.

[39] Bai D L Pure & Applied Chemistry, 1993, 61:1103.

[40] Kozikowski A et al. J Org Chem, 1991, 56:4637.

[41] He XC et al. Tetrahedron: Asymmetry, 2001, 12(23):3213.

[42] Zhu DY et al. International Patent, WO 96/20176, 1.

Yang YE and Ren-Sheng XU

CHAPTER 5

Sesquiterpenoids

Terpenoids constitute a large, widespread and diverse group of natural products derived from more than two isoprene units. Sesquiterpenes are a large and ubiquitous family of C_{15} isoprenoid natural products and are widely distributed in plants, microorganisms, marine organisms and insects. Some sesquiterpenes, especially sesquiterpene lactones, show a wide range of biological activities, such as anti-bacterial, anti-tumor, anti-virus, cytotoxic, immune inhibition, phyto-toxic, insect hormone, insect anti-feedant, plant growth regulation and neurobiological activities. In the last thirty years, many sesquiterpenoids with special structures that contain halogens, sulfur, cyanide and thiocyanide have been isolated from marine organisms. These novel sesquiterpenes exhibit very interesting and peculiar structures and biological activities. Therefore, work on sesquiterpenoids is one of the most active research areas in natural products chemistry to date. Many sesquiterpenes are major components found in the higher boiling point parts of essential oils, and are very important for controlling the odor of fragrances used in the perfume industry.

Although the sesquiterpenes consist of only fifteen carbon atoms, they are able to form many surprising skeletons as well as complex stereo-structures in the plant metabolic process. Chemical and spectral studies of sesquiterpenoids have culminated during the last three decades and have been reviewed by Fraga et al.[1−8]. This chapter mainly introduces the isolation, spectral analysis, structure determination and some recent progress in sesquiterpene chemistry.

Section 1   Chemical Properties, Isolation and Purification

Unlike the alkaloids and flavonoids, sesquiterpenes have no special colour test or UV absorption for detection, nor a specific isolation method. For the detection of sesquiterpenes, TLC methods with vanilline-sulfuric acid as coloration are most commonly used. In TLC, the Rf values of sesquiterpenes are very closely related to their polarities. The TLC values of sesquiterpenes are developed by using the solvent system hexane:benzene (9:1) or (19:1). Derivatives of alkane, aldehyde, ketone, ether, ester, alcohol and acid have different Rf values, which are valuable for the determination of the chemical types of sesquiterpene compounds.

Sesquiterpenes are the main components of the higher boiling point parts of essential oils. The fragrant herb is steam distillated to give an essential oil, a mixture of terpenes and aromatic compounds, which then are further fractionally distillated. The lower boiling point parts are mainly monoterpenes, and the final higher boiling point parts, including most sesquiterpenes, are further isolated by chromatographic methods to yield pure sesquiterpenes.

Systematic isolation and purification of sesquiterpenes were designed by F. Bohlmann[9].

The plant materials are ground and then extracted using ether. The ether extract is dissolved in methanol and cooled to −20°C, and then filtered to remove most impurities including oils, fats, wax as well as steroids and triterpenes. The methanol soluble part is separated with a silica gel column eluted with petroleum ether, petroleum ether:ether (9:1, 3:1 and 1:1), ether:methanol (9:1, 1:1) and then methanol.

The petroleum ether fractions contain mainly sesquiterpene alkane and alkene hydrocarbons. The petroleum ether:ether (9:1) fractions include sesquiterpene ethers, cyclo-oxygen compounds, aldehydes and ketones, etc. The petroleum ether:ether (3:1) fractions contain sesquiterpene esters, lactones and hydroxy compounds, etc. The ether fractions include hydroxy and polyhydroxy compounds. The ether:methanol (9:1) fractions contain polyhydroxy compounds and carboxy compounds. As a rule, the chromatographic properties of organic natural compounds depend on their polarities. The elution order is alkanes, alkenes, ethers, esters, aldehydes, ketones, hydroxyl compounds, and acids. Steroid and triterpene molecules are larger than those of sesquiterpenes, therefore, their esters can be eluted by petroleum ether:ether (9:1) and their hydroxyl derivatives can be eluted by petroleum ether:ether (3:1).

The crude fractions mentioned above are separated by using column chromatography and further isolated by TLC and HPLC to give pure sesquiterpenes. A sample of about 200 g plant materials can be used for the isolation and identification of more than thirty compounds, as long as the content is no less than 0.001% in the plant. In most cases, enough material can be isolated for biological tests, especially for in vitro screening.

For the isolation of sesquiterpene lactones, the ethanol is typically used as the extraction solvent. As a preliminary treatment, the ethanol extracts of medicinal plants are partitioned in different solvents. The procedure is mainly designed on the basis that the sesquiterpenes have mild polarity, and are insoluble in petroleum ether, and also have low solubility in water. To remove the water soluble impurities, the ethanol extract is partitioned in chloroform and water. The chloroform extract is further partitioned with petroleum ether and 10% aqueous methanol. The petroleum ether soluble substances, such as most of the fatty acids, fatty alcohols and their esters, are removed and the sesquiterpenes remain in the aqueous methanol layer. The methanol extract is separated by silica gel or alumina column chromatography and further purified by HPLC. Thin layer chromatography is also widely used for sesquiterpene isolation, owing to the simple and quick ability to isolate 10 to 100 mg pure compounds or purified fractions for in vitro screening or further chemical investigations. Recently, liquid chromatography coupled with MS and NMR detectors has been successfully applied for the analysis of natural substances. These LC-MS and LC-NMR techniques can be directly used for isolation and detection of some natural products from plant extracts.

[**Example 5-1**]   The sesquiterpenes in *Saussurea lappa*∗

A powdered *Saussurea lappa* root was extracted with ether, and the ether extract was initially separated on silica gel to yield six crude fractions, A-F. Each fraction was further separated by column chromatography or TLC and HPLC. The chemical components of the fractions were then analyzed with the help of UV, IR, MS and NMR methods.

Fraction	Eluting solvents	Compounds
A.	pt-ether	**1** 180 mg
B.	pt-ether:ether (9:1)	**2** 60 mg, **3** 24 mg, sterol and triterpene esters
C.	pt-ether:ether (3:1)	**4** 175 mg, **5** 100 mg, sterols and triterpenes
D.	pt-ether:ether (1:1)	**6** 40 mg, **7** and **8** mixture 5 mg
E.	ether	**9** and **10** mixture 5 mg, **11** 14 mg, **12** 15 mg **13** 8 mg, **14** 6 mg
F.	ether: methanol(9:1)	**15** 20 mg, **16** and **17** mixture 8 mg

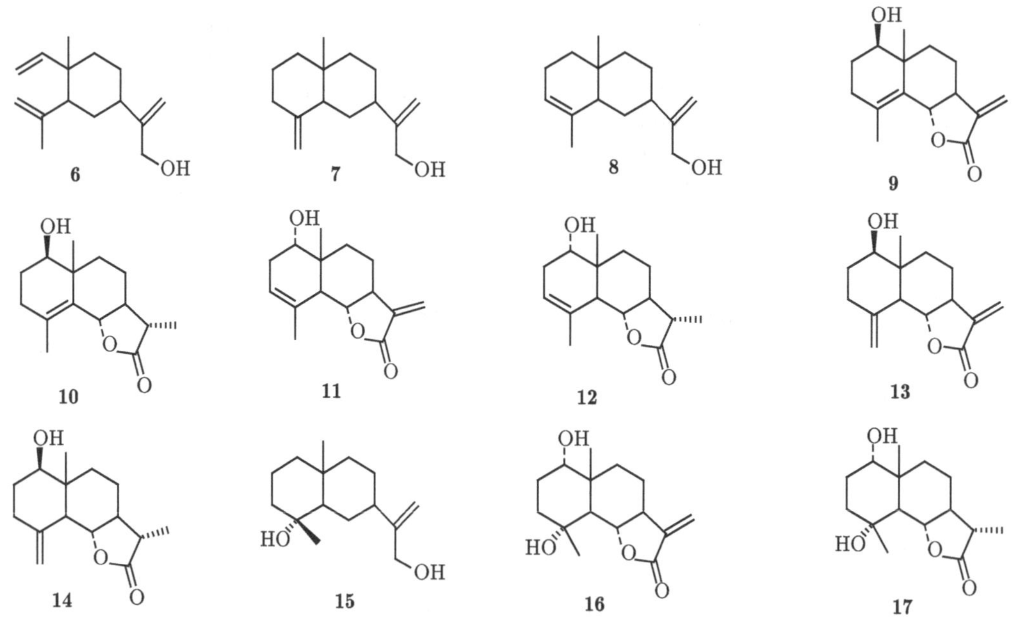

Structures of **1-17**

* Bohlman F. and Chen Z. unpublished work.

[**Example 5-2**]   The antitumor constituents of *Eupatorium rotundifolium*[10]

The concentrated alcoholic extract of *E. rotundifolium* from 17 Kg of dried plant (A 500 g) was partitioned between 4 L of chloroform and 2 L of water. The water layer was washed with 1 L of chloroform, and the combined chloroform extract was washed with water. The water layer was evaporated to give red oil (B 150 g). The evaporation of the chloroform fraction gave a dark green residue (C 283 g). The chloroform fraction was further partitioned between 3 L of petroleum ether and 3 L of 10% aqueous methanol to give the petroleum ether (D 85 g) and methanol (E 170 g) fractions. The cytotoxic activity of each fraction was tested with a KB cell culture to give results of $ED_{50}(\mu g/ml)$(A) 2.6, (B) 100, (C) 2.1, (D) 30, (E) 1.9. See Table 5-1.

The most active fraction (E) 170 g was separated by silica gel column chromatography (3 Kg), and eluted with $CHCl_3$: CH_3OH at different ratios. Then, their activities were tested again with a KB cell culture. Eight cytotoxic sesquiterpene lactones were obtained. The yield and cytotoxic activity results were as follows:

Table 5-1   Yield and Cytotoxic Activities of Sesquiterpene Lactones

Compound	Yield (g)	ED50 ($\mu g/ml$)
[1mm] Euparotin **18**	1.2	2.9
Euparotin acetate **19**	6	0.29
Eupachlorin **20**	0.95	2.6
Eupachlorin acetate **21**	7	
Eupatoroxin **22**	1.5	
Eupatundin **23**	3.9	0.80
Eupachloroxin **24**	0.25	2.4
10-Epi-Eupatoroxin **25**	0.6	

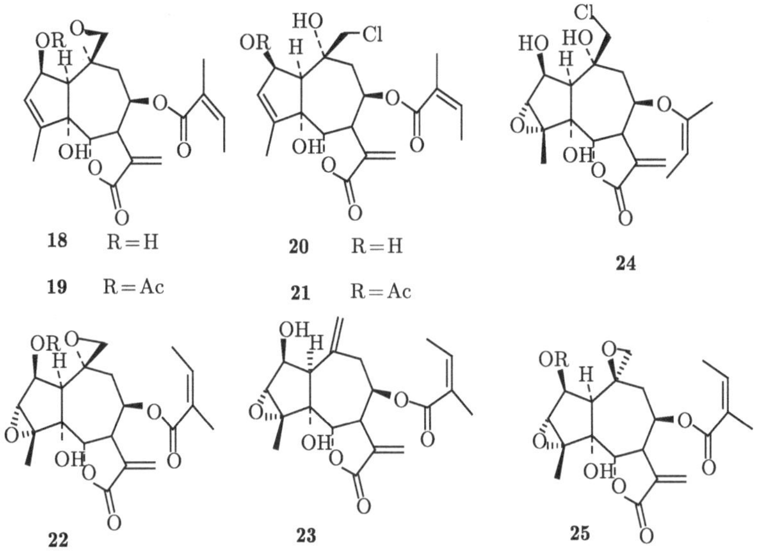

Structures of **18-25**

This research was the earliest report of sesquiterpene chloride isolation from plants. It is worthwhile to point out that the epoxy ring of sesquiterpene molecules can be opened to form the chlorine derivative when using silica gel chromatography eluted with chloroform. A trace amount of hydrochloric acid may be formed during the chromatographic process.

This study is a typical example of bioactive compound isolation by following bioactive directed separation and isolation. Bioassays, such as the inhibition of specific enzymes, are widely used as preliminary tests for directing the isolation of active components from medicinal plants.

Molecules found in naturally occurring compounds can be ionized to form a molecular ion peak in a mass spectrum, and the molecular ion can be further fragmented. Different types of natural products show different fragmentation patterns. Therefore, these different fragmentations can be used for the determination of chemical structures. Until about forty years ago, the GC-MS technique was successfully used for analysis of volatile natural compounds, but the combination of liquid chromatography and mass spectrometry was more complex than GC-MS. Since about twenty years ago, LC-MS techniques have also been used in chemical laboratories. The main methods were electrospray LC-MS (ES LC-MS), thermospray LC-MS (TSP LC-MS) and Frit-FAB LC-MS, etc. To date, the TSP LC-MS is most widely used for the analysis of natural compounds in the 200 to 1000 Dalton range, such as flavonoids, coumarins, xanthones, catechins, quinones, phenolic compounds, lignans, monoterpenes, sesquiterpenes, diterpenes, triterpenes, steroids, lipids, saponins, monosaccharides, oligosaccharides and alkaloids. This technique provides a very useful and powerful tool for the analysis of most organic natural products except polymers[11].

Sesquiterpene lactones are not readily vaporized and decompose easily at high temperatures. They are rather difficult to analyze by GC-MS. Most of the sesquiterpene lactones have no characteristic UV absorption peak. Therefore, TSP LC-MS may be a good choice for the analysis of sesquiterpene compounds.

Artemisinin **42** is a potential antimalaria sesquiterpene lactone. Owing to the fact that artemisinin lacks a suitable chromophore, it is difficult to detect with conventional UV detectors. And, as the natural phenolic compounds occur together with artemisinin in the same extract, such compounds may strongly interfere during the HPLC-UV analysis. Furthermore, artemisinin contains a peroxy functional group, which is readily thermally decomposed during GC-MS analysis. Therefore, TSP LC-MS is a suitable method for direct analysis of plant extracts[12]. TSP LC-MS can be performed on the dichloromethane extract of the *Artemisia annua* plant using the positive ion mode with ammonium acetate. The TSP mass spectrum of artemisinin **42** showed a number of quasimolecular peaks at m/z 283 [M+H]$^+$, 300 [M+NH$_4$] and 341 [M+CH$_3$CN+NH$_4$]$^+$. The HPLC column used was a RP-8 Nucleosil and the extract was eluted with the following gradient and solvents: CH$_3$CN:H$_2$O (0.05% TFA) (50:50) to (75:25) in 30 min (1 ml/min).

As another example[13], the Compositae plant *Centaurea solstitialis* was extracted with chloroform and then analyzed for its sesquiterpene lactone components using TSP LC-MS. The HPLC column was a RP-18 Novapak, and the elution gradient was as follows: CH$_3$CN: H$_2$O (15:85) to (85 to 45:55). Three sesquiterpene lactones m/z 380 **26**, m/z 418 **27**, and m/z 452 **28** were identified.

Structures of **26-28**

Recently, a combination of LC-MS and LC-NMR for the isolation and identification of secondary metabolites of plants, marine organisms and micro bacteria has been successfully used. Kraus et al. reported that they have isolated nine antibacterial active sesquiterpene lactones from *Vernonia fastigiata*[14]. This plant material was extracted with ethyl acetate and the extract was isolated using rotatory locular countercurrent chromatography. The elution solvent was petroleum ether:ethyl acetate:methanol:H$_2$O (9:5:5:1). The eluted fractions were screened for antibacterial activity against *Bacillus subtilis*, and the active fractions were further isolated using LC-UV with MeOH:H$_2$O to give nine compounds. Their molecular ions were 379, 381, 437, 421, 423, 435, 463, 421 and 468, respectively. The antibacterial fractions were analyzed with an LC-MS method, RP-18 chromatographic column, gradient elution using CH$_3$CN:H$_2$O with the acetonitrile concentrations increased from 15% to 40% in 30 min, and then from 40% to 100% in 30 min. The flow rate was 0.7 ml/min.

From the analysis of these fractions, nine vernosistifolide sesquiterpene lactones have been isolated and their ester side chains identified as Methac, *i*-Bu and Ac. LC-NMR was carried out by liquid chromatography with the same conditions as the LC-MS. The above nine vernosistifolide compounds were analyzed using mono and dimensional NMR spectroscopy, and their structures were elucidated as **29-37**.

The combination of LC-MS and LC-NMR is now established as a rapid and practical technique for analyzing the chemical composition of naturally occurring secondary metabolites from plants, marine organisms and micro bacteria. It can also provide some important chemical information on every compound, for identification of known compounds, and for structural elucidation of new compounds.

This method can be carried out by using crude plant extracts and without complex isolation and purification. It is an important innovation in modern phytochemistry, but there are some limitations. The main problem is that the NMR spectrum provides relatively low sensitivity, and it may be difficult to identify trace substances from the plants. We anticipate that it will be widely used as a routine technique in future phytochemical study.

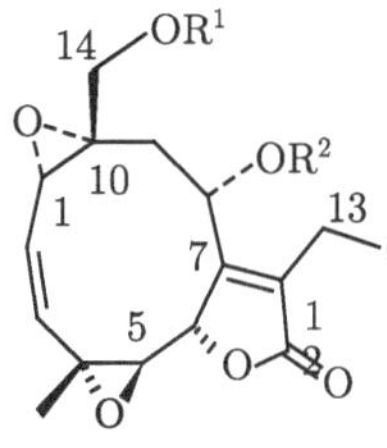

	R^1	R^2	R^3
29	H	Methac	H
30	H	i-Bu	H
31	H	Methac	Ac
32	H	i-Bu	Ac
33	H	ANG	Ac
34	H	Methac	Ac

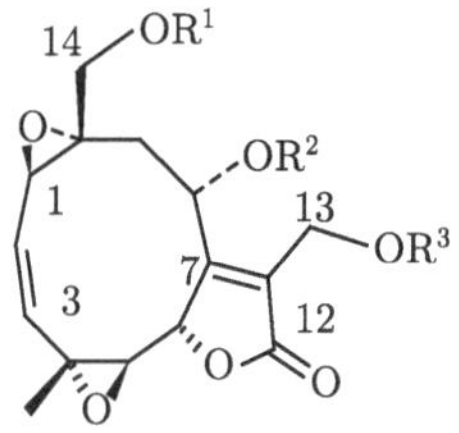

	R^1	R^2	R^3
35	H	Methac	Ac
36	H	i-Bu	Ac

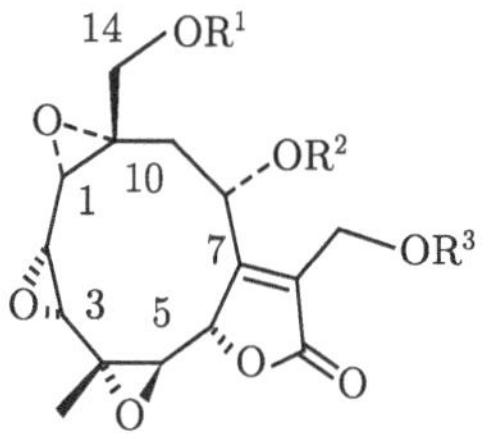

	R^1	R^2	R^3
37	H	Methac	Ac

Structures of **29-37**

Section 2   Spectroscopic Analysis of Sesquiterpenes

2.1   UV, IR and MS spectra

The skeletons of sesquiterpenes lack suitable chromophores and have no characteristic UV absorptions. Sesquiterpene lactones, although having the lactone ring, only show absorption at short wavelengths (200 to 220 nm). If the sesquiterpene molecule presents a conjugated double bond or aromatic ester side chain, it will exhibit UV absorption.

The IR spectra of sesquiterpenes are important for detection of carbonyl groups. The different carbonyl groups present different absorptions as shown in Table 5-2.

Table 5-2   IR Absorption of Carbonyl Groups in Sesquiterpenes

Compound	absorption (cm^{-1})	Compound	absorption (cm^{-1})
[1mm] General esters	1735 to 1750	Carboxylic acid	1700 to 1725
δ−lactone	1735 to 1750	Cyclopentanone	1735 to 1750
γ−lactone	1760 to 1780	Cyclohexanone	1700 to 1725
α, β−unsaturated-γ-lactone	1740 to 1760	Cyclopentenone	1690 to 1710
β−lactone	About 1840	Cyclohexenone	1665 to 1690
Aldehyde	1720 to 1740	α, β-unsaturated aldehyde	1680 to 1710

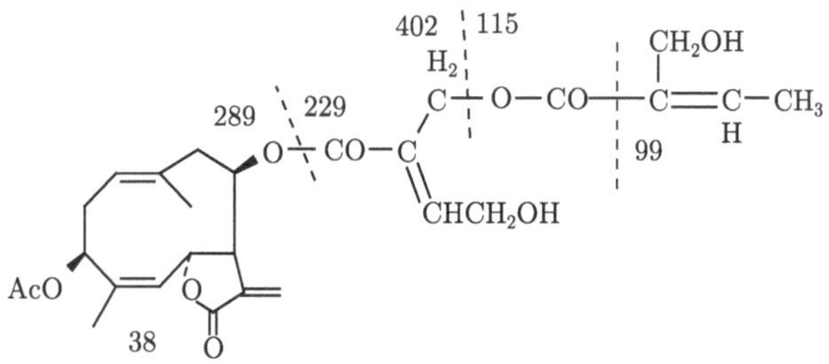

Structures of **38**

Owing to the different functional groups in sesquiterpene skeletons, the mass spectra fragmentation may lead to different EI mass spectra fragment peaks. If there are hydroxyl or ester substitutions on sesquiterpene molecules, these functional groups may easily split off from the molecular ion, resulting in the loss of water or a carboxylic acid first. The molecule is then cleaved in different ways from the different substituted positions. The FAB mass spectrum is commonly used for the detection of molecular ions and identification of acyl substitutions to these compounds. For instance, the fragmentation of provincialin **38** shows the following.

2.2 ^{1}H-NMR spectra of sesquiterpenes

From the spectroscopic analysis to the structural elucidation of naturally occurring compounds, the ^{1}H-NMR spectrum is very useful and can be sequentially analyzed as follows:

CH—OR   3.4-6.2,   —CH—OH   3.4-4.6,   =C—CH—OH 4.5-6.0

—CH—OCOR   4.9-5.9,   =C—CH—OCOR shift to lower magnetic field

=C—CH—OR 4.5-6.0,   —C—CH— 1.6-2.8,
    | \ /
   OH   O

2.2.1   Detection of acyl side chain

In general, sesquiterpenes contain fifteen carbon atoms. When more than fifteen carbons are found from the MS or ^{13}C-NMR spectra, a compound may include an acyl side chain.

The acyl group RCOO- in sesquiterpene molecules is rather unstable in EIMS analysis, which can subsequently easily eliminate a RCOOH from its molecule. We can find the quasimolecular ion from FAB-MS, and the fragmentation of a lost RCOOH peak [M-RCOO]$^+$ from EIMS, and further find the partial molecular formula of a RCOO group. Most of the acyl group of sesquiterpenes commonly occur in natural products, and are easily found and elucidated from ^{1}H-NMR spectra.

2.2.2   Detection of skeleton type

Sesquiterpenes are composed of three isoprene molecules which are connected from head to tail with each other. Therefore, it is very important to find the methyl and/or methylene signals in the ^{1}H-NMR spectra, and it may be helpful to analyze their chemical skeleton.

2.2.3   Analyzing patterns of oxygen atoms

In oxygen substituted sesquiterpenes, the different signals of oxygen vicinal protons are easily distinguished in ^{1}H-NMR spectra. The chemical shift δ of vicinal protons from different substitutions is shown in the following sections.

2.2.4   Analysis of all vicinally connected protons

Due to the complexity of sesquiterpenes, two-dimensional (2D) NMR needs to be used for structural determination. Analysis of ^{1}H-^{1}H COSY spectrum may start from low magnetic field protons, which are vicinally connected to hydroxyl, ester or vinyl groups. The connection order of carbon atoms can be found by analyzing HMQC spectrum. All proton signals will then be assigned. Afterwards, by using the HMBC technique to find the long range connection of all hydrogen and carbon atoms, the whole planar structure can be described.

2.2.5   Stereochemical elucidation

The high resolution ^{1}H-NMR spectrum is valuable in order to resolve the stereochemistry of sesquiterpenes. It is based on the following spectra data: (1) Coupling constant (J) of proton signals: The coupling constant (J) in the ^{1}H-NMR spectrum is closely related to the two dimension angle from both vicinal protons. Therefore, the coupling constant may be used for detecting the related configuration of natural compounds. (2) Chemical shield effect deduced from functional groups: The chemical shield effect on stereo space can also cause the proton signals to change to a lower magnetic field. (3) Nuclear Overhauser effect (NOE): The functional group is connected to neighboring protons, which may appear as the NOE if the distance between the functional group and proton is as near as 6 Å. The NOE can be detected by NOE differential spectrum and NOESY spectrum.

In the ^{1}H-NMR spectrum, two-dimensional angle, and chemical shield effect are helpful to determine the stereochemistry of natural organic compounds.

The above mentioned plane structure can be further analyzed for its NOESY and ^{1}H-NMR spectrum, and then the chemical structure (including stereochemistry) will finally be established.

2.3 ^{13}C-NMR of sesquiterpenes

The ^{13}C-NMR spectrum is also a very useful and powerful tool for chemical structure elucidation of natural products. In general, the ^{13}C-NMR spectrum is used with a proton noisy decoupling technique to obtain the simple singular peak of every carbon atom. Then the INEPT and DEPT techniques are used for distinguishing the primary CH, secondary CH$_2$, tertiary CH$_3$ and quaternary carbons as d, t, q, and s peaks, respectively.

Accordingly, all chemical shift signals can be distinguished from each carbon atom's chemical environment. Owing to the ^{13}C-NMR, the spectrum shows a wide range, from 0 to 140 ppm. Its anisotropic effect is smaller than ^{1}H-NMR. Therefore, the structural relationship of chemical shifts is more regular than ^{1}H-NMR.

An important way for determining the structure of natural compounds, is by comparing the spectral data of an unknown sample with those of known compounds.

[**Example 5-3**]   Structure elucidation of spathulenol[15]

Spathulenol **39** was isolated from the essential oil of *Citrus grandis* peel. Its structure was elucidated using only NMR techniques by Fuyuhiko Inagaki.

From the ^{1}H-NMR spectrum, eighteen proton signals representing a total of twenty four protons were found. They are (Figure 5-1a) q, p, o, n, m, l, k, j, i, h, g, f, OH, e, d, c, b, a. In these protons, the f, e, and d are methyl protons and q and p are vinyl protons.

From the ^{13}C-NMR spectrum, fifteen signals representing a total of fifteen carbon atoms were found. They are (Figure 5-1b) O, L, N, K, J, I, H, G, F, E, D, C, B, M, A. Signal O (152 ppm) and signal L (106 ppm) are vinyl carbons. Signal N (81 ppm) may be connected to an OH group.

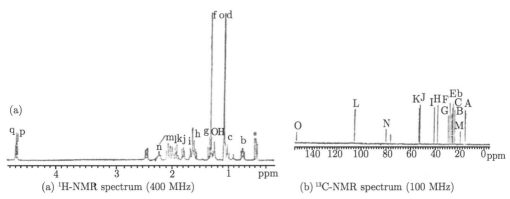

(a) ^{1}H-NMR spectrum (400 MHz)   (b) ^{13}C-NMR spectrum (100 MHz)

Figure 5-1   NMR spectrum of spathulenol

The fifteen carbon signals were differentiated with the insensitive nuclei enhanced by polarization transfer (INEPT) technique to give:

5 CH_2 with t peaks I, H, D, B, L

3 CH_3 with q peaks A, C, F

4 CH with d peaks K, J, G, E

3 quaternary carbons with s peaks M, N, O

From the heteronuclear shift-correlated 2D ^{1}H-^{13}C-COSY NMR spectrum, the cross peaks of Jn, Kg, Ih, Ij, Ho, Hm, Ga, Fe, Cf, Di, Dk, Bl, Bc, Ad, Eb can be found.

The ^{1}H detected heteronuclear multiple-quantum coherence spectrum (HMQC) is used for detection of the H-C connection.

From the homonuclear shift-correlated 2D H-H-COSY NMR spectrum, the cross peaks as proton-proton coupling relations: ab, ag, bc, bl, cl, cm, co, gn, hi, hj, hk, ik, in, ji, jk, lo, lm, (nk) can be found.

From the above results, the following structural fragments were deduced:

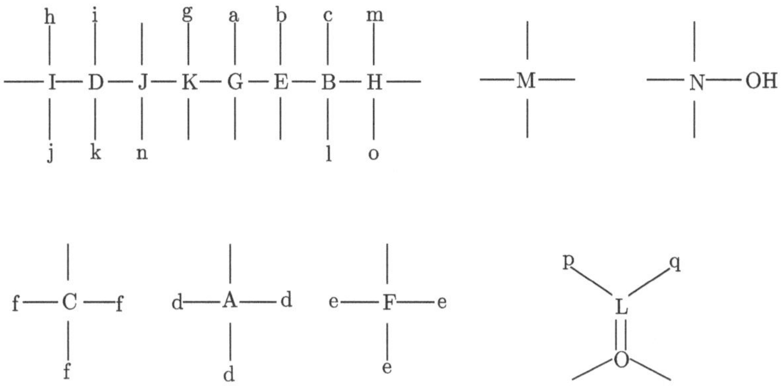

From the long-range ^{1}H-^{1}H COSY NMR spectrum, the following long-range proton coupling relations were found: ae, be, co, qj, mp, nq and hf.

HMBC spectroscopy is now used for detection of long-range proton coupling relationships, instead of long-range ^{1}H-^{1}H COSY NMR spectra.

According to the long-range coupling of mp and nq, the L=O double bond may be inserted between carbons J and H. The long-range coupling of hf indicates that a quaternary carbon F may be inserted between carbons I and K. Given that the gj peak was found, there was only one carbon, not two, found between carbons I and K.

The long-range coupling of ae and ab shows that carbons G and E may be connected to another quaternary carbon M, forming a tricyclic ring. Carbon M may be connected to carbons A and F.

Given the above results, with total connection of C-C and H-C chemical linkages, structure **39** was deduced.

The stereochemistry of spathulenol **39** can be elucidated by a NOE differential NMR spectrum or a NOESY NMR spectrum. The NOESY NMR spectrum of **39** shows the cross peaks: ab, ae, af, be, dg, hf, hj, ik, jk, el, om, kg, and op. According to the results of the nuclear Overhauser effect cross peaks: d with g; e with a, b, l; f with a, h; a with b, e, f, n, the stereochemical structure of spathulenol was deduced as **39**[15].

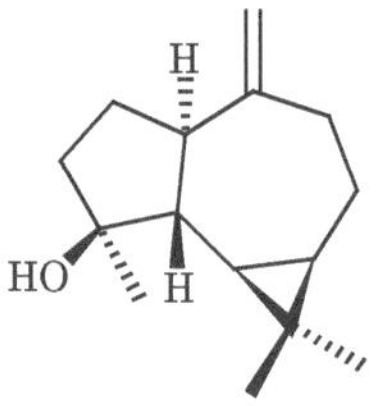

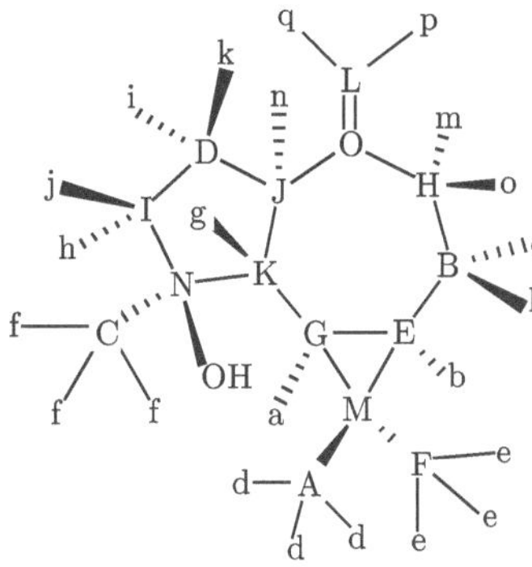

Structure of **39**

The structural determination of spathulenol is a typical example of structural elucidation using only NMR techniques. Although this work was finished in an early stage, it indicated that the NMR technique is extremely useful for structural elucidation of natural products.

In general, the combination of main spectra UV, IR, MS and NMR is always used for determining complex structural compounds.

For the last twenty years, 2D NMR techniques have greatly progressed. The following are very useful for the structural elucidation of natural products:homonuclear shift correlated 2D spectroscopy (^{1}H-^{1}H COSY), heteronuclear shift correlated 2D spectroscopy (^{1}H-^{13}C-NMR), NOE enhancement spectroscopy (NOESY), ^{1}H detected heteronuclear multiple-quantum coherence spectroscopy (HMQC) and ^{1}H detected heteronuclear multiple bond connective spectroscopy (HMBC). For HMQC and HMBC NMR, please refer to the spectral analysis of artemisinin below.

Section 3   Artemisinin–Chemistry, Pharmacology, and Clinical Uses

The Qinghao plant, *Artemisia annua* L. (Compositea), is the main component of the Chinese antimalaria prescription (qing-hao-bie-jia-tang), which is a decoction of qinghao plant and turtle shell. In the Chinese herbal market, *Artemesia annua* and *A. apiacea* are both commonly used as qinghao. From 1965 to 1970, more than five thousand plants have been screened using animal malarial tests. It was found that the alcoholic extract of *A. annua* was one of the most active samples against mice malaria (*Plasmodium berghai*), and the active component, artemisinin, was isolated. Artemisinin, a colorless sesquiterpene lactone with the formula $C_{15}H_{22}O_5$, mp 156 to 158, and $[\alpha]_D + 69$ (C = 0.5, CHCl$_3$), was previously summarized by Luo[16] and Ziffer[17].

3.1   Chemical properties of artemisinin

Artemisinin **42** is readily soluble in chloroform, methanol, ethanol, and ethyl acetate and sparingly soluble in water and ether. It is insoluble in petroleum ether. It is rather stable in neutral solutions, but very sensitive in an alkali solution. It is easily reduced by FeSO$_4$ or P(C$_6$H$_5$)$_3$ and its mass spectrum shows a peak at m/z 250 [M-32]$^+$, which may be the loss of an oxygen atom at the peroxide linkage.

Artemisinin **42** was reduced by Zn+HCl or Pd/CaCO$_3$ catalytic hydrogenation to form deoxyartemisinin **43**. If **42** was reduced by NaBH$_4$, its lactone carbonyl group was reduced to hydroxyl to give a mixture of two enantiomeric dihydroartemisinins α-**45** and β-**44**. It is important to note that the peroxy group was preserved. Dihydroartemisinin **45** was treated with BF$_3$catalyst in a methanol solution, and converted to α-artemether **47** and β-artemether **46**. The β-artemether **46** was readily crystallized in solution and could be separated. When the NaBH$_4$ reduction reaction of **42** was finished, HCl was added to destroy the excess NaBH$_4$. The acidic solution was left at room temperature for 4 h and methylation occurred directly. This is a simple process for the industrial manufacture of artemether.

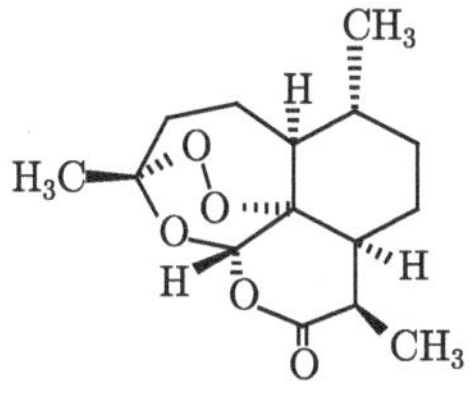

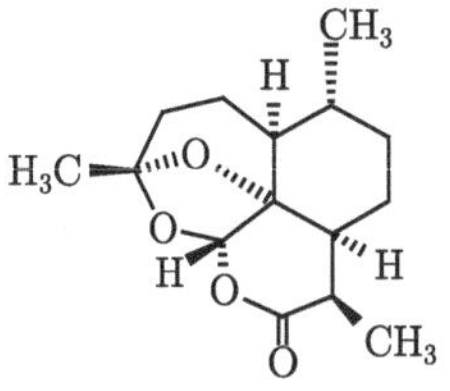

Structures of **42-47**

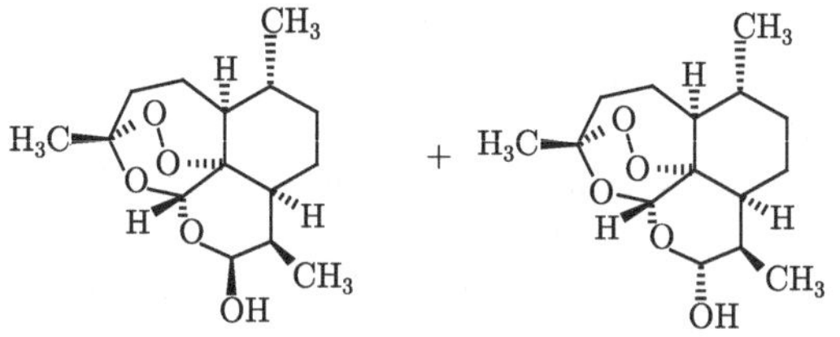

3.2   Spectra analysis of artemisinin

Artemisinin, **42** with the molecular formula C$_{15}$H$_{22}$O$_5$,has only one lactone ring. Its UV spectrum shows no absorption above 220 nm. The IR spectrum showed δ-lactone absorption at 1745 cm^{-1}, peroxy absorptions at 831, 881, 1115 cm^{-1}, and HRMS m/z 282.1472. Its ^{1}H-NMR spectrum showed the following signals in DMSO-d_6:

H-1α 1.29 m; H-2α 1.92 ddd (3.4, 7.1,10.5); H-2β 1.32 m; H-3α 2,66 ddd (3.9, 9.0,10.5,); H-3β 2.06 ddd (4.2, 4.2, 13.3); H-5β 6.11 s; H-7α 1.78 ddd (3.7, 3.5, 13.2); H-8α 1.77 dddd (3.3, 3.3, 3.3, 13.2); H-8β 1.14 dddd (3.0, 13.1, 13.1, 13.1); H-9α 1.0 dddd (3.5, 12.2, 12.2, 12.2); H-9β 1.63 dddd (3.3, 3.3, 3.3 12.5); H-10β 1.51 m; H-11α 3.15 dq (8.7); H-13 1.06 d (7.3); H-14 0.91 d (3.4); H-15 1.34 s.

3.2.1 ^{1}H-NMR and ^{13}C-NMR spectra of artemisinin

The ^{1}H-NMR spectrum of artemisinin **42** only showed a 6.11s (H-5) signal in the lower magnetic field. All signals found in the upper field are shown in Figure 5-2. In the molecular model of artemisinin **42,** the band 1α, 2β, 3α, 7α, 8β, 9α, and 10β are axil protons and 2α, 3β, 8α, and 9β are equatorial protons respectively. Therefore the axial-axial torsion angles of 1α,2β; 2β,3α; 7α,8β; 8β, 9α, and 9α, and 10β are present at about 180 and splitting constant J value = 13 to 14 Hz. The equatorial-equatorial torsion angle of 1α,2α; 2α,3α; 2α,3β; 7α, 8α; 8α,9β; and 9β, 10β present at about 45 and J value = 4 to 5 Hz. The 2α, 2β; 3α, 3β; 8α, 8β; and 9α, 9β are bonded on the same carbon and the splitting constants J are as big as 14 to 15 Hz. All proton signal peaks were very typical, as shown in Figure 5-2.

Fifteen carbon signals are present in the ^{13}C-NMR spectrum of artemisinin **42**, including three s peaks, five d peaks, four t peaks and three q peaks. The signals are assigned as follows: C-1 49.468 d, C-2 24.352 t, C-3 35.468 t, C-4 104.709 s, C-5 93.277 d, C-6 79.564 s,

C-7 43.881 d, C-8 22.414 t, C-9 33.126 t, C-10 36.076 d, C-11 32.515 d, C-12 171.493 s, C13 12.404 q, C-14 19.560 q, C-15 24.927 q.

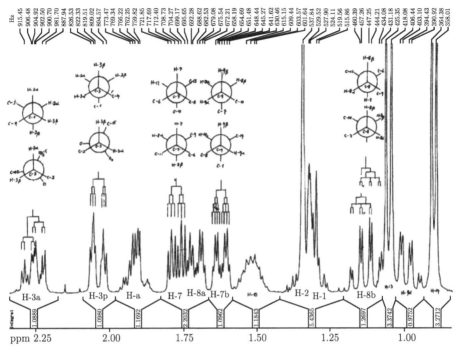

Figure 5-2 ^{1}H-NMR spectrum of artemisinin

3.2.2   HMQC and HMBC spectra

HMQC and HMBC NMR spectra are very useful and widely used to determine natural product structures. The HMQC technique is similar to ^{1}H-^{13}C-COSY 2D NMR spectra for the assignment of the H-C combinations. The HMBC technique is similar to long rang H-C coupling 2D NMR for the 4J coupling relationships between H and C atoms.

Table 5-3   Correlation Signals in HMQC and HMBC Spectra of Artemisinin

Proton Signal		HMQC (H to C)		HMBC (H to C)
H-1α	1.29	C-1	49.486	C-14, C-7, C-6, C-5
H-2α	1.92	C-2	24.362	C-1, C-4, C-3, C-5
2β	1.32			C-3, C-4
H-3α	2.26	C-3	36.076	C-2, C-4
3β	2.04			C-2, C-4, C-1
		C-4	104.709	
H-5β	6.11	C-5	93.277	C-1, C-6, C-4, C-12
		C-6	79.584	
H-7α	1.78	C-7	43.881	C-8, C-9, C-1, C-6, C-5, C-12
H-8α	1.00	C-8	22.414	C-6
8β	1.63			C-7
		C-9	35.464	
H-10β	1.51	C-10	38.874	C-14, C-2
H-11α	3.15	C-11	33.126	C-13, C-8, C-7, C-12
		C-12	171.493	
H-13	1.06	C-13	12.404	C-11, C-7, C-12
H-14	0.91	C-14	19.560	C-9, C-10, C-1
H-15	1.34	C-15	32.515	C-3

The relationship between the HMQC and HMBC two-dimensional NMR spectra of artemisinin **42** is shown in Table 5-3.

3.3   Pharmacology and clinical uses of artemisinin and its derivatives

Human malaria is caused by protozoa of the genus *Plasmodium*, such as *P. falciparum, P. malaria, P. ovale and P. vivax*, which are injected into the blood stream by the bite of an infected female mosquito. In many parts of the world, the malignant malaria strains are responsible for cerebral malaria and can cause the patient to lapse into a coma and ultimately lead to death. Today, malignant malaria strains have become resistant to most of the synthetic antimalaria agents such as chloroquine, mefloquine, etc. Artemisinin is a novel class in anti-malarial agents, in that it can overcome resistant strains. Pharmacological studies show artemisinin is a well-tolerated drug which rapidly destroys malaria parasites.

In vitro testing indicated that artemisinin **42** at the dose of 20 to 50 μg/ml directly inhibits the growth of parasites in mice infected with chloroquine resistant strains of *P. berghai*. Artemisinin exhibits its anti-malaria activity at a dose of 6 mg/kg and suppresses parasite infection in 90% of all cases (ED_{90}). Artemisinin inhibits the malaria protozoa at the erythrocytic stage, but is inactive at the exo-erythrocytic and pre-erythrocytic stages. Its chemical structure and biological mechanism are quite different from those of synthetic anti-malaria agents. Therefore, artemisinin is not resistant to chloroquine and mefloquine, and can inhibit the resistant chloroquine strains much effectively than normal malaria protozoa[18]. The injection of artemisinin or its derivatives can be easily absorbed, widely distributed, and rapidly metabolized. They easily cross the blood-brain barrier. Cell biology studies on the distribution characteristics of artemisinin found that the concentration of artemisinin and derivatives distributed in malaria protozoa infected RBC were much higher (by about a factor of 100) than those of normal RBC. The higher drug concentrations are collected in the target cells of malaria protozoa, therefore artemisinin and its derivatives are very suitable for use in the emergency treatment of patients with malignant brain malaria. See Table 5-4.

Artemisinin **42** and its derivatives are metabolized rapidly and maintain a short half-life in blood. Hence, they show a higher recurrence rate[19−21]. In general, primaquine or pyrimethamine is used in the later stage of therapy for recurrent treatments.

3.4   Chemical modification and structural-activity relationship of artemisinin

Owing to its poor solubility in water or oil, artemisinin **42** is difficult to prepare as a clear solution for parenteral administration. The suspended injection of artemisinin causes localized pain at the injection area. Most brain malaria patients went into a coma, and the intravenous administration form is the best choice. Therefore, structural modification of artemisinin may be a better way to find new anti-malaria derivatives with low toxicity for clinical use.

Artemisinin **42** only contains peroxy, lactone and acetal functional groups in its structure. Most chemical reactions may attack the peroxy group first, but in converting the peroxy to the deoxyartemisinin derivative **43**, it loses anti-malarial activity also. Artemisinin can also be reduced by $NaBH_4$ to form another peroxy containing compound, dihydroartemisinin **44+45**. Dihydroartemisinin showed four -fold anti-malarial activity compared to artemisinin itself. Modification of artemisinin to dihydroartemisinin was recently finished and its main skeletal derivatives have been totally synthesized.

Reduction of artemisinin **42** with $NaBH_4$ reduces its 12-carbonyl group to a hydroxyl group and forms two isomers, α-**45** and β-**44**. Both the α and β forms show similar anti-

malarial activity and can be further reacted to prepare their ether, ester, and other derivatives. The pure or mixed dihydroxyartemisinin was methylated by adding a catalyst BF_3 to
form the ether derivative **48**. Dihydroxyartemisinin was acylated with anhydride $(RCO)_2O$
or acyl chloride RCOCl in pyridine, Et_3N or DMAP, to form its ester derivatives **49**. If
dihydroartemisinin was acylated with dicarboxylic anhydride, **50** or **52** can be synthesized.
If dihydroartemisinin was acylated with ROCOCl in pyridine, the carboxy ester derivatives
51 were formed.

Earlier, the Shanghai Institute of Materia Medica, Chinese Academy of Science synthesized a series of ether, ester, and carboxy ester derivatives. These were screened for their
anti-malarial activity on mice malaria (*Plasmodium berghai*)[18,22].

At a dose of 6.20 mg/kg, artemisinin shows inhibitory effects against malarial protozoa in
90% of all cases (ED_{90} = 6.20 mg/kg)

Of the dihydroxyartemisinin ether derivatives, artemether (**48-1** R=CH_3) has the most
potency, being twice as active as artemisinin against mice malaria *P.berghai*. (SD_{90}=1.02
mg/kg). In general, the longer chain ether derivative may be less potent in anti-malarial
activity than shorter ones.

Of all ester derivatives, dihydroxyartemisinin benzoate (**49-53**, R = C_6H_5) is most potent.
It is 13 times as active against mice malaria. (SD_{95}= 0.48 mg/kg).

Among carboxy ester derivatives the **51-54** R=n-C_3H_7 has most potency, being 12 times
as active as artemisinin itself. (SD_{90} = 0.50 mg/kg).

Structures of **48-54**

Among the dicarboxylic acid diester derivatives, compound **52-52**, $R=(CH_2)_3$ has the most potency. It is nine times as active as artemisinin. $(SD_{90}=0.65$ mg/kg).

In modern investigations on anti-malarial activity, the in vitro screening is frequently used as the first choice. Investigators at Walter Reed Army Institute and universities in the United States have performed primary tests against two drug resistant clones of *P. falciparum*. The W-2 clone from Indochina is resistant to chloroquine but sensitive to mefloquine, whereas the D-6 clone from Africa (Sierra Leone) is resistant to mefloquine but sensitive to chloroquine. Activity data are reported as artemisinin indices: IC_{50} artemisinin/ IC_{50} compound.

A series of synthetic artemisinin **42** derivatives were synthesized and screened, such as acylamides **53**[23], water soluble esters **54** [24], C-4 alkyl C-11demethyl artemisinin derivatives **55**[25], C-4 alkyl C-11 alkyl derivatives **56**, C-12 decarbonyl C-4 alkyl artemisinin derivatives **57**, C-12 decarbonyl C-11 alkyl artemisinin derivatives **58**[26], C-12 N-containing artemisinin derivatives **59**[28] tricyclic artemisinin derivatives and C-11 alkyl artemisinin derivatives[27]. The main outstanding active compounds are listed as follows:

Table 5-4   Anti-malaria Activities of Artemisinin Derivatives

Compound	Number	Structure R=	Anti-malarial activity (artemisininindices)	
			W-2	D-6
Artemisinin	**42**		1.00	1.00
Dihydroartemisinin	**44+45**	H	3.58	4.75
Artemether	**48-1**	CH_3	1.92	2.34
Arteether	**48-2**	C_2H_5	2.08	2.53
Artemisunate	**54-1**	$CO(CH_2)_2COONa$	3.86	4.83
	54-36	$(S)\text{-}CH_2CH(CH_3)COOCH_3$	6.00	1.73
	54-44	$(R)\text{-}CH(C_6H_5)COOC_2H_5$	10.8	8.0
	54-49	$(S)\text{-}CH(C_6H_5)COOC_2H_5$	20.7	9.3
	53-1	CH_3	4.00	2.13
	55-2	C_2H_5	21.2	6.73
	56-6	$CH_2(C_2H_5)_2$	13.82	22.84
	58-5	$n\text{-}C_4H_9$	58.26	20.90
	58-7	$(CH_2)_3C_6H_5$	50.73	24.06
	58-8	$(CH_2)_3C_6H_4-p\text{-}Cl$	69.90	33.17
	57-2	$n\text{-}C_3H_7$	7.22	6.85

From these screened *in vitro* results, compound **58-8** is the most active among artemisinin derivatives. It merits further research on anti-malarial agents.

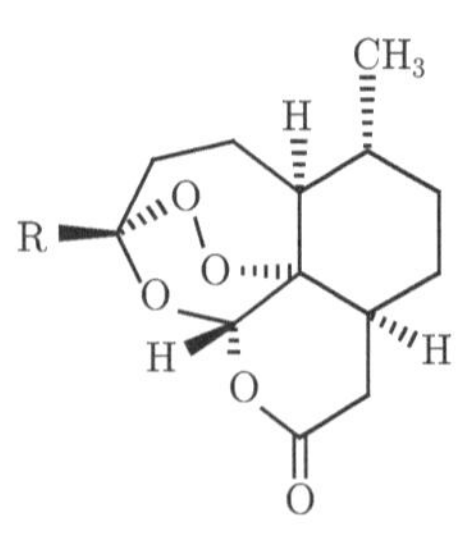

55 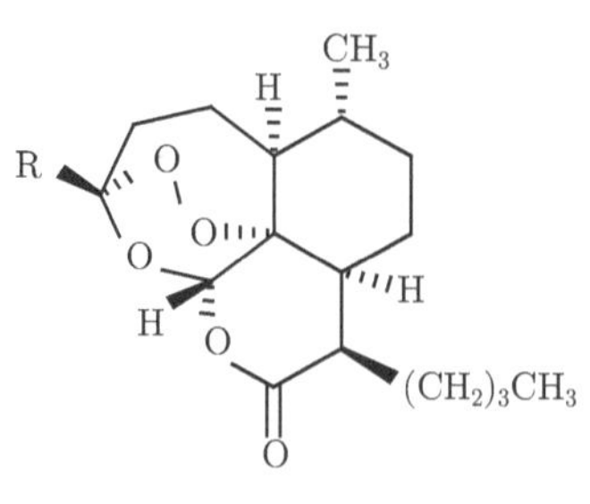   56 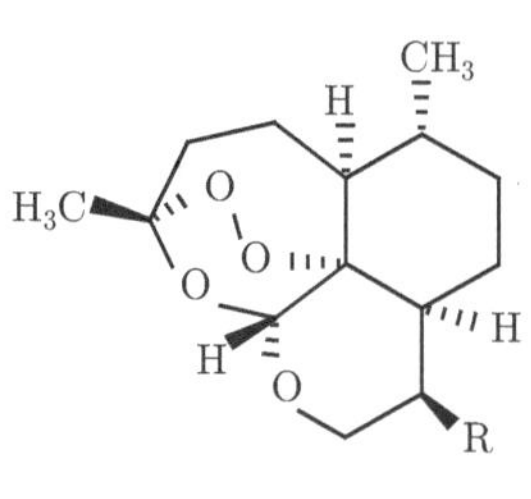   58

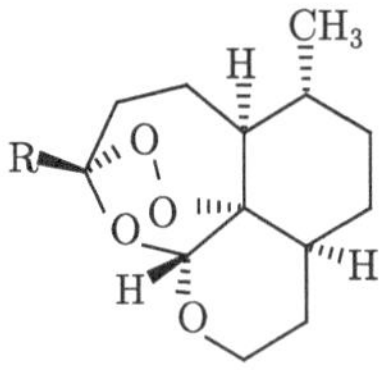

Structures of **55-59**

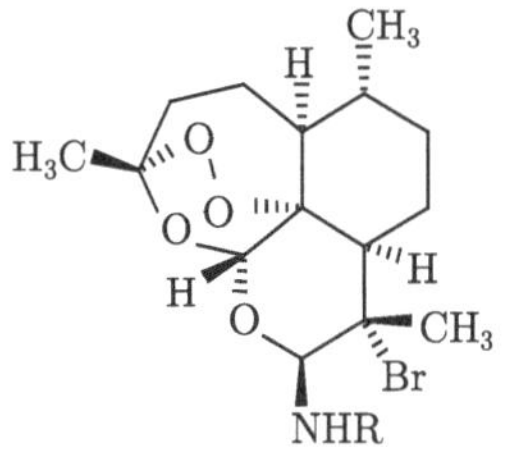

Section 4   Recent Progress of Specific Sesquiterpenes

Over the last twenty years, natural product sesquiterpenes have been actively researched. Owing to the discovery of the anti-malarial drug artemisinin, the naturally occurring peroxide has attracted profound attention from organic chemists and pharmacologists. The natural peroxides include common hydroxyperoxides and cycloperoxy ethers. The former are very unstable and the latter are rather stable.

From *Artemisia abrotannum*, some peroxy compounds, such as arteincultone **60**, epi-arteincultone **61** and compounds **62-65,** were isolated[29]. Novel compounds **62-65** possessed a peroxy spiro ketal structure, and arteincultone **60** exhibited a moderate anti-malarial activity against *P. falciparum*.

Artabotrys uncinatus is an Annonaceae plant found in the Hainan province of southern China. During anti-malarial screening, it was found to exhibit anti-malarial activity against *P. berghai*, and the active components yinzuaosu A **66** and B **67** were isolated[30]. Their structures and total synthesis were finished in China. Although their anti-malarial activities were found earlier than artemisinin, they were very difficult to prepare from natural sources, and research on the anti-malarial activity of their derivatives is inprogress[31].

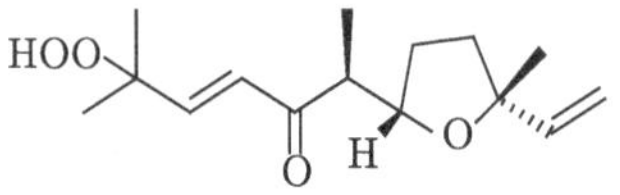

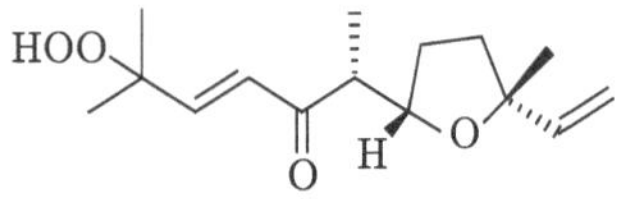

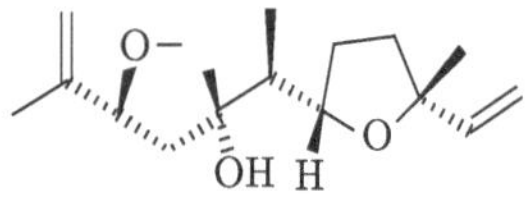

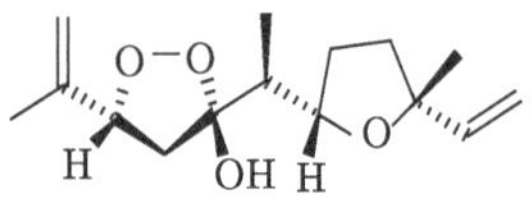

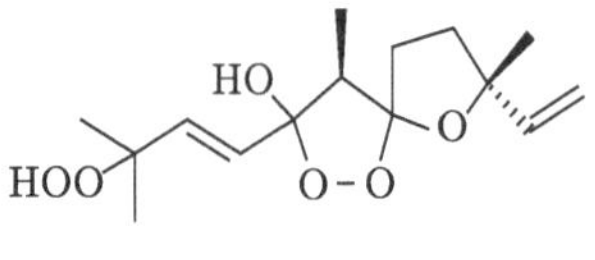

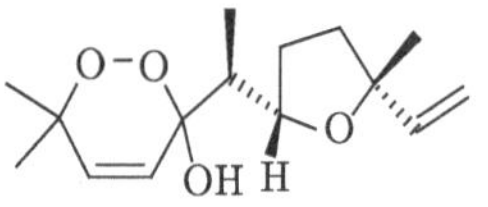

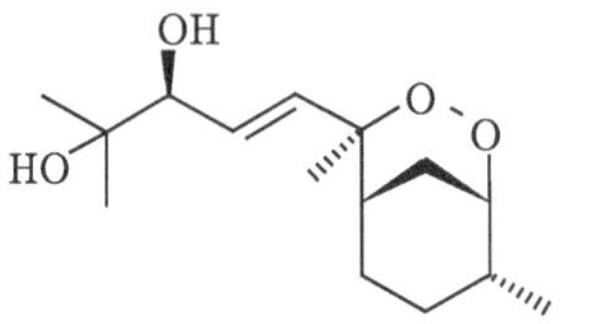

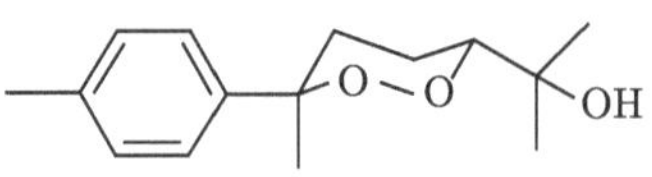

Structures of **60-67**

For twenty years, some dimer and trimer sesquiterpenes have been discovered in nature. Previously, a dimeric sesquiterpene lactone, artanomaloid **68** was isolated from the Chinese medicinal plant *Artemisia anomala*[32]. Arteminolides A-F and artselenoid **69** were isolated from *Artemisia selengensis*. These *Artemisia* dimeric sesquiterpene lactones are formed by a Diels-Alder reaction between two different sesquiterpene lactones.

From the Japanese plant *Chloranthus japonica*, the dimeric compound shizukaol **70** and trimer trishizukaol **71** were found[33].

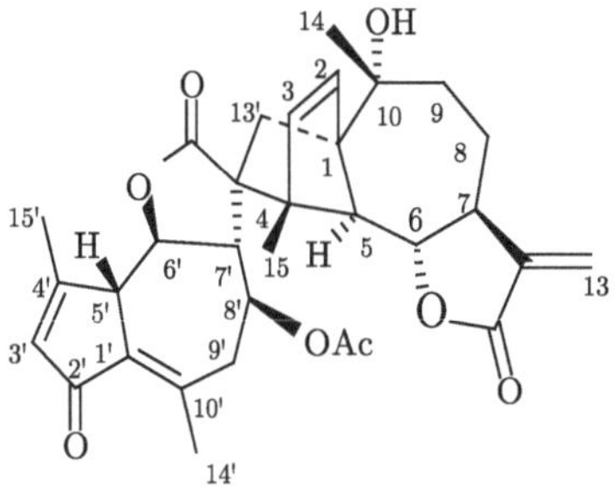

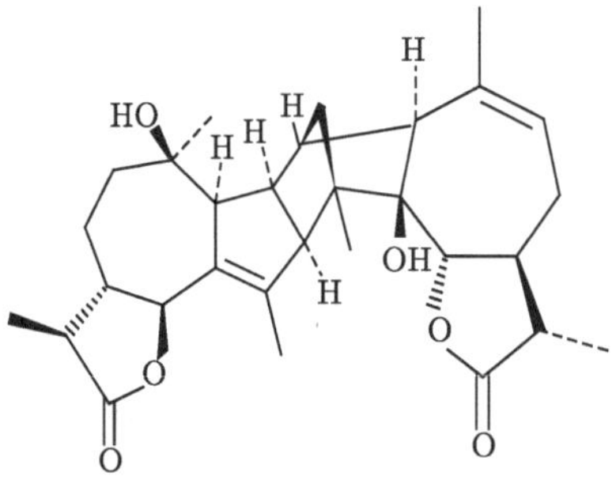

Structures of **68-71**

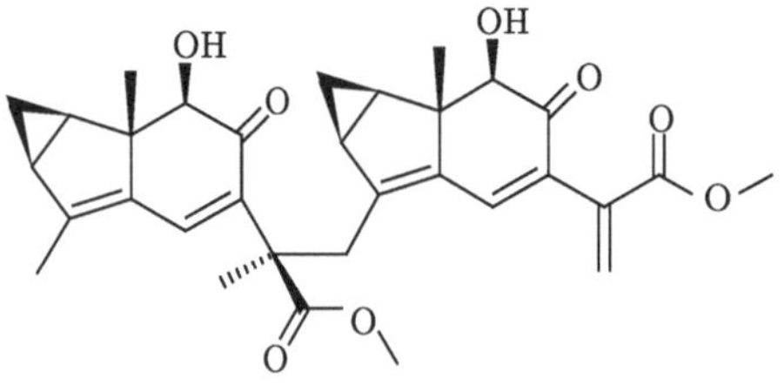

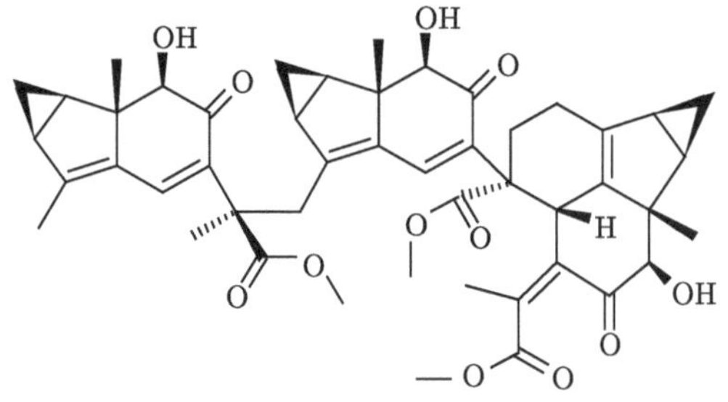

Many combinations of sesquiterpene compounds with phenolic compounds have been found in marine organisms and fungal products. Simplicissin **72** was isolated from fungal *Penicillium simplicimum* and exhibited inhibition of pollen growth. The antibiotic substances hongoquercin A **73**, B **74**, and (+)-hyatel laquinone **75** were isolated from fungus LL-23G227[34,35]. Recently, some combinations of sesquiterpene-phloroglucinol compounds were isolated from ferns. For example, dryofragin **40**, which was isolated from *Dryopteris fragrans*, and phloroglucinol A **41a** and B **41b** isolated from *D. atrata*[36].

The natural nitrogen compounds are biosynthesized from amino acids and defined as alkaloids. From *Stachybotrys* sp. the nitrogen containing sesquiterpenes **76**, **77** and dimeric compound **78** were isolated. **77** and **78** showed endothelin receptor antagonistic activity.

The sesquiteroene alkaloids polyalthenol **79**, polyveoline **80**, isopolyalthenol **81** and neopolyalthenol **82** were isolated from the Annonaceae plant *Polyalthia olieri*[37]. Additionally, a series of drimentine alkaloids were isolated from *Actinomycetes* sp. MS8651 strain, and these compounds exhibit anti-bacterial, cytotoxic and anti-neoplasma activities[38].

Atractylodes macrocephala is a famous Chinese medicinal plant that contains several biologically active sesquiterpene lactones. A lactam, atractylenolactam **83**, was found in its rhizome[39]. Another similar sesquiterpene, lactam **84**, was isolated from the marine soft coral *Clavularia inflate*[40].

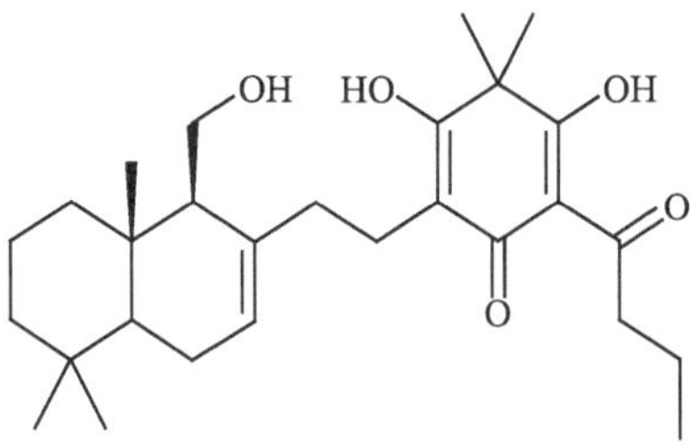

40

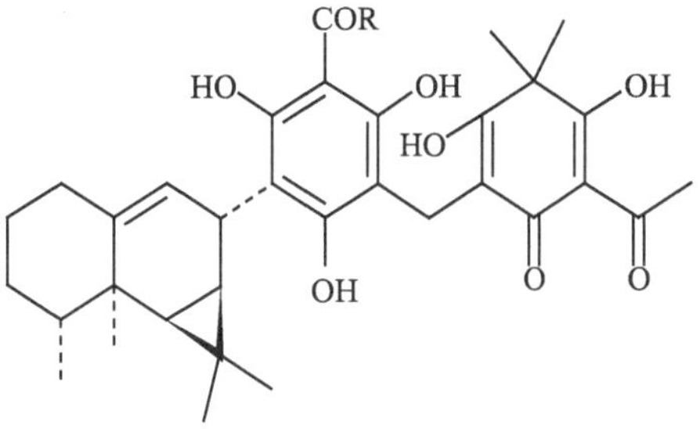

41a R=Me
41b R=CH$_2$CH$_2$-Me

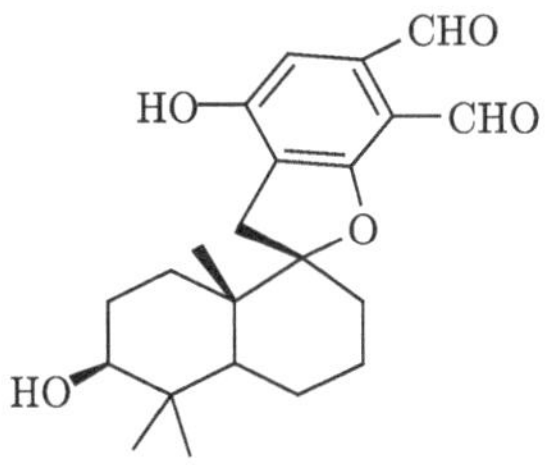

72

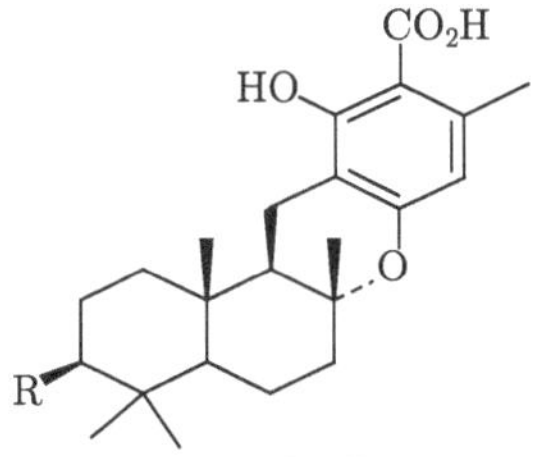

73   R=H
74   R=OAc
Structures of **72-75**

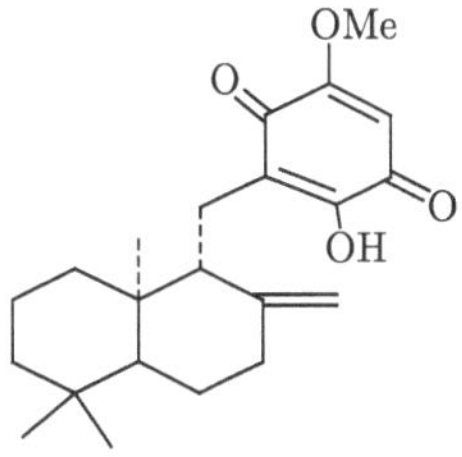

75

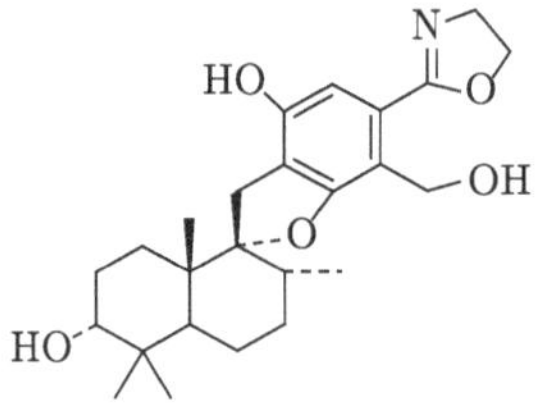

76 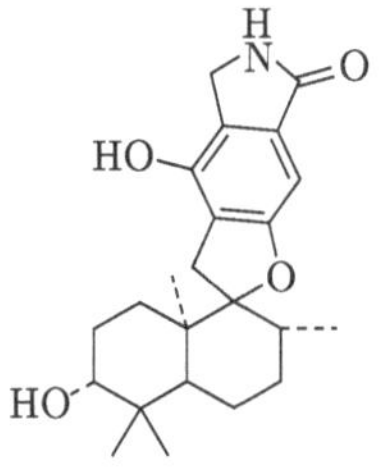   **77** 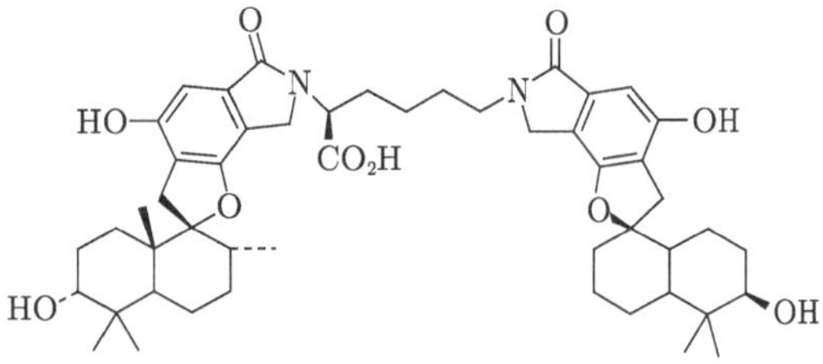   **78**

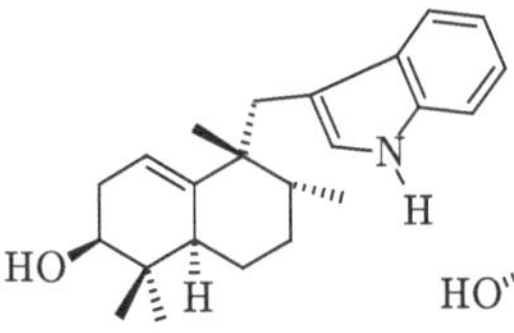

79 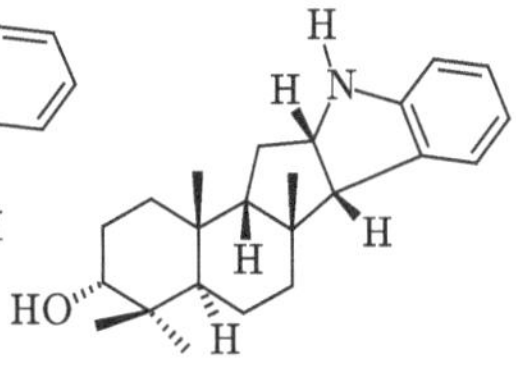   **80** 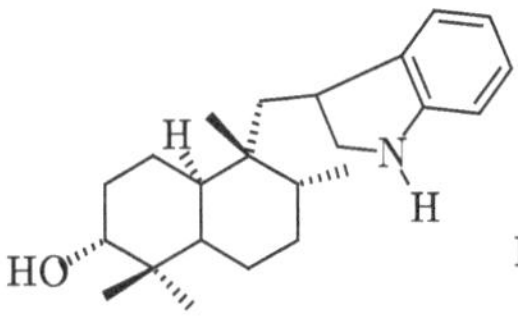   **81** 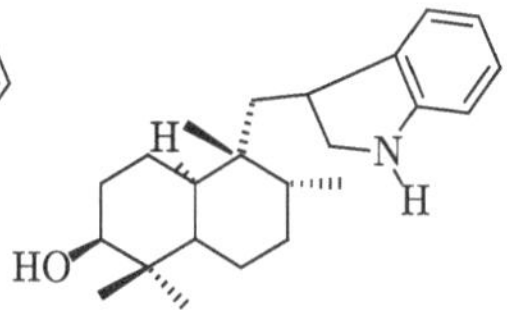   **82**

Structures of **76-82**

Some marine organisms contain sesquiterpene alkaloids. 4-Methylaminoavarone **85** was isolated from *Dysidea avara*. Several sulfonic acid derivatives of sesquiterpenes were isolated from marine biological resources. These acidic compounds can form water-soluble salts and may be important for pharmacological research. From *Callyspongia* sp., a pholpation enzyme inhibitor, akaterpene **86**, was isolated[41]. Ispidospermidin **87** was isolated from *Chaeronema hispidulum*, and was created from a new type of pholpation enzyme C inhibition.

Many sesquiterpene-amino acid combination compounds occur in marine organisms. Naki-jiquinone A **88** and B **89** were isolated from Spongiidae, and both compounds show anti-fungal activities[43].

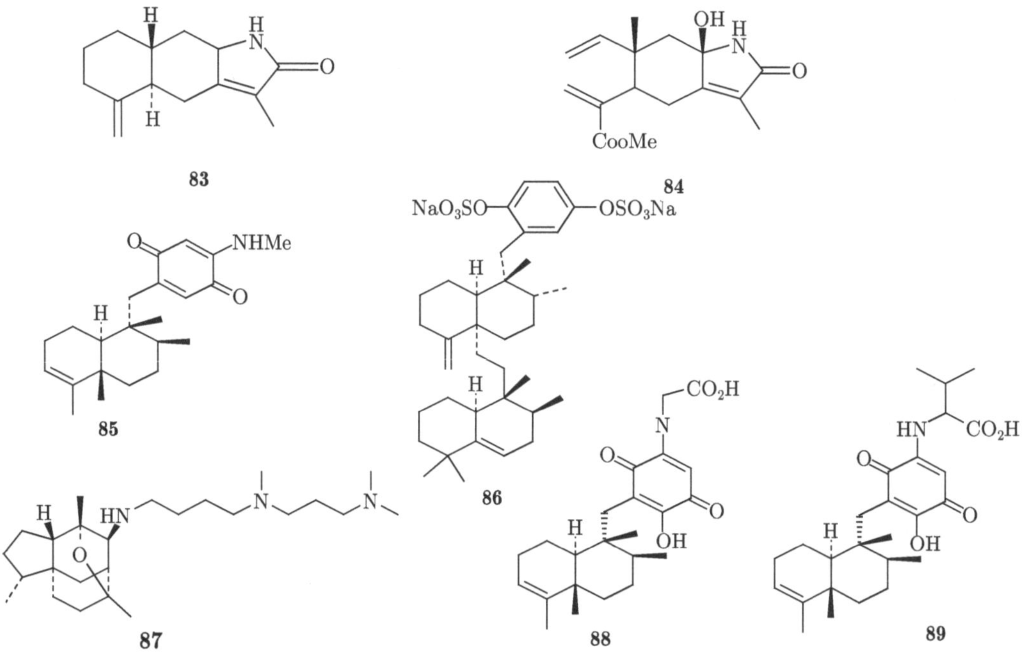

Structures of **83-89**

The above-mentioned novel sesquiterpenes occur in plants, marine organisms and micro bacteria. As a matter of fact, many very complicated biosynthetic processes can take place in nature, which cannot yet be emulated by modern synthetic chemistry. For the discovery of new drugs, natural compounds with both complex structures and potential biological activities can be used as lead compounds for further modification. Some may be developed for clinical uses. To overcome the inherent problems, great effort will be required by natural product scientists.

References

[1]   Fischer N H et al. Prog Chem Nat Prod, 1979, 38: 47.

[2]   Fraga B M. Nat Prod Rep, 1995, 12:303.

[3]   Fraga B M. Nat Prod Rep, 1996, 13: 307.

[4]   Fraga B M. Nat Prod Rep, 1997, 14: 145.

[5]   Fraga B M. Nat Prod Rep, 1998, 15: 73.

[6]   Fraga B M. Nat Prod Rep, 1999, 16: 21.

[7]   Fraga B M. Nat Prod Rep, 1999, 16: 711.

[8]   Fraga B M. Nat Prod Rep, 2000, 17: 483.

[9]   Chen Z. Huaxue Tongbao chemistry, 1984: 57.

[10] Kupchan S M et al. J Org Chem, 1969, 34: 3876.

[11] Wolfender J L et al. Phytochem Analysis, 1994, 5: 153.

[12] Mailard M P et al. J Chromatogr, 1993, 647: 147.

[13]   Hamburger M et al. Nat Toxins, 1993, 1: 315.

[14]   Vogler B et al. J Nat Prod, 1998, 61: 175.

[15]   Fuyuhiko Inagaki Kagaku no Ryoiki (J. Jpn. Chem.) 1983,141 (Supp.): 1.

[16]   Ziffer H et al. Fortschritt Chem Naturstoff, 1997, 72: 121.

[17]   Luo X D, Shen C C. Med Res Rev, 1987, 7: 29.

[18]   Qinghaosu Antimalaria Coordinating Research Group. Chin Med J, 1979, 92: 811.

[19]   Zhu D et al Aeta Pharmaceutica Sinica, 1980, 15: 509.

[20]   Zhu D et al Aeta Pharmacologica Sinica, 1983, 4: 194.

[21]   Gu H M et al. Trans R Soc Trop Med Hyg, 1984, 78: 265.

[22]   Chin Cooperative Research Group on Quinhaosu and Its Derivatives as Antimalarials. J Trad Chin Med, 1982, 2: 45.

[23]   Avery M A et al. J Med Chem, 1995, 38: 5038.

[24]   (a) Lin A Getal. J Med Chem, 1989, 32: 1247.
       (b) Lin A Getal. J Med Chem, 1995, 38: 764.

[25]   Avery M A et al. J Med Chem, 1996, 39: 2900.

[26]   Avery M A et al. J Med Chem, 1996, 39: 4149.

[27]   Ma J et al. Bioorg. Med Chem Lett, 2000, 10: 1601.

[28]   Lin A J et al. J Med Chem, 1990, 33: 2610.

[29]   Marco J A et al. J Nat Prod, 1994, 57: 939.

[30]   Liang X. et al Acta Chemica Sinica, 1979, 37: 215.

[31]   (a) Casteel D A. Antimalarial Agents. In: Wolf M E (Ed). Burger's Medicinal Chem istry and Drug Discovery. New York: Wiley, 1997.
       (b) Hofheinz W et al. Trop Med Parasitol, 1994, 45: 261.

[32]   Jakopovic J et al. Phytochem, 1987, 26: 2777.

[33]   Kawabata J et al. Phytochem, 1998, 47: 231.

[34]   Roll D M et al. J Antibiot, 1998, 51: 635.

[35]   Abbanal D A et al. J Antibiot, 1998, 51: 708.

[36]   Fuchino H et al. Chem Pharm Bull, 1997, 45: 1101.

[37]   Kunesch N et al. Tetrahedron Lett, 1985, 26: 4937.

[38]   (a) Lacey E et al. PCT Int. Appl.,WO 1998, 9809968.
       (b) Lacey E et al. PCT Int. Appl., CA 1998, 128, 217535.

[39]   Chen C Z et al. Phytochem, 1997, 45: 765.

[40]   Patit A D et al. J Nat Prod, 1997, 60: 507.

[41]   Fukami A et al. Trahedron Lett, 1997, 38: 1201.

[42]   Hashimoto T et al. Tetrahedron Lett, 1994, 35: 4787.

[43]   Shigemori H et al. Tetrahedron, 1994, 50: 8347.

Zhong-Liang CHEN

CHAPTER 6

Diterpenes

Diterpenes are composed of C_{20} hydrocarbons and have a general molecular formula $C_{20}H_{32}$. Diterpenes are distributed widely in nature (particularly in plants) and are frequently present in the families of Labiatae, Verbenaceae, Celastraceae, Euphorbiaceae, Thymelaeaceae as well as many genera such as *Rhododendron* and *Taxus*. In addition to plants and fungi, diterpenes are also found in many marine and insect species. According to "The Combined Chemical Dictionary," more than 10,000 diterpenoids of over 100 skeletons have been characterized with new diterpene-derived frameworks emerging at an accelerating speed. Due to modern spectroscopic techniques, diterpene related research has greatly advanced with many complicated structures determined successfully in a short amount of time. More importantly, in-depth bioactivity research has demonstrated that some diterpenoids are valuable for treating severe diseases. For example, taxol showed significant cytotoxic and antileukemic actions. It was approved as an anticancer drug for treating ovarian, breast, lung and squamous cancers. Taxol has become one of the most prescribed cancer-treating medicines. Andrographolides, isolated from *Andrographis paniculata*, exhibited anti-inflammatory and anti-bacterial activities to permit applications for treating gastritis, enteritis and pneumonia. It is anticipated that the utilization of diterpene compounds in medicine is still at its early stage, as more promising bioactivities of diterpenes need to be explored. The main diterpene skeletons are illustrated below[1].

Section 1   Main Diterpene Skeletons

Acyclic and mono-cyclic diterpenes

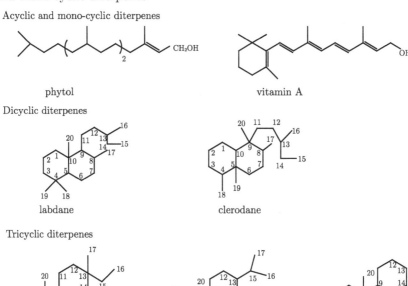

Acyclic and mono-cyclic diterpenes

phytol

vitamin A

Dicyclic diterpenes

labdane

clerodane

Tricyclic diterpenes

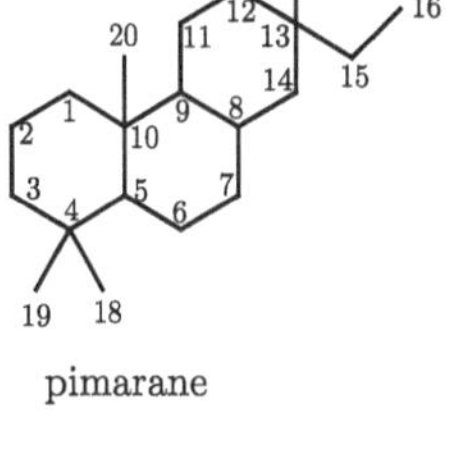

pimarane

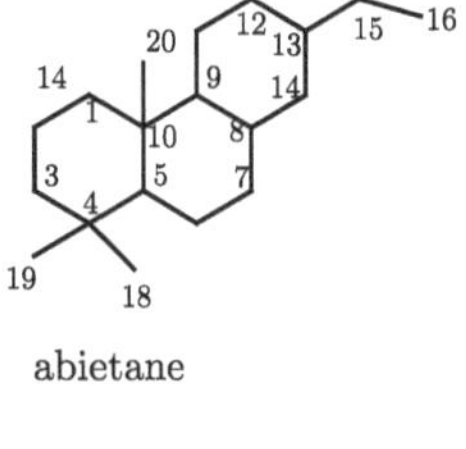

abietane

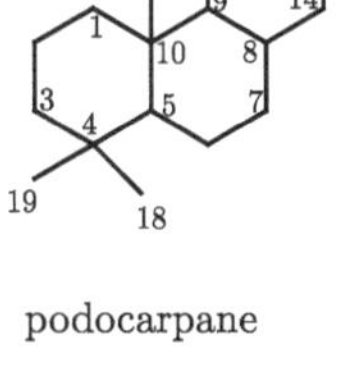

podocarpane

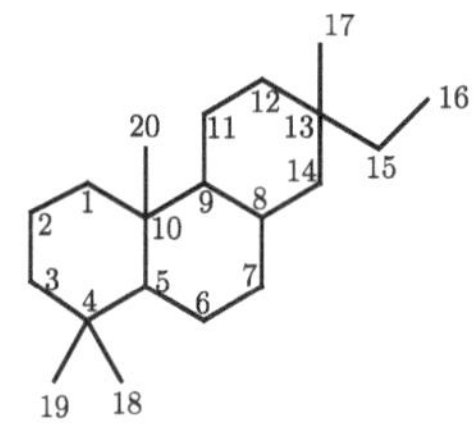

cassane

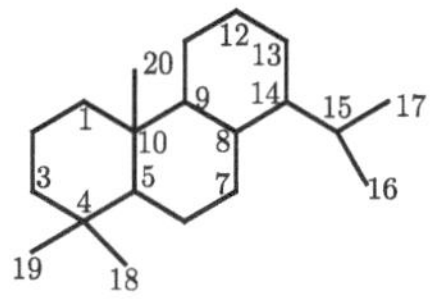

totarane

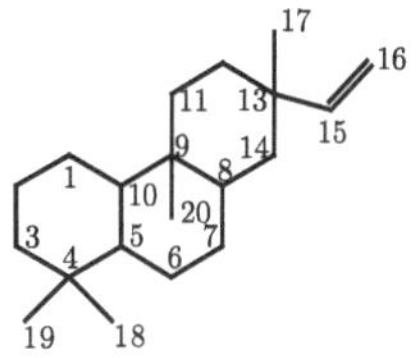

rosane

Tetracyclic diterpenes

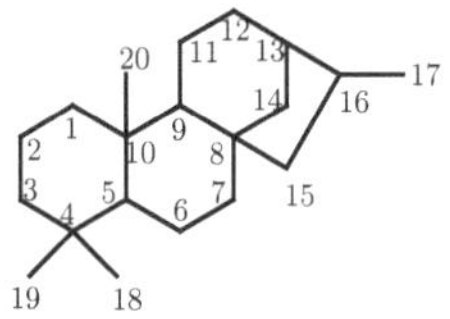

kaurane

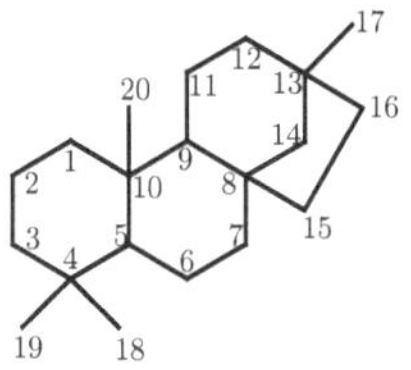

beyerane

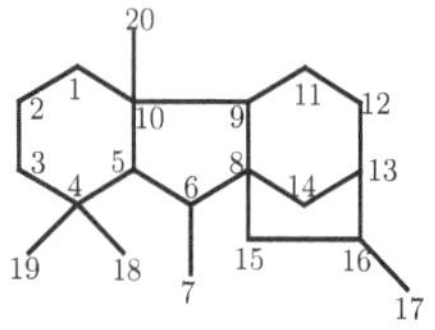

gibberellane

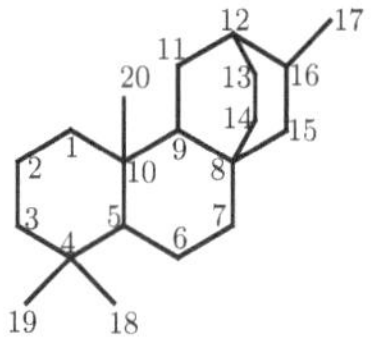

atisane

Main macrocyclic diterpene skeletons

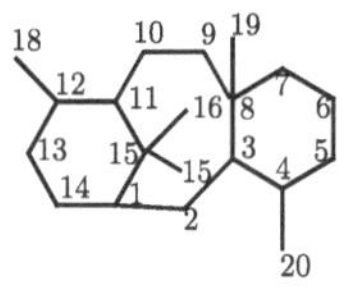

taxane

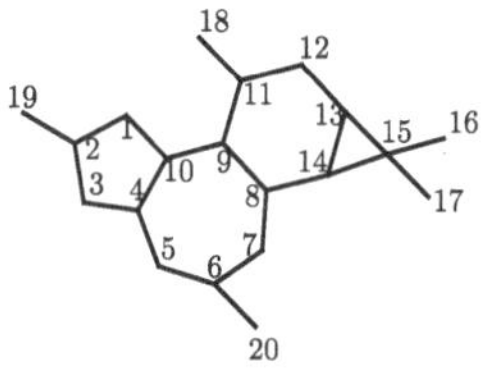

tigliane

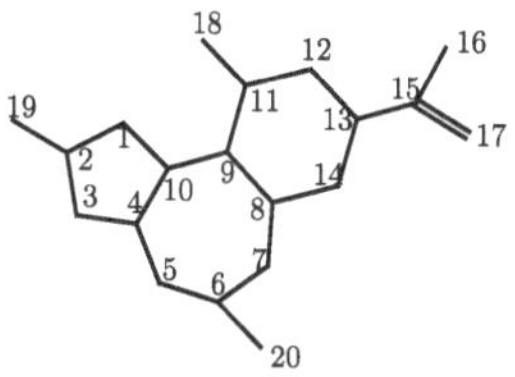

daphnane

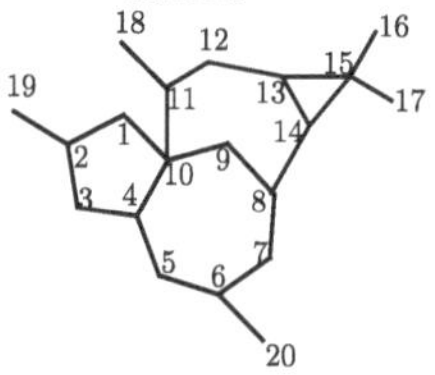

ingenane

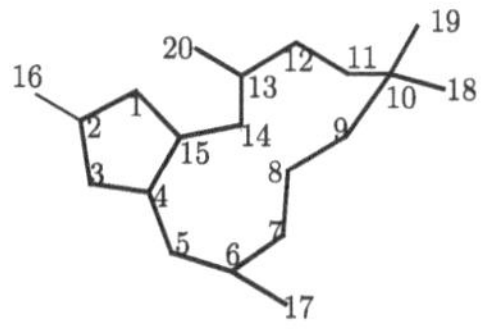

jatrophane

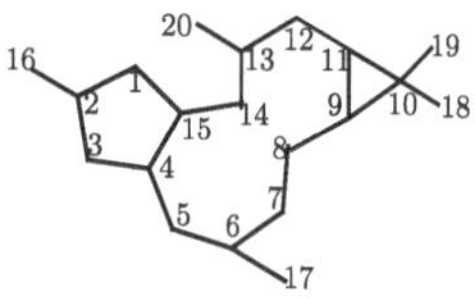

lathyrane

Section 2   Biogenesis

From the biogenetic viewpoint, the classification of diterpenes agrees well with the cycliza-tion pattern of the common precursor geranygeranyl pyrophosphate. The enzyme-driven cyclization reaction first produces labdane pyrophosphate and then forms labdane after eliminating pyrophosphate. Clerodane is produced if the first-order cyclization processes

are followed and/or accompanied by further skeletal rearrangements. Second-order cyclization of the labdane pyrophosphate intermediate allows generation of the tricyclic diterpene
pimarane, which gives abietane upon rearrangement. The tetracyclic diterpenes are thought
to be formed by cyclization of pimaradienes to yield an ionic intermediate that is prone to
"collapse" to afford diterpenes such as kaurane, atisane, etc.[2]

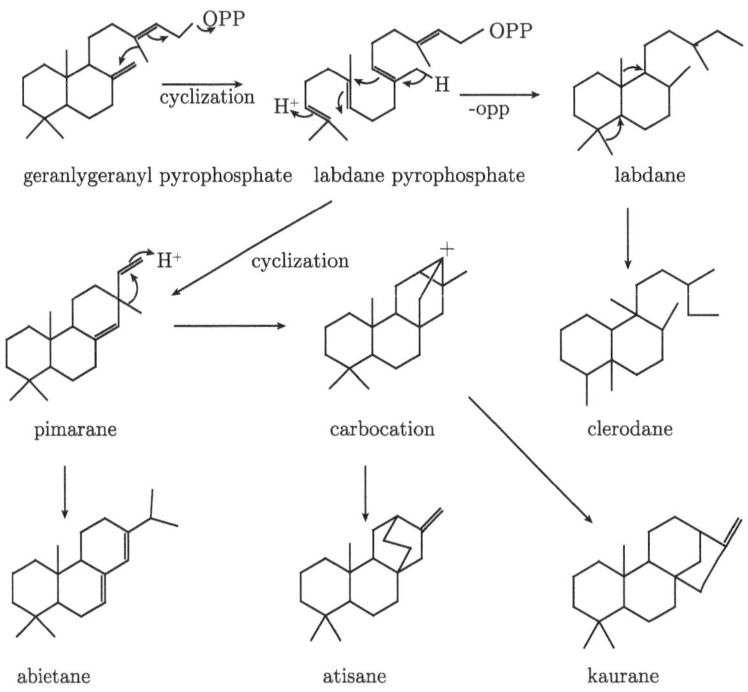

geranlygeranyl pyrophosphate   labdane pyrophosphate   labdane

pimarane   carbocation   clerodane

abietane   atisane   kaurane

Macrocyclic diterpenes are also formed from the cyclization of geranylgeranyl pyrophosphate
as exemplified by the biogenic route of casbane and jatrophane, and are shown as follows:

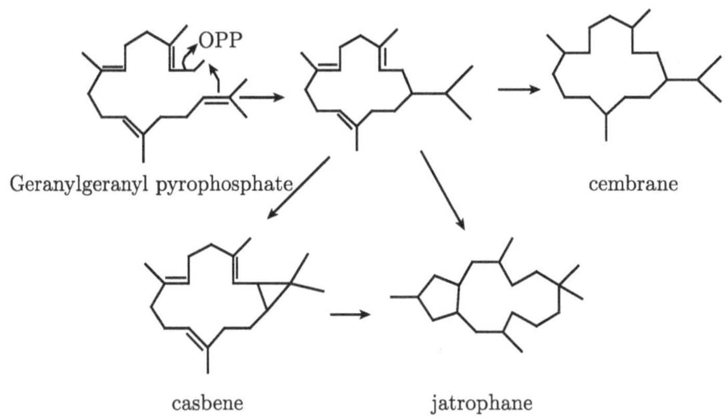

Geranylgeranyl pyrophosphate   cembrane

casbene   jatrophane

Section 3   Labdanes

The labdane represents a family of bicyclic diterpenes that are biosynthetically generated
from the enzyme-driven cyclization of the acyclic precursor geranylgeranyl pyrophosphate

(1). A key factor governing the cyclization is most presumably a proton's attack on the read-ily accessible C-14 atom of the 14,15-vinyl moiety since it leads to sequential polarizations of 10,11- and 6,7-double bonds. The polarization arouses the illustrated bond formation to finally form labdane pyrophosphate after the elimination of a H-17. It was found that the 10-methy group is always opposite to attachments on C-5 and C-9 in the normal series (10β-methyl accompanied by 5α-H and 9α-H) and *ent*-series (10α-methyl coexisting with 5β-H and 9β-H), suggesting that the linear precursor adopts one of two possible conformations at the onset of cyclization. In comformation **1** , the electrons were attached from underneath the 18-methyl group, therefore the cyclization of **1** formed normal labdane. In conformation **2**, the *ent*-labdane formed if the electrons were attached from above the 18-methyl group[3].

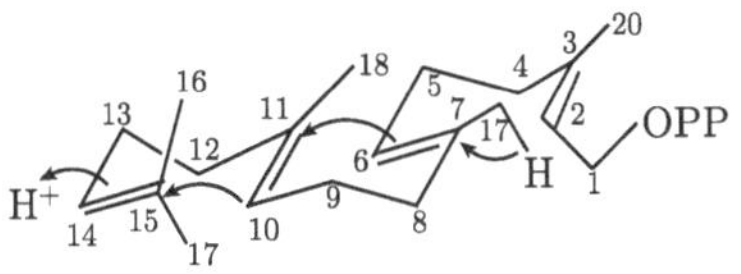

labdane pyrophosphate (**2**)

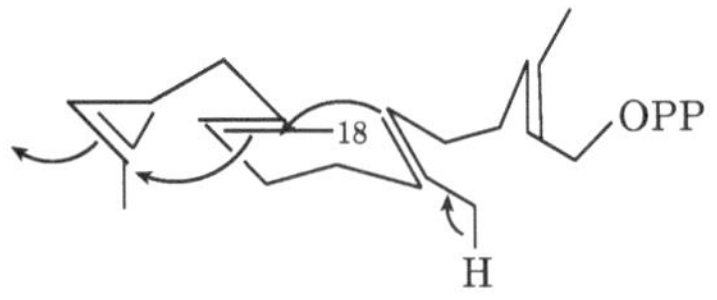

ent-labdane pyrophosphate (**4**)

This rationally explains why the labdane diterpenoids derived from geranylgeranyl py-rophosphate naturally fall into either normal or *ent* series. In literature, the labdane diter-penes belonging to the normal series have been named "netly" [without mentioning "nor-mal", e.g., 8α,13α-dihydroxyl-labda-14(15)-ene (sclareol, **5**) is a "normal" one]. However, for the *ent* series, diterpenes with an "*ent*-" symbol have to be added at an appropriate postion in their names. For example, the diterpene **6** could be named either as *ent*-8β,13β-dihydroxyl-labda-14(15)-ene or as 8α,13α-dihydroxyl-*ent*-labda-14(15)-ene, but the configu-ration of the former must be contrary to and the configuration of the latter has to be similar to the structural scheme.

As to the distribution, a number of important or interesting labdane diterpenes have been characterized from nature. Sclareol (**5**), with a 13*S*-configuration, was isolated from *Salvia sclarea* and shown to have anti-bacterial function. Its *ent* form, *ent*-sclareol (**6**), was obtained from *Gnaphalium undulatum*[4,5].

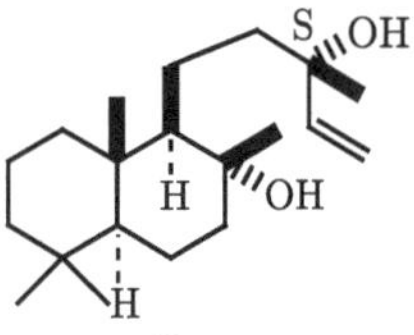

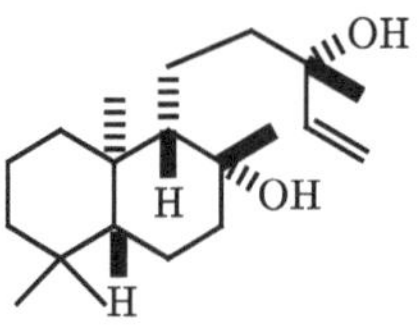

Andrographolide (**7**), obtained from *Andrographis paniculata*, displayed anti-inflammatory and anti-bacterial activities that could be of value for treating gastritis, enteritis and pneumonia[6]. A highly oxygenated labdane forskolin (**8**), isolated from *Coleus forskohlii*, was demonstrated to possess anti-hypochondria, cardiotonic, and antihypertensive actions. Its possible biogenic route could be written as follows:[7]

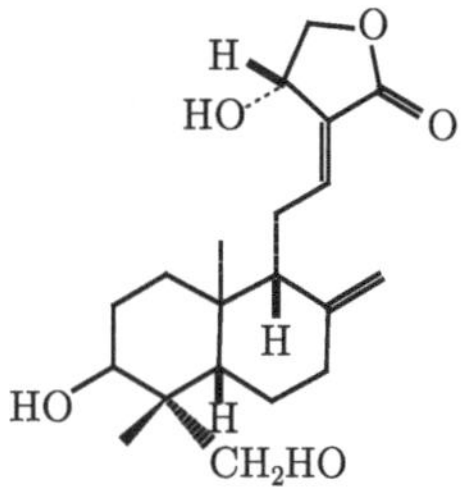

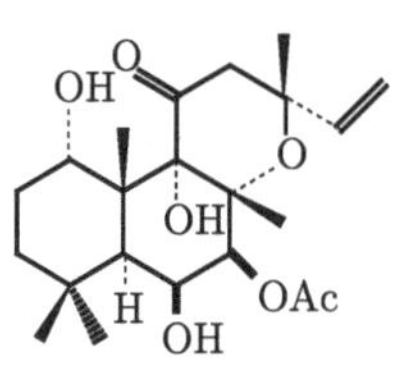

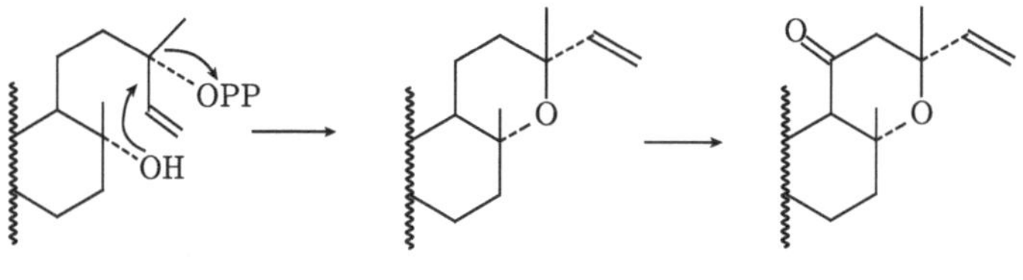

The NMR data of andrographolide and forskolin are given below, since they could be of reference value:

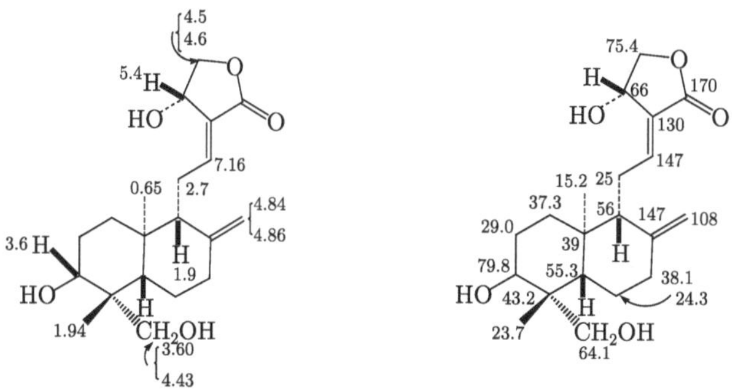

^{1}H and ^{13}C NMR data of andrographolide (pyridine-d_5)

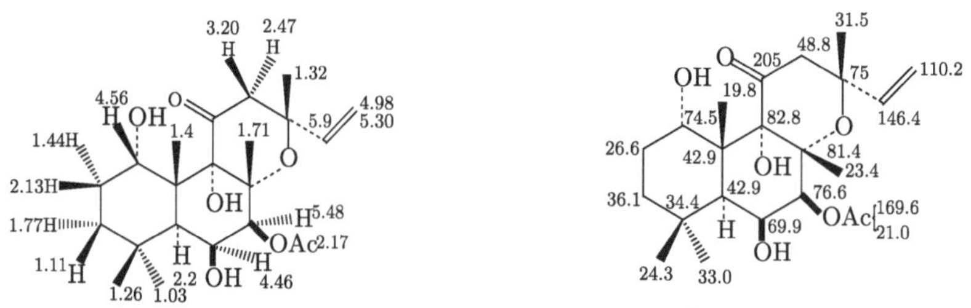

^{1}H and ^{13}C NMR data of forskolin (CDCl$_3$)

A number of other types of diterpenes have been characterized as rearranged labdanes. For example, pallavicinia 3 (**9**), from *Pallavicinia subuciliata*, belongs to abeo8(7→2) labdane[8]. The 5,10-seco labdane diterpene, seconidorella lactone (**10**), was reported to be from *Nidorella hottentotica*[9]. The nor-labdanes, such as agallochin (**11**)[10], and cyclic labdanes, like 14,17-cyclo-8,13 (16)-labdadien-19-ol (**12**), were isolated from *Excoecania agallocha* and *Vellogia flavicans* respectively[11].

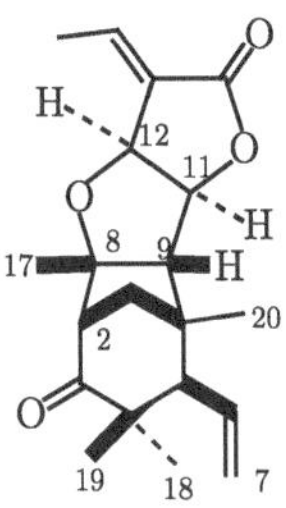

9 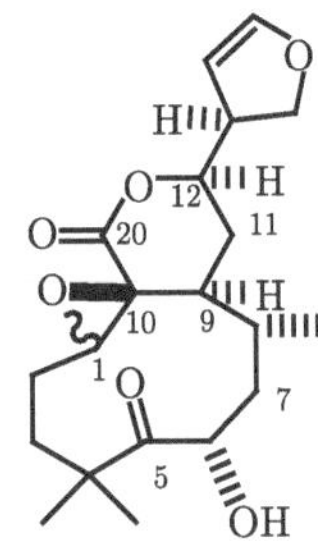   **10** 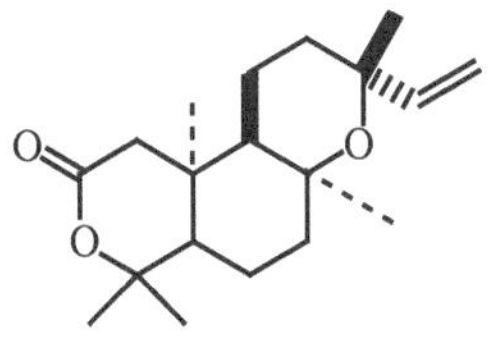   **11** 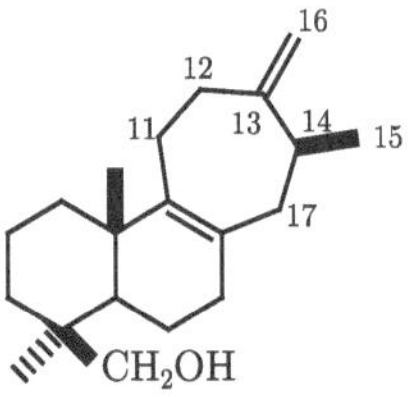   **12**

Section 4   Clerodanes

The clerodanes suggest that they are derived from the rearrangement of labdane, which was

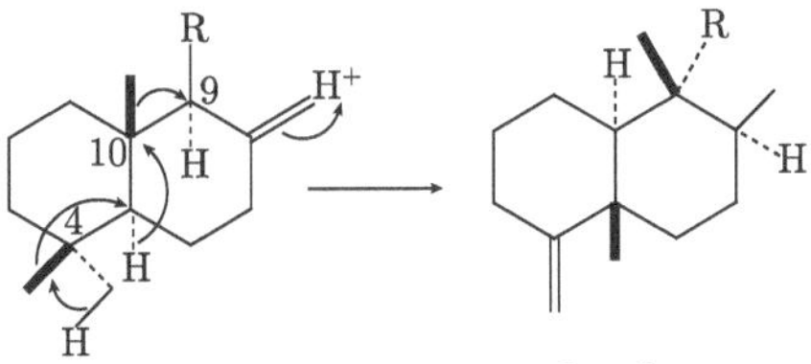

labdane   clerodane

intiated by protonation of the C-8 double bond and a hydride shift from C-9 to C-8 followed by a methyl migration from C-10 to C-9. A further hydride shift from C-5 to C-10 and a methyl migration from C-4 to C-5 finished the rearrangement. The stereochemistry of clerodane has been related to the absolute configuration of clerodin (**13**), which was established by x-ray[12]. Comprision of the configuration of clerodane with compound **13** has a same configuration as neo-clerodane (normal series) and their configuration revealed the opposite which was attributed to *ent*-neo-clerodane (*ent*-series)[2].

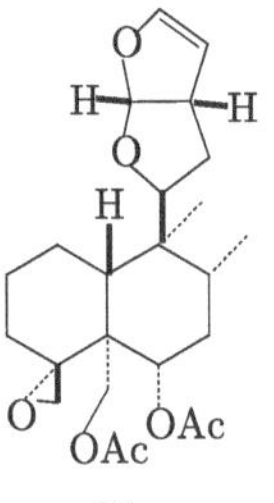

13 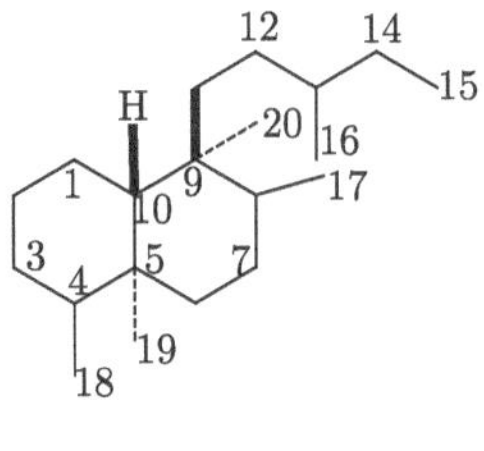   neo- clerodane 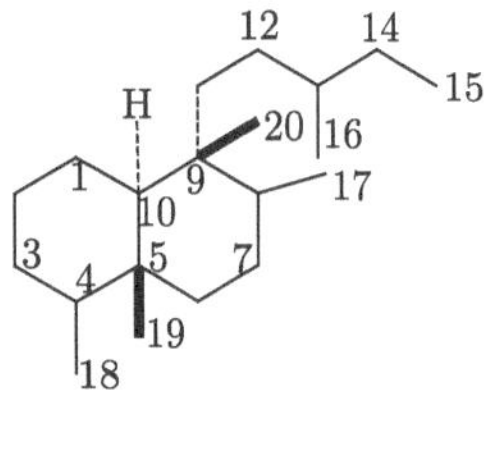   *ent*-neo-clerodane

The clerodanes were usually found in Labiatae plants, which displayed insect antifeedant activity. Many neo-clerodanes were isolated from *Ajuga decumben* as ajugacumbins A through G (**14-18**)[13].

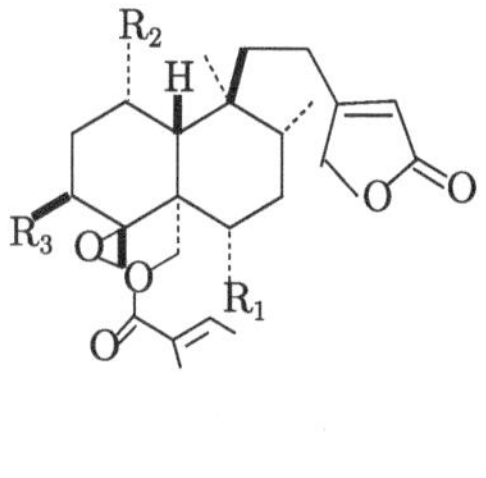

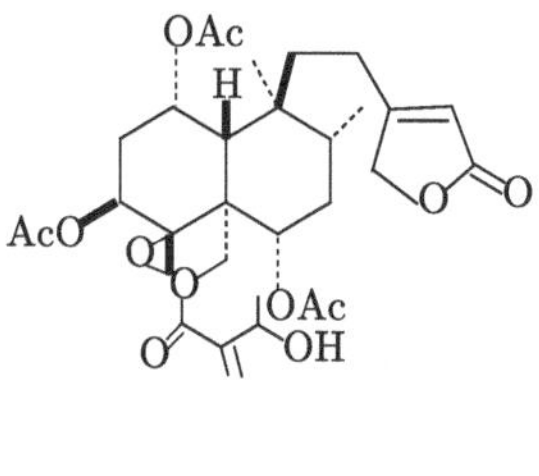

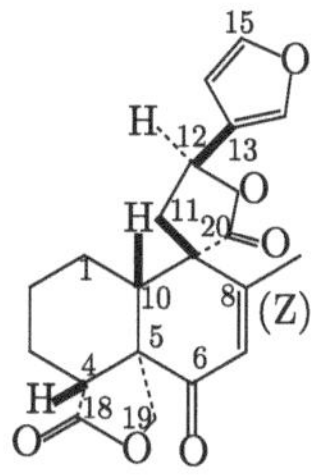

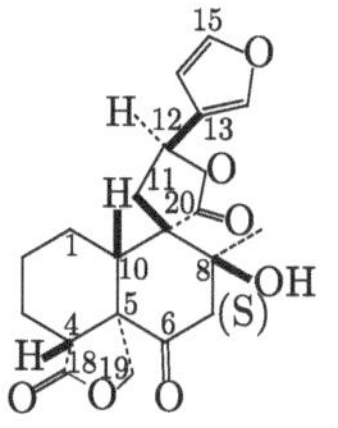

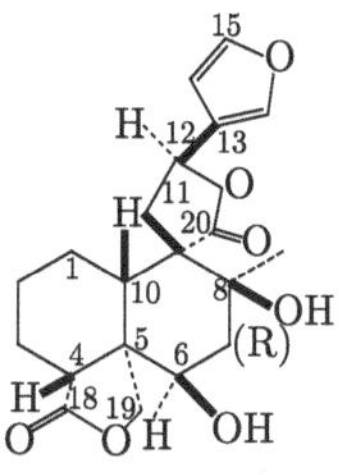

14 $R_1 = OAc$, $R_2 = R_3 = H$ **18**   **19**   **20**   **21**
15 $R_1 = OH$, $R_2 = R_3 = H$
16 $R_1 = R_2 = R_3 = OAc$
17 $R_1 = OAc$, $R_2 = H$, $R_3 = OH$

Teupernins A, B and C (**19-21**) were isolated from *Teucrium pernyi*, which belongs to the 18,19-olide of neo-clerodane[14]. Teaponin (**22**) and teupernin D (**23**) were obtained from *Teucrium japonicum*[15].

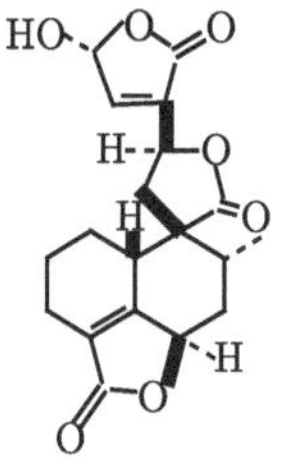

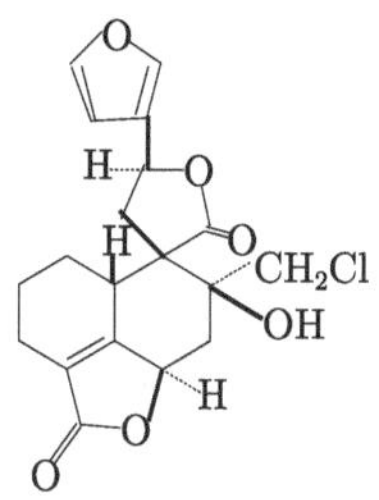

22   **23**

Clerodin (11,16:15,16-diexpoxy clerodane, **13**) is a model reference compound of clerodane and extracted from *Clerodendrum infortunatum*. The brominated derivative of **13** was subjected to x-ray analysis to determine the absolute configuration[12]. Successive separation of the same type of diterpenes, such as ajugavensin (**24**), was achieved from *Ajuga genevensis*[16]. Scuterepenin F_1(**25**) and F_2 (**26**) were obtained from *Scutellaria repens*[17]. The clerodanes bearing cycloptopane, such as blepharolide A (**27**), and containing a seven-cyclic ring such as blepharolide B (**28**), were isolated from *Salvia blepharophylla*.[18]

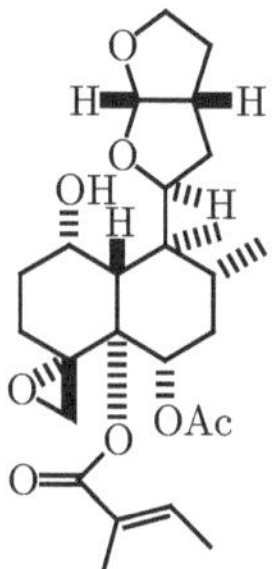

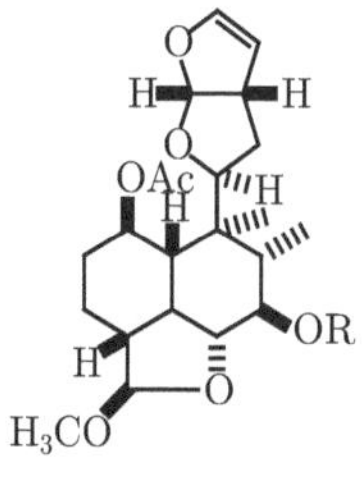

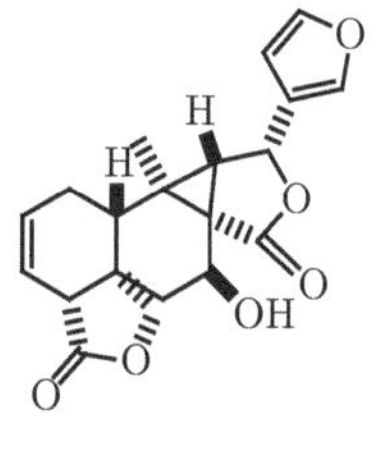

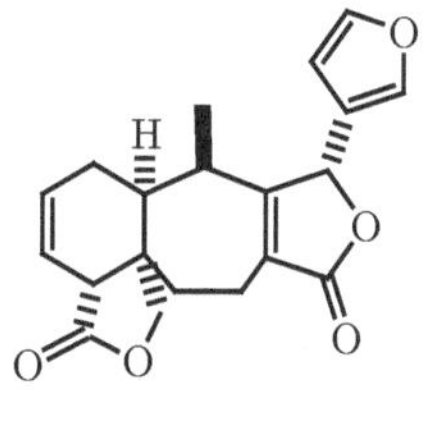

24   **25** R = cinnamoyl (E-)   **27**   **28**
                    26 R = cinnamoyl (Z-)

Section 5   Pimaranes

The pimaranes are tricyclic diterpenes which can be found in the resin of pine trees. The pimaranes are formed through the cyclization of the labdane pyrophosphate. The pimarane skeleton consists of the hydrophenanthrine. The epimers at C-13 exist as the two epimeric forms, e.g. pimarane and iso-pimarane, of which the configuration of 13-vinyl and 13-methyl group are β and α- oriented and assigned as the pimarane or *ent*-pimarane. On the contrary, the vinyl and methyl groups are α and β-configurations assigned as the isopimarane or *ent*-isopimarane[2].

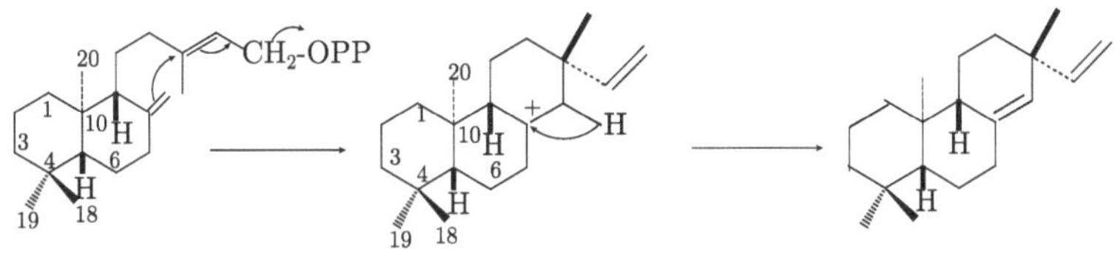

labdane pyrophosphate   pimarane

Examples include pimaric acid (**29**) and oryzalexins (**30**), which belong to pimarane type diterpenes. Isopimaric acid (**31**), sandaracopimaric acid (**32**), and compound **33** are assigned as the isopimarane type diterpenes. Compounds **29, 31** and **32** are normal series. Compound **30** and **33** are *ent* series. Compounds **34** and **35** are characterized as 9-epi-*ent*-pimarane[19].

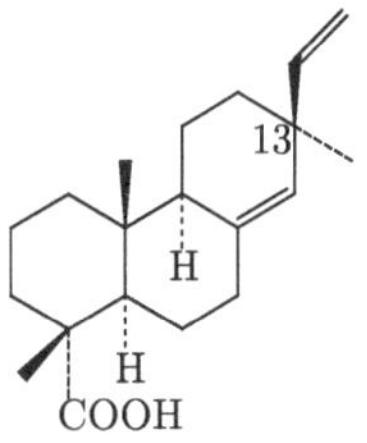

29 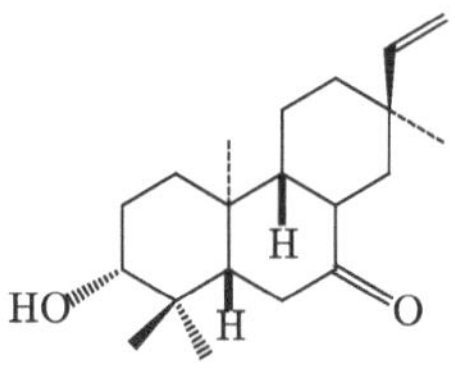   **30** 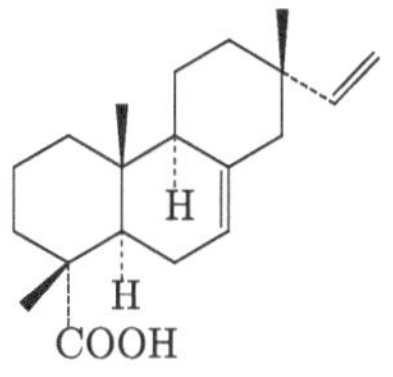   **31**

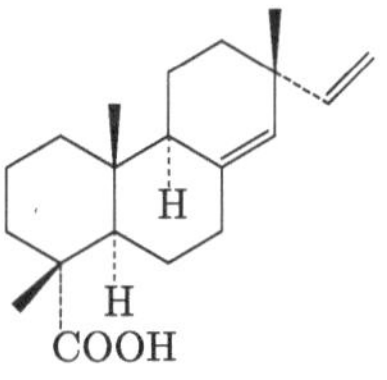

32 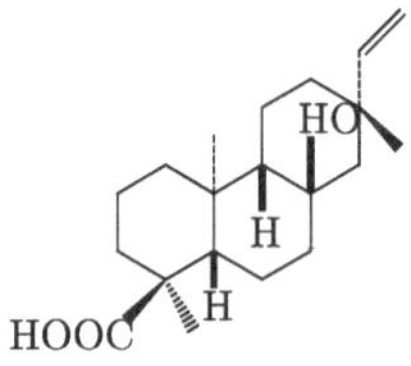   **33** 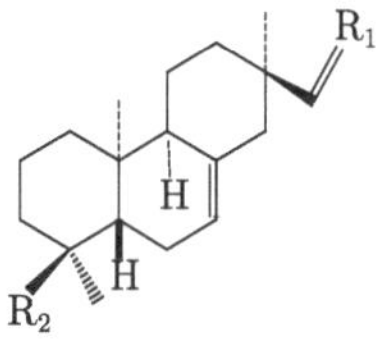   **34** R$_1$=CH$_2$OAc, R$_2$=Me
                                                                35 R$_1$=Me, R$_2$=CH$_2$OH

Section 6   Abietanes

The abietane is presumed to be formed by rearrangement of the pimarane. The rearrangement reaction starts with protonation of the C-15 double bond to produce 16-methyl followed by the 17-methyl group migration from C-13 to C-15 to form an isopropanyl group at C-13.

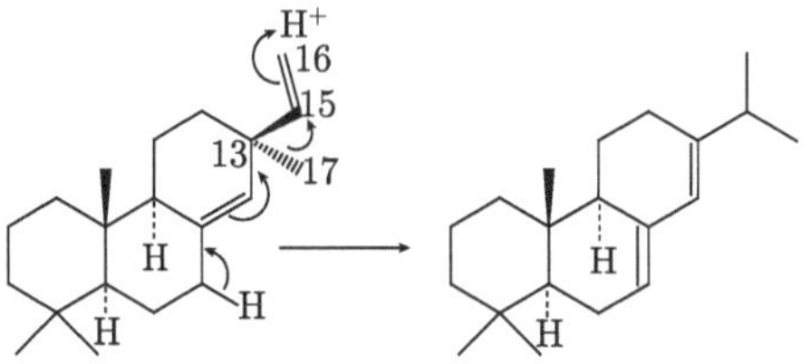

pimarane   abietane

In nature, abietanes react with various degrees of oxidation and hydrogenation to produce many new abietanes, which show various physiology activities.

The simple structural abietanes include abietic acid (**36**), from the resin of *Pinus* spp., and ferruginol (**37**), from *Podocarpus ferrugineus*.

Ring A experiences the rearrangement reaction as that of triptolide (**38**) and its many derivatives from *Tripterygium wilfordii*, which display strong anti-inflammatory and decreased immunity function activity. The triptolide has a LD_{50} value of 0.82 mg/kg (rat /intravenous injection) and was used for the treatment of rheumatoid arthritis, lupus erythematosus, and nephritis in clinical trials[20]. The rearrangement reaction is illustrated below:

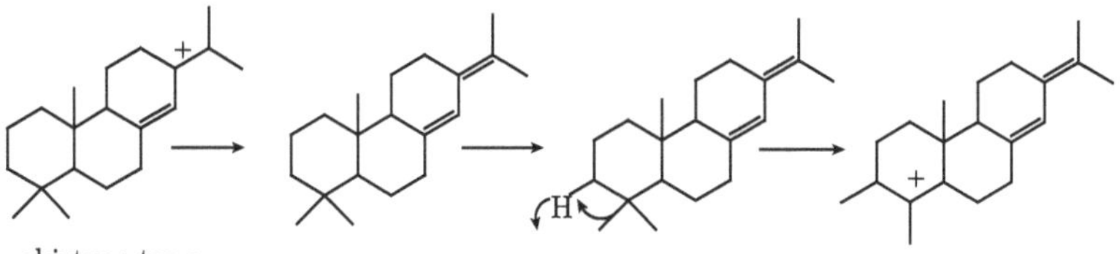

abietane type

The ring C, bearing the α,β-unsaturated-γ-lactone diterpenes, such as Yuexiandajisu D (**39**), E (**40**) and F (**41**), were obtained from *Euphorbia ebracteolata*[21].

The C ring of the abietanes was substituted onto the *o*-quinone or *p*-quinone as tanshinone II_A(**42**) and isotanshinone (**43**), which can be isolated from *Salvia miltorrhiza* and used to treat coronary disease[22].

Ring C undergoes aromatization and rearrangement of the abietanes isolated from *Clerodendrum mandrainorum*[23].

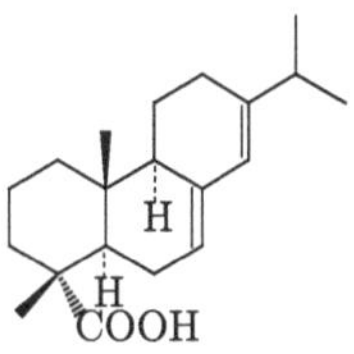

36 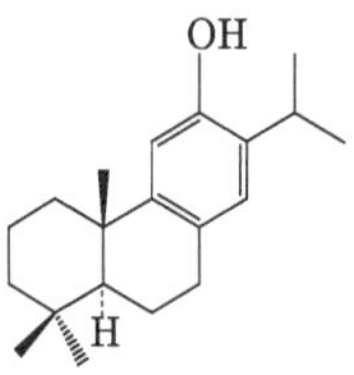   37 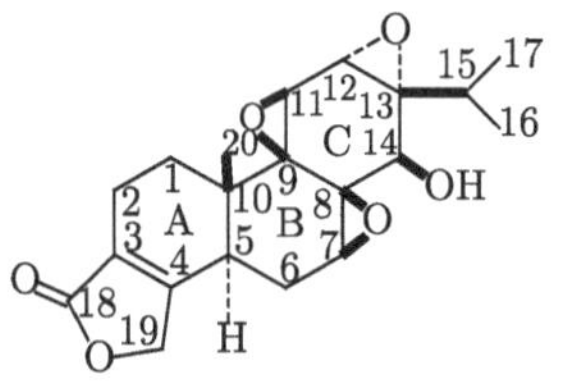   38 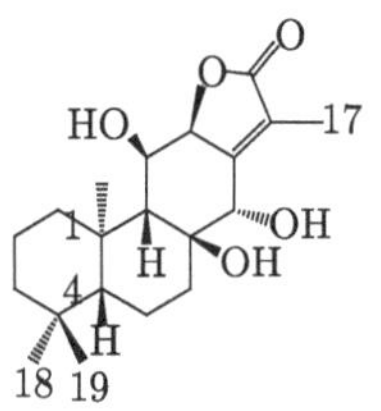   39

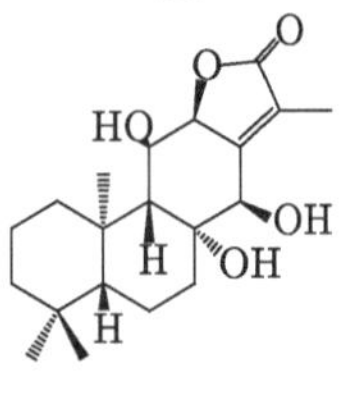

40 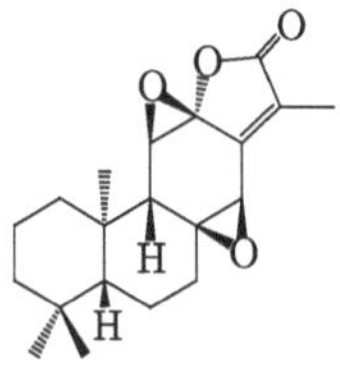   41 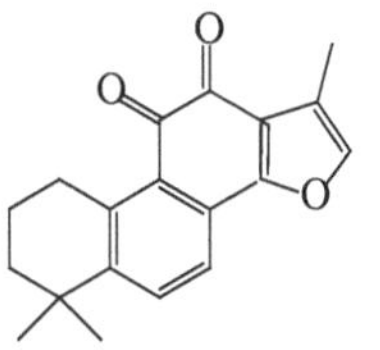   42 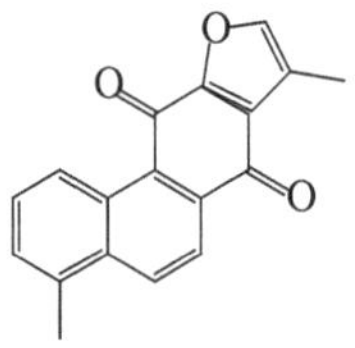   43

The rearrangement products of abietanes were reported in the literature. The abeo9(10→20) abietane, named icetexane, has resulted in over 40 derivatives. Taxamairin A (**44**) and B (**45**) were isolated from *Taxus chinensis* and showed inhibited hepatoma cell activity (IC$_{50}$ of compound 40 = 30.21μg/mL: IC$_{50}$ of compound 41 = 26.78 μg/mL)[24]. The rearrangement reaction is described below:

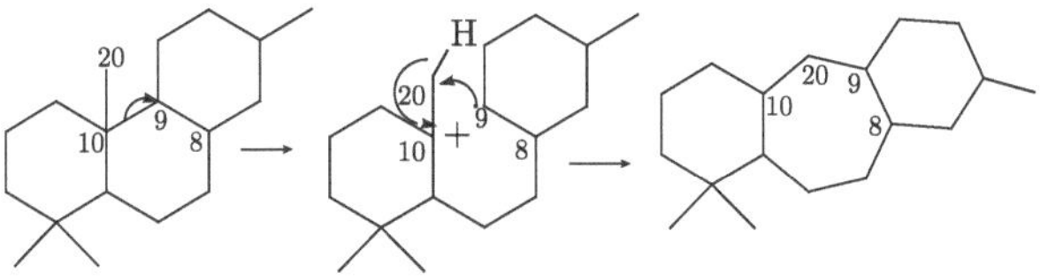

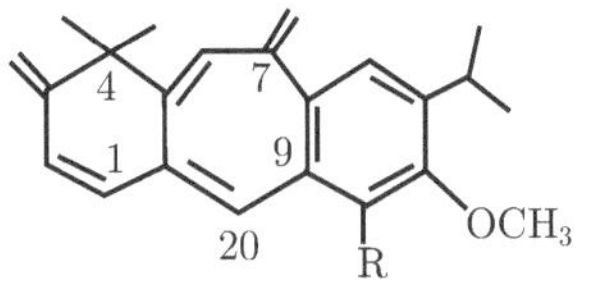

44 R=OCH$_3$
45 R=OH

The abeo9(8→7) abietane, larikaemferic acid (**46**), was obtained from *Laix kaempfen*[25]. The abeo 5(6→7) abietanes, such as taiwaniaquinol A (**47**) and taiwaniadduci A (**48**), were obtained from *Taiwania crytomerioides*[26]. The abeo20(10→5) abietanes, such as abeo20(10→5)-1(10),6,8,11,13-abietapentaene-11,12,16-triol (**49**),were isolated from *Salvia apiana*[27]. The triabeo14(8→7),11(9→8), 20(10 →9)abietane, such as aegyptinone (**50**), was isolated from *Salvia aegyptiaca*[28].

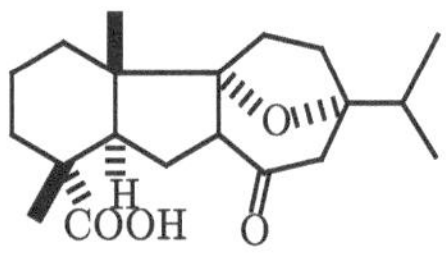

46 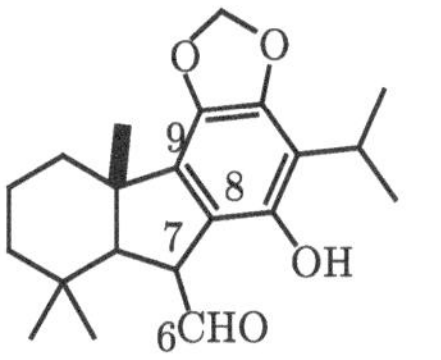   47 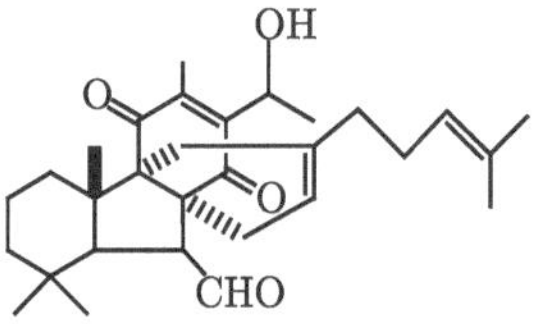   48

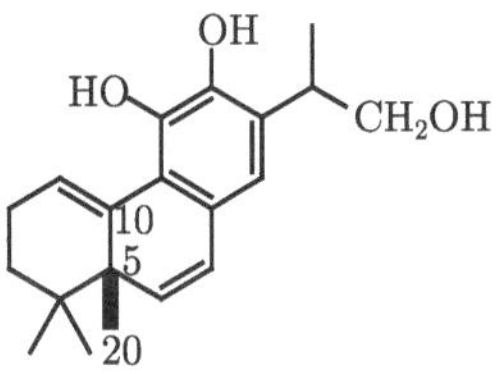

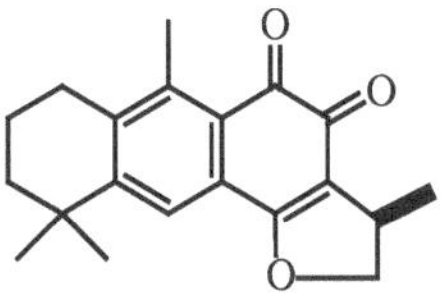

Section 7   Cassanes and Totaranes

7.1   Cassanes

Cassanes are derived from the migration of the 13-methyl group to C-14 of the pimarane. The migration route is shown below.

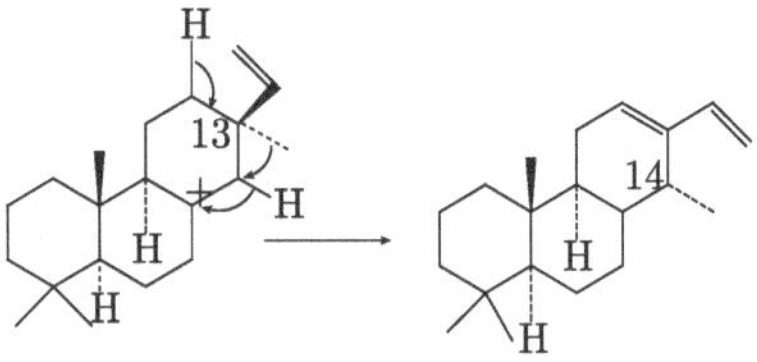

pimarane   cassane

Cassaine (**51**) and cassaidine (**52**) were isolated from *Erthrophleum guineenise, E. ivorense, E. suaveolens and E.couminga*, which show strong cardiac simulant, antihyptertensive and local anaesthetic activities[29]. The cassanes containing a furan ring, for example ε-caesalpin (**53**), were obtained from the seed of *Caesalpinia bonducella*. The nor-cassanes, such as nor-caesalpinin A (**54**) and B (**55**), were isolated from *Caesalpinia crista*[30].

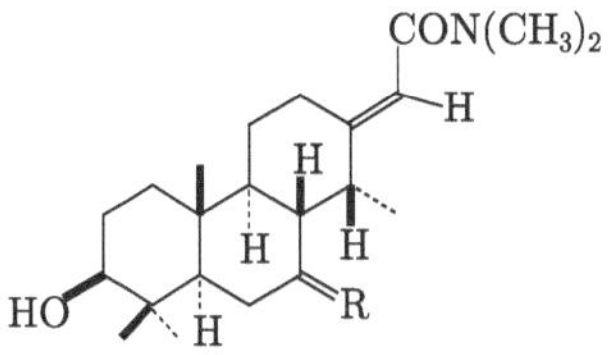

51 R=OH, H
52 R=O

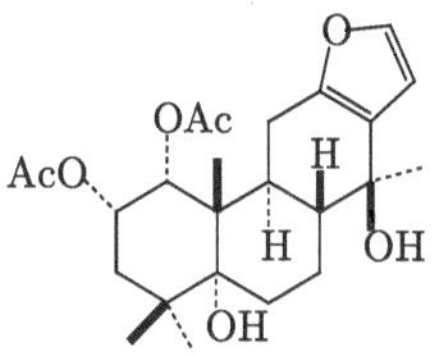

53

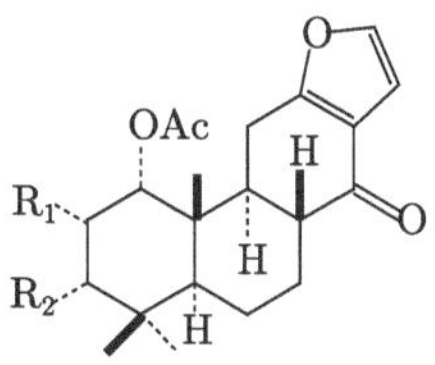

54 R₁=OAc, R₂=H
55 R₁=H, R₂=OAc

7.2   Totaranes

The totarane has an isopropanyl group at C-14 which is derived from the migration of the 13-isopropanyl group of the abietane. For the sake of elucidation of the configuration of totarol (**56**), the Clemmensen reaction of totarolone (**57**) produces totarol (**56**). The Cotton effect of compound **57** is identified with the lanost-8-en-3-one. Above results confirm that totarol belongs to normal series. Totarol was isolated from *Podocarpus totara*, and totarolone from *P. nagi*; alilactone C (**58**), from *P. nagi*, and inunakilactone (**59**), from *P. macrophyllus*, also belong to totarane[31].

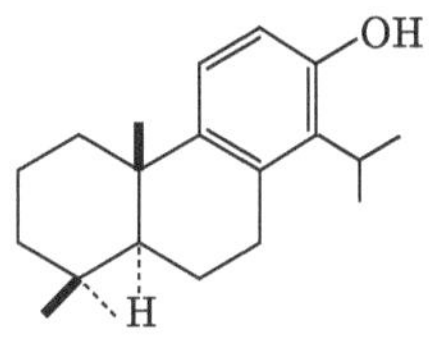

56 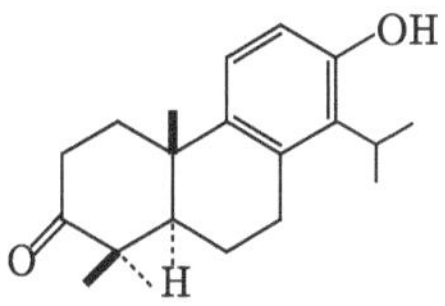   **57** 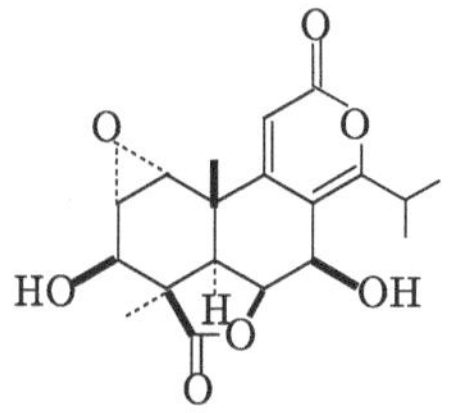   **58** 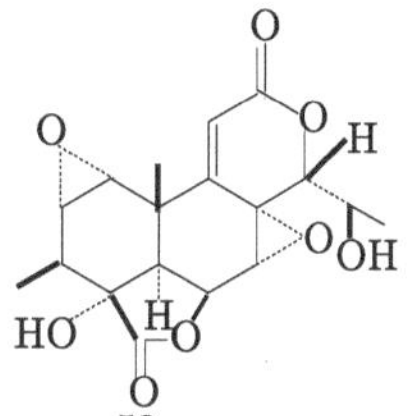   **59**

Section 8   Rosanes

Rosanes are believed to be formed by rearrangement of the 20-methyl group to C-9 of pimarane. The reaction is described below.

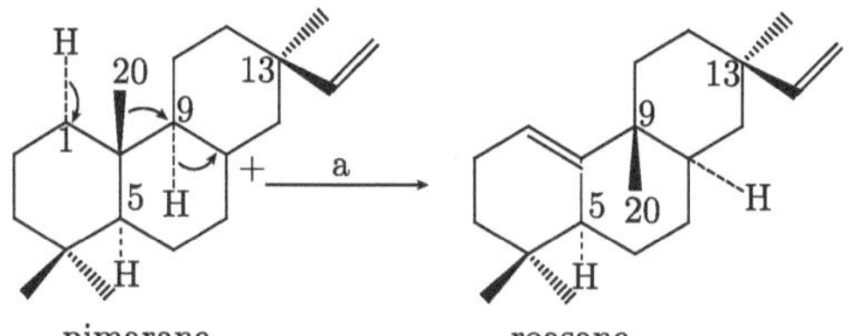

pimarane   roasane

The 20-methyl group of rasane placed at C-9, which is β-oriented, is assigned as normal series and the 20-methyl group, which is α-oriented, is named as *ent* series.

6,10-Dihydroxy-15-rosen-19-oic acid (**59**) and 7,10-dihydroxy-15-rosen-19-oic acid (**60**) were obtained from *Trichothecium roseum*,[32] and belong to normal series. The sagittine A (**61**) B (**62**) and C (**63**) were isolated from *Sagittaria sagittifolia* which, according to the biogenic relationship, are between pimarane and rosane. The structures of sagittines should be characterized as *ent*-iso-rosane type[33,34].

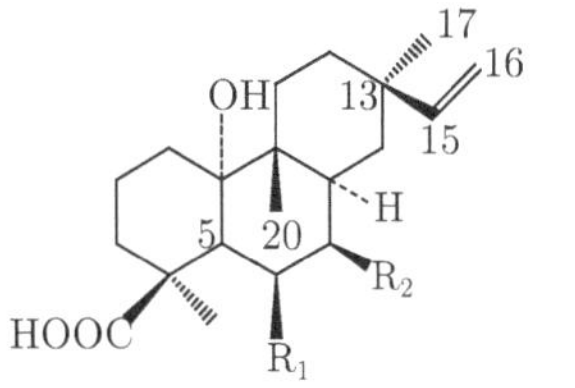

59 R₁ = OH, R₂ = H   **61** 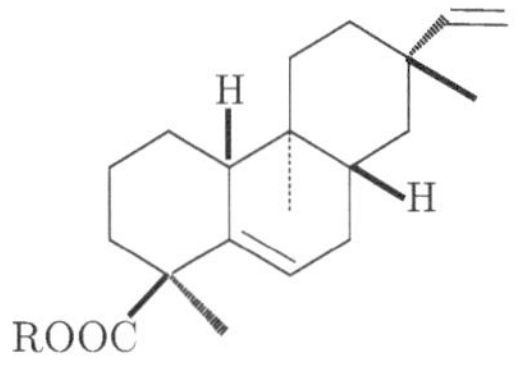   **62** R = arabinofuranosyl
60 R₁ = H, R₂ = OH   **63** R = 5'-acetoxyarabinofuranosyl

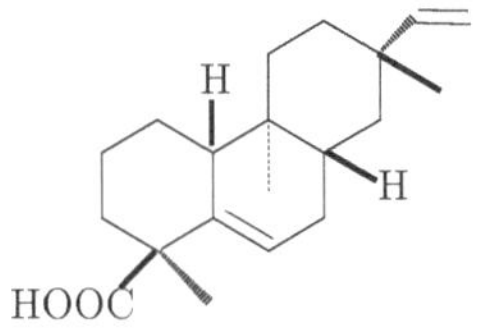

Section 9   Kauranes

The tetracyclic diterpenes are thought to be formed by cyclization of pimaradienes to an intermediate non-classical carbocation to create diterpenes such as kaurane. The dehydrogenation reaction of kaurane with Se resulted in the formation of retene and pimarthrene derivatives, which confirmed the hydrophenanthrine skeleton of kaurane. The kaurane structures were characterized as **64** and **65**. Compound **64** consisted of a *trans*fused A/B ring, B/C and C/D were *cis*-fused ring junctions, the C-20 was a β-configuration, and the 5-H and D ring were α-oriented, respectively. These types of kauranes belong to normal kaurane. Compound **65** consisted of a *trans*-fused A/B ring, B/C and C/D were *cis*-fused at the ring junction, C-20 was α-oriented, and the 5-H and D rings were β-oriented, respectively. These types of kauranes are assigned as *ent*-kauranes[35].

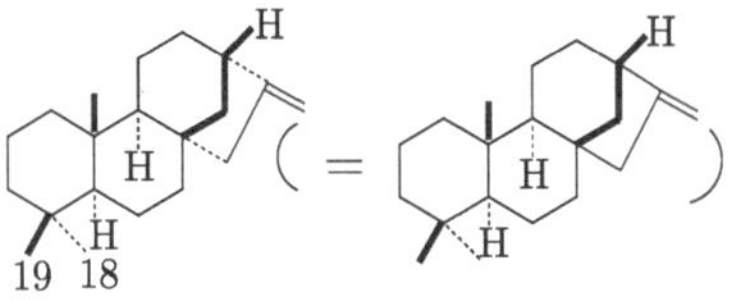

kaurane(64)[= (+)-kaurene] 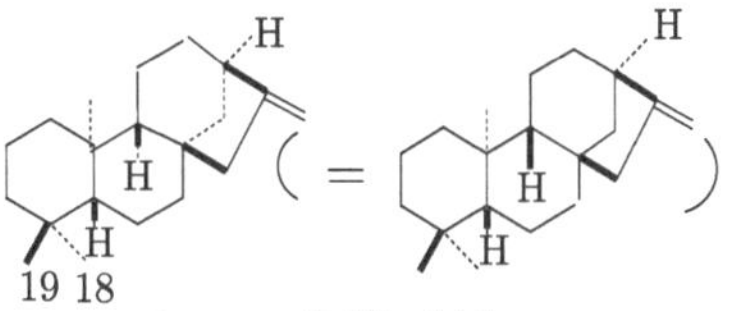   *ent*-kaurane(65)[= (-)-kaurene]

The *ent*-kauranes were extensively studied and many of them were isolated mainly from *Rabdosia* plants. Dehydrogenation of the C-17 methyl group from the *ent*-kauranes forms exocyclic methylenes at C-16 (named *ent*-kaurenes).

9.1   C-20-non-oxygenated-*ent*-kauranes[35]

These types of *ent*-kauranes are distributed extensively in *Rabdosia* plants, and include over 150 compounds. For example, kamebanin (**66**) was isolated from *Isodon kameba* and lushanrubescensin A (**67**) from *Rabdosia rubescens*.

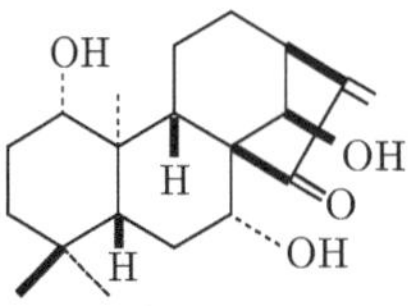

66 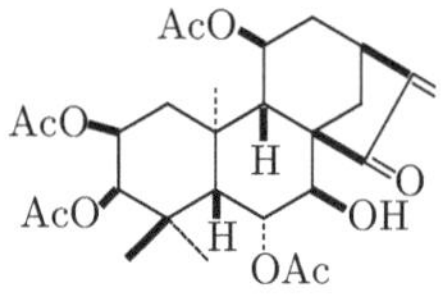   **67**

The very sweet glucoside stevioside has continued to attract attention as a sugar substitute and is extensively used as a food or beverage sweetener for diabetes or hypertension patients. The commercial product sterioside is an extract of *Stevia rebaudiaua*. The main component is the sterioside (**68**). Recently, a new diterpene, rebaudioside A (**69**), has been isolated from the same plant. It is sweeter than stevioside and the taste is similar to that of sugar[36].

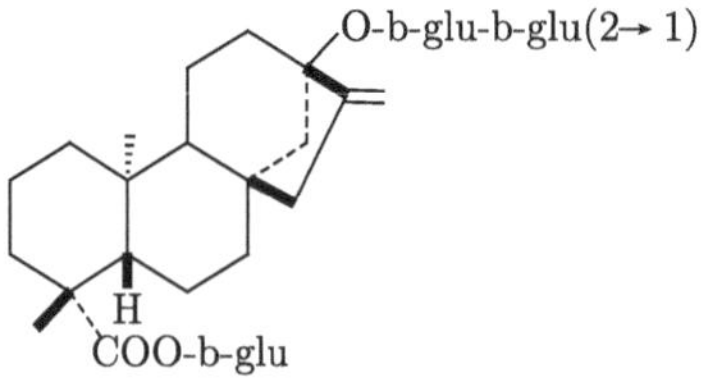

68 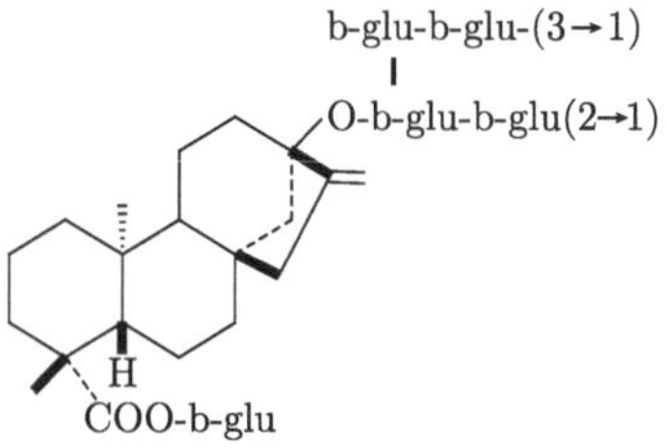   **69**

9.2   C-20 oxygenated-*ent*-kauranes[35]

The structures of C-20 oxygenated-*ent*-kaurane were derived from the oxygenation reaction of C-20 non-oxygenated-*ent*-kauranes, followed by semiacetal or an oxygen bridge formed between C-20 and C-7. Examples of this group include oridonin (**70**), ponicidin (**71**) and rubescensin C (**72**), which can be isolated from *Isodon rubescens*. The content of **70** was reported as 0.34% and **71** as 0.05% in the plant. Compounds **70** and **71** show anti-cancer activity.

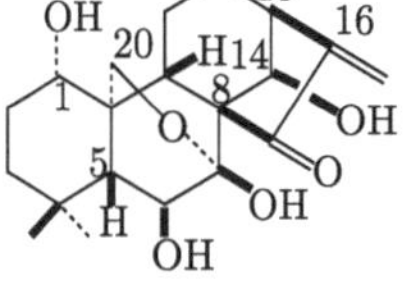

70 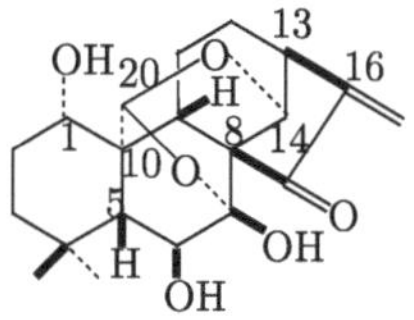   **71** 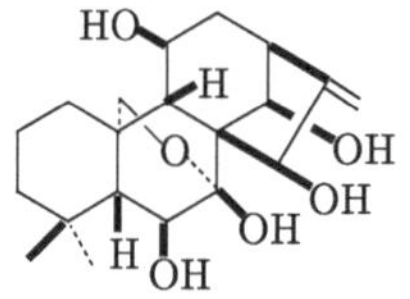   **72**

9.3   Seco-kauranes[35]

Examples of 6,7-seco-*ent*-kauranes include enmein (**73**), obtained from *I. trichocarpus*, and rabdosin A and B (**74** and **75**), isolated from *R. japonica*. 8,9-Seco-*ent*- kauranes, such as epoxyshikoccin (**76**) and shikodomedin (**77**), were obtained from *R. shikokiana*. Spiro-lactone-*ent*-kauranes, such as ludongnin A (**78**) and ludongnin B (**79**), were isolated from *I. rubescens*.

Other types of tetracyclic diterpenes, such as beyerane, gibberellane, atisane, helifulvane, trachylobane, aphidicolane leucothol and grayanotoxane, may be produced from the rearrangement reaction of kaurane and are shown as follows:

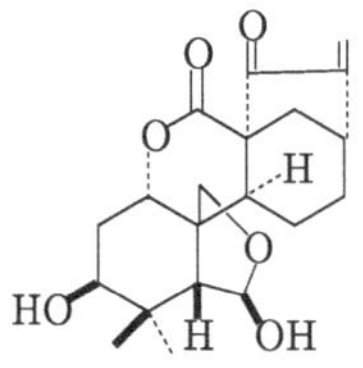

73 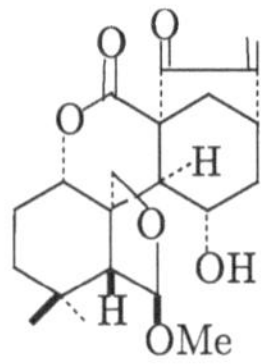   **74** 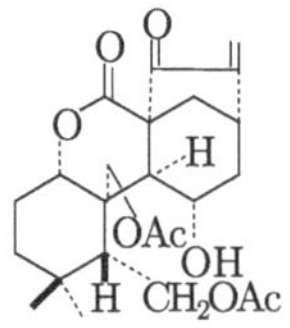   **75** 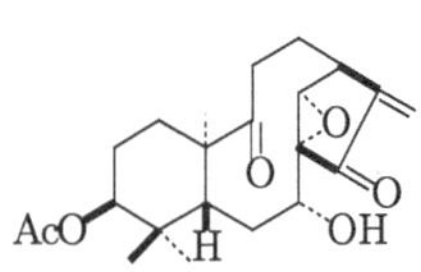   **76**

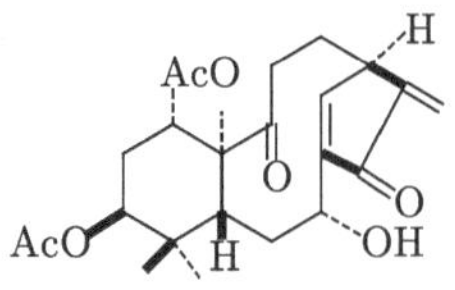

77 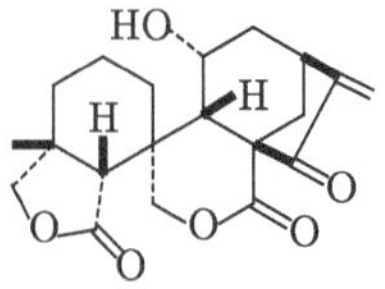   **78** 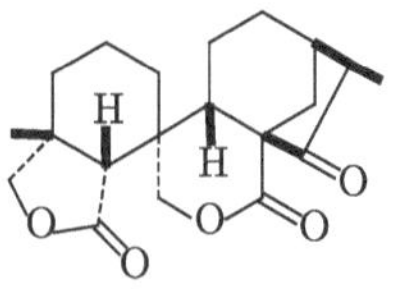   **79**

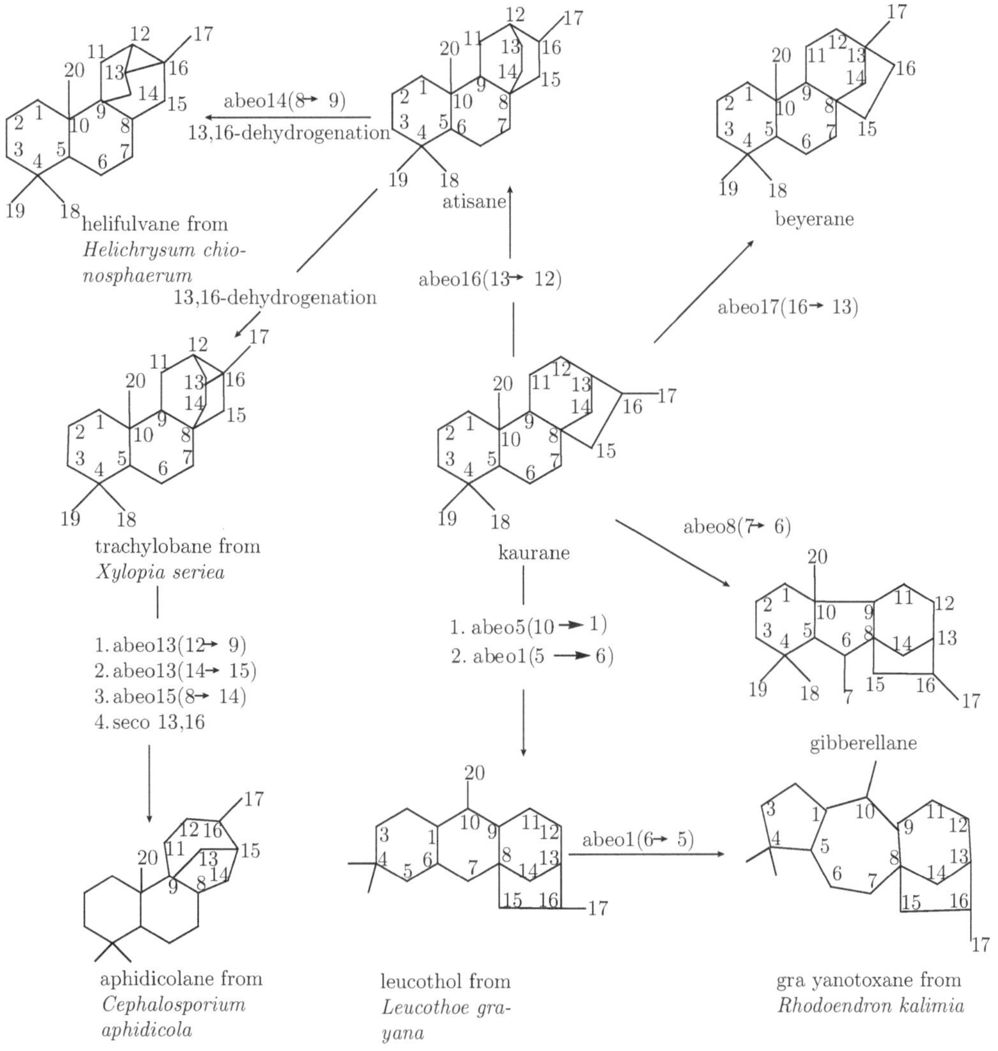

abeo14(8→ 9)
13,16-dehydrogenation

helifulvane from
Helichrysum chio-
nosphaerum

atisane

beyerane

13,16-dehydrogenation

abeo16(13→ 12)

abeo17(16→ 13)

trachylobane from
Xylopia seriea

kaurane

abeo8(7→ 6)

1. abeo13(12→ 9)
2. abeo13(14→ 15)
3. abeo15(8→ 14)
4. seco 13,16

1. abeo5(10 → 1)
2. abeo1(5 → 6)

gibberellane

aphidicolane from
Cephalosporium
aphidicola

leucothol from
Leucothoe gra-
yana

abeo1(6→ 5)

gra yanotoxane from
Rhodoendron kalimia

Figure 6-1   The rearrangement reaction of kaurane diterpene, etc.

Section 10   Taxanes

Taxane-based phytochemicals, such as taxol (**80**), are secondary metabolites exclusively found in *Taxus* species (Taxaceae). Taxol, significantly cytotoxic and antileukemic, was approved in 1992 by the Food and Drug Administration (USA) for treating drug-resistant ovarian cancer, and in 1994 for curing breast cancer. Clinical use of taxol is steadily increasing since it is also prescribed for the treatment of non-small-cell and small-cell lung cancers, squamous cancers on heads and necks, and other tumors. Today, taxol is one of the major anti-cancer drugs[37−39].

Taxol was isolated originally from the *Taxus brevifolia* bark by Wani and Wall in 1971[37]. So far, a total of over 400 taxane-based compounds have been characterized from *Taxus* plants. Biosythetically, the taxane skeleton was generated upon the cyclization of geranyl-geranyl pyrophosphate as illustrated below.

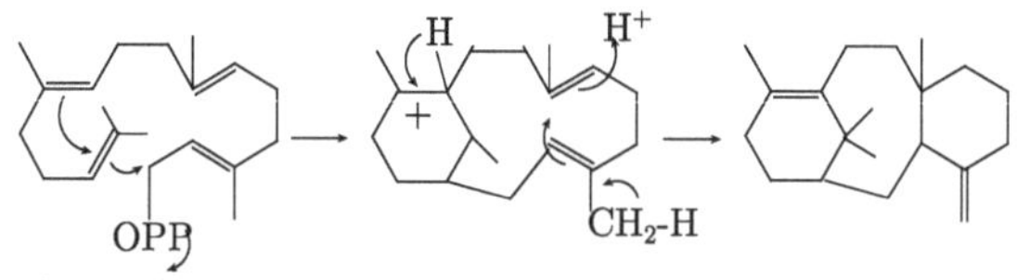

geranylgeranyl pyrophosphte   taxane

The taxane framework is unique in the fusion of its 6/8/6/4-membered rings that usually carry a $C_{4,5}$-oxetane and C_9-ketone along with several acyl residues, such as acetyl and benzoyl groups. The structure of taxol is unique in the presence of 1,7-dihydroxyls and a benzyl amino ester moiety anchoring on C-13 (**80**).

Taxotere (**81**), a semisynthetic analogue as effective as taxol, was also approved for clinical use in 1996.

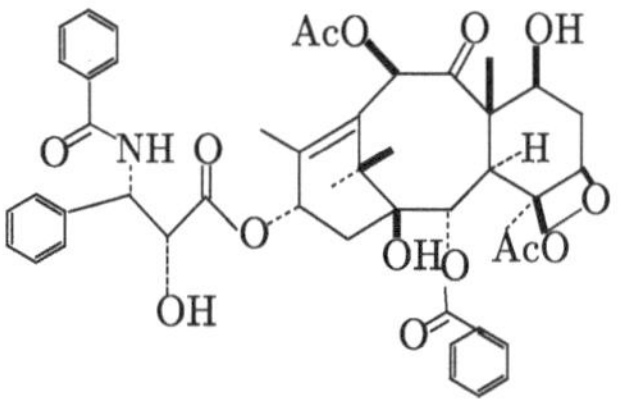

80 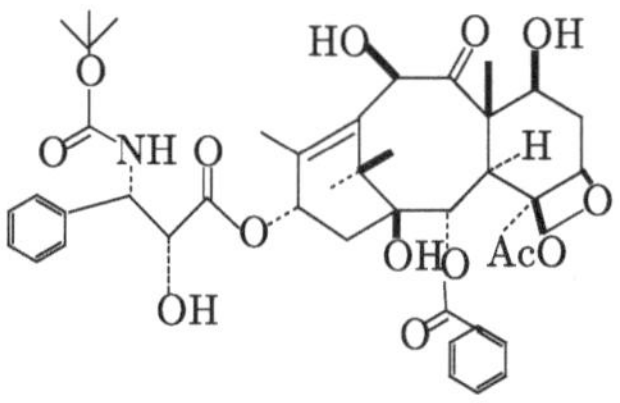   **81**

Although the total synthesis of taxol has been reported[40], the straightforward semisyntheses of taxol and taxotere are more feasible for industrial production.

10.1   Taxol[41]

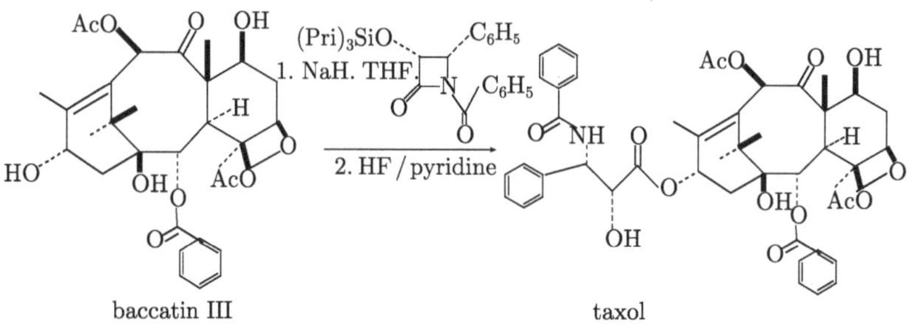

baccatin III   taxol

10.2   Taxotere[42]

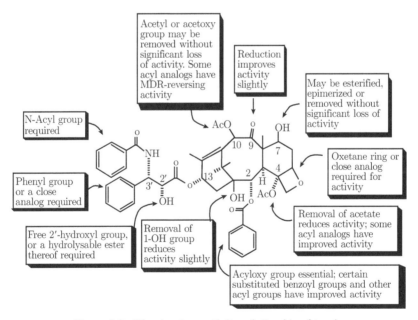

The structure-activity relationship of taxol has been well investigated as summarized below[37].

Figure 6-2   The structure-activity relationship of taxol

The functionality of taxol is essential for anticancer action and includes the oxetane, the cyclic framework, the side chain bearing a 2'-hydroxyl group at C-13, and the ester groups at C-2 and C-4. The structural change in the segment from C-6 through C-12 insignificantly affects its antitumor activity.

Owing to its poor solubility in water, taxol is very difficult to administer by injection. It was reported that this solubility issue could be partially overcome by the esterification of the 2'-hydroxyl with acids such as malic acid or butyric acid.

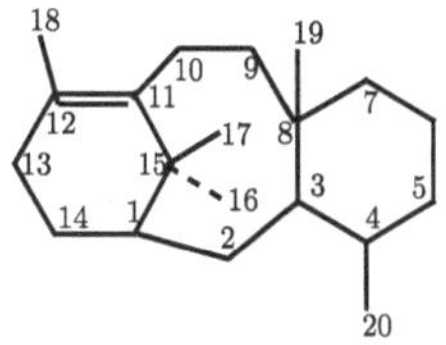

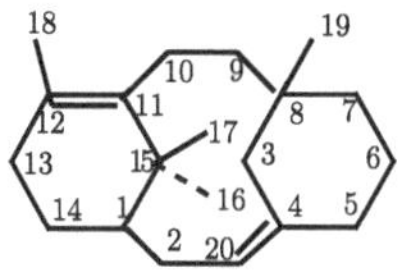

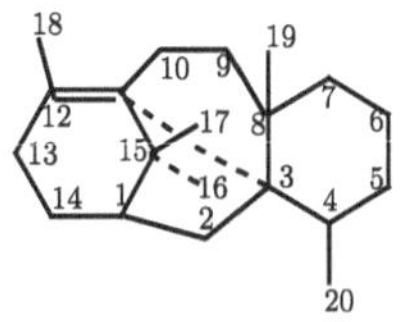

In terms of the phytochemistry, some new taxane-related skeletons like II through VI have been discovered from *Taxus* species.

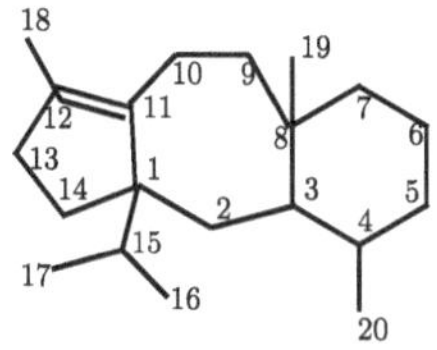

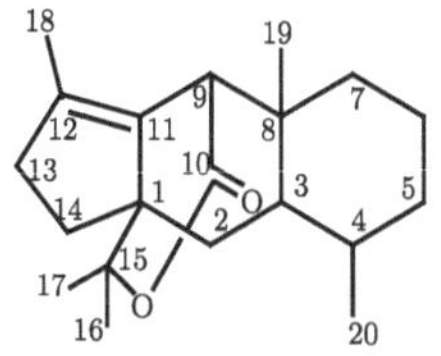

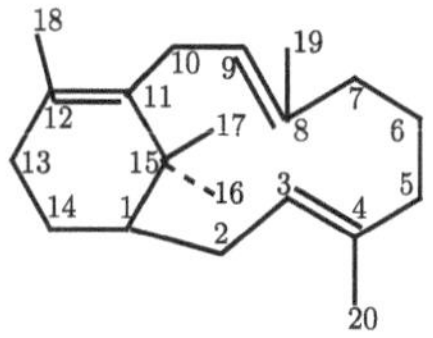

Type II belongs to the abeo11(15→1)-taxane skeleton, which was confirmed by x-ray diffraction analysis to be taxchinin A (**82**)[43], and the rearrangement reaction of the abeo 11(15→1) taxane resulted in **83**. Type III was assigned as an abeo2(3→20)-taxane (taxumairone A(**84**)[44]), and type IV (3,11-cyclotaxane skeleton) was found to be taxuspine C (**85**)[8]. Type V is actually a bisabeo 11(15→1),11(10→9) taxane known as 13-O-acetylwallifoliol (**86**)[45]. Type VI is a 3,8-seco-taxane named verticillane and is represented by taxuspine U (**87**)[44].

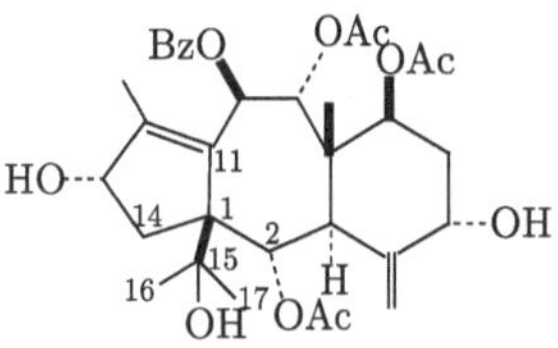

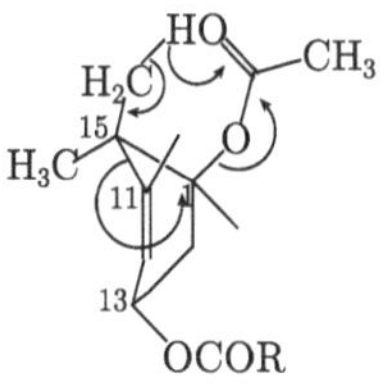

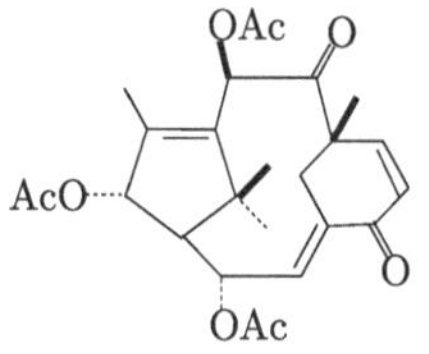

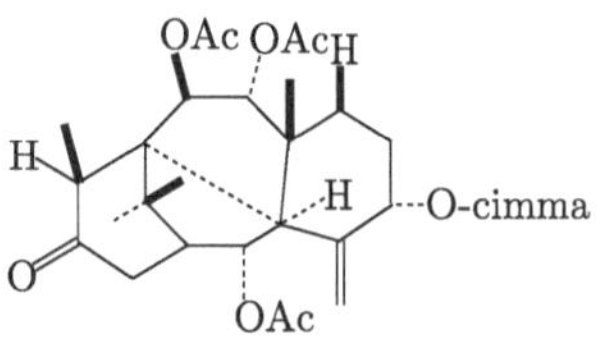

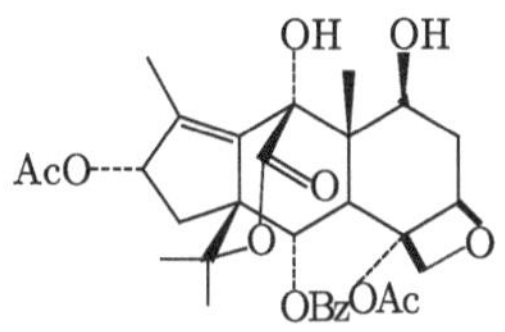

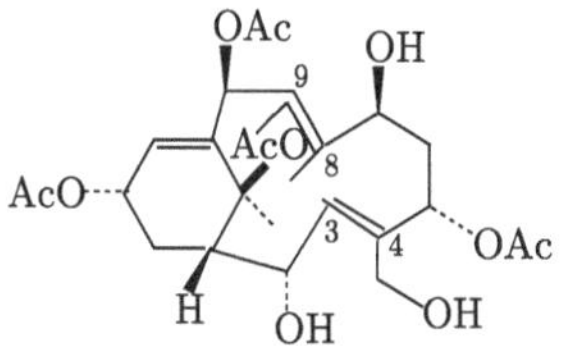

The ^{1}H and ^{13}C NMR data of taxol are given here for readers' reference.

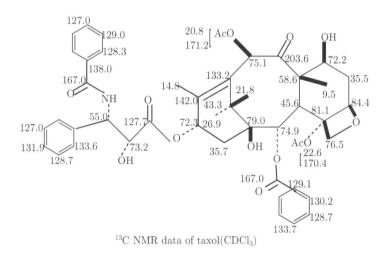

^{1}H NMR data of taxol(CDCl$_3$)

^{13}C NMR data of taxol(CDCl$_3$)

Section 11   Tiglianes, Ingenanes and Daphnanes

The toxic diterpenes of the families Euphorbiaceae and Thymelaeaceae are based upon tigliane, ingenane and daphnane. Among the diterpenes of these types, many of them exhibit tumor-promotion and intiation.

Tiglianes are usually ester derivatives of diterpene alcohols which include a tetracyclic ring that consists of the 4/7/6/3 ring junction. Those such as 12-O-tetra-decanoylphorbol-13-acetate (TPA,**88**), from *Euphorbia tiglium*, exhibit a strong cocarcinogenesis. However, phorbol (**89**) shows no biological toxicology. When either ester group of 12,13-diesters is highly consitituted of fatty acids, the ester derivatives of phorbol will reveal cocarcinogenesis[46].

Ingenanes The characteristic structure is revealed to be a tetracyclic diterpene consisting of the 5/7/6/3 ring fusions, and a keto-bridge between C-8 and C-10 such as 3-O-hexadecanoyl ingenol (3HI, **90**). The toxic components of *Euphorbia kansui* have been isolated and also found to belong to an ingenane diterpene[47,48].

Daphnanes The characteristic skeleton is shown to be a tricyclic diterpene, which is found in the 5/7/6 ring junction, and an orthoester benzoate between C-9, C-14, and C-13, such as daphnetoxin (**91**) from *Hura crepitans*. Yuanhuacine (**92**) and yuanhuadine (**93**) were isolated from *Daphne genkwa*, which was shown to induce abortion[49].

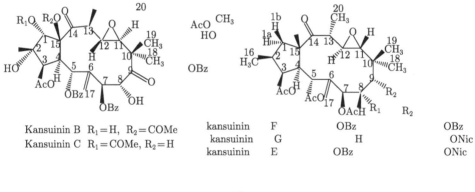

88 $R_1 = OAc$, $R_2 = OOC(CH_2)_{12}CH_3$
89 $R_1 = R_2 = OAc$

90

91 $R_1 = H$, $R_2 = C_6H_6$
92 $R_1 = PhCO$, $R_2 = CH_3(CH_2)_4(CH=CH)_2$
93 $R_1 = CH_3CO$, $R_2 = CH_3(CH_2)_4(CH=CH)_2$

Section 12   Jatrophanes and Lathyranes

Jatrophanes The skeleton of Jatrophane macrocyclic diterpenes consist of a two ring junction (5/12 ring). These exhibit various physiological activities, such as anti-tumor and NGF inducement. Kansuinins A through H were isolated from the roots of *Euphorbia kansui* The kansuinins A, D and especially E, exhibited a specific survival effect on fibroblasts that expresses TrkA, which is a high-affinity receptor for nerve growth factor ($ED_{50} = 0.23 \ \mu g/mL$)[50−52].

Kansuinin B $R_1 = H$, $R_2 = COMe$
Kansuinin C $R_1 = COMe$, $R_2 = H$

	R_1	R_2
kansuinin F	OBz	OBz
kansuinin G	H	ONic
kansuinin E	OBz	ONic

	R_1	R_2	R_3
kansuinin	OH	Bz	Ac
kansuinin	H	Bz	Nic
kansuinin	H	Bz	Ac

The other types of jatrophane diterpenes have been isolated from Euphorbiaceae plants, such as euphorins A to D, euphoscopins A to D, and eupholeins A to H isolated from *Euphorbia helioscopia* and which exhibit anti-tumor activity[53]. Terrchinolides A to H were isolated from *E.terracina*. These structures possess a six-cyclic lactone at C-5 and C-6[54−56].

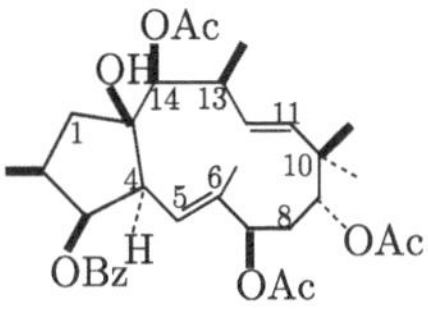

euphornin A 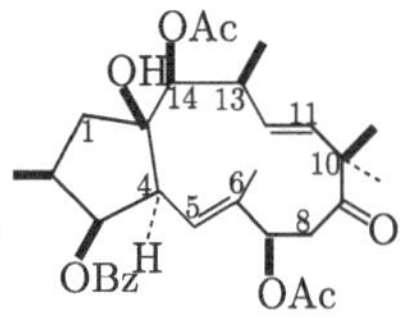   euphoscopin A 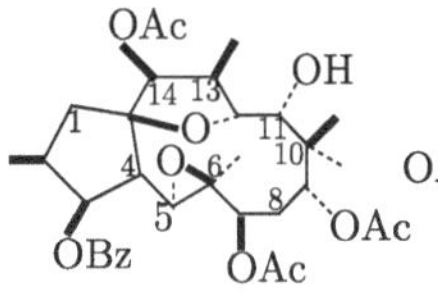   eupholelin A 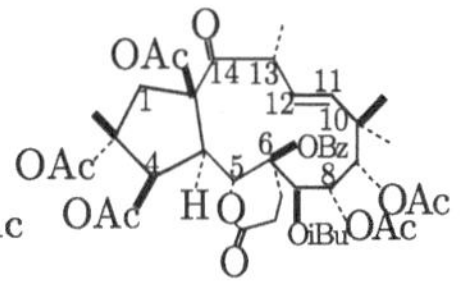   terrchinolide A

Lathyranes   The lathyrane skeleton consists of 5/11/3 ring junctions and is derived from the rearrangement reaction of jatrophane. The cyclic propane ring of lathyrane was fused to the macrocyclic ring at C-9 and C-11, with H-9α and H-11α. The cyclic pentane and macrocyclic ring are linked *trans* with H-4α.

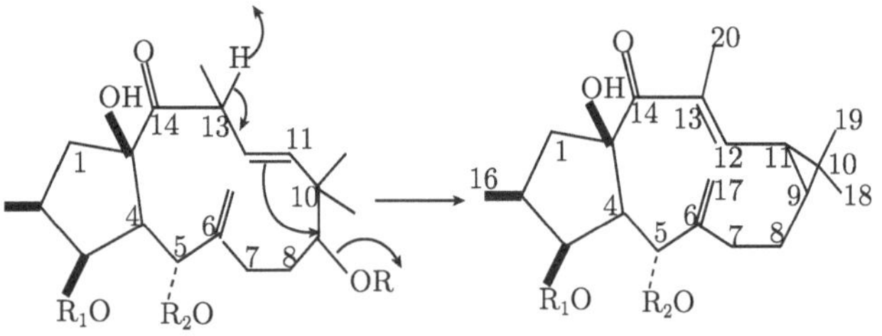

jatrophane   lathyrane

Lathyranes (**94**) and their ester derivatives were isolated from *Euphorbia lathyris*, which is said to possess the 4α-lathyrane skeleton and show slight cytotoxic activity. Ingol (**95**) and its ester derivatives were obtained from *E. ingens*, *E. kamerunnica*, and *E.antiquorum*, etc.,and exhibit anti-neoplastic and cytotoxic activity. Jolkinols A through D were isolated from *E. jolkini*.

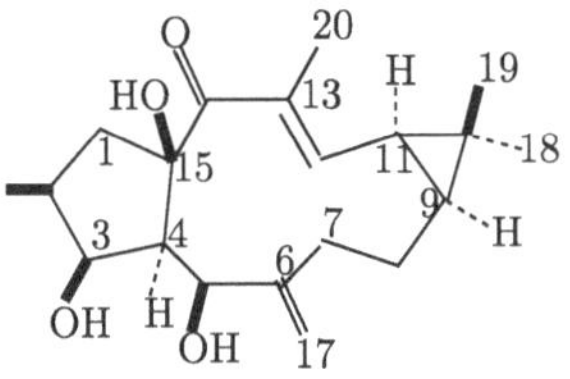

94 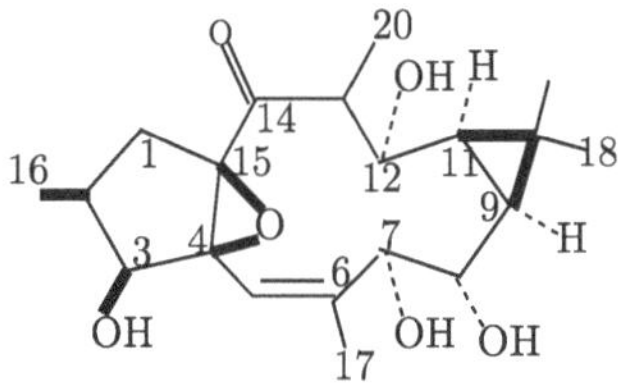   **95** 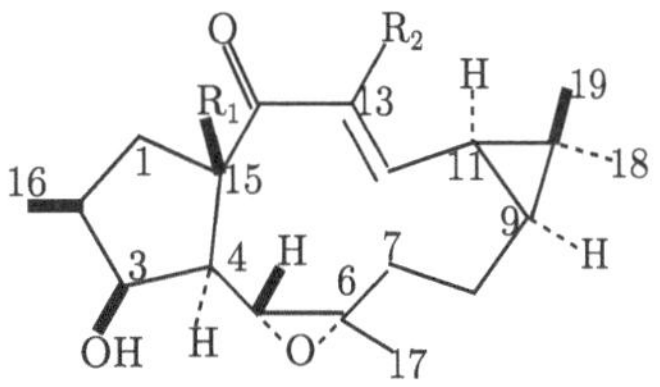   jolkinol A R$_i$=cinnamate,
                                                            R$_2$=CH$_2$OH
                                                 jolkinil B R$_i$=cinnamate,
                                                            R$_2$=CH$_3$

Section 13   Myrsinols and Euphorsctines

Non-irritant macrocyclic diterpenes, such as esters of myrsinol (**96**), have been isolated from *Euphorbia myrsinites*. The skeleton consists of a 5/7/6 ring junction bearing a 3,17-ether bridge. The analogous myrsinol ester diterpenes, such as esters of euphorappinol (**97**) were isolated from *E. aleppica*. Sprol (**98**) esters were isolated from *E. seguieriana* and *E. prolifera*. Sprol is derived from the rearrangement reaction of myrsinol. The partial myrsinol diterpenes exhibited a moderate degree of activity towards HIV-1[59−61].

Euphorsctines A to E and eupphactin A to D were obtained from *Euphorbia micractina*, and were formed from the rearrangement reaction of lathyrane. Euphorsctine A (**99**) exhibited anti-tumor activity[62].

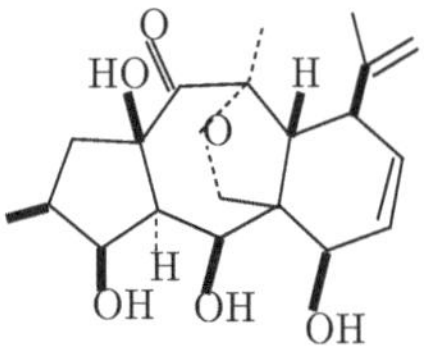

96 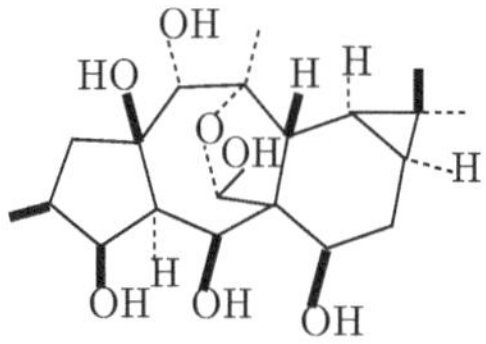   97 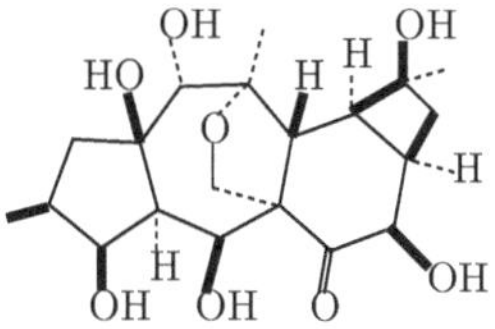   98 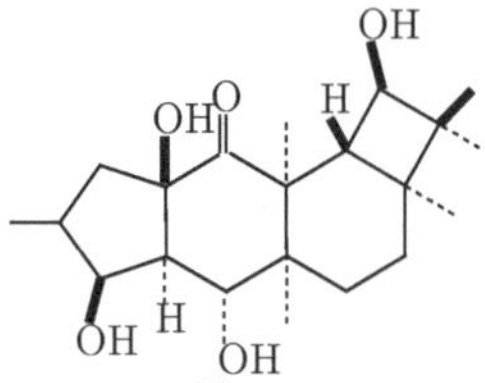   99

Section 14   Ginkgolides and Pseudolaric Acid

Ginkgolides A, B, C, J, M and bilobalide were obtained from the leaves of *Ginkgo biloba*. The ginkgolide skeleton is a very rigid structure consisting of hexacyclic C_{20} trilactone. The *cis*-fused F/A/D/C ring junction forms an empty semi-ball hole, the D ring contains a cage form tetrahydrofuran ring which occupies the center of the empty hole, and the oxygen atoms of the D,C and F ring and 10-hydroxyl group consist of an analogous crown ether structure. Ginkgolides are good PAF antagonists. Ginkgolide B showed an ED_{50} of 10^{-7} to 10^{-8} M in various tests. A mixture of ginkgolides A, B, and C was used to treat asthma[63].

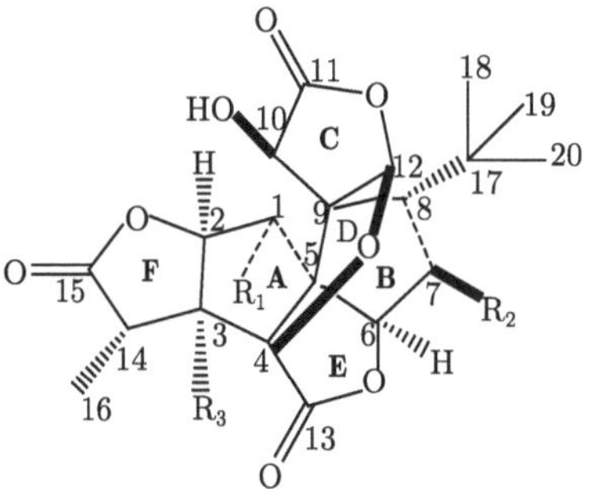

ginkgolide A, $R_1 = R_2 = H, R_3 = OH$
ginkgolide B, $R_1 = OH, R_2 = R_3 = H$
ginkgolide C, $R_1 = R_2 = R_3 = OH$
ginkgolide J, $R_1 = H, R_2 = R_3 = OH$
ginkgolide M, $R_1 = R_2 = OH, R = H$

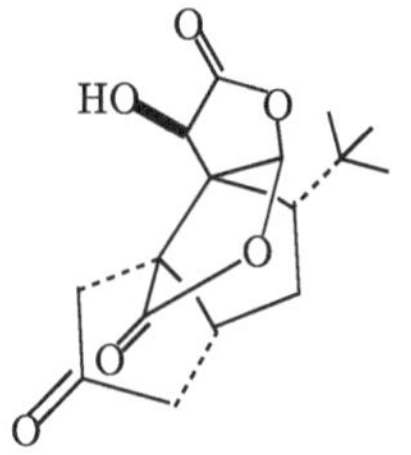

bilobalide

The antibacterial constituents pseudolaric acid A, B and C were isolated from the bark of *Pseudolarix kaempferi* and used for the treatment of skin diseases[64].

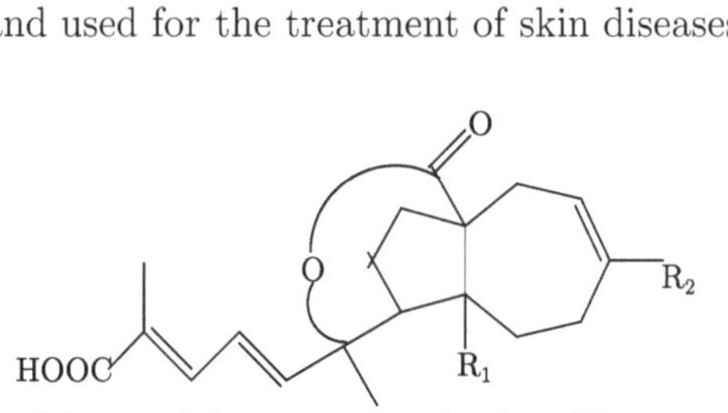

pseudolaric acid A $R_1 = Ac, R_2 = CH_3$
pseudolaric acid B $R_1 = Ac, R_2 = COOCH_3$
pseudolaric acid C $R_1 = H, R_2 = COOCH_3$

References

[1]   Misuhashi H et al., Natural Products Chemistry, Tokyo, Nankoda Press, 1985, 102.

[2]   Dry P M et al., Method in Plant Biochemistry, New York, Academic Press, 1991, 265.

[3]   Seaman F et al., Diterpenes of flowering of plants compositae, New York, Springer-Verlag, 1990: 385.

[4]   Ulubelen A et al., Phytochem, 1994, 36: 971.

[5]   Torrenegra R et al.,Phytochem, 1994, 35: 195.

[6]   Mattsuda T et al., Chem Pharm Bull, 1994, 42: 1216.

[7]   Blhat S V, Prog Chem Org Nat Prods, New York, Springer-Verlag, 1993, 1.

[8]   Nagashima et al., Tetrahedron, 1999, 55: 9117.

[9]   Bohlman F et al., Phytochem, 1982, 21: 1109.

[10]   Anjaneyulu A S R et al., Phytochem, 2000, 55: 891.

[11]   Pinto A C, Phytochem, 1966, 42: 767.

[12]   Ragers D et al., J Chem Soc Chem Comm, 1974, 97.

[13]   Min Z D et al., Chem Pharm Bull, 1989, 37: 2505, 1990, 38: 3167, 1995, 43: 2253.

[14]   Xie N etal., Phytochem, 1991, 30: 1.

[15]   Xie N etal., Chem Pharm Bull, 1992, 40: 2193.

[16]   Jolad S D et al., Phytochem, 1988, 27: 1211.

[17]   Kizu H et al., Chem Pharm Bull, 1998, 46: 988.

[18]   Al-Yahya M A et al., J Nat Prod, 1993, 56: 30.

[19]   Hanson J Q et al. Nat Prod Reports, 1991, 8: 5.

[20]   Wang X W et al., Drug of the Future, 1999, 40: 2193.

[21]   Shi H M et al., Planta Medica, 2005, 71: 349.

[22]   Liang Y et al., Chinese Tradition and Herbal Drug, 2000, 31: 304.

[23]   Fan T P, J Asian Nat Prod Research, 2000, 2: 237.

[24]   Liang J Y et al., Chem Pharm Bull., 1987, 35: 2613.

[25]   Ohtsu H et al., Planta Med, 1999, 61: 907.

[26]   Lin W H et al., Phytochem, 1997, 46: 169.

[27]   Gongalag A G et al., Phytochem, 1992, 31: 1691.

[28]   Sabri N M et al., J Org chem., 1989, 54: 409.

[29]   Cronlund A, planta Med, 1976, 76: 126.

[30]   Arijun H et al., Tetra Lett. 2003, 54: 6809.

[31]   Nakanish K et al., Nat Prod Chem, New York, Academic press, Vol. 1, 1974: 231.

[32]   Djerssi C et al., J Chem Soc, 1966, 624.

[33]   Liu X T et al., J Nat Prod, 2006, 69: 255.

[34]   Liu X T et al., Planta Med, 2007, 73: 84.

[35]   Sun H D et al., Diterpenoids from Isodon Species, Beijing, Science Press, 2001.

[36]   Fujita E et al., Prog Chem Org Nat Prods, 1984, 46: 78.

[37]   Kingston D G I et al., J Nat Prod, 2000, 63: 726.

[38]   Kingston D G I et al., J Nat Prod, 1999, 62: 1448.

[39]   Virinder S P et al., Phytochem., 1999, 50: 1267.

[40]   Holton R A et al., J Am Chem Soc, 1994, 116: 1597: 1599.

[41]   Holton R A et al., J Am Chem Soc, 1994, 116: 1599.

[42]   Mangatal M T et al., Tetrahedron, 1989, 45: 4177.

[43]   Fuji K et al., Tetra Lett, 1992, 33: 7915.

[44]   Kobayashi J et al., Med Res Rev, 2003, 22: 305.

[45]   Shen Y C et al., J Nat Prod, 2002, 65: 1848.

[46] Hiroho I Bioactive Molecular, Vol. 2, Naturally Occurring Carcinogens of Plant Origin. Toxicogy, Pathology and Biochemistry, Rodansha Ltd. 1987, 181.

[47] Wang L. Y. et al., J Nat Prod, 2002, 65: 1246.

[48] Wang L. Y. et al., Chem. Pharm Bull, 2003, 51: 935.

[49] Ying B P et.al., Acta Chemica Sinica, 1977, 35, 103.

[50] Pan Q et al., J Nat Prod, 2004, 67: 1548.

[51] Wang L Y et al., J Nat Prod, 2002, 65: 1246.

[52] Wang L Y et al., Chem Pharm Bull, 2003,51: 935.

[53] Yamamura S et al., Phytochem, 1989, 28: 3421.

[54] Jakupovic J et al.,Phytochem, 1998, 47: 1583.

[55] Marco J A et al., Phytochem, 1997, 45: 137.

[56] Marco J A et al., J Nat Prod, 1999, 62; 110.

[57] Evans F J et al., Prog Chem Org Nat Prod, 1983, 44: 1.

[58] Schroeder G etal., Planta Med., 1979, 35: 234.

[59] Shi Y P et al., Natural Product and development, 1998, 11: 85.

[60] Wu D G et al., J Nat Prod., 1995, 58: 408.

[61] Shi J D et al., Phytochem, 1995, 38: 1445.

[62] Shi J D et al., J Nat Prod, 1995, 58: 51.

[63] Kitai M et al., Drug, 1991, 42: 9.

Zhi-Da MIN

CHAPTER 7

Saponins

Section 1   Introduction

The saponin group of glycosides contain triterpenoid and spirosteroid as their aglycones. It can be further divided into triterpenoid saponin and steroidal saponin. A saponin may contain one, two, or three saccharide chains, and also possibly an acyl group attached to the saccharide moiety. Saponins are mainly distributed in terrestrial higher plants. For example, steroidal saponins are found primarily in the Dioscoreaceae, Liliaceae, and Scrophulariaceae families and triterpenoid saponins in Araliaceae, Leguminosae, Polygalaceae, and Cucurbitaceae plants. In addition, several marine organisms, like starfish and sea cucumbers, were also found to contain saponins.

Most saponins are natural surfactants due to their aglycone lipophilic properties and hydrophilic effects of their sugar chains. Thus, there may be a lasting bubble effect after oscillation. Certain plant extracts rich in saponins were used to produce emulsifiers, detergents and blowing agents. Saponins are also able to destroy cell membranes and show hemolytic, spermicidal, and cytotoxic activity in fish and snails.

Section 2   Extraction and Isolation of Saponins

Saponins are typically highly polar compounds and are necessarily extracted with alcohol, water, or their mixtures. After evaporation of alcohol in vacuo, the residue could be partitioned between chloroform or ethyl acetate and water to remove low polarity components. This is then followed by partition between water saturated n-butanol and water to enrich the saponins in the n-butanol fraction. It should be noted that certain saponins are lipophilic due to the acylation of sugar hydroxy groups, and are therefore soluble in chloroform or ethyl acetate.

In earlier methods, precipitation was often used to purify saponins. Saponins may be precipitated after pouring a large amount of acetone or ethyl ether into a methanol or ethanol solution containing saponins. However, macroporous resin has become widely preferred for the isolation, investigation, and industrial production of saponins.

2.1   Chromatography using macroporous resin

Macroporous resins are polymers, formed from styrene and other small molecules, and can be divided into different types. Diaion HP-20 macroporous resin (Japan) and D-101, DA-201 (China) are suitable for the separation of saponins

Newly purchased macroporous resin should be repeatedly washed with acetone, methanol, or ethanol to remove organic impurities, and then thoroughly exchanged with water before application. Saponin-rich fractions obtained by the solvent partition method could be dissolved in water and then subjected to the macroporous resin column. Elution of the column with water may remove proteins, polysaccharides, sugars, and also certain small molecules such as catechins and iridoid glycosides. Refined saponin fractions may be obtained from elution with gradient alcohol and require further purification by other chromatographic methods.

2.2   Chromatography using silica gel

Silica gel chromatography is often used following macroporous resin chromatography as an additional purification method. However, silica gel may adsorb compounds with high polarity, such as saponins, and lead to sample loss. Therefore, chromatography over silica gel is not suggested to purify trace saponin samples. Mixtures of chloroform-methanol-water are often found to be a satisfactory solvent system in silica gel chromatography. Addition of a suitable amount of water to chloroform and methanol mixtures can help avoid the chromatographic "tailing" effect and give better resolution.

2.3   Reversed-phase chromatography

Since the 1980s, the development and universal application of reversed-phase adsorbants significantly promoted the purification of saponins. At present, the most adopted reversed-phase adsorbant is C_{18} bonded silica gel (octadecylsilane, ODS). Mixtures of methanol and water or acetonitrile and water are often used as eluents of reversed-phase HPLC due to their low viscosity. However, the cheaper and less toxic ethanol-water system may also give good isolation and has been used in certain low pressure and medium pressure chromatography.

A RP-18 Lobar column (Merck), a glass column prepacked with 40-63 μm C_{18} silica gel, can be used for the purification of saponins. The author of this chapter has isolated a series of triterpenoid saponins from the plant *Mussaenda pubescens* (Rubiaceae) by using an RP-18 Lobar column with ethanol-water and acetonitrile-water as eluent[1]. RPTLC should be used for the selection of elution solvent and as a detection tool. Certain routine RPTLC doesn't work when a low concentration of alcohol solvent has to be used (~50% in water) for the TLC development. In this case, RP-18 TLC WF254 (Merck), with hydrophilic substance added in the adsorbant, which can be developed even by using pure water, should be a good choice.

2.4   Liquid-liquid partition chromatography

Liquid-liquid partition chromatography has the advantage of no irreversible adsorption for high polarity compounds and is suitable for the isolation of saponins. Diallo et al. has reported the separation of asiaticoside and madecassoside from the extract of the medicinal plant *Centella asiatica* (Umbelliferae) using MLCCC equipment with a chloroform-methanol-isobutyl alcohol-water (7/6/3/4) solvent system. The two compounds are different in only one hydroxy group[2].

Section 3   Structure Determination of Saponins

The development of modern spectroscopic methods, especially NMR, has enabled rapid structure identification of saponins of several milligrams or less without sample depletion. Structural determination of the aglycone and sugar portions, and also their linkage, are necessary steps in elucidation of the saponin structure. Although most triterpenoid and steroid aglycones of saponins are known compounds and can be identified by comparing their ^{13}C NMR data with those reported in literature, their structures can also be established unambiguously by using modern spectroscopic methods. This section will focus mainly on the elucidation of the structure of the sugar moiety, including the determination of the type and number of sugar units, the linkage site and sequence of each sugar unit, and the conformation and configuration of each sugar unit.

3.1   Cleavage of glycosidic bond

The cleavage of the glycosidic bond may not only yield the aglycone and sugar(s) composed in the saponin, but also possibly produce secondary saponins by controlling the condition and time of hydrolytic reaction, which is important in determination of the saponin structure. The following methods are widely used for the hydrolysis of the glycosidic bond.

3.1.1   Acidic hydrolysis

The glycosidic bond is a part of the acetal functional group, which is vulnerable to acid hydrolysis. The reaction rate of acidic hydrolysis is related to the sugar and aglycone structures, and also to the linking position of sugar to aglycone. Deoxy sugars, furan sugars, and five-carbon sugars normally have faster reaction speeds than non-deoxy sugars, pyran sugars, and six-carbon sugars. The general acidic hydrolysis method is to dissolve the saponin in a diluted HCl or H_2SO_4 solution and then heat for a period of time. The aglycone can be extracted from the hydrolysate or collected by filtration. The filtrate containing the sugar may be neutralized with alkaline or anion ion exchange resin, and then detected by co-TLC on silica gel with authentic sugar samples. Absolute configuration of each sugar can also be identified by GLC analysis after derivatization into one of the two low polarity diastereoisomers. For saponins containing two or more sugar units, the stability of each glycosidic bond may be different toward acid. Therefore, secondary saponins could be obtained by changing the acid concentration, reaction temperature, and time.

3.1.2   Two-phase acid hydrolysis (mild acid hydrolysis)

Severe acidic hydrolysis conditions may result in structure deviation and make further investigation complicated. Two-phase acidic hydrolysis refers to the addition of a liquid organic phase, such as benzene and toluene, into the acidic aqueous solution. Upon generation of the aglycon, it is transferred into the organic phase, and thus avoids further acid contact side reactions.

3.1.3   Smith degradation

Sodium periodate may selectively oxidize two adjacent hydroxy groups without affecting other neighboring groups. The sugar moiety containing two adjacent hydroxy groups in a saponin may be oxidized by sodium periodate into an intermediate with two aldehyde functions. Reduction of the intermediate with sodium borohydride and then treatment with acid at room temperature may yield the aglycone or secondary saponin with a sugar unit but without adjacent hydroxy groups. Such a method is called Smith degradation[3]. It is a mild reaction, and can be used to obtain the real aglycons that are unstable toward acidic environments.

3.1.4   Enzymatic hydrolysis

Glycosidases are the enzymes that catalyze the biosynthesis of glycosides. Under suitable conditions, the process can also result in the decomposition of glycosidic bonds. Because enzymes catalyze the chemical reaction of substrates at the same condition as in the body, they can minimize the structural derivatization of the aglycones. Glycosidases most often used include amygdalin (emulsion), maltase, cellulase, takdiastase, and crude hesperidase. Enzymatic reaction of saponins is usually conducted at 37°C by mixing the saponin with an enzyme in water. TLC analysis may be used to detect the reaction progress, and it may take several days for a thorough reaction. The aglycone can be filtered or extracted with ethyl acetate from the aqueous solution when the hydrolysis reaction is completed. After simple chromatographic purification, or even without further purification, the aglycone can be subjected to spectroscopic analysis. The residual aqueous solution can be used for the identification of sugar components, their absolute configuration, proportion of each sugar moiety, and so on.

3.1.5   Alkaline hydrolysis

Esteric glycosidic bonds often exist in the pentacyclic triterpenoid saponins, and they can be selectively cleaved by alkaline hydrolysis without affecting the etheric glycosidic bond. Alkaline hydrolysis is usually undertaken by heating the aqueous solution containing sodium hydroxide and saponin in a water bath for some time. Reaction progress and reaction conditions, such as the alkaline concentration, reaction time, and temperature, should also be determined by TLC analysis.

3.2   Structure determination of saponins by chromatography

Chromatographic methods may play an important role in the determination of sugar composition, absolute configuration, and proportion of each kind of sugar.

3.2.1   Silica gel thin-layer chromatography

Using silica gel thin-layer chromatography, the sugars obtained from hydrolysis of saponins can be identified by comparison with authentic sugar samples. Developing solvents can be mixtures of chloroform-methanol-water, ethyl acetate-n-butanol-water, acetone-butanol-water, or ethyl acetate-methanol-water-acetic acid. Kong, et al. developed both the acidic hydrolysis sample of bolbastemmosaponin A and standard sugar samples on silica gel TLC using acetone-butanol-water (5/4/1), and rhamnose, xylose, arabinose, and glucose were identified with R_f values of 0.62, 0.49, 0.35, and 0.21, respectively[4].

3.2.2   Gas chromatography

Gas chromatography has been an important method in the investigation of the structure of saponin since the late 1950s owing to its high sensitivity and separation ability, as well as its ability for online qualitative and quantitative analysis. GC was used mainly in the identification of the types of sugars, linking sites and determination of the absolute configurations of sugar samples. The sugar samples, due to their low volatility, have to be prepared into the methyl ether, trimethylsilyl ether, or acetate derivatives before GC analysis. By standard comparison of the GC chromatogram, the types of sugar and sugar linking sites can be determined.

It was previously believed that sugar components of natural saponins only exist in one form of the two isomers, namely D or L configuration, and the absolute configuration of sugars was not usually confirmed. However, as other sugar enantiomers were also found in some organisms, it became necessary to determine the absolute configuration of sugar components. Hara et al. reported a method to identify the absolute configuration of sugar samples, in which the sugar sample was reacted with L-cysteine methyl ester hydrochloride and then subjected to peracetylation or trimethylsilylation before GC analysis. The method is simple, almost all aldose enantiomers can be separated satisfactorily, and analysis can be performed with ordinary GC conditions[5].

3.3   Structure determination of saponins by spectroscopy

3.3.1   Mass spectrum

Saponins are less volatile natural products with high polarity. Various kinds of soft ionization mass spectrometry, such as electrospray ionization mass spectrometry (ESIMS), fast atom bombardment mass spectrometry (FABMS), and matrix-assisted laser desorption/ionization time-of-flight mass spectrometry (MALDI-TOFMS), can be applied to determine the molecular weights of saponins. In the negative ion mode detection of MS, [M-

H]$^-$ and [M+HCOO]$^-$ are often observed, while [M+H]$^+$, [M+Na]$^+$, and/or [M+K]$^+$ can be found in the positive ion detection mode. Combinational analyses of both MS detection modes are helpful for the identification of the saponin molecular weight.

A group of fragments with the loss of sugar units can sometimes be observed in the soft ionization mass spectrometry of saponins. According to the mass differences of the fragments and quasimolecolar ion peaks, the existence of five-carbon sugars (-132), six-carbon sugars (-162), 6-deoxy-glucoses (-146), and uronic acids (-176) and sequence of sugars in the saccharide chain can be inferred.

3.3.2   Nuclear magnetic resonance spectrum

It was necessary to undertake a series of chemical works in the structural elucidation of saponins before high field NMR facilities were available. The structural identification of saponins seemed to benefit more from the technological advances of NMR when compared to other small molecular natural products. In the high field NMR spectra of saponins, signals due to aglycone and sugar moieties can be better resolved. Because NMR does not destroy the sample, the structure of trace saponin may be identified especially when the sample amount is very small. The following is a brief description about the structural identification of saponin sugar chains using NMR methods.

3.3.2.1   1D NMR

^{1}H-NMR spectra of saponins are generally recorded in deuterated pyridine, methanol or water due to their high polarity. An anomeric proton signal of a sugar moiety normally appears at δ 4.3 to 6.5 ppm, isolated from other oxygen-connecting sugar protons, and anomeric proton signals with esteric glycosidic bond appear downfield from those with ether glycosidic bonds. The number of anomeric proton signals could be used to identify the number of sugar units. In a certain glycopyranoside, the six-membered sugar ring is generally fixed in a chair conformation, and the relative configuration of an anomeric proton can be deduced by the coupling constant. For α-glucose and α-galactose, the coupling constants of the anomeric protons are 2 to 4 Hz, while those of β-glucose and β-galactose are 7 to 9 Hz. For mannose and rhamnose, it is hard to establish the configuration of the anomeric protons due to the small H-1 and H-2 coupling constants[6]. The 6-deoxy sugar may be identified by the doublet methyl signal at δ 1.5 to 1.7 ppm and with a coupling constant of 6 to 7 Hz. Other sugar proton signals are normally overlapped between δ 3.5 to 4.5 ppm, especially when there are many sugar units such as those in many saponins.

Compared with ^{1}H-NMR spectrum, the ^{13}C-NMR signals are distributed in the range from 0 to 200 ppm, and therefore, with less possibility of signal overlap. For oxygen-bond glycosides, the anomeric carbon signals appear between δ 90 and 112 ppm, and those with an ether glycosidic bond are located in the lower region, generally more than δ 98 ppm, while those with an esteric glycosidic bond are located in the higher region, around δ 95 ppm. According to the number of anomeric carbon signals, and in combination with the carbon resonances between δ 60 and 85 ppm, the number of sugar units in a saponin could be confirmed. Comparison of the chemical shifts of sugar carbons with those of a known compound is helpful for the identification of the kind of sugars if only one or two sugar units exist. In addition, uronic acid and 6-deoxy sugars can be confirmed by their specific carbon signals. The size of a sugar ring can be determined by the chemical shifts of sugar carbons. For example, the resonances at C1, C2, and C4 of a furan-aldose downshift at about 4 to 14 ppm, while C5 upshifts around 4 to 7 ppm, when compared to pyran-aldose. The resonances at C1, C2, C3, and C5 of α conformer of pyran sugar upshift 2 to 7 ppm compared to β conformer types, except D-mannose and L-rhamnose, which can be distinguished by the difference of chemical shifts between C-3 and C-5.

Another main function of ^{13}C-NMR of saponin is to confirm the location of a sugar residue. The difference between carbon resonances of sugars and signals of corresponding methyl glycoside results in the location of glycosydation. C_α of the glycosydation position downshifted around 4 to 10 ppm, while C_β upshifted around 0.9 to 4.6 ppm with other carbons almost unchanged. When two ortho hydroxyl groups of sugar are glycosidated, shift values of glycosidation will differ greatly and the linkage sites of sugar units have to be established by the observation of NOE or ^{13}C-^{1}H long-range correlation signals.

3.3.2.2   2D NMR

In many cases, only anomeric proton signals and methyl signals of 6-deoxy sugars can be identified in the ^{1}H-NMR spectra of saponins possessing several sugar units. With the development of high megahertz NMR instruments, the resolution of ^{1}H-NMR spectra are improved, which enables the *ab initio* structural elucidation by using two-dimensional (2D) NMR techniques.

The first step of ab initio structural elucidation of a saponin is to assign all the proton signals of the sugar units. The linear proton coupling system in each sugar unit makes it suitable to assign all proton signals by using ^{1}H-^{1}H COSY and TOCSY (2D HOHAHA) experiments. General assignment steps begin with the isolated anomeric proton signals or the doublet methyl signals of 6-deoxy sugars. In order to clearly identify the structure of a sugar chain, peracetylation of saponin is sometimes necessary. Sugar proton signals attached to acetoxy groups appear in the down field, while the signals of protons at the glycosidation position as well as other protons attached to carbons without free hydroxyl groups are located in the high field. The chemical shifts of acetoxymethine protons (CHOAc) normally appear at δ 4.7 to 5.4 ppm, and the signals for CH_2OAc, CHOR and CH_2OR can be observed at δ 3.0 to 4.3 ppm. Peracetylation enables the sugar proton signals to be distributed in a broad range, and then these signals can be assigned more easily by using ^{1}H-^{1}H COSY and TOCSY experiments[7].

The carbon signals of each sugar unit can be assigned by using HMQC and HSQC experiments after assignment of all proton signals.

HMBC and NOESY (or ROESY) spectra are widely used in the structural determination of a sugar chain. In the HMBC spectrum of saponin, ^{13}C-^{1}H long-range correlation signals between anomeric proton of one sugar unit and the carbon that the sugar unit is connected to could be observed. In the NOESY or ROESY spectrum of saponin, NOE correlation signals between the anomeric proton of one sugar unit and the proton on which carbon the sugar unit is linked could be found. Some saponins contain mid-size molecules (relative molecular weight of about 1000 Dalton), and certain NOE signals may not be observed in the NOESY spectrum. In such case, ROESY experiments should be adopted[8].

Section 4   Biological Activity of Saponins

Many saponins are biologically active compounds of medicinal plants and play a variety of physiological functions. In a recent review by Rao and Yu, the biological activities of saponins included anti-tumor activity, immunomodulatory activity, anti-inflammatory activity, cholesterol-lowering activity, antihepatotoxic activity, hypoglycemic activity, antimicrobial activity, and cardiovascular activity, etc.[9,10]. A brief introduction of the major biological activities of saponins is listed below.

4.1   Anti-tumor and cytotoxic effects

Many saponins exhibit cytotoxic and anti-tumor activities. Tubeimoside I, a triterpenoid glycoside isolated from traditional Chinese medicine *Bolbostemma paniculatum*, and ginseniside Rg3, isolated from the famous plant *Panax ginseng* C. A. Meyer, revealed anti-tumor effects[11,12]. The anti-tumor effects of some saponins are related to their immunomodulatory function[13]. Intravenous injection of some saponins may reduce serious hemolysis due to their surfactant effect.

4.2   Immunomodulatory activity

The immune system is critical to human health. Medicines promoting immune system ability can be used for the treatment of cancer, AIDS, and other infectious diseases, and also reduce anti-aging effects. Many saponins, such as ginsenosides and gypenosides, possess the ability to modulate human immune system function.

Some saponins are also used as auxiliary agents for disease prevention and therapeutic vaccine research and development in addition to promoting immune system functions. Combined with antigens, saponins can help integrate these macromolecules through the mucosal membrane and promote absorption, which increases the effectiveness of injected and oral vaccines. Saponin QS-21, derived from the South American plant *Quillaja saponaria*, promotes the immune system without cytotoxic effects and is currently being developed as an auxiliary AIDS vaccine[14].

Esculentoside A, isolated from the *Phytolacca esculenta* Van Houtte, exhibited immunosuppressive functions, which may be related to the significant anti-inflammatory effect of the plant[15].

4.3   Antimicrobial effects

Many saponins have been found to exhibit activities against plant pathogens and human pathogens.

4.3.1   Antiviral activity

Saikosaponins can inhibit a variety of DNA and RNA viruses, irreversibly inactivate herpes virus, and also exhibit inhibitory activity toward HIV. Glycyrrhizin may inactivate measles and herpes simplex viruses at 5 μM *in vitro*. Saponins isolated from the stems and leaves of Chinese ginseng and American ginseng showed a protective role against HSV-1 infection, and further investigation revealed that the active ingredients were ginsenoside Rb series saponins, especially ginsenoside Rb2[16].

4.3.2   Antifungal activity

Many saponins have been reported to possess antifungal activity. One antifungal saponin mechanism was considered to be a complex formed between saponin and sterols in the fungal serosa that damaged the permeability of the cell membrane. Takechi et al. studied the relationship between the structure of α-hederins and their antifungal activity, as well as hemolytic activity, and found that the terminal rhamnopyranosyl unit is important to the antifungal activity, while the free carboxylic group is important to the hemolytic effect[17]. Certain saponins with antimicrobial activity have been used in cosmetics.

4.4   Cardiovascular activity

Cardiovascular diseases are among the most dangerous to human health. Many saponins possess various bioactivities related to cardiovascular diseases, including lowering cholesterol, anti-hypobaric hypoxia, anti-arrhythmia, positive inotropic effect, and capillary protection activity. Notoginsenosides isolated from *Panax notoginseng* exhibited protective effects on experimental myocardial injury induced by ischemia and reperfusion in rats[18]. The anticerebral ischemia effect of ginsenoside Rb1 isolated from *Panax notoginseng* was found to be related to its calcium antagonistic role, and saponins with panaxatriol type aglycones of the same plant may act against ischemia-reperfusion arrhythmia induced by coronary artery ligation[19,20]. Astragalosides can significantly improve myocardial contractility, reduce coronary flow, and protect myocardial function[21]. Some diosgenins are effective in the treatment of coronary heart disease, reducing angina, and promoting metabolism regulation[22].

4.5   Anti-inflammatory, anti-exudative, and anti-edema effects

The early stage inflammation is often manifested as an increase in vascular permeability, release of histamine, blood amine, and alkaline polypeptide accompanied by hyperemia and hematoma. Some saponins were found to possess anti-exudation and anti-edema activities, and thus display anti-inflammatory effects. For example, escins, a series of triterpenoid saponins isolated from the dried ripe seeds of *Aesculus wilsonii*, possessed anti-inflammation and anti-edema effects. They may also increase the vein tension and have been used clinically to treat diseases such as cerebral edema.

4.6   Other effects

In addition to the above mentioned biological effects, saponins also possess a variety of other activities. For example, soy saponins were found to lower body cholesterol levels by binding sterols into insoluble forms. Saikosaponin possessed antifever and liver protection effects and saponins isolated from *Aralia elata* and *Polygala tenuifolia* may inhibit the absorption of glucose and ethanol. Glycyrrhizin was found to possess adrenocorticotropic hormone (ACTH)-like activity and has been used as an anti-inflammatory agent for the treatment of ulcer disease. Licorice saponins and mogrosides are used in the food industry as sweeteners.

Section 5   Triterpenoid Saponin

5.1   Triterpenoid

A triterpenoid is composed of 30 carbon atoms. Triterpenoid compounds are widespread in the plant kingdom, including both monocot plants and dicotyledonous plants. Plants of the families Araliaceae, Leguminosae, Campanulaceae, and Scrophulariaceae often contain a high content of triterpenoid components. Triterpenoids are also found in animals, such as lanolin from wool, fungi, such as *Ganoderma lucidum*, and marine organisms, such as sea cucumbers and soft corals.

5.2   Main structural skeletons of triterpenoid saponins

5.2.1   Tetracyclic triterpenoids

5.2.1.1   Lanostane (1) type triterpenoid

The lanostane type tetracyclic triterpenoids are characteristic of the *trans* junction of rings A/B, B/C, and C/D, β-oriented methyl at C-10 and C-13, α-oriented methyl at C-14, β-oriented side chain at C-17, and the R configuration for C-20.

Astragalus membranaceus is a traditional Chinese medicine possessing tonic and diuretic effects. A series of triterpenoid saponins with the aglycone cycloastragenol [(20R, 24S)-3β, 6α, 16β, 25-tetrahydroxy-20, 24-epoxy-9, 19-cyclolanostane, (**2**) have been isolated. The C-3 and C-6 of the glycone were linked to a sugar unit in the structure of astragaloside I (**3**) also with the substitution of acetyl groups in a sugar moiety. Astragaloside I (**3**) was reported to exhibit immunomodulatory activity[23]. Astragaloside IV (**4**) was the major bioactive component in *Astragalus membranaceus*, while astragaloside VII (**5**) was the first triterpenoid saponin with three saccharide chains found in nature[24],

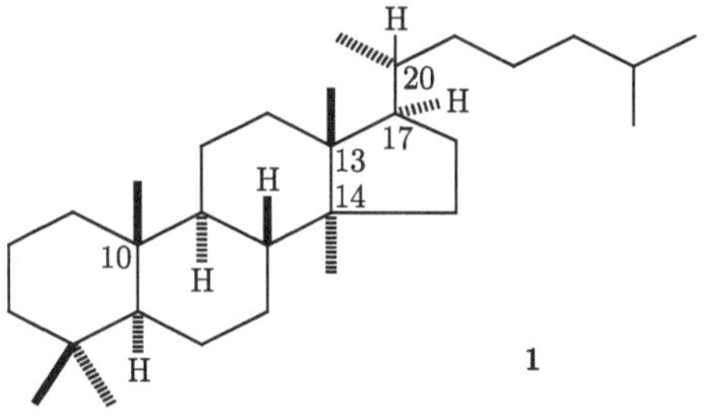

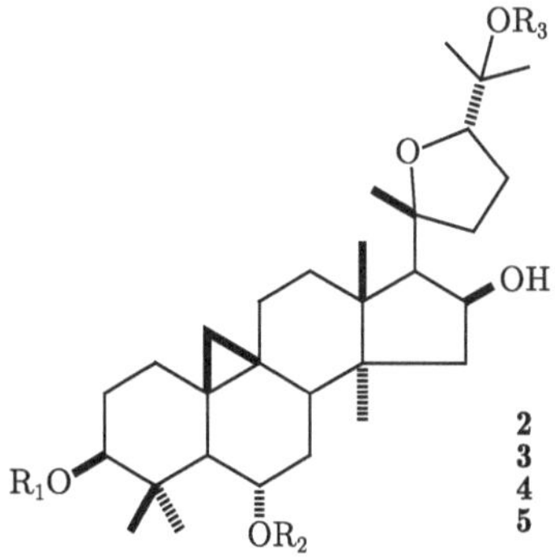

	R1	R_2	R_3
2	H	H	H
3	xyl(2,3-diAc)	glu	H
4	xyl	H	H
5	xyl	glu	glu

5.2.1.2   Dammarane (6) type triterpenoid

Dammarane type tetracyclic triterpenoids are characterized by a β-methyl at C-8, β-H at C-13, α-methy at C-14, and β side chain at C-17.

The real saponin aglycones protopanaxadiol (**7**) and protopanaxatriol (**8**), from the famous medicinal plants *Panax ginseng* (Araliaceae) and *P. notoginseng*, both belong to dammarane-type tetracyclic triterpenoids. The structures of protopanaxadiol and protopanaxatriol are very similar except for one more 6-OH in the latter compound. In most of these saponins, the absolute configuration of C-20 is S. The number of the saponins with dammarane type aglycones is only next to that of oleanane type saponins in all triterpenoid saponins. The sugar chain may be located at C-3, C-6, and C-20 of the dammarane type aglycone saponins, as in the case of ginsenoside Rb1 (**9**), where two saccharide chains consisting of a glucose moiety were linked to C-3 and C-20, respectively.

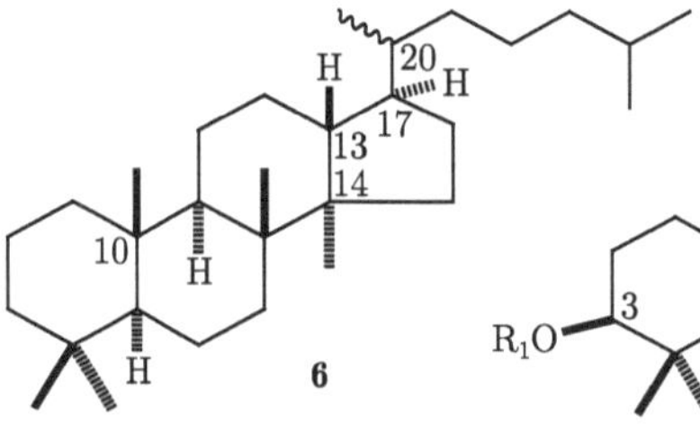

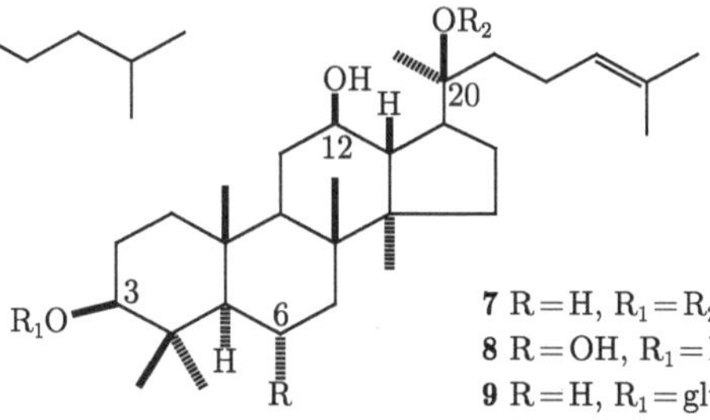

7 R=H, R_1=R_2=H
8 R=OH, R_1=R_2=H
9 R=H, R_1=glu(1-2)glu,
     R_2=glu(1-6)glu

5.2.1.3   Cucurbitane (10) type triterpenoids

Cucurbitane type triterpenoids are a category of highly oxidized tetracyclic triterpenes with bitter taste. Unlike lanostane type triterpenoids, the substitutions in A/B rings of cucurbitane type triterpenoids are C8β-H, C9β-CH$_3$, and C10α-H.

Cucurbitane type triterpenoids exist mainly in the plants of the family Cucurbitaceae. Some of these triterpenoids exhibit significant cytotoxic, anti-tumor, liver protection, and anti-inflammatory effects. Cucurbitacins IIa (**11**) and IIb (**12**), isolated from the roots of *Hemsleya amabilis* (Cucurbitaceae), were tested clinically for the treatment of acute dysentery, tuberculosis, and chronic bronchitis with good results. Cucurbitane type triterpenoids mainly exist in the form of glycosides in plants. Rui et al. reported the isolation of a new bitter substance, cucurbitacin IIa glucoside (**13**) from *Hemsleya amabilis* collected in Guizhou[25]. It is not easy to obtain a cucurbitane type triterpenoid saponin in its original form if β-glycosidase exists in the plant.

11 R = H, R? = Ac
12 R = R? = H
13 R = glu, R? = Ac

5.2.1.4   Meliacane (14) type triterpenoids

The skeleton of a meliacane type triterpenoid is composed of 26 carbons. Therefore, meliacane type triterpenoids are also known as tetranortriterpenoids. Meliacane type triterpenoids are widely distributed in *Melia* species of the family Meliaceae and exhibit a bitter taste and insect antifeeding activity. The extract of the fruits of *Melia azedarach* L. have been commercialized as insect antifeedant. Toosendanin (**15**), isolated from the fruits, root barks, and barks of *Melia toosendan*, has been used as a lumbricide with more than 90% efficiency. Azadirachtin (**16**), obtained from the seeds of neem (*Azadirachta indica*), displayed very strong insect antifeeding, insect molting, and growth inhibitory activities. Extracts of neem fruits, with azadirachtin as the major biologically active component, have been developed as insecticides. The structure of azadirachtin was wrongly determined, and then revised by Kraus et al.[26]. The skeleton of meliacane type triterpenoids is usually highly oxidized, and single crystal x-ray diffraction is often used to confirm their complicated structures. Sawabe et al. reported the identification of meliacane type triterpenoid glycosides, such as methyl nomilinate 17-O-β-D-glucopyranoside (**17**), from the plant *Citrus unshiu*[27].

5.2.1.5   Quassinoids

Quassinoids are highly oxygenated triterpenes, which were mainly isolated as bitter principles from the plants of the Simaroubaceae family. According to their basic skeleton, quassinoids are categorized into five distinct groups: C-18, C-19, C-20, C-22, and C-25. The most prevalent quassinoids have a C-20 picrasane (**18**) skeleton. These compounds were classified as diterpenoids. However, they are now believed to be derived from apoeuphol (**19**) or its isomer, apotirucallol (**20**).

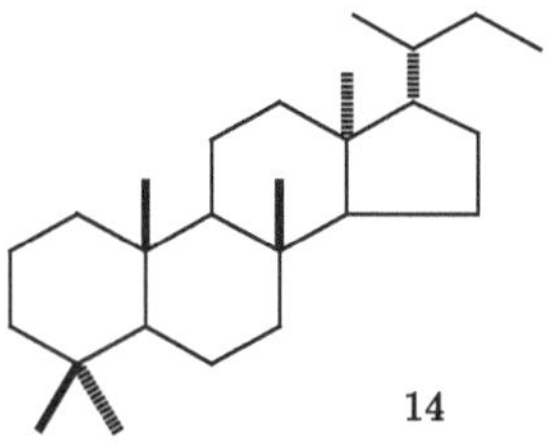

14

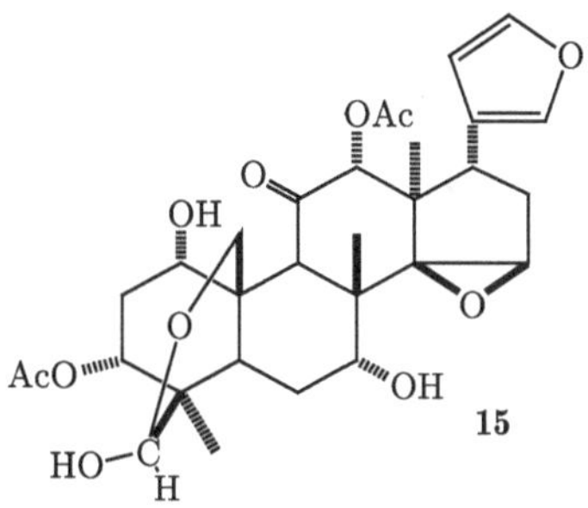

15

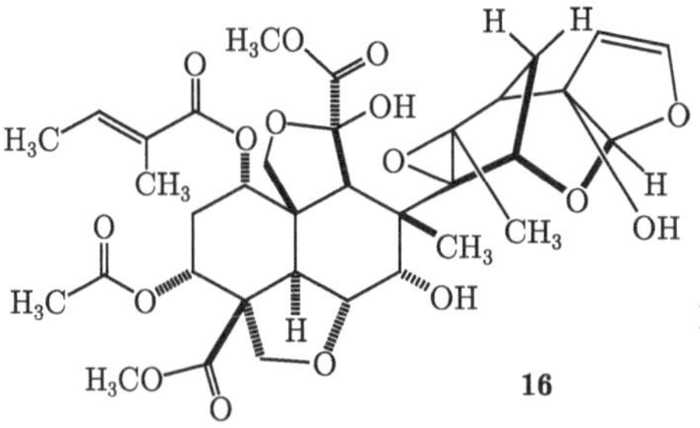

16

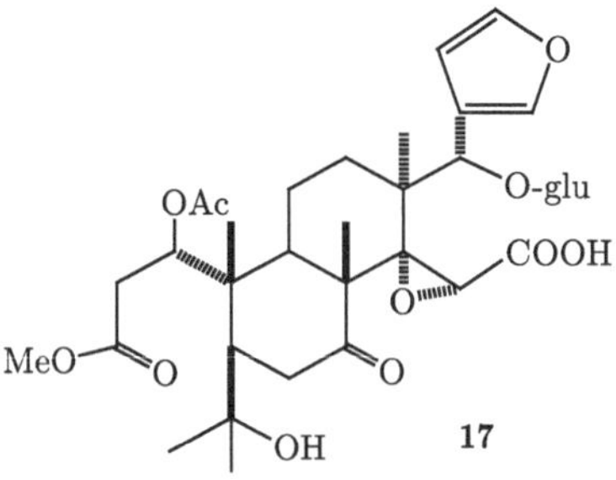

17

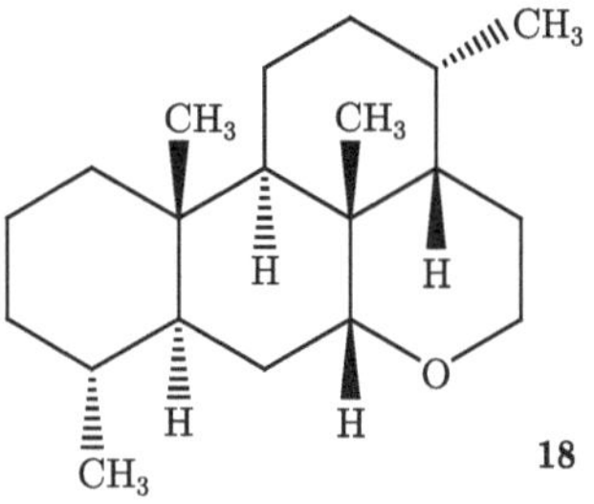

18

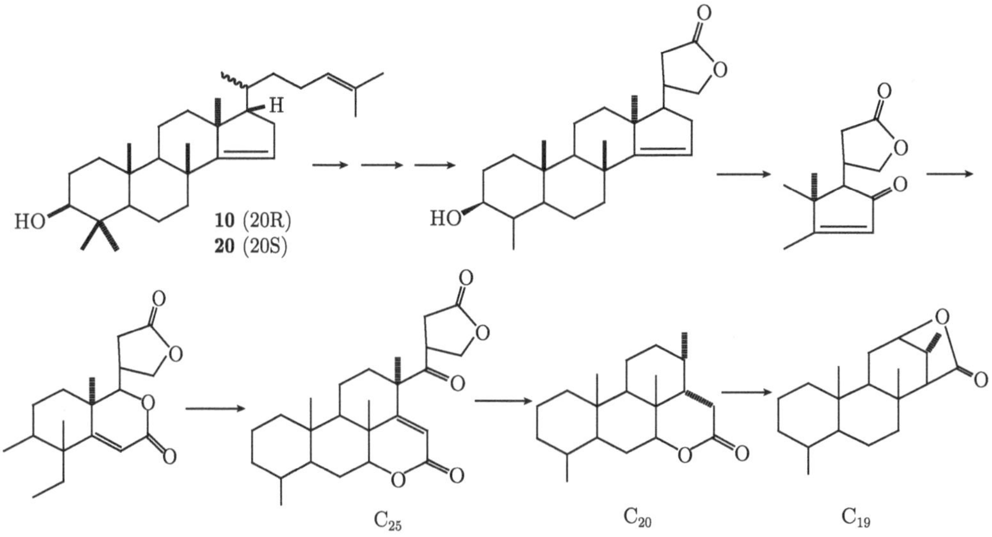

10 (20R)
20 (20S)

C_{25} C_{20} C_{19}

Many quassinoids have been isolated from *Brucea javanica* by Zhang et al.[28,29]. Bruceantin (**21**), obtained from *B. javanica*, exhibited significant anticancer activity. However, its clinical application was limited due to its severe toxicity. The total syntheses of compounds with quassinoid skeletons were also reported[30,31]. Glycosides with quassinoid types of aglycones were identified from plant sources, and examples include 2-*O*-glucosylsamaderine (**22**) from *Quassia indica* and bruceatin E 2-O-β-D-glucopyranoside (**23**) from *Brucea javanica*, respectively[32,33].

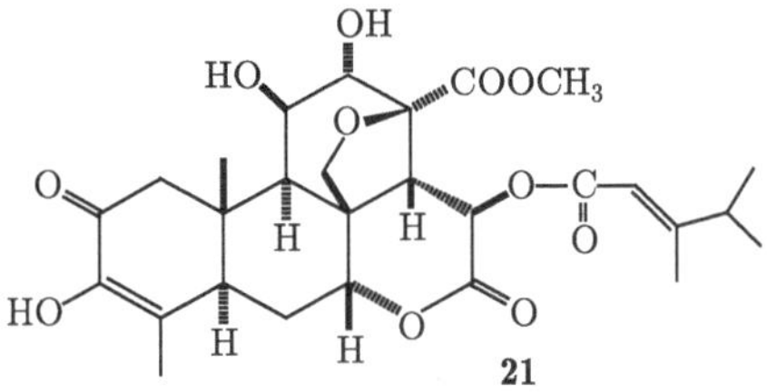

5.2.2   Pentacyclic triterpenoids

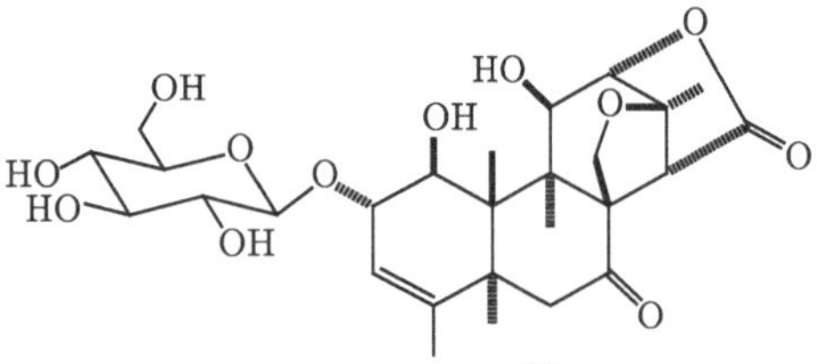

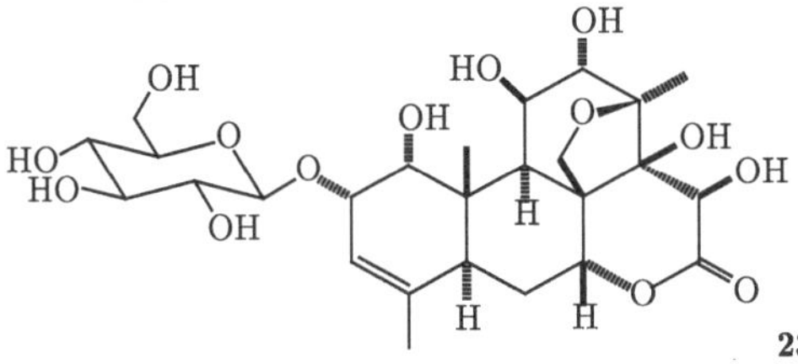

5.2.2.1   Oleanane (24) type triterpenoids

Oleanane type triterpenoids, also known as β-amyrin type triterpenoids, consist of five six-membered rings with the *trans* junction of rings A/B, B/C and C/D, and *cis* junction of rings D/E. Eight methyls were linked at C-4, C-4, C-8, C-10, C-14, C-17, C-20, and C-20, respectively. Oleanane type triterpenoids are widely distributed in the plant kingdom.

Oleanolic acid (**25**) exists in *Olea europaea* and *Ligustrum lucidum* from the Oleaceae family and many other plants. This compound may reduce the aminotransferase level, and has been used for the clinical treatment of acute jaundice hepatitis. Glycyrrhetinic acid (26) and glycyrrhizic acid (**27**), isolated from *Glycyrrhiza uralensis* (Leguminosae), possess ACTH-like activity, are also used clinically as anti-inflammatory drugs and for the treatment of gastric ulcers.

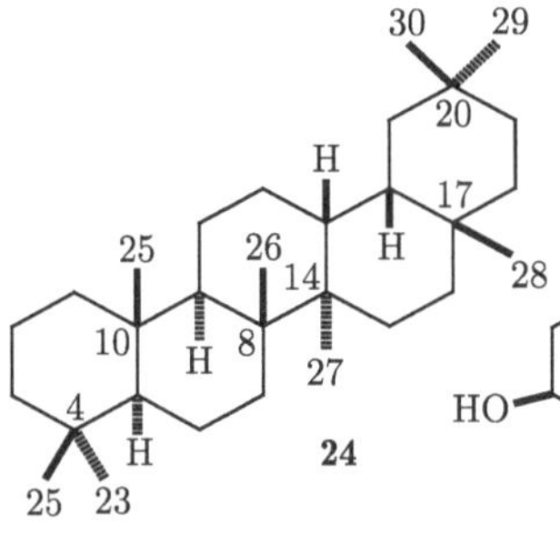

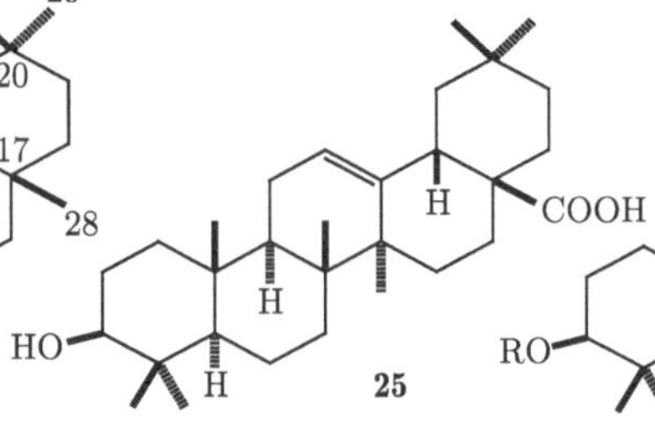

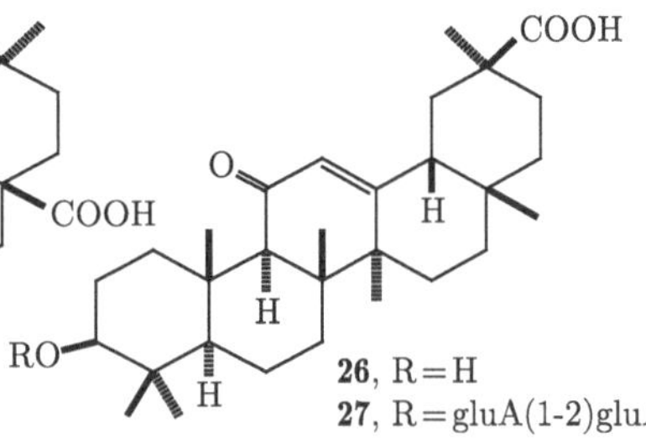

Oleanane type saponins are the most common triterpenoid saponins and most have a hydroxyl group at C-3, $\Delta^{12,13}$, and a carboxylic group at C-28.

5.2.2.2   Ursane (28) type triterpenoids

Ursane type triterpenoids, also known as α-amyrin type triterpenoids, contain five identical six-membered rings as compared to those of oleanane type triterpenoids with *trans* junction of rings A/B, B/C, and C/D, and *cis* junction of ring D/E. Their difference lay in the location of eight methyls at C-4, C-4, C-8, C-10, C-14, C-17, C-19, and C-20, respectively, in the former.

Free ursolic triterpenoids and ursolic triterpenoid glycosides are both common in plants. Saccharide chains are normally linked at C-3 and C-28 of aglycones. Asiaticoside (**29**) exists in high content in the plant *Centella asiatica*, and this compound may promote wound healing.

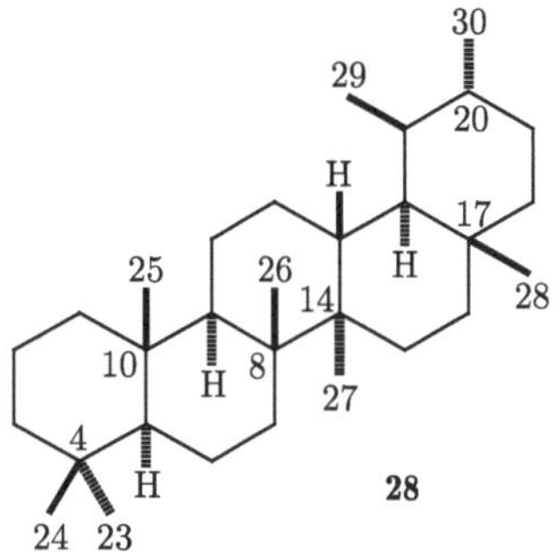

5.2.2.3   Lupane (30) type triterpenoids

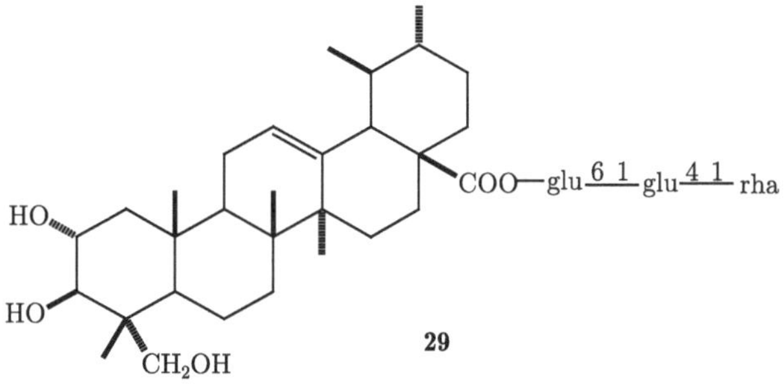

Lupane type pentacyclic triterpenoids possess five-membered E rings substituted by α-isopropyl at C-19, and the A/B, B/C, C/D, and D/E rings are all *trans* arranged.

Betulinic acid (**31**), of the lupane type triterpenoid, exists in the barks of *Betula platyphylla* Suk., barks and leaves of *Punica granatum* Linn., seeds of *Ziziphus jujuba* Mill. *var. spinosa* (Bunge) Hu *ex* H. F. Chou, and *Asparagus cochinchinensis* (Lour.) Merr., etc. The compound has been found to possess anti-HIV activity and selective cytotoxicity against tumor cells[34]. Ye et al. reported the isolation of pulsatilloside C (**32**), a lupane type glycoside, from the traditional Chinese herb *Pulsatilla chinensis*[35].

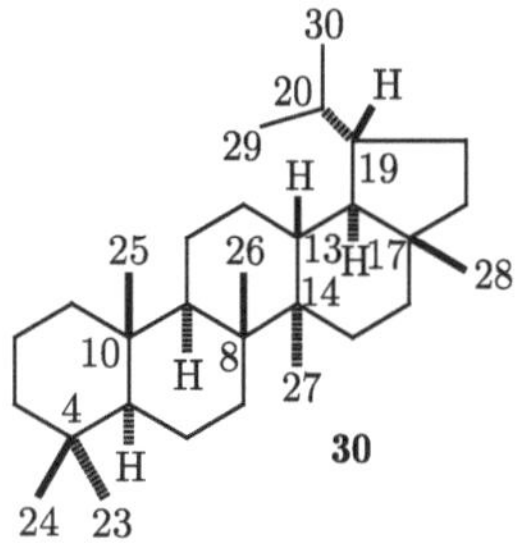

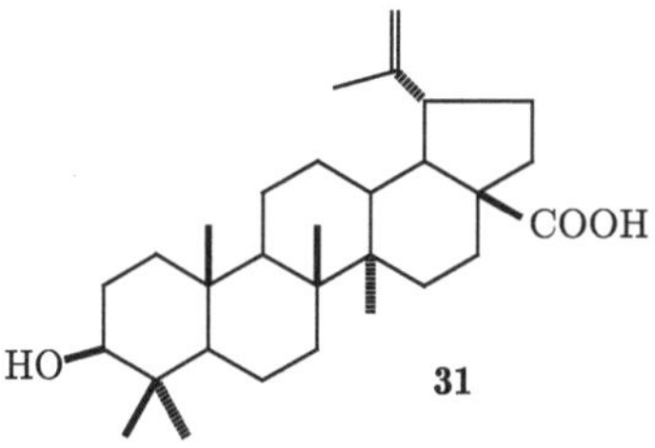

5.2.2.4   Hopane (33) and isohopane (34) type triterpenoids

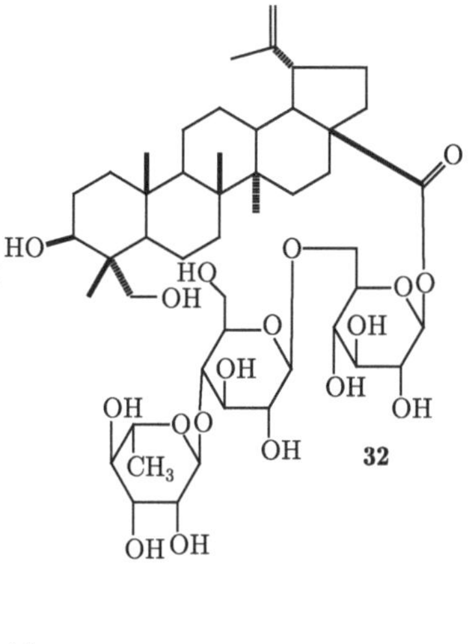

Hopane and isohopane type triterpenoids both possess five-member E rings substituted by isopropyl at C-21 with rings A/B, B/C, C/D, and D/E *trans* arranged. The structural difference between hopane and isohopane type triterpenoids lay in the orientation of the isopropy at C-21. Tanaka and Inatomi, et al. reported the identification of 17, 24-dihydroxyhopan-28, 22-olide (**35**) and diplazioside VI (**36**) from the fern *Diplazium subsinuatum*[36,37].

33 $R_1 = H$, $R_2 =$ isoproply
34 $R_1 =$ isopropyl, $R_2 = H$

35 R = H
36 R = glu(1-2)gluO-

5.3   Spectroscopic analysis of triterpenoids

5.3.1   Ultraviolet spectrum

Most triterpenoids have no strong UV absorption due to the lack of conjugated functional groups. Furan rings of toosendanin-type compounds and some triterpenoids with α, β-unsaturated carbonyl functional groups show characteristic UV absorption peaks. When two or more unsaturated groups exist in triterpenoids, UV spectroscopy can be used to determine whether these groups are conjugated.

5.3.2   Infrared spectrum

IR spectroscopy can be used to distinguish oleanane type, ursane type, and tetracyclic triterpenoids according to the absorption of hydrocarbons in regions A (1355 to 1392 cm^{-1}) and B (1245 to 1330 cm^{-1}). Oleanane type triterpenoids exhibit two bands (1392 to 1379 cm^{-1} and 1370 to 1355 cm^{-1}) in region A and three bands (1330 to 1315 cm^{-1}, 1306 to 1299 cm^{-1}, 1269 to 1250 cm^{-1}) in region B. Three bands can be found in region A (1392 to 1386 cm^{-1}, 1383 to 1370 cm^{-1}, and 1364 to 1359 cm^{-1}) and region B (1312 to 1308 cm^{-1}, 1276 to 1270 cm^{-1}, and 1250 to 1245 cm^{-1}), respectively, for ursolic type triterpenoids. For tetracyclic triterpenoids, there is only one band in region A and B, respectively.

5.3.3   Mass spectrum

Some characteristic fragments can be found in the EI-MS of pentacyclic triterpenoids. When an endo-cyclic double bond exists, ion signals due to typical RDA fragmentation can be observed. If there is no endo-cyclic double bond, fragments are often found due to the breakage of ring C. RDA fragmentation and breakage of ring C may happen at the same time in some cases. The functional group positions in a pentacyclic triterpenoid skeleton can be deduced according to the analyses of m/z fragments. Characteristic fragment ions in EI-MS of tetracyclic triterpenoids are due to the loss of side chains.

5.3.4   Nuclear magnetic resonance spectrum

Since the 1980s, high field NMR instruments and a series of new 2D NMR techniques have been applied in natural product chemistry research, made it possible to assign all hydrogen and carbon signals, and ensured the accuracy and increased the speed of structural identification of triterpenoid saponins.

The most distinguished ^{1}H NMR signals of a triterpenoid are protons from methyl groups, olefinic functional groups, and oxygen-connected carbons. All methyl proton signals of oleanane type triterpenoids are singlets. Doublet methyl proton signals can often be observed in the ^{1}H NMR spectra of ursane and tetracyclic type triterpenoids. The allyl methyl proton signal of a lupene type triterpenoid normally appears as a broad singlet at δ 1.63 to 1.80 ppm. The chemical shifts of endo-cyclic olefinic proton signals were at a lower field than 5 ppm, such as H-12 of oleanolic acid and ursolic acid between δ 4.93 and 5.50 ppm, while the chemical shifts of exo-cyclic olefinic proton signals were at higher field than 5 ppm, such as those in lupene and hopene between δ 4.30 and 5.00 ppm. The C-3 of most tetracyclic or pentacyclic triterpenoids is oxygenated. In some of them the H-3 appears as dd signals in the ^{1}H NMR spectrum if C-2 is not substituted. The two coupling constants can be used to identify the relative configuration of C-3.

Compared to ^{1}H NMR spectra, signals from ^{13}C NMR spectra are distributed in a broader range, and the phenomena of signal overlap are less frequent. Although the application of modern 2D NMR techniques enables the assignment of all proton and carbon signals of a triterpenoid, it is easier to identify the structures of known compounds by comparison of the acquired ^{13}C NMR data with those reported in literature. Mahato et al. collected ^{13}C NMR data of 396 pentacyclic triterpenoids with various skeletons;[38] Kalinovskii and Agrawal summarized ^{13}C NMR data of oleanane type triterpenoids, respectively;[39,40] Ageta and Chakravarty reported the assignment of signals in the ^{13}C NMR spectra of some hopane type triterpenoids;[41,42] Zuo et al. reviewed the application of ^{13}C NMR data in the structural determination of dammarane type triterpenoids[43].

Section 6   Steroidal Saponins

Aglycones of steroidal saponins are C-27 steroidal compounds, mainly distributed in the plants of Liliaceae, Dioscoreaceae, and Solanaceae. Plants in the family Scrophulariaceae, Amaryllidaceae, Leguminosae, and Rhamnaceae may also contain steroidal saponins. Several traditional Chinese medicines, such as Zhi-Mu (rhizome of *Anemarrhena asphodeloides* Bge), Mai-Dong (radix of *Ophiopogon japonicus* (Thunb) Ker-Gawl.), and Qi-Ye-Yi-Zhi-Hua (rhizome of *Paris chinensis* Franch.) contain abundant steroidal saponins. Steroidal aglycones are important raw materials in the production of progesterones, sex hormones, and glucocorticoids.

6.1   Steroidal aglycones

The most common steroidal aglycones are spirostane (**37**) derivatives with characteristic spiral ketal functional groups in the side chain. The C-5 and C-25 of natural spirostane derivatives may be isomeric, while the other chiral carbons are unchanged.

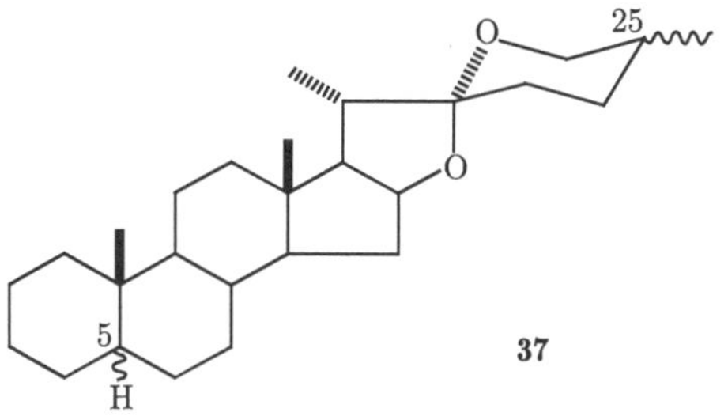

Three chiral centers exist in the side chains of spirostane derivatives, C-20 and C-22 are in S and R configurations, respectively, while C-25 isomers are distributed widely in plants. The two C-25 isomers can be transformed into each other. Using a hydrochloric acid-ethanol solution, more sarsasapogenin with C-25 axial methyl groups can be transformed into the isosarsasapogenin with a C-25 equatorial methyl group due to the poor stability of the former.

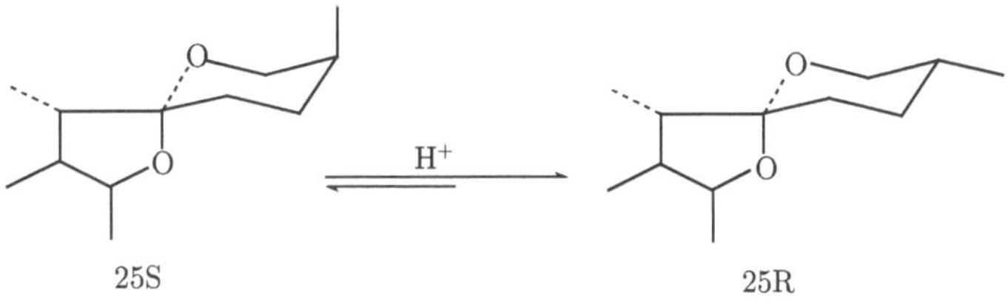

Pseudodiosgenin diacetate (**39**) can be obtained by reflux of diosgenin (**38**) in acetic anhydride. Further oxidation of pseudodiosgenin diacetate with Cr_2O_3 and elimination of the β-carboxyl ester group yields 3-acetoxypregna-5,16-dien-20-one (**40**), which may be used to produce progesterone[44]. A similar degradation reaction can also be applied to other steroidal aglycones to produce raw materials for large-scale production of progesterone, sex hormones, and glucocorticoid.

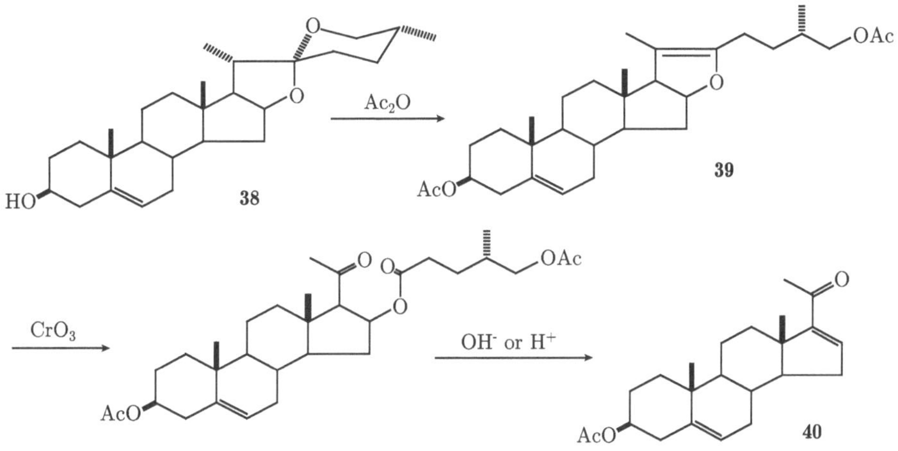

6.2   Spectroscopic analysis of steroidal aglycones

6.2.1   Ultraviolet spectrum

Saturated aglycones of steroidal saponins show no absorption between 200 and 400 nm. Introduction of isolated double bonds, carbonyl groups, α, β-unsaturated ketones, or conjugated double bonds may result in UV absorption at 205 to 225 nm (ε 900), 285 nm (ε 500), 240 nm (ε 11000), and 235 nm, respectively.

6.2.2   Infrared spectrum[45,46]

Steroidal saponins containing a ketal side chain may show four characteristic absorption bands at around 980 cm^{-1} (A), 920 cm^{-1} (B), 900 cm^{-1} (C), and 860 cm^{-1} (D). Among them, band A is the strongest. As for 25S-steroidal saponins or their aglycones, band B is stronger than band C, while for 25R-steroidal saponins or their aglycones, band C is stronger than band B. Such results could be used to distinguish the two C-25 stereoisomers.

6.2.3   Mass spectrum[47]

In the EI-MS of steroidal aglycones, fragmented peaks derived from the ketal side chain can be observed at m/z 139 (strong intensity), 115 (medium intensity), and 126 (weak intensity). The fragmentation pathways can be explained as below:

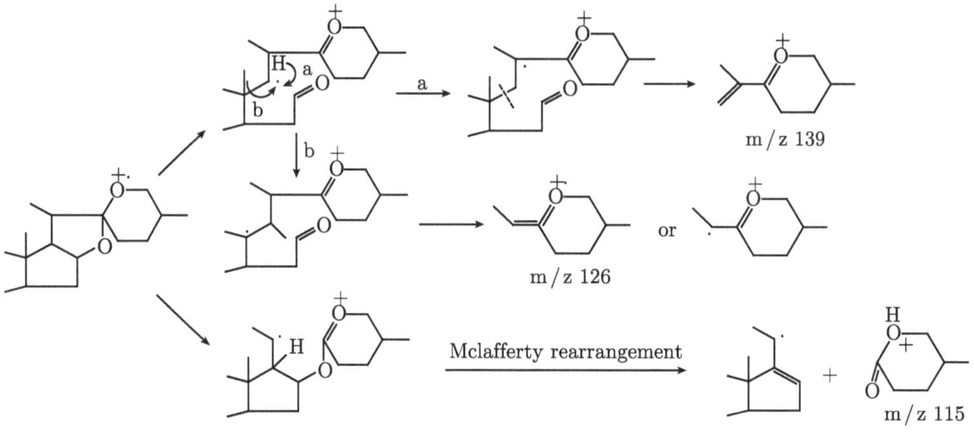

If C-25 or C-27 was substituted by a hydroxyl group, the above three fragmented peaks will shift up to m/z 155, 131, and 142, respectively. If an olefinic bond exists between C-25 and C-27, the above three fragmented peaks will exhibit at m/z137, 113, and 124. If a hydroxyl group exists at C-23, the base peak at m/z139 will disappear and no peak corresponding to the substitution can be found.

6.2.4 Nuclear magnetic resonance spectrum

Four characteristic methyl (18, 19, 21, and 27-Me) signals can be observed in the high field of the ^{1}H-NMR spectra of steroidal aglycones. Among them, 18 and 19 methyls are singlets with the former in the higher field, while 21 and 27 methyls are doublets with the latter in the higher field. If C-25 is substituted with a hydroxyl group, the 27-Me proton signal will exhibit as a singlet, and shifts to low field. H-16 and H-26 proton signals display at low field due to the linkage of oxygen with C-16 and C-26. The above proton signals can be easily distinguished, while other aliphatic proton signals are seriously overlapped and difficult to assign without using other NMR techniques.

In the ^{1}H NMR spectrum, the chemical shift of 27-Me in α orientation (equatorial, 25R) is in a higher field than its isomer of 25S configuration. Chemical shifts of the two signals of H-26 with 25R configuration are closer than those of the 25S isomer, and the difference in chemical shifts can be used to identify the 25R and 25S isomers[48].

The 27 carbon signals of the steroidal saponin aglycones are typically well separated in ^{13}C-NMR spectra compared to ^{1}H NMR spectra. The ^{13}C-NMR data of many steroidal saponins have been reported, and the structure of an aglycone can be confirmed by comparison of the data with those of known compounds[49−51].

6.3 Spirostanol saponins

Spirostanol saponins usually contain a saccharide chain at C-3, and in some cases, C-1, C-2, C-5, and C-11 can also be glycosylated. In recent years, many saponin glycosides with an open F-ring and connected by a sugar chain at C-26 were also isolated from plants. They will be discussed in section 6.5 due to their unique properties.

Dioscorea is the largest genus in the Dioscoreaceae family, and most *Dioscorea* plants contain steroidal saponins. In the 1930s, the aglycone of steroidal saponindiosgenin (**38**) was first isolated by Fukuda et al. from the plant *D. tokoro*. After the successful transformation of diosgenin into hormone drugs, the compound has become a major raw material for the synthesis of hormones. Professor Ren-Hong Zhu obtained diosgenin for the first time from the plant *D. septemloba*, which broke new ground in the investigation of steroidal saponins in

China. So far, a large number of steroidal saponins have been identified from the *Dioscorea* plants. *D. zingiberensis* and *D. nipponica* are the main sources of diosgenin in China. Trillin (**41**), gracillin (**42**), and diosgenin-diglucoside (**43**) were the three major saponins isolated from dried stems of *D. zingiberensis*, while *D. nipponica* mainly contained dioscin (**44**) and gracillin[52,53]. *D. althaeoides* contains a higher content of diosgenin and possesses potential economic value. The main steroidal saponins contained in *D. althaeoides* are diosgenin (**38**) and gracillin (**42**)[54].

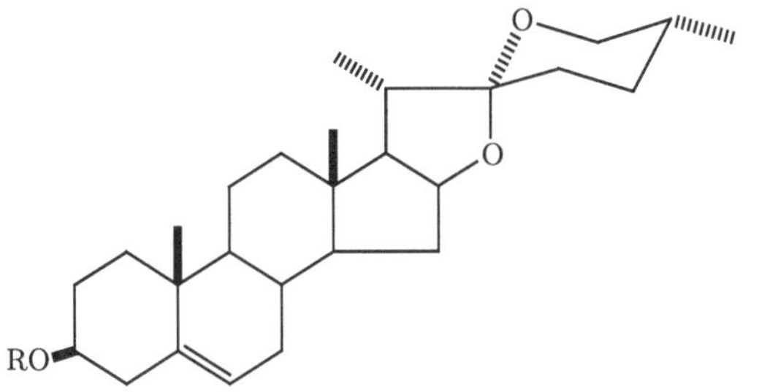

41 R = D-glu

42 R = D-glu $\xrightarrow{1\quad 3}$ D-glu-
                                             $\underset{\underset{\text{L-rha}}{1}}{\overset{2}{|}}$

43 R = D-glu $\xrightarrow{1\quad 4}$ D-glu-

44 R = L-rha $\xrightarrow{1\quad 2}$ D-glu-
                                             $\underset{\underset{\text{L-rha}}{1}}{\overset{4}{|}}$

6.4   Furostanol saponins

Studies showed that some spirostanol saponins do not exist in fresh plants, but are actually produced in the process of drying and storage of the plant material. Furostanol saponins were not obtained during the early research work because of purification difficulties and also the failure to recognize the enzymatic hydrolysis of the original saponins. Re-examination revealed that furostanol saponins are widespread in steroidal saponin-containing plants.

Dioscorea plants contain many kinds of saponins. Protozingberenssaponin (**45**) and pro-togracillin (**46**), the two major original saponins, were obtained from the fresh plant material of *D. zingiberensis*. Only three secondary saponins could be isolated from the dried plant[55].

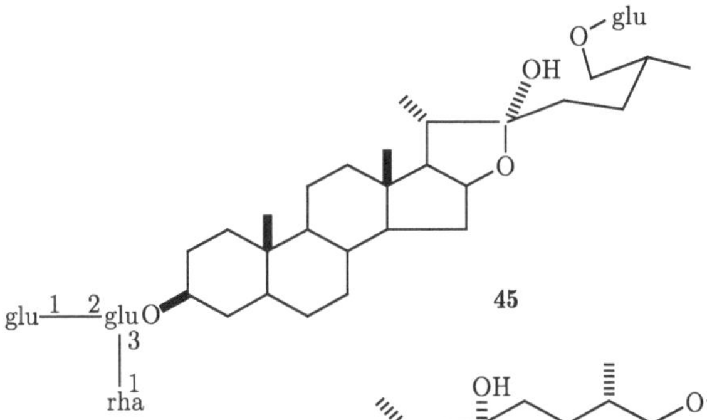

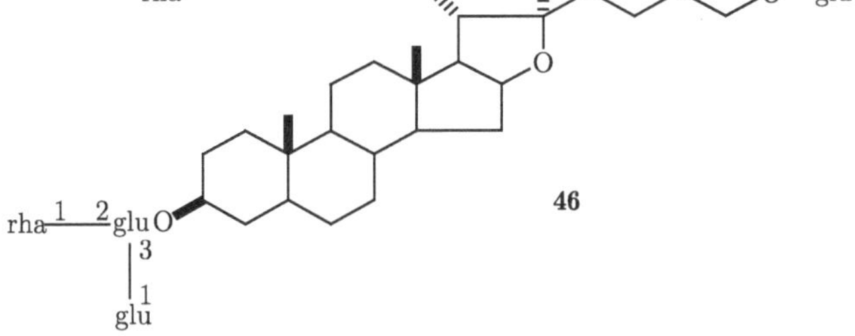

6.5   Furospirostanol saponins

Furospirostanol saponins possess a furan F-ring instead of a pyran F-ring in the spirostanol aglycones. Aculeatisides A (**47**) and B (**48**), obtained from *Solanum aculeatissimum*, are disaccharides of the aglycone nuatigenin (**49**). Acidic hydrolyses of 47 and 48 afforded not only nuatigenin (**49**), but also isonuatigenin (**50**). Baeyer-Villiger degradation of **47** and **48** yielded **51** and (S)-4, 5-dihydroxypentanoic acid 5-O-β-D-glucopyranoside (**52**), which confirmed the existence of only one glucose moiety linked to C-26 of the aglycone[56].

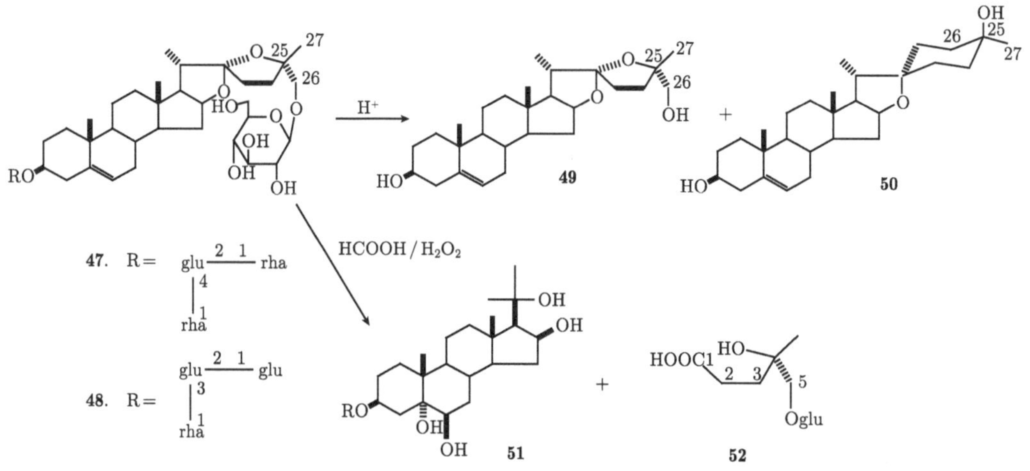

Bibliography

[1] Rensheng Xu. Natural Products Chemistry. 1st ed, Beijing: Science Press, 1989.

[2] Xinsheng Yao. Chemistry of Natural Medicines. 2nd ed, Beijing: People's Medical Publishing House, 1997.

[3] Xiankai Wang. Chemistry of Natural Medicines. Beijing: People's Medical Publishing House, 1988.

[4] Waller G R, et al.. Saponins Used in Food and Agriculture. New York: Plenum Press, 1996.

[5] Waller G R, et al.. Saponins Used in Traditional and Modern Medicine. New York: Plenum Press, 1996.

References

[1] Zhao W M, et al., J. Nat. Prod. 1994, 57: 1613.

[2] Diallo B, et al., J. Chromatogr. 1991, 558: 446.

[3] Smith F, et al., Methods in Carbohydrate Chem. 1965, 5: 361.

[4] Kong F H, et al., Acta Chim. Sinica 1988, 46: 775.

[5] Hara S, et al., Chem. Pharm. Bull. 1987, 35: 501.

[6] Agrawal P K, Phytochemistry 1992, 31: 3307.

[7] Massiot G, et al., J. Chem. Soc. Chem. Commun. 1986, 1485.

[8] Bothnerby A A, et al., J. Am. Chem. Soc. 1984, 106: 811.

[9] Rao A V, et al., Drug Metab. Drug Interact. 2000, 17: 211.

[10]   Biotechnology Research Institute of The Hong Kong University of Science and Technology, Traditional Chinese Medicine Research and Development, Reviews by Biotechnology Research Institute Visiting Scholars (1997-1999). Beijing: Science Press, 2000: 255.

[11]   Kong F H, et al., Tetrahedron Lett. 1986, 27: 5765.

[12]   Yu L, et al., Planta Medica 1994, 60: 204.

[13]   Kim W, et al., Koryo Insam Hakhoechi 1989, 13: 24.

[14]   Kensil C R, et al., Adv. Exp. Med. Biol. 1996, 405: 165.

[15]   Ju D W, et al., Acta Pharmaceutica Sinica, 1994, 29: 252.

[16]   Li J B, et al., Chinese Traditional and Folk Medicine, 1992, 23: 249.

[17]   Takechi M, et al., Phytochemistry 1990, 29: 451.

[18]   Li X, Chen J X, Sun J J, Acta Pharmacologica Sinica 1990, 11: 26.

[19]   Li L X, et al., Chinese Pharmacological Bulletin 1991, 1: 56.

[20]   Gao B Y, et al., Acta Pharmaceutica Sinica 1992, 27: 641.

[21]   Lei C L, et al., Journal of Norman Bethune University of Medical Science 1994, 20: 326.

[22]   Tang S R, et al., Acta Botanica Sinica 1983, 25: 556.

[23]   Bedir E, et al., Biol. Pharm. Bull. 2000, 23: 834.

[24]   Xinsheng Yao, Chemistry of Natural Medicines. 2nd ed, Beijing: People's Medical Publishing House, 1997: 327.

[25]   Rui H K, et al., Acta Pharmaceutica Sinica, 1981, 6: 445.

[26]   Kraus W, et al., Tetrahedron Lett. 1985, 26: 6435.

[27]   Sawabe A, et al., Carbohydr Res. 1999, 315: 142.

[28]   Zhang J S, et al., Planta Medica 1980, 39: 265.

[29]   Zhang J S, et al., Acta Chim. Sinica 1984, 42: 684.

[30]   Vanderroest J M, et al., J. Am. Chem. Soc. 1993, 115: 5841.

[31]   Grieco P A, et al., J. Am. Chem. Soc. 1994, 116: 7606.

[32]   Kitagawa I, et al., Chem. Pharm. Bull. 1996, 44: 2009.

[33]   Fujioka T, et al., J. Nat. Prod. 1994, 57: 243.

[34]   Zuco V, et al., Cancer Lett. 2002, 175: 17.

[35]   Ye W, et al., J. Nat. Prod. 1998, 61: 658.

[36]   Tanaka N, et al., Chem. Pharm. Bull. 1982, 30: 3632.

[37]   Inatomi Y, et al., Chem. Pharm. Bull. 2000, 48: 1930.

[38]   Mahato S B, et al., Phytochemistry 1994, 37: 1517.

[39]   Kalinovskii A I, Chem. Nat. Compd. (Engl Transl) 1992, 28: 1.

[40]   Agrawal P K, et al., Prog. Nucl. Magn. Reson. Spectrosc. 1992, 24: 1.

[41]   Ageta H, et al., Chem. Pharm. Bull. 1994, 42: 39.

[42]   Chakravarty A K, et al., Tetrahedron 1994, 50: 2865.

[43]   Zuo G Y, et al., Chinese J. Org. Chem. 1997, 17: 385.

[44]   Marker R E, et al., J. Am. Chem. Soc. 1939, 61: 3592; 1940, 62: 518, 648, 898, 2532.

[45]   Rolandeddy C, et al., Anal. Chem. 1953, 25: 266.

[46]   Jones R N, et al., J. Am. Chem. Soc. 1953, 75: 158.

[47]   Djerassi C. et al., Org. Mass Spectr. 1970, 3: 1187.

[48]   Williams D H, et al., Tetrahedron 1965, 21: 1841.

[49]   Garcia J A R, et al., Magn. Reson. Chem. 1987, 25: 831.

[50]   Yang R Z, et al., Acta Botanica Yunnanica 1987, 9: 374.

[51] Agrawal P K, et al., Phytochemistry 1985, 24: 2479.

[52] Liu C L, et al., Acta Botanica Sinica 1984, 26: 283.

[53] Fang Y W, et al., Acta Pharmaceutica Sinica 1982, 17: 388.

[54] Liu C L, et al., Acta Pharmaceutica Sinica 1984, 19: 799.

[55] Liu C L, et al., Acta Botanica Sinica 1985, 27: 68.

[56] Saijo R, et al., Phytochemistry 1983, 22: 733.

Wei-Min ZHAO

CHAPTER 8

Amino Acids and Peptides

Proteins, saccharides, and nucleic acids are three common polymers and are vital substances found in all living organisms. Among them, proteins have the most diversified physiological functions. As building blocks, they are present in all living cells and involved in the construction of all tissues such as skin, muscle, bone, nerve and blood. As enzymes or hormones, they catalyze and regulate universal chemical reactions in a living body. As antibodies, they help to defend against attack by many pathogens. Almost every protein has a different function, although there is some overlapping of jobs. For example, hemoglobin in red blood cells can transport oxygen, a substance vital for the continuance of life, to every cell in the body.

The word "protein" is derived from the Greek word "proteios", which means primary or holding first place. Proteins are polymers of α-amino acids. There are only 20 α-amino acids used to make proteins in nature. These 20 α-amino acids are called common amino acids and they are all L-α-amino acids. Proteins are highly complicated biomolecules which usually have molecular weights larger than 10,000. As a small protein, insulin has a molecular weight of 6,000. When dimerized, its molecular weight is doubled to 12,000 and the dimer has a more complicated high-order structure than the monomer has. Peptides are smaller than proteins. They are defined as molecules consisting of two or more amino acids. They also have stable high-order structures and diversified physiological functions. Peptides have very important roles in life activities, which include the transformation of signals, regulation of metabolism, and harmonization of different physiological pathways.

Section 1 Amino Acids

1.1 Structure and classification of amino acids

The 20 natural α-amino acids which are used to make up proteins are listed in Table 8-1. According to the properties of their R groups, amino acids are divided into three subgroups (acidic, basic and neutral). Except for glycine, which is not a chiral molecule, amino acids are all L- amino acids. In other words, their absolute configurations are related to L-glyceraldehyde.

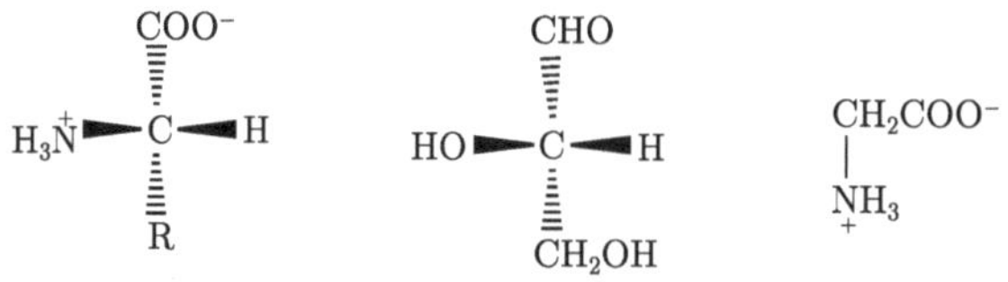

Following Cahn-Ingold-Prelog's rule, most L-α-amino acids have the S configuration because the carboxyl groups in natural α-amino acids are usually prior to the R groups.

To date, there are only L-α-amino acids found in proteins, but D-α-amino acids also exist in nature. For example, D-serine was found in the forebrains of mammals and D-aspartic acid was found in nerve endings[1]. Some D-amino acids have been found in invertebrates and amphibious organisms such as frogs, crawdads and lobsters. Among them, an interesting example is D-Alanine, which was found in dermorphin in 1981[2]. In sea biomass, we could also find some D-α-amino acids. Certain D-α-amino acids are involved in some marine natural products, such as valinomycin, tyrocidine and gramicidine[1].

Table 8-1   Common Amino Acids

Side chain group	Name	Ab.	pK$_{\alpha1}$α-COOH	pK$_{\alpha2}$α-NH$_3$	pK$_{\alpha3}$RGROUP	pI
Neutral amino acids						
-H	Glycine	G or Gly	2.3	9.6		6.0
-CH$_3$	Alanine	A or Ala	2.3	9.7		6.0
-CH(CH$_3$)$_2$	Valine	V or Val	2.3	9.6		6.0
-CH$_2$CH(CH$_3$)$_2$	Leucine	L or Leu	2.4	9.6		6.0
- CH CH$_3$CH$_2$CH$_3$	Isoleucine	I or Ile	2.4	9.7		6.1
-CH$_2$C$_6$H$_5$	Phenylalanine	F or Phe	1.8	9.1		5.5
-CH$_2$CONH$_2$	Asparagine	N or Asn	2.0	8.8		5.4
-CH$_2$CH$_2$CONH$_2$	Glutamine	Q or Gln	2.2	9.1		5.7
-CH$_2$C$_8$H$_5$NH	Tryptophan	W or Trp	2.4	9.4		5.9
HOOC-CHCH$_2$CH$_2$CH$_2$NH	Proline	P or Pro	2.0	10.6		6.3
-CH$_2$OH	Serine	S or Ser	2.2	9.2		5.7
-CHOHCH$_3$	Threonine	T or Thr	2.6	10.4		6.5
-CH$_2$-C$_6$H$_4$-OH	Tyrosine	Y or Tyr	2.2	9.1	10.1	5.7
HOOC-CHCH$_2$CHOHCH$_2$NH	Hydroxyproline	Hyp	1.9	9.7		6.5
-CH$_2$SH	Cysteine	C or Cys	1.7	10.8	8.3	5.0
-CH-S	Cystine	Cys-Cys	1.6	7.9		5.1
-CH-S	Cystine		2.3	9.9		
-CH$_2$CH$_2$SCH$_3$	Methionine	M o rMet	2.3	9.2		5.8
Acidic amino acids						
-CH$_2$COOH	Aspartic acid	D or Asp	2.1	9.8	3.9	3.0
-CH$_2$CH$_2$COOH	Glutamic acid	E or Glu	2.2	9.7	4.3	3.2
Basic acids						
-CH$_2$CH$_2$CH$_2$CH$_2$NH$_2$	Lysine	K or Lys	2.2	9.0	10.5	9.8
-CH$_2$CH$_2$CH$_2$NHCNHNH$_2$	Arginine	R or Arg	2.2	9.0	12.5	10.8
-CH$_2$C$_3$N$_2$H$_3$	Histidine	H or His	1.8	9.2	6.0	7.6

Besides the common amino acids, hundreds of types of non-protein amino acids have been found in nature and they have multiple functions in living organisms. These uncommon amino acids cannot be used to synthesize protein because there is no specific *t*RNA or codon for them. Neither could they be incorporated into protein by post-translational modification. Non-protein amino acids are structurally diversified and widely distributed in nature. There are more than 240 non-protein amino acids found in plants[3], and most of them are multi-functionalized aliphatic amino acids. 2, 6-Diaminopimelic acid **1** is a neutral amino acid and 4-carboxyl-4-hydroxy-2-amino adipic acid **2** is a basic amino acid.

(1)   (2)

Besides open-chain aliphatic amino acids, there are also heterocyclic amino acids found in nature. 3-Chloro-5-keto-3,4-pyrrolidene-2-carboxylic acid, **3**, was isolated from fungus[4] and imidazole 2-carboxylic acid, **4**, was isolated from a sponge *Tedamia anhelans*[5]. 4-Methylaeruginoic acid is a cytoxic amino acid, which was isolated from the microbe *Streptomyces* KCTC9303[6].

(3)   (4)

Some non-protein amino acids like β-alanine and γ-amino butyric acid (GABA) are widely distributed in plants, possibly because they are closely involved in the metabolism of protein amino acids and other primary amines.

All the above mentioned common amino acids and non-protein amino acids occur naturally. However, there are also many synthetic amino acids which are widely used in place of natural amino acids during the synthesis of peptides. The peptides bearing uncommon synthetic amino acids may retain their normal physiological functions. They may also have some additional physiological properties which may be of medical significance and have clinical applications. Synthetic amino acids, such as naphthylalanine (Nal), pyridyl alanine (Pal), 1,2,3,4-tetrahydroisoquinoline-3-carboxylic acid (Tic), β-(2-thienyl)alanine (Thi) and β-cyclohexylalanine (Cha), etc., are widely used in synthesis of peptidic medicines including antagonists of luteinizing hormone-releasing hormone (LH-RH), inhibitors of rennin, inhibitors of HIV protease, and so on.

1.2   Physical properties of amino acids

In a crystalline state, an amino acid exists as a zwitterion, a substance which has both positive and negative charges, due to the coexistence of a basic group (-NH_2) and an acidic group (-CO_2H). In aqueous solution, because amino acids are both proton acceptors and proton donors, they are balanced in a cationic, zwitterionic and anionic state.

$$\overset{+}{H_3}NCHCOOH \underset{+H^+}{\overset{-H^+}{\rightleftharpoons}} \overset{+}{H_3}NCHCOO^- \underset{+H^+}{\overset{-H^+}{\rightleftharpoons}} H_2NCHCOO^-$$
$$\underset{R}{\qquad} \qquad \underset{R}{\qquad} \qquad \underset{R}{\qquad}$$

cation   zwitterion   anion

(in acidic medium)   (in basic medium)

The ionization properties of amino acids can be expressed as Ka (or pKa). The constant Ka is called the ionization constant and could be determined from a titration curve. The pKa represents the relative proton-donating ability and every amino acid has its own unique pKa value. When an amino acid has more than one proton to donate, the ionization constant of the most acidic proton would be defined as pKa_1, followed by pKa_2, pKa_3, etc. Usually, for α-amino acids, the α carboxylic acid group is the strongest proton-donating group. Therefore, it has the pKa_1, the β carboxylic acid group has pKa_2, and the α-N^+H_3 has pKa_3. For example, L-aspartic acid has a pKa_1 2.10, a pKa_2 3.86 and a pKa_3 9.83.

Each amino acid has an isoelectric point (PI), which is a pH value at which the amino acid is neutral. In other words, at its PI, the amino acid carries no net electrical charge since the negative charge originated from carboxylic dissociation equals the positive charge from the protonation of the amino group. Although still in a dissociated state, an amino acid would not migrate in an electric field at its PI due to its zero net charge. For glycine, from its titration curve, the ionization constant can be determined as pKa_1 2.34 and pKa_2 9.69. Then, the PI of glycine can be calculated as follows to be 6.02.

$$pI = 1/2(pKa_1 + pKa_2)$$

For amino acids which have more than two ionizable groups (-COOH, -N^+H_3), such as Asp, Glu, and Lys, their PI values can be calculated from the pKa values at both sides of the zwitterion. For example, lysine has a PI value of 9.8 which is the average of its pKa_2 9.0 and pKa_3 10.5.

$$\frac{pKa_2 + pKa_3}{2} = \frac{9.0 + 10.5}{2} = 9.8$$

$$\overset{+}{H_3}N(CH_2)_4CHCOOH \underset{H^+}{\overset{OH^-}{\rightleftharpoons}} \overset{+}{H_3}N(CH_2)_4CHCOO^- \underset{H^+}{\overset{OH^-}{\rightleftharpoons}}$$

NH$_3$ NH$_3$
+   +

pka$_1$=2.2   pka$_2$=9.0

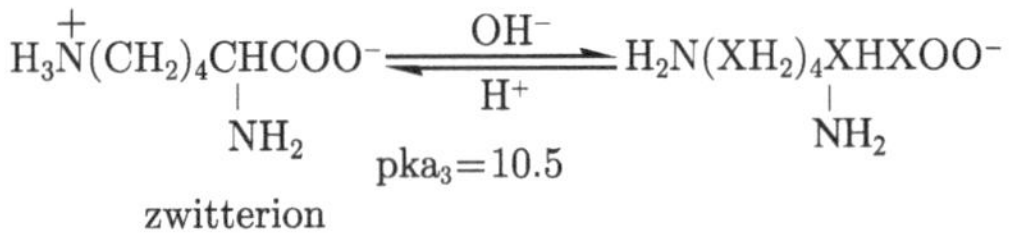

$$\overset{+}{H_3}N(CH_2)_4\underset{\underset{NH_2}{|}}{C}HCOO^- \underset{\underset{pka_3=10.5}{H^+}}{\overset{OH^-}{\rightleftharpoons}} H_2N(XH_2)_4XH\underset{\underset{NH_2}{|}}{X}OO^-$$

zwitterion

1.3   Chemical properties of amino acids

All the functional groups including the amino group, carboxyl group, and side chain group in an amino acid can take part in their respective specific chemical reactions. The carboxy acid can be ionized, esterified, converted to amide, etc., and the amino group can be acylated, converted to halide or hydroxyl group, etc. Different side chains may contain different functional groups which would have different reaction abilities. For example, the side chain of cysteine has a mercapto group which can be alkylated or oxidated. The most important chemical property of amino acids is the formation of peptide bonds which lays the foundation of peptide and protein chemistry.

1.3.1   Acylation

Under acidic conditions, acyl halide reacts with an amino acid to produce the O-acylated product while under weak basic conditions. A N,O-diacylated product could be obtained. Under strong basic conditions, only the N-acylated product could be obtained. Preparation of N-acylated derivatives to protect the amino group of an amino acid is an important step during peptide synthesis.

Acylation is also a method for detection of amino acids. For example, the amino group of an amino acid can react with dansyl chloride (DNS-Cl) to produce DNS-amino acid, a strong fluorescent substance which could be applied to detect an amino acid[7].

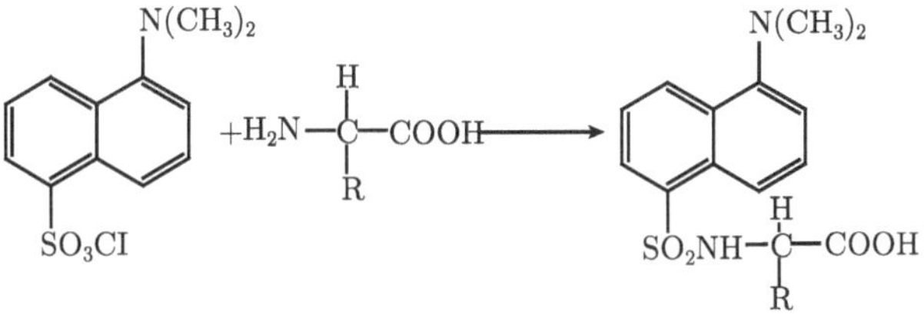

Amino acids can react with phenylisothiocyanate under mild basic conditions to produce a N-acylated product. Then, under strong acidic conditions, the N-acylated product can be converted to a very stable PTH-amino acid (3-phenyl-2-thiohydantoin-amino acid). PTH-amino acid can be isolated and identified by chromatographic methods. This process is called Edman degradation. In the Edman process, the amino acid's residues of a peptide or protein will be released sequentially from the N-terminus which allows the determination of the primary structure of a given peptide or protein[7].

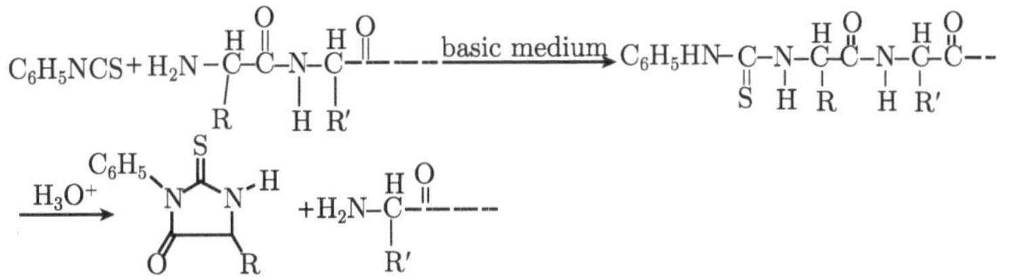

1.3.2   Reaction with CO_2[8,9]

The amino group of an amino acid can reversibly react with carbon dioxide to produce carbamate.

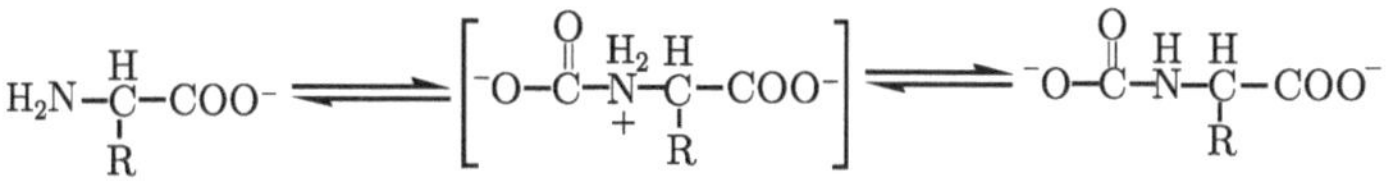

Besides the α-amino group of an amino acid, any other primary amino group, such as the N-terminal amino group of a polypeptide, can also be converted to carbamate when reacted with carbon dioxide. However, a cyclic secondary amino group, such as in proline, would not undergo this reaction.

1.3.3   Formation of Schiff base

A Schiff base can be formed by a carbonyl group and a primary amino group. It has geometric isomers (Z/E) due to the presence of a carbon nitrogen double bond.

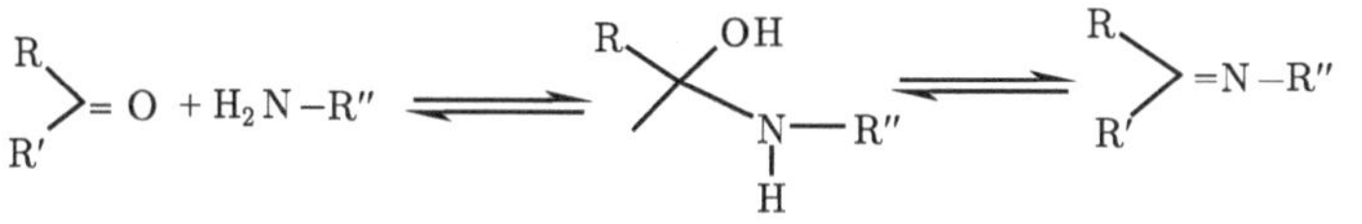

Under basic conditions, a Schiff base may tautomerize via a deprotonated intermediate. During the tautomerization, with the migration of hydrogen, a chiral carbon bonded to the amino group will be racemized.

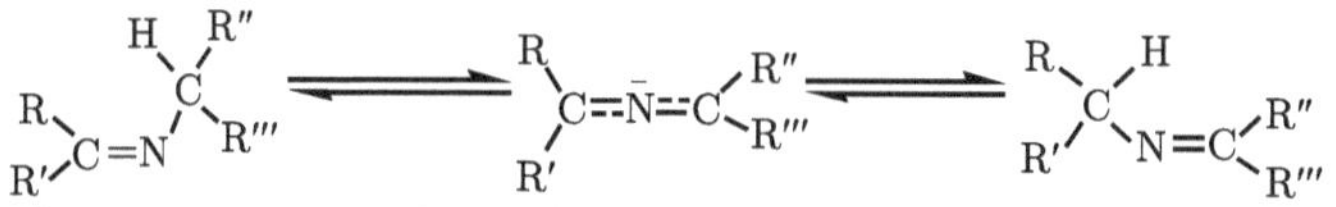

The presence of a metal ion will facilitate the formation of Schiff bases of amino acids and pyridoxal. For example, compound **5** was obtained from tryptophan and pyridoxal[10]. This class of Schiff bases lays the general foundation for enzyme-catalyzed amino acid biotransformations including transamination, decarboxylation, racemization, etc. The condensation of amino acids or other amines with aldehydes is also an important pathway for the biosynthesis of alkaloids.

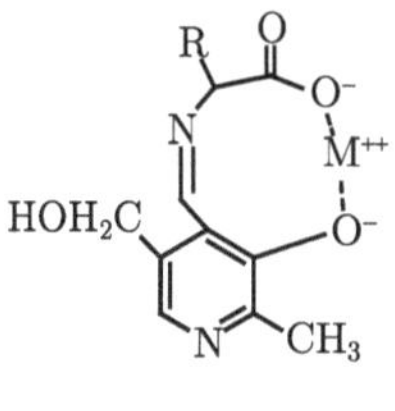

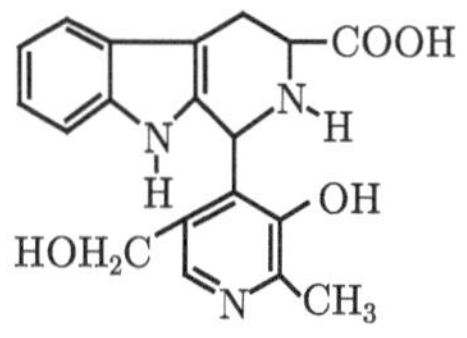

(5)

1.3.4   Alkylation

The amino group and side chain functional group, such as a hydroxyl or mercapto group of an amino acid, can be alkylated. The alkylated amino acids are less polar than unalkylated ones. Therefore, alkylation of an amino acid will facilitate its isolation and identification. Alkylation is also an important method for the preparation of amino acid derivatives, which leads to structural diversity. The alkylation of a cysteine's mercapto group by α-iodoacetic acid can be utilized to study the cysteine residues in a protein.

$$R\text{-}SH + ICH_2COOH \longrightarrow R\text{-}S\text{-}CH_2COOH + HI$$

Under weak basic conditions, the amino group of an amino acid can react with 2,4-dinitroflorobenzene (DNFB) to produce a yellow substance, N-dinitrobenzyl amino acid (DNP-amino acid). Treated with DNFB, followed by hydrolysis in 20% aqueous HCl, a protein can be degraded into free amino acids. Among these degraded amino acids, α-DNP amino acids are from the N termini of the protein while other DNP amino acids, such as ε-NH-DNP lysine, O-DNP tyrosine, etc., are from amino acid residues which have an additional amino group on their side chains. Therefore, DNP derivatization is widely used in N-terminal sequence analysis of peptides[7].

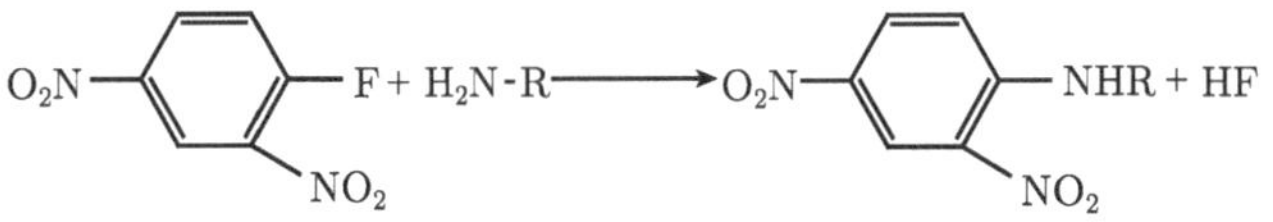

N-alkyl amino acids are widely distributed in proteins and peptidic antibiotics from plants or some sea biomass. For example, cyclosporin is enriched in N-methyl amino acid. Alkylation of the amino group of amino acids is a common method to prepare N-alkyl amino acid. N-alkyl amino acids are also often used to synthesize peptidic medicines.

1.3.5   Reactions involving both amino and carboxyl groups

N-carboxy-anhydride (NCA), also known as Leuchs anhydride, can be obtained by reacting an amino acid with phosgene.

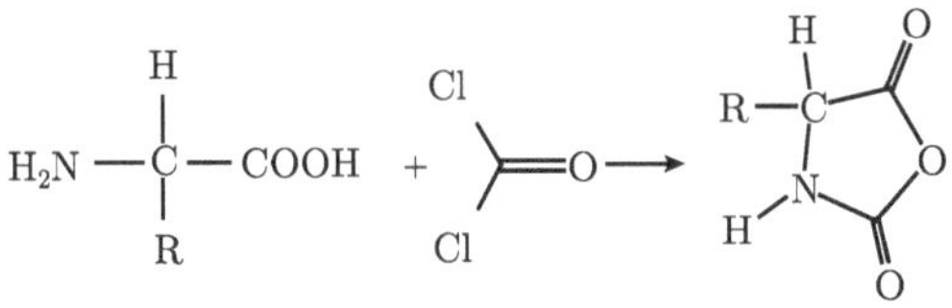

Leuchs anhydrides are very reactive and can polymerize to a peptide; thus they have been widely used for the preparation of polypeptides. However, the peptides prepared from NCAs are often randomly sequenced. Recently, Goodman and coworkers[11] developed a new method to synthesize peptides using protected N-carboxy anhydrides' UNCAs as starting materials. Goodman's method has the advantages of fast reaction and easy operation. It can be applied both in liquid and solid phase synthesis.

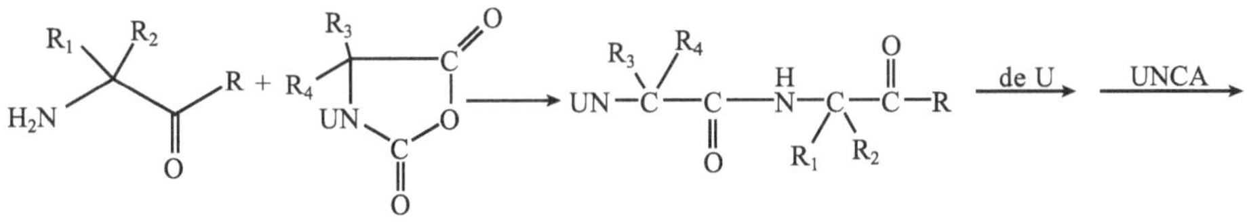

1.3.6   Interaction with metal ions

Amino acids can chelate with metal ions such as Na^+, K^+, Mg^{++}, Ca^{++}, Ba^{++}, etc. by forming chelating compounds. Metal amino acid chelate has a rather high stability constant. When the lone pair electrons of an amino nitrogen atom chelate with metal, the alkaline character of the amino group disappears. In other words, in a neutral aqueous solution of amino acids when the zwitterionic amino acids are chelated with metal, the solution will become acidic. The chelated amino acids and free amino acids have a great difference in their chromatographic and electrophoresis behaviors. Based on different dissolving ability of metal amino acid chelates, some amino acids can be purified through their metal chelated form by crystallization. In some cases, the metal chelated amino acids can even be used in the kinetic resolution of racemic amino acids. For instance, DL-aspartic acid can be resolved by their copper chelates[12].

Copper chelation can also facilitate α alkylation of an amino acid. For example, threonine can be prepared from copper-chelated glycine[13].

1.4   Purification and Characterization of Amino Acids

1.4.1   Color reaction

If an amino acid exists in its free form, it can be directly identified by many kinds of chromogenic reagent. So If the amino acid is bonded to another structural unit through its amino or carboxyl group, it might need to be hydrolyzed before being subjected to color reagents. Among numerous chromogenic reagents, ninhydrin is the most important one and is widely used for detecting an amino acid. Ninhydrin can react with the free amino groups of an α-amino acid to produce a purple or violet colored substance **6**[14].

(6)

The colors observed from the reactions of non-protein amino acids with ninhydrin differ greatly due to the varied structures of non-protein amino acids. Except for tertiary amines, ninhydrin can also react with other primary amines, secondary amines, oligopeptides, and amino alcohols to produce specifically colored substances.

Besides the substrate structure, the ninhydrin test is also influenced by factors such as the pH value of the reaction medium, metal ions, etc. With the presence of an appropriate metal ion, such as Ca^{++} or Cu^{++}, different amino acids would produce more differentiatable colored solutions while reacting with ninhydrin[15]. Isatin is another widely used chromogenic reagent for detecting an amino acid. The isatin test is not as sensitive as the ninhydrin test and the color produced by the isatin test is not stable. However, different amino acids would generate more differentiable colors in the isatin test, which will lead to identification of a specific amino acid.

Some reagents react with amino acids to produce fluorescent substances and can be used to detect amino acids. For example, fluorescamine can react with amino acids to produce fluorescent derivatives **7** (Ex = 390 nm, Em = 475 nm), which are easily detected by a fluorescence spectrophotometer.

(7)

Other applicable chromogenic reagents for detection of amino acids are listed in Table 8-2.

1.4.2   Techniques in isolation and analysis

There are a variety of applicable methods for the isolation and analysis of amino acids. Paper chromatography and paper electrophoresis have been widely used. However, due to their complicated operation and low efficacy, these two methods are not commonly used nowadays. Instead, thin layer chromatography (TLC) techniques using silica gel, cellulose, and polyamide as stationary phases are the most commonly used methods.

From a TLC plate, amino acids can be detected by a variety of chromogenic reagents and ninhydrin is one of the most widely used. A UV lamp is also applicable to detect amino acids, such as Phe, Tyr, Trp, which have an aryl group in their side chains.

Ion exchange chromatography (IEC) was first applied in quantitative analysis of amino acids by Moore and Stein in 1958. Although the isolatical, isolational, and analytical techniques have advanced greatly since then, IEC is still the most important. There are a number of other existing methods for quantitative analysis of amino acids nowadays. In addition to ICE, gas chromatography (GC), high performance liquid chromatography (HPLC) and reverse-phase HPLC (RP-HPLC) are all applicable. The amino acid analyzer, which is based on post-column derivetization of amino acids with ninhydrin, has also advanced greatly. By decreasing the resin diameter, increasing column pressure, changing isocratic elution to gradient elution, and promotion of automation, the detection speed and sensitivity of an amino analyzer are both greatly enhanced (see Figure 8-1). While the analysis of the amino acid composition of a protein took up to a day decades ago, it takes only about half an hour with only P mol of samples[15].

Table 8-2   Chromogenic Reagents for Detection of Amino Acids

Reaction name	Reagent	Color	Amino acid
Sakaguchi reaction	alpha-naphthol and hypochlorous acid	red	arginine and other guanide group containing compounds
Nitroprusside reaction	sodium nitroferricyanide in diluted aqueous ammonia	red	cysteine and other mercapto group containing compounds
Sullivan reaction	1,2-naphthoquinone-4-sulfonic acid and sodium sulfite	red	cysteine
Pauly reaction	diazotized sulfanilic acid	red	tyrosine, histidine and other imidazole derivatives
Ehrlich reaction	p-dimethylaminobenzaldehyde and conc. HCl	blue	tryptophan and other indole derivatives
Hopkins-Cole reaction	glyoxalic acid and conc. H_2SO_4	red and purple	tryptophan and other indole derivatives
Folin-Ciocalteau reaction	phosphomolybdium tungstic acid	blue	tyrosine
Millon reaction	mercurous nitrate and mercuric nifrate in conc. HNO_3	red	tyrosine and other phenolic compounds
Xanthoproteic reaction	conc. HNO_3	yellow	phenylalanine, tryptophan, tyrosine
Phthalaldehyde	o-phthaldehyde alcohol solution	dark green	glycine

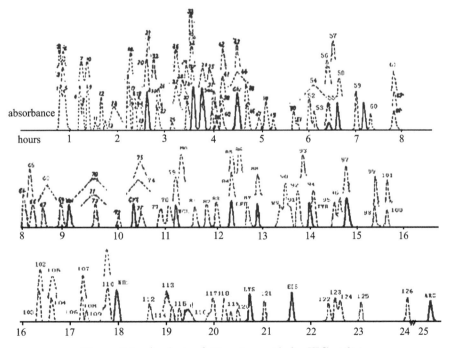

Figure 8-1   Analysis of 18 amino acids by IEC utilizing post-column derivatization with ninhydrin[7]

1.4.3 Infrared spectroscopy[16]

Amino acids can exist in zwitterionic form as well as cationic and anionic forms. Both the amino group and the carboxyl group of an amino acid can exist in their free form ($-NH_2$, $-COOH$) or dissociated form ($-^+NH_3$, $-COO^-$).

When the amino acid exists in its free form or anionic form, it has a protonated amino group ($-^+NH_3$) instead of a free amino group. There is no characteristic infrared absorption band in the area of 3300 to 3500 cm^{-1} representing N-H stretching vibration. Instead, $^+NH_3$ asymmetric stretching vibration appears at 3030 to 3130 cm^{-1}. Symmetric stretching vibrations of $^+NH_3$ will appear in the area of 3000 to 2000 cm^{-1}. For proline, the protonated imino group $^+NH_2$ displays one absorption band near 2900 cm^{-1} representing the asymmetrical stretching vibration of $^+NH_2$. The metal salts of amino acids have a characteristic $-NH_2$ infrared absorption band at 3390 to 3260 cm^{-1}. Amino acid esters will display N-H stretching vibrational absorption at lower frequencies.

A protonated primary amino group, $-^+NH_3$, has two additional characteristic absorption bands within the range of 1660 to 1485cm^{-1}. One is at 1660 to 1610cm^{-1} (W) representing its asymmetrical bending vibration and often appears as shoulder absorption to the absorption peak for $-COO^-$. Another band which appears at 1550 to 1485 cm^{-1} (6.45 to 6.73 μm) represents the symmetrical bending vibration of $-^+NH_3$.

When the $-NH_2$ group was mono-alkylated, such as N-methyl glycine, the resulting –NH group will display stretching vibration as absorption at 3500 to 3300cm^{-1}.

The number of carbons between the amino group and carboxyl group seems to have no influence on the zwitterionic character of an amino acid. For $H_2N(CH_2)_{10}COOH$, even though there are ten carbons between the amino group and the carboxyl group, characteristic absorption bands for $-NH_3^+$ are still observable.

Zwitterionic as well as anionic forms of amino acids have strong absorbance at 1600 to 1560 cm^{-1} due to the existence of a deprotonated carboxyl group ($-COO^-$). For the cationic form of amino acids, the absorption bands for the stretching vibration of a carboxyl group are related to the position of the amino group. When the amino group is α positioned to the carboxyl group, ν_{co} appears at 1750 to 1740 cm^{-1} or it will appear at 1730 to 1700 cm^{-1}.

Comparison of IR spectra of different forms is necessary for detecting an amino acid. Certainly, for an amino acid which has two carboxyl groups and one amino group, there will be one more band for ν_{co} and for an amino acid which has two amino groups and one carboxyl group there will be extra bands indicating the presence of another amino group.

1.4.4 Mass spectroscopy

Mass spectroscopy is indispensable for the structural determination of peptides and proteins. Under electron impact (EI) ionization conditions, the molecular ion peak of an amino acid is usually very weak or even undetectable due to the highly polar nature of free amino acids[17]. Some derivatization methods, like acylation and TMS alkylation, can decrease the polarity of amino acids. Thus, they are often utilized in mass spectroscopic analysis of amino acids[18].

$$CH_3-CH-CH_2-CH_2-COOH \qquad TMS-O-CH_2CH_2CH_2-CH--COOH-TMS$$
$$CH_3 \qquad NH_2 \qquad\qquad NH$$
$$\qquad\qquad\qquad\qquad\qquad\qquad TMS$$
$$M^+=159 \qquad\qquad\qquad\qquad M^+=349$$

Mass spectroscopy can also be used in determination of absolute configuration of amino acids. By Marfey's method, amino acids are derivatized with 1-fluoro-2,4-dinitrophenyl-5-

L-alanine amide (FDAA) and then subjected to LC/MS analysis for determination of the absolute configuration[19].

Mass spectroscopy is prevalently used in the determination of a peptide or protein mainly because only a tiny amount of sample is needed for the test[20]. With the great advances in ionization technologies, a number of ionization methods such as chemical ionization (CI), field desorption ionization (FD) and fast atom bombardment ionization (FAB) can be used to detect protonated amino acids in mass spectroscopy. Electron spray ionization (ESI) and matrix assisted laser desorption ionization (MALDI) techniques developed at the end of 1980s especially promoted the application of mass spectroscopy in structural determination of peptides and proteins.

1.4.5   Identification of succinamopine[21]

Succinamopine **9** is an imino acid produced by tobacco crown gall tumors induced by *Agrobacterium fumefaciens*. Succinamopine is prone to lactamization under acidic conditions. For example, at pH 2, succinamopine **9** would be lactamizated to **11** in a few days at room temperature. Succinopine lactam **12** would be obtained at a higher temperature.

9 X = NH₂

10 X = OH

11 X = NH₂

12 X = OH

Compound **9** is negative to ninhydrin testing. It can be detected by bromocresol green, silver/mannitol, or ferric thiocyanate. Compounds **9** and **10** are positive to hypochlorite/tolidine test, which indicates the presence of an imino group (>NH). Potentiometric titration provided dissociation constants of compound **11**: $pK_{a1} = 2.3$, $pK_{a2} = 4.7$, which showed the presence of two carboxyl groups. Although succinamopine gives no ninhydrin color reaction, it moves very slightly in the cationic direction in pH 1.8 electrophoresis and is retained from 50% ethanol on a cation exchange resin. These facts also suggest the presence of the imino diacid group, and the facile lactamization strongly points to a glutamyl residue in the imino polyacid succinamopine.

The NMR spectrum of succinopine lactam **12** in D_2O contains a set of two coupled signals (d, 2.98 ppm; t, 4.88 ppm) in the ratio 2:1 and a second set of broad signals (2.48 and 4.55 ppm) in the ratio 4:1. The 4.88 ppm triplet is the signal of the α-H adjacent to the most acidic carboxyl (aspartyl α-carboxyl) while the 4.55 ppm multiple is the signal of the α-H adjacent to the glutamyl α-carboxyl. The NMR spectrum of succinopine lactam **12** methyl ester has methoxyl resonances at 3.51 and 3.58 ppm (ratio 1:2), indicating the presence of three esterifiable carboxyls in compound **12**.

The NMR spectrum of succinopine lactam **12** trimethyl ester in carbon tetrachloride had signals of appropriate areas at 2.20 (m, glutamyl β, γ-H), 2.80 (d, J = 6 Hz, aspartyl β,β'-H), 3.54 (s, 1 methoxy), 3.60 (s, 2 methoxys), 4.30 (m, glutamyl α-H), and 4.85 ppm (t, J = 6 Hz, aspartyl α-H). The infrared spectrum of the ester in carbon tetrachloride shows a strong ester band at 1745 cm⁻¹, a tertiary γ-lactam at 1705 cm⁻¹, and the absence of any NH or OH stretch above 3000 cm⁻¹.

The fast atom bombardment (FAB) positive-ion mass spectrum of succinamopine lactam **11** had the strongest peak at 245 $[M + H]^+$, with a weaker, ammoniated peak at 262 $[M + NH_4]^+$ and a cluster ion at 489 $[2 M + H]^+$. Weaker cluster ions involving succinamopine lactam were also observed. Succinamopine gave a positive FAB quasi-molecular ion at 263 $[M +H]^+$ and a negative FAB ion at 261 $[M–H]^-$. The high resolution MS of succinopine lactam triethyl ester (molecular ion 329.1464) is in excellent agreement with the elemental composition of $C_{15}H_{23}NO_7$ required for the triethyl ester of structure 2b. Major electron-impact fragments can be observed for the loss of ethoxy (284), loss of carbethoxy (256), and extrusion of CO (182).

Synthetic succinamopine was prepared by reductive condensation of α-ketoglutaric acid and (S)-asparagine. The product, after elution from a cation exchange resin, showed the expected presence of diastereomers barely resolvable by electrophoresis at pH 2.8. Each synthetic diastereomeric pair (succinamopine, succinamopine lactam, and succinopine lactam) has a greater difference between first dissociation constants than between second dissociation constants. This difference in pK was used to fractionate synthetic succinopine lactam and isosuccinopine lactam on an anion exchanger by suppression of ionization with a formic acid gradient. As expected, the weaker acid isosuccinopine lactam emerged first at ca. 3.5 M formic acid and succinopine lactam at ca. 4.8 M.

Succinopine lactam was prepared by methods involving one center of known chirality and creation of one new center, leading to diastereomeric pairs in each case. For convenience, the two chiral centers of succinopine lactams are referred to as the aspartyl and the glutamyl chiral centers. The method of synthesis and selectivity of fermentation clearly indicate that natural succinopine lactam has (R)-glutamyl and (S)-aspartyl chirality; hence, natural succinamopine and succinamopine lactam have this chirality also.

Section 2   Peptides

Both peptides and proteins are basic substances of life. From a chemical point of view, they both are polymers of amino acids and the basic amino acid units are linked together through an amide bond (peptide bond). There are only 20 amino acids consisting of natural peptides or proteins and they are all L-α-amino acids. Occasionally, some non-natural D-amino acids can be found incorporated into microbic metabolites, plants, and occasionally low grade animal originated peptides.

Peptides and proteins have no difference in their basic structures. They just differ in their molecular mass. A peptide is conventionally defined to be a molecule which has a molecular mass less than 10,000 (containing about 100 amino acid residues), while a protein is defined to be a molecule which has a bigger molecular mass. Due to their bigger molecular mass, proteins obviously have more internal structural stabilizing factors, such as secondary bonds like hydrogen bonds, salt bonds, and interactions between aryl groups. So, generally, proteins have more stable and complicated conformational structures than peptides.

Peptides and proteins are extremely important for life sciences, biology and medical sciences, and endocrinology. They are extensively related to every physiological and pathological event including cell differentiation, immune defense, carcinogenesis, aging, reproduction, and so on. Some natural peptides, as well as some synthetic peptides, are well known for their application as biological medicines. Insulin is used to treat diabetes. Calmodulin is used to treat osteoporosis. Luteinizing hormone-releasing hormone (LH-RH) is used to treat some reproductive system diseases and some peptide antibiotics are able to inhibit tumors.

As enzymes, proteins are undoubtedly vital for life activities. Research on the chemistry of peptides and proteins will improve our understanding of life processes and help to promote life quality[22].

2.1   Structures and properties of peptides and proteins

Peptides and proteins are a group of amino acids linked together through a peptide bond (-CONH-). The structure of a tripeptide consisting of an alanine, a cysteine, and a valine is shown below. There are two peptide bonds in a tripeptide.

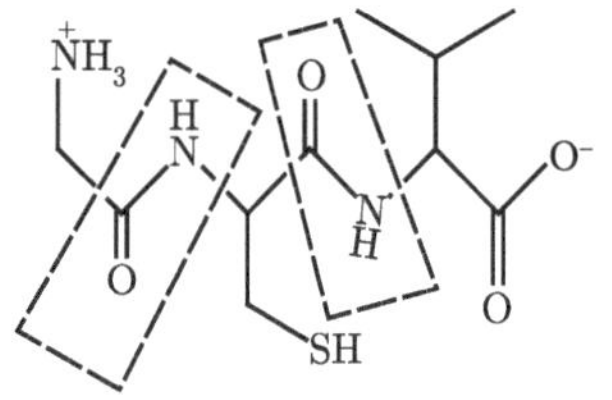

The amide structure has two resonance contributors:

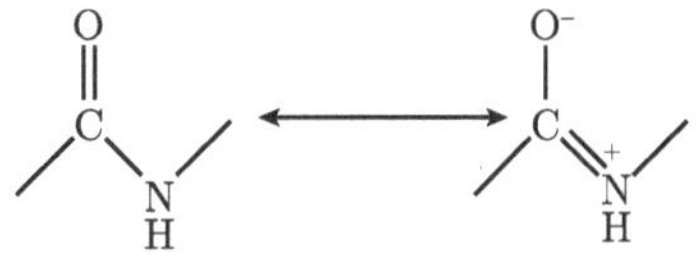

The resonance suggests that the amide group has a partial double bond character so that rotation around this bond is restricted. As a result, the four atoms of a peptide bond lie in a single plane, which is a crucial factor for the stable conformations of peptides and proteins.

Based on their structures, peptides can be divided into two major classes: linear peptides and cyclic peptides. Luteinizing hormone-releasing hormone (LHRH), as an example, is a linear decapeptide.

Glp-His-Trp-Ser-Tyr-Gly-Leu-Arg-Pro-Gly-NH$_2$

Besides linear peptides, cyclic peptides are also widely distributed in nature. As seen in many cyclic peptide antibiotics, cyclic peptides may be formed through head-to-tail cyclization. Sometimes, certain unnatural amino acids, such as cyclosporine A, are found in these cyclic peptides.

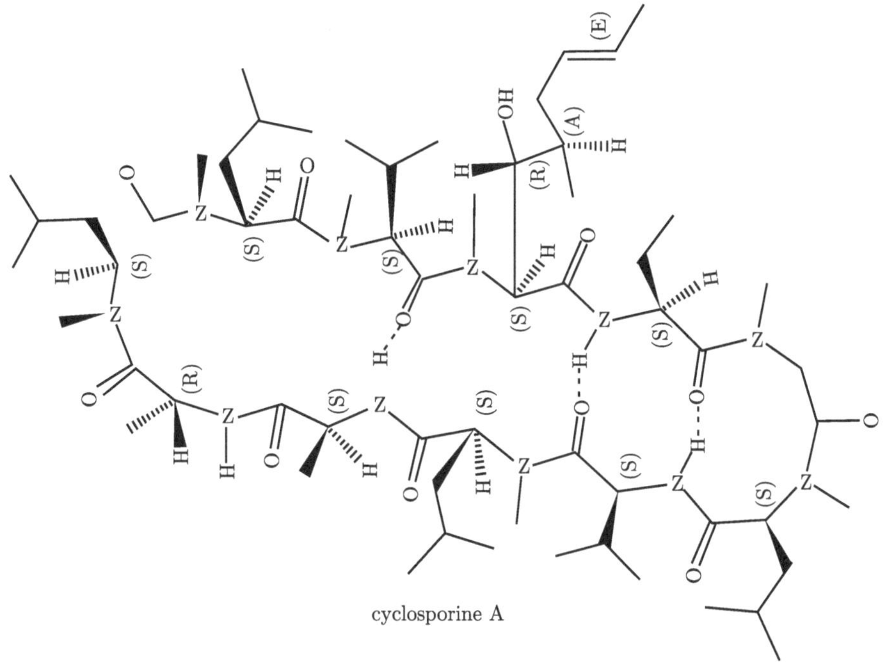

cyclosporine A

Many cyclic peptides are formed from a linear peptide by a disulfide bond between two cysteine residues. For an example, oxytocin is a cyclic peptide of this class. Formation of a macrocyclic lactone is another commonly used method for construction of a cyclic peptide.

$$\text{Cys-Tyr-Ile-Gln-Asn-Cys-Pro-Leu-Gly-NH}_2$$

oxytocin

As science and technology develop, more and more natural cyclic peptides have been found in plants. These cyclic peptides can be further divided into two groups: cyclopeptides and cyclotides. Cyclopeptides are plant secondary metabolites and usually have fewer than 14 amino acid residues, some of which are modified amino acid residues[23]. Cyclotides are recently characterized molecules found in plants and they are small disulfide-rich proteins which usually consist of 28 to 37 amino acid residues[24]. Kalata B1 is a cyclotide of 29 amino acid residues from the Rubiaceae family[25]. Plant cyclotides are usually found in the Rubiaceae, Violaceae, and Cucurbitaceae families and possess multiple bioactivities[26,27].

2.2 Natural bioactive peptides

To date, the research of natural bioactive peptides has focused on those from animals, especially from mammals. In mammals, like human beings, all endocrine organs contain plenty of natural bioactive peptides.

There are numerous peptidic hormones widely distributed in mammals. Although they often exist in trace amounts, these peptidic hormones have their physiological functions. For example, luteinizing hormone-releasing hormone (LH-RH), a hormone that is produced by adenohypophysis, is essential for the maturation of sex organs. In the early 1970s, from hundred thousands of pigs and goats, only micrograms of LH-RH were isolated and purified. Based on this small amount of sample, the structure of LH-RH was determined by Schally and Guillenin[28].

According to their different secretive glands and target organs, as listed in Table 8-3, peptide hormones can be classified as targeting the hypothalamus, hypophysis, digestive tract, pancreas hormone, etc.[29]. Studies have demonstrated that most peptide hormones are widely distributed rather than present in a specified tissue or organ. In regards to the distribution of bioactive peptide hormones, Erspamer[30] proposed a concept of trigonal (brain, intestine, skin) distribution. Recently, some proposed that neuropeptides should be independent from other bioactive peptides to form a new principle because they are synthesized and released by neural cells and specifically act on neurons[31]. Some bioactive peptides, like kinins, are not only present in high grade animals but also widely distributed in lower animals. For example, magainins, a class of antimicrobial peptides from amphibian skin, have similar physiological functions and highly homologous sequences as some bioactive peptides from high grade animals.

While few known bioactive peptides originated from plants, plenty of them were found in microorganisms and fungi. Two well known antibiotics, penicillins and cephamycins, are both lactam tripeptides.

Valinomycin is a cyclic peptide antibiotic consisting of 12 amino acid residues. It can accommodate a naked potassium ion, and carries it across the membrane[32]. Nisin has a more complicated cyclic structure which was determined in 1971 and later synthesized by Shiba[33]. It is typically used as an antiseptic for food preservation. Cyclosporine is a D-amino acid and non-protein amino acid enriched cyclic peptide which can be isolated from the fungi *Trichoderma polysporum* and *Tolypocladium inflatum*. It is an important immune inhibitor clinically used after transplant surgery[34]. Cyclosporin can also inhibit the growth of fungi and yeasts and exhibits anti inflammatory activity. More than 100 antibiotic peptides have been isolated from insects. Silkworm antimicrobial peptide D and snake venom peptide 13 are very active against both Gram positive and Gram neqative bacteria. A 33 peptide obtained from tarantulas displays strong analgesic activity[35].

Table 8-3   Distribution and Physical Functions of Common Peptide and Protein Hormones

Endocrine gland	Peptide hormone	Target organ	Biological function
Adenohypophysis	Luteinizing hormone (LH) or interstitial cell stimulating hormone (ICSH)	Ovary, testes	Essential to sex gland development, sex hormone production and secretion, and ovulation
	follicle stimulating hormone (FSH)	Ovary, testes	Stimulates the growth of Graafian follicles, enhances spermatogenesis and controls the production and secretion of sex hormones with LH
	Thyroid stimulating hormone (TSH)	Thyroid gland	Stimulates thyroid gland development and production and release of thyroid hormones T3 and T4
	Prolactin (PRL)	Galactophore, mammary gland, corpus luteum	Stimulates lactation and mammary glands and the production of progesterone
	Human growth hormone (HGH)	Whole body	Anabolic effect on the growth and metabolism of body and tissues.
	Adrenocorticotrophic hormone (ACTH)	Adrenal cortex	Stimulates the secretion of corticosteroid hormones from adrenal cortex
	Lipotropic hormones (LPH)	——	Precursors of prohormones β-endorphin, β-melanotropin, etc.
Neurohypophysis	Oxytocin	Uterine smooth muscle, breast smooth muscle	Stimulates the contraction of uterine smooth muscle at birth, stimulates milk secretion after birth
	Vasopressin	Small artery, renal tubule	Regulates water retention and raises blood pressure
Pancreas	Insulin	Liver, adipose tissue, muscle	Controls cellular intake of glucose, decrease blood glucose level and enhances lipogenesis
	Glucagon	Liver, adipose tissue, muscle	Increases the breakdown of glycogen, raises blood glucose level and lipolysis of triglyceride
Digestive tract	Gastrin	Stomach	Stimulates the secretion of gastric acid
	Cholecystokinin (CCK)	Gallbladder, pancreas	Stimulates the secretion and delivery of pancreatic enzymes and bile
	Vasoactive intestinal peptide (VIP)	Vascular smooth secretin muscle	Similar to that of glucagon and inhibits the secretion of gastric acid
	Gastric inhibitory polypeptide (GIP)	Stomach, pancreas	and induces insulin secretion
Parathyroid	Parathyroid hormone (PTH)	Bone, kidney, intestinal tract	Raises calcium level of blood and regulates phosphate metabolism
	Calcitonin (CT)	Bone, blood	Reduces blood calcium level and regulates phosphate metabolism
Corpus luteum	Relaxin	Pubic symphysis, uterine muscle	Relaxes birth canal muscles
Thymus gland	Thymus hormone	Lymphatic system	Stimulates T-lymphocyte development and enhances immune system

2.2.1   Purification and identification of peptides

Generally speaking, methods for peptide purification are similar to those used in the purification of amino acids and proteins. Chromatography, electrophoresis, and ion exchange all can be used in peptide isolation. Here, we will explore some practical methods, such as gel filtration chromatography, SDS PAGE electrophoresis, and RP-HPLC.

1. Gel filtration chromatography

Gel filtration chromatography is also known as molecular sieve chromatography or molecular exclusion chromatography. A porous macromolecule is used as stationary phase and peptides can be separated according to their molecular sizes. As shown in Figure 8-2, larger peptides only move across the external volume, whereas smaller peptides move across the external and internal volumes. As a result, larger peptides usually are eluted out earlier than smaller ones.

Dextran and agarose are two stationary phases mainly used for gel filtration chromatography. Dextran provided by Pharmacia is known as Sephadex. According to their different crosslinking degree and applications, Sephadex has subtypes from G10 to G200. Sepharose and Biogel are two common commercially available types of agarose.

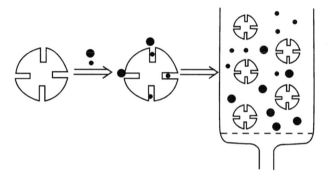

Figure 8-2   Peptide movement in gel piltration chromatography

2. SDS PAGE electrophoresis

Polyacrylamide is also a kind of gel, but it is more often used as a matrix for electrophoresis. Sodium dodecyl sulfate polyacrylamide gel electrophoresis (SDS-PAGE) is widely used in the purification and analysis of peptides and proteins.

3. HPLC

HPLC refers to high performance (or high pressure) liquid chromatography. A number of polymers, such as ion exchange resin and gel are applicable as stationary phases for HPLC. Silica gel is most commonly used, especially when it is modified to be hydrophobic (reverse phase silica gel). Reverse phase HPLC is common for peptide separation. As a stationary phase, reverse phase silica gel is a hydrophobic alkylated silica gel. The mobile phases often used in reverse phase HPLC are polar aqueous solution and buffers with different pH values. Organic solvents, like acetonitrile and methanol, can be mixed into the mobile phase to enhance the eluting power. In RP-HPLC, higher polarity peptides will be eluted earlier than lower polarity peptides. Under appropriate conditions, peptides with similar properties and structure can be effectively isolated. An example is shown in Figure 8-3.

Some other methods often used in the purification of peptides, like isoelectric focusing, capillary electrophoresis and SDS-PAGE, are also widely used in the purification and identification of proteins.

4. Examples of isolation, purification, and identification of bioactive peptides

Because most bioactive peptides exist in tiny amounts in animal tissues, their isolation and purification processes are much more complicated. Usually, a great deal of hard work needs to be done in order to obtain a pure bioactive peptide.

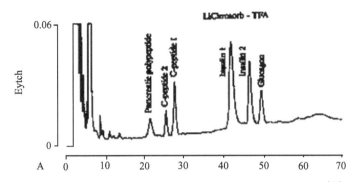

Figure 8-3   Isolation of synthetic GnRHs by RP-HPLC[36].
0.2 μg of each GnRH was loaded on a VydacC$_{18}$ (5 μm), 0.21×15 cm column, TEAP buffer (pH 6.5) as mobile phase, flow rate 0.2 ml/min.

Isolation, purification, and identification of gonadotropin releasing hormone (GnRH) and Leuteinizing hormone releasing hormone (LH-RH) are typically complicated processes. GnRH is a linear decapeptide. It is mainly present in the hypothalamus of mammals and can promote the secretion of LH and FSH. Although McCann found that crude hypothalamus extract can promote ovulation, many years passed before Schally and Guillemin obtained the pure LH-RH and identified its structure.

Schally and his coworkers reported this research in 1971[37]. Though it is a 12-step operation, not more than 1 mg pure GnRH was obtained from the hypothalamus (2.5 kg) of 165,000 pigs. The obtained pure GnRH showed high activity to stimulate the release of LH and FSH.

pig hypothalamus $\xrightarrow[\text{2N HOAc}]{\text{powdered, decreased}}$ freeze extract $\xrightarrow{\text{glacial acetic acid}}$ freeze-dry powder
165,000(2.5kg)   (665 g)

$\xrightarrow[\substack{\text{Sephadex G25}\\\text{1 M HOAc}}]{\text{Gel filtration chromatography}}$ freeze-dry powder that can stimulate release of LH and FSH (380 g) $\xrightarrow[\text{with phenol}]{\text{desalt, extract}}$ 83 g (LH release activity 30 to 150 μg, FSH 100 ug)

$\xrightarrow{\text{CMC, NH}_4\text{OAc}}$ 277 mg (LH 0.1 ug, FSH 0.5 ug) $\xrightarrow{\text{electrophoresis (EFE)}}$ 94.8 mg (LH 50 ng, FSH 0.3 ug)

$\xrightarrow[\text{Distribution}]{\text{Counter Current}}$ 14.2 mg (LH 2 to 25 ng, FSH 10 to 30 ng) $\xrightarrow[\substack{\text{partition chromatography}\\\text{Sephadex G25}}]{}$ 2.3 mg

$\xrightarrow{\text{zone electrophoresis}}$ 830 ug

The above mentioned process is typical for biological active peptide isolation from animal tissues. With the development of modern isolatical technology, especially the wide application of HPLC, the isolation process can be greatly simplified with high efficacy.

2.2.2   Synthesis of peptides

The first successful synthesis of a peptide was achieved by E. Fischer and T. Curtius in the 1880s. More than one hundred years later, with the fast development of synthetic organic chemistry and modern analytical technology, peptide synthetic processes have greatly advanced. The application of solid phase synthesis and mechanical automation facilitated the synthesis of peptides. While most natural peptides are rare and not stable enough, synthetic peptides can be prepared in large amounts and also can be modified to more stable structures. Many synthetic peptides are applied as medicines and play indispensable roles in life sciences.

A bipeptide can be obtained from the condensation of two amino acids. However, from a thermodynamic point of view, the condensation of two amino acids and release of a H_2O molecule is unfavorable.

$$^+NH_3CHR'COO^- + \; ^+NH_3CHR^2COO^- \longrightarrow \; ^+NH_3CHR'CONHCHR^2COO^- + H_2O$$

To fulfill the synthesis of a bipeptide, usually we need to protect the amino group of one amino acid and the carboxyl group of another amino acid in order to avoid unwanted condensations.

$$^+NH_3CHR'COO^- \longrightarrow XNHCHR'COOH$$
$$\text{amino protection}$$

$$^+NH_3CHR^2COO^- \longrightarrow NH_2CHR_2COOY$$
$$\text{carboxyl protection}$$

$$XNHCHR'COOH + NH_2CHR^2COOY \longrightarrow XNHCHR'CONHCHR^2COOY$$

The carboxyl group which participated in peptide bond formation often needs to be activated before being subjected to condensation. It can be activated by formation of acyl halide, acid anhydride, and many other active ester derivatives.

1. Protection of amino groups

Among the various amino protecting strategies, carbamates are those most widely used. Benzyloxycarbonyl (Z)[38], t-butyloxycarbonyl (BOC)[39] and 9-fluorenylmethoxycarbonyl (Fmoc)[40] are three representative protecting groups for carbamate formation.

Benzyloxycarbonyl (Z) was widely used years ago. It has the advantages of easy preparation and easy product purification by crystallization. The deprotection of Z can be achieved under acidic conditions or catalytic hydrogenation. The BOC protecting group applied later is much more practical, especially when it is used in solid phase peptide synthesis. The BOC group is sensitive to acid and can be removed under acidic conditions, such as CF_3COOH and HCl containing an organic solvent.

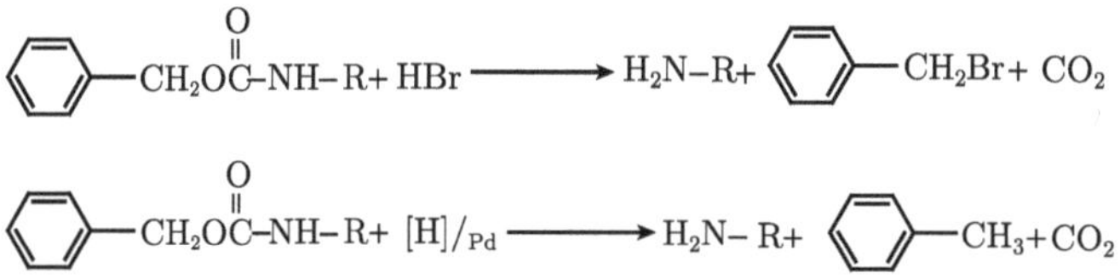

$$((CH_3)_3COCO)_2O + NH_2CHRCOOH \longrightarrow (CH_3)_3COCONHCHRCOOH$$

$$(CH_3)_3COCONHCHRCOOH \xrightarrow{\;H^+\;} \; ^+NH_3CHRCOOH + (CH_3)_2C=CH_2 + CO_2$$

The Fmoc group is sensitive to basic conditions and stable under acid conditions. Basic reagents, such as morpholine, piperidine, and diethyl amine, can be utilized to remove a Fmoc group.

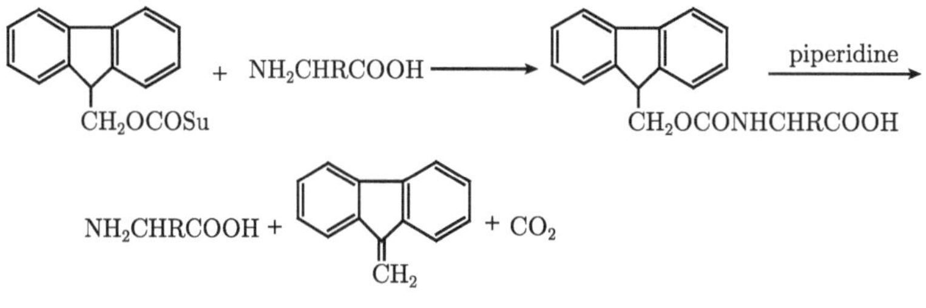

2. Protection of carboxyl groups

Compared with protection of an amino group, protection of a carboxyl group is relatively easy. The simplest method for carboxyl protection is the formation of a salt with a metal ion, but practically, the carboxyl group is often protected by esterification, such as the formation of methyl, ethyl, tert-butyl and benzyl esters. The deprotection of a methyl ester of ethyl ester can be achieved by simple saponification. Benzyl ester protection can be deprotected by catalytic hydrogenation. Tert-butyl ester can be hydrolyzed under acidic conditions (HCl, CF_3COOH, etc.) to release a carboxyl group.

3. Protection of side chain functional groups

Before peptidic condensation, the functional groups in the side chain of an amino acid often needed to be protected too. These side chain functional groups include the mecapto group in a cysteine, the hydroxyl group in a serine or tyrosine, the guanidyl group in an arginine, the imidazole group in a histidine, and so on. A good protecting group for side chain functional groups should be compatible with the selected amino protecting group and carboxyl protecting group.

4. Construction of peptide bond

As mentioned above, the key step for construction of a peptide bond is the activation of the carboxyl group in order to facilitate the nucleophilic attack from an amino group. Converting the carboxyl group to an electrophilic acyl chloride (RCOCl) or acyl azide ($RCON_3$) was used in earlier times to activate the carboxyl group. But these activated carbonyl groups suffer from the disadvantages of unstable and harsh reaction conditions. Methods involving activated ester, acid anhydride, and condensation agents are more commonly used nowadays to construct a peptide bond.

The electron withdrawing group-substituted alkyl or aryl amino acid ester is electrophilic enough to attack the amino group to form a peptide bond. p-Nitrophenol[41], N-hydroxy succinimide[42], and pentafluorophenol[43] are three common reagents used to form an activated ester with amino acids.

$$XNHCHR'COOR + NH_2CHR^2COOY \longrightarrow XNHCHR'CONHCHR^2COOY + ROH$$

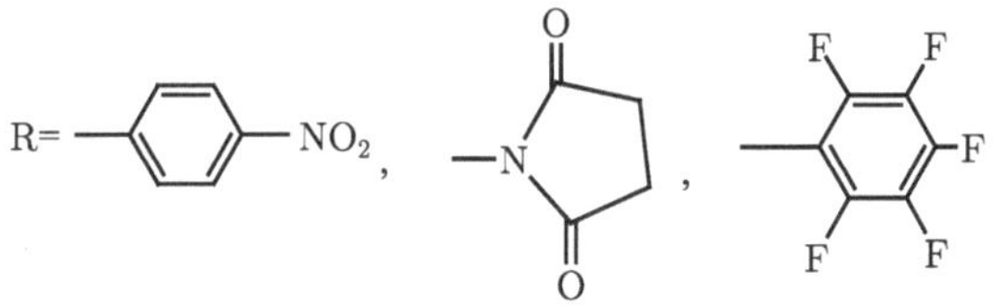

Formation of anhydrides, especially formation of mixed anhydrides with a chloro-carbonic ester, is another widely used method to activate the carboxyl group of N-protected amino acids during peptide synthesis. The peptide bond formed by this mechanism can be carried out under mild conditions and the reaction is often clean without any side reaction.

Compared to the methods mentioned above, the method involving utilization of a condensation agent is much more convenient. Due to its easy manipulation, N,N'-dicyclohexylcarbodiimide (DCC) is the most widely used condensation agent both for liquid and solid phase peptide synthesis[44].

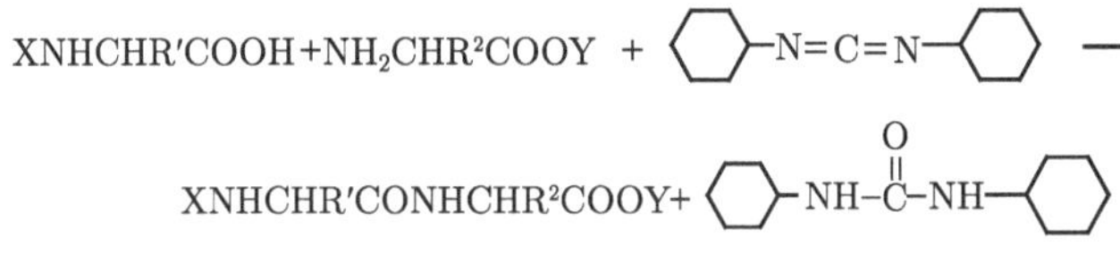

During the past decades, a number of powerful condensation agents have been invented. Generally, they have powerful condensation ability and produce few unwanted side reaction. Among them, BOP reagent, first used by Castro, is the most practical one[45]. These condensation agents may differ greatly in structure and can be classified as phosphorus cations, carbon cations, nitrogen cations, and so on[46,47]. For example, pyridine salt condensation agents (BEP, FEP, etc.) are more appropriate for synthesis of bulky peptides[47].

5. Solid phase synthesis

The methods for peptide bond formation discussed above are applicable not only in liquid phase synthesis, but also in solid phase synthesis. For solid phase synthesis, an undissolvable support is used and the condensation proceeds in a semi-heterophase. Excess reactants are used to ensure completion of reaction and the unreacted reactants are removed by a simple wash operation. The peptide attached to the undissolvable macromolecule can be elongated by repeated operation, which adds one amino residue a time. For instance, to synthesize a hexpeptide of A-B-C-D-E-F, we suggest a procedure as below:

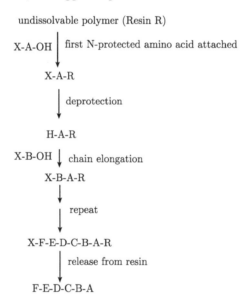

Solid phase synthesis, which introduced a novel concept to peptide synthesis, is much more convenient and operationally simple. Since Merrifield[48] developed this method in 1963, solid phase synthesis was extensively explored. A number of supportive resins with pronounced properties have been invented. More importantly, solid phase synthesis is so greatly automated today that it is predominantly used in peptide synthesis.

References

[1] Kreil G. A. Rev Biochem, 1997, 66: 337.

[2] Kuroda Y, et al. J Antibiotics, 1980, 33: 259.

[3] Rosenthal G A. Plant Nonprotein Amino and Imino acids, New York: Academic Press, 1982.

[4] J Nat Prod, 1997, 60: 802.

[5] J Antibiotics, 1997, 50: 256.

[6] Gray N R. Methods in Enzymology. New York: Academic Press, 1972, 25, Part B: 121.

[7] Xia, Q.C. et al. Development of Research Technology in Protein Chemistry, Beijing: Science Press, 1997 (in Chinese), 80.

[8] Caplow M. J Am Chem Soc, 1968, 90: 6795.

[9] Morrow J S, et al. J Biol Chem, 1974, 249: 7484.

[10] Schott H F, et al. J Biol Chem, 1952, 196: 449.

[11] Goodman M, et al. Biopolymers, 1996, 40: 183.

[12] Harada, et al. Bull Chem Soc Japan, 1983, 56: 653.

[13] Barretl G C. Chemistry and Biochemistry of the Animo Acids, New York: Chapman and Hall, 1985, 365.

[14] Dawson R M, et al. Data for Biochemical Research. London: Oxford University Press, 1969.

[15] Chang, B.Y. et al, Anal. Chem. (Chinese) 1993, 21: 1220.

[16] Bellamy L J. The Infrared Specfra of Complex Molecules, New York: John Wiley, 1975, 3(1): 263

[17] Liu, S.Y. et al, Acta Physico-Chimica Sinica 1997 (in Chinese), 13: 252.

[18] Biemeun K, et al. J Am Chem Soc, 1961, 83: 3795.

[19] Ken-ichi Harada, et al. Tertrahedron Lett, 1995, 36: 1515.

[20] Xia, Q.C. et al. Development of Research Technology in Protein Chemistry, Beijing: Science Press, 1997 (in Chinese), 165.

[21] Chilton W S, et al. Biochemistry, 1984, 23: 3290.

[22] Wang, Y. et al. Trichosanthin (second edition), Beijing: Science Press, 2000 (in Chinese).

[23] Mu Q et al., J Pharmazie, 2003, 58, 10.

[24] Witherup K M et al. J Nat Prod, 1994, 57: 1619.

[25] Saether O et al. Biochemistry, 1995, 34: 4147.

[26] Dou, H. et al. Natural Product Research and Development 2004 (in Chinese), 16, 76.

[27] Claeson P et al. Nat Prod, 1988, 61: 77.

[28] Ye, Y.H. et al, Acta Scientiarum Naturalium Universitatis Pekinensis 1996 (in Chinese), 26: 207.

[29] Gong Y.T., Bioactive Peptide, Beijing: Science Press, 1985 (in Chinese), 40.

[30] Erspamer V. Trends Neurosci, 1983, 6: 200.

[31] Du, Y. C. Neuropeptide and Brain Function, Shanghai: Shanghai Scientific and Technological Education Publishing House, 1998 (in Chinese), 7.

[32] Brockma H, et al. Chem Ber, 1955, 8857.

[33] Shiba T, et al. Peptide Chemistry, 1988, 337.

[34] Wieland, T.; Bodanszky, M. The World of Peptides, Heidelberg: Springer-Verlag, 1991, 211.

[35] Wu, W.T. et al. Chin J Nat Med, 2004 (in Chinese), 2, 70.

[36] Milles C.; Rivies J. Biopolymers, 1996, 40: 265.

[37] Schally A V, et al. Biochem Biophys Res Commun, 1971, 43: 393.

[38] Bergmam M.; Zervas, L. Chem Ber, 1932, 65: 1192.

[39] Carpino L A. J Am Chem Soc, 1957, 79: 98.

[40] Carpino L A, et al. J Am Chem Soc, 1970, 92: 5748.

[41] Bodanszky M. Nature, 1955, 175: 685.

[42] Anderson G W, et al. J Am Chem Soc, 1963, 85: 3039.

[43] Kovacs J, et al. J Am Chem Soc, 1967, 89: 183.

[44] Sheehan J C, et al. J Am Chem Soc, 1955, 77: 1067.

[45] Castro B. Tetrahedron Lett, 1975: 1219.

[46] Li P, et al. J Peptide Res, 2001, 58: 129.

[47] Li P, et al. Tetrahedron, 2000, 56: 8119.

[48] Merrifield R B. J Am Chem Soc, 1964, 86: 304.

Chang-Qi HU and Jie-Cheng XU

CHAPTER 9

Flavonoids

Section 1   Overview

Flavonoids are group of natural products that are widely present in plants. Most flavonoids are present in the form of O- or C- glycosides and a small portion are in the form of aglycone. Flavonoids exist in almost all studied species in the plant kingdom, and play an important role in plant growth, blossoming, fruit bearing, and disease and bacteria defence, etc. Many monographs about flavonoids have been published[1-7]. In the last 10 years, flavonoid research has turned more to the development of medical value, extraction and isolation technology, assay determination, and production of finished dosage forms.

1.1   General structures and categories of flavonoids

In earlier days, flavonoids were defined as compounds which contained the basic structure 2-phenyl-chromone. Nowadays, the term generally refers to a series of compounds which contain two benzene rings (ring A and ring B) connected via the central three carbon atoms.

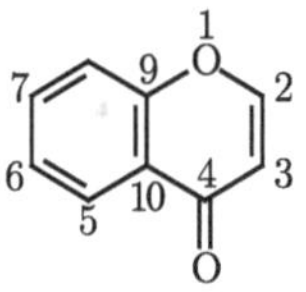

Chromone 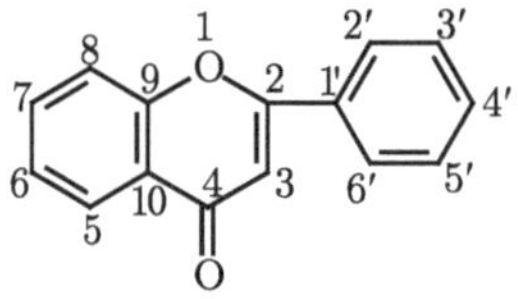   2-phenyl-chromone 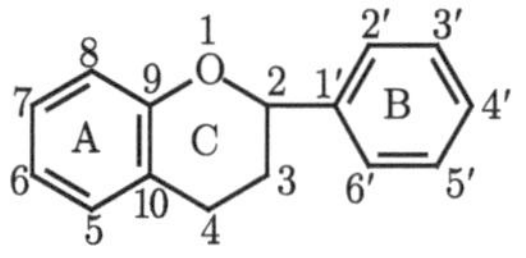 C_6—C_3—C_6

According to the status of the central three carbon atoms including their oxidation level, looping or not, and substitution pattern (3-or 2- position), flavonoids can be classified into several structure types (see Table 9-1). Most natural flavonoids are derivatives of these basic structures, with common substituent groups such as –OH, –OCH$_3$, –OCH$_2$O–, isopentenyl and caffeoyl, etc.

Table 9-1   Basic Structural Types of Flavonoids

Type	Basic structure	Type	Basic structure
flavones	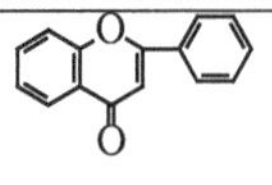	flavan-3-ols	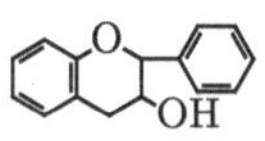
flavanones	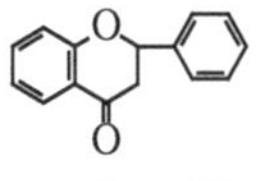	flavan-3,4-diols	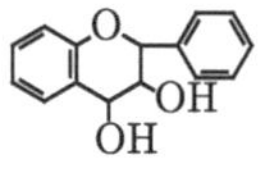
flavonols	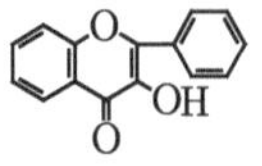	anthocyanidins	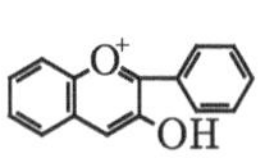
flavanonols	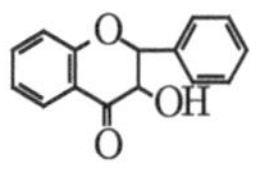	xanthones	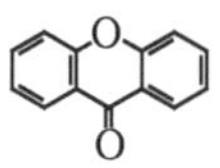
isoflavones	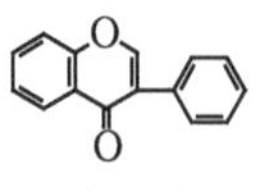	aurones	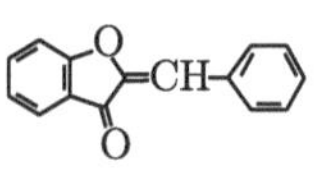

continued

Type	Basic structure	Type	Basic structure
isoflavanones		furanochromones	
homoisoflavones		phenylchromones	
chalcones		dihydrochalcones	

The flavonoid glycosides are mostly O-glycoside and C-glycoside. Besides these, there are also other structure types, such as biflavonoids, triflavonoids and homoflavonoids, etc.[4,5], which are formed by flavonoid dimers or trimers.

1.2 Physical and chemical properties of flavonoids[6,7]

Generally, flavonoids are crystalline solids, and only a few of them are amorphous powders. Flavones, flavonols, and their glycosides, generally, are grayish yellow to yellow. Chalcones are yellow to orange yellow, flavanones and flavanonols have no cross conjugation system, and thus, are generally colorless. However, a characteristic color and fluorescence can be seen under ultraviolet light, and the color would change under ammonia gas. Isoflavones only have a few conjugations, thus are pale yellow. The color of anthocyanidin is dependent on pH value, generally red (pH < 7), purple (pH = 8.5), and blue (pH > 8.5), etc.

Flavonoid aglycones that contain an asymmetric carbon, such as flavanone, flavanonol, isoflavanone and their derivations, have optical activity. Other aglycones have no optical activity, but all flavonoid glycosides have optical activity and are generally levorotatory (-) because monosaccharides or oligosaccharides are bonded in their structure.

Flavonoid aglycones are sparingly soluble or insoluble in water, freely soluble in organic solvents such as methanol, chloroform, etc.. However, flavonoid glycosides are freely soluble in strong polar solvents such as water, methanol, etc.and sparingly soluble or insoluble in organic solvents such as benzene, chloroform, etc. Because most flavonoids contain a phenolic hydroxyl group and thus show acidity, they can be dissolved in alkaline solutions (Na_2CO_3, etc.) and alkaline organic solvents (pyridine, formylamine and DMF). Because of the two free electrons at the 1-oxygen of the pyrone ring, flavonoids exhibit weak alkalinity and can form salts with strong acids, like concentrated HCl and H_2SO_4. However, the obtained salt is extremely unstable and will be hydrolyzed immediately with the addition of water.

Flavonoids can show characteristic color when reacted with Mg powder (or Zn powder)/HCl, $NaBH_4$, $AlCl_3$, NaOH, and other reagents that can be used for their identification.

1.3 The presence of flavonoids in plants

The flavonoids in plants are derived from chalcones, which are synthesized by hydroxylcinnamic acid and tri-acetic acid, that is, they are formed by glucose via shikimic acid and HAc-malonic acid pathways, respectively. Chalcone and (-)-flavanone can be mutually transformed, and both can further form flavanonols, isoflavones, aurones and dihydrochalcones. The flavanonols can further form flavonols, anthocyanidins, (+)-catechins and (-)-epicatechins. The isoflavones can further form pterocarpines and rotenones. For the origin and derivative relationships, see Figure 9-1. Some flavonoids are centered in specific plant families.

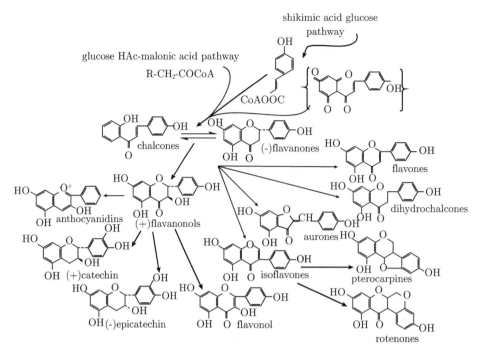

Figure 9-1   The origins and derivatives of flavonoids

1.3.1   Flavones and flavanones

Flavones are widely present among angiosperms, especially in the families of Rutaceae, Compositae, Scrophlariaceae, Labiatae, Umbelliferae, Acanthaceae, Gesneriaceae and Leguminosae. 6-Hydroxy-flavones mostly exist in the herbaceous plants. 8-Hydroxy-flavones are generally present in the ligneous plants. Flavanones are extensively present, especially in angiosperms, such as in the Rosaceae, Rutaceae, Compositae, Zingiberaceae, Ericaceae and Leguminosae families.

1.3.2   Flavonols and flavanonols

Flavonols are widely present in dicotyledons, especially in flowers and leaves of some ligneous plants. Kaempferol and quercetin are the most common flavonols. Flavanonols are present in gymnosperms, a few plants of monocotyledon Zingiberaceae, and widely in dicotyledons, Leguminosae and Rosaceae.

1.3.3   Chalcones and dihydrochalcones

Chalcones extensively exist in the families of Compositae, Leguminosae and Gesneriaceae. They are also found in the families of Scrophlariaceae and Valerianaceae. Dihydrochalcones are seldom found in the plant kingdom.

1.3.4   Isoflavones and isoflavanones

Isoflavones are mainly present, in angiosperms. for instance, in Leguminosae (about 70%), and the rest are in the Iridaceae and Moraceae families. Pterocarpin, trifolirhizin, and maackiain are contained in *Sophora tonkinensis* Gagnep. (a traditional Chinese herb), are the derivatives of isoflavanone, and are known to have anticancer activity. The activity of glycosides is more potent than that of aglycons. As the soybean isoflavone and puerarin were found to have anticancer and coronary heart disease improvement effects, respectively, isoflavones research has gained more attention.

1.3.5   Anthocyanidins

Anthocyanidins make flowers, fruits, leafage and stems to appear different colors, such as blue, purple and red, etc. They are widely present in angiosperms, and the most common anthocyanidins are cyanidine, delphinidin, pelargonidin and malvidin.

1.3.6   Flavanols

The derivatives of flavan-3-ol, known as catechins, are widely present in the plant kingdom, mainly in the ligneous plants, which contain tannins.

The derivatives of flavan-3,4-diol, known as leucoanthocyanidins such as leucocyanidin, leucodelphinidin and leucopelargonidin, etc., are also widely present in regnum vegetables, especially in ligneous and fern plants which contain tannin. These compounds may be polymerized and have similar properties as those of tannins.

1.3.7   Aurones

Aurones are different from other flavonoid compounds. Aurones are seldom used in traditional Chinese medicine. However, aurones are commonly present in the plants of Compositae, Scrophlariaceae, Gesneriaceae, and in monocotyledonous cyperus plants.

1.3.8   Biflavonoids

Biflavonoids are centered in gymnosperms, such as in the Ginkgoaceae, Taxodiaceae, and Pinaceae families. They also can be found in ferns, such as selaginella.

1.3.9   Other flavonoids

1. Xanthones

Xanthones, commonly found in Gentianaceae and Guttiferae, are a special kind of flavonoids. They can also be found in the Liliaceae family.

2. Furanochromones and phenylchromones

Furanochromones and phenylchromones are seldom found in the plant kingdom. Khellin, isolated from the fruit and seed of *Ammi visnaga* Lam, is a furanochromone.

Section 2   Extraction and Isolation of Flavonoids

The extraction and isolation process can be divided into two phases. Phase 1 is extraction, and the main issue in this phase is how to select extraction solvent, which is dependent on whether the target compounds are glycosides or aglycones, and also what the part of the plant is to be used. Phase 2 is isolation, and the goal of this phase is to separate the flavonoids from other co-existing compounds. When necessary, further separation and purification will be performed to obtain different flavonoid components. As these two phases are interrelated, it is very difficult to clearly define them during actual operations.

2.1   Extraction

As flavonoid solubility varies greatly, no single extraction solvent has been found that can suit all flavonoids. The selection of an extraction solvent is dependent on the properties of the target compounds and the impurities. The common extraction methods are listed below.

2.1.1 Hot water extraction

Hot water extraction is the most common method used for flavonoid glycosides. When using this method, raw materials are put in boiling water to inactivate enzymes. This method can also be used to extract relatively highly polar flavonoid aglycones, such as flavanol, flavandiol and proanthocyanidin. During the extraction, one will also need to take the quantity of water added, immersion time, decocting time, extraction time, and other factors into consideration[8].

2.1.2 Methanol or ethanol extraction

Methanol and ethanol are the most common extraction solvents used for flavonoids. Highly concentrated alcohol (90 to 95%) is suitable for extracting aglycones, and alcohol with about 60% concentration is suitable for extracting glycosides. The extraction process often takes two to four sequences. Percolation, reflux and merceration can also be used for extraction.

2.1.3 Succession solvent extraction

For this extraction process, different solvents from low to high polarity are used. For example, first, petroleum benzin or hexane would be used to degrease. Then, benzene follows to extract poly-methoxyl flavonoids or isopentenyl and methyl flavonoids. After that, ether, chloroform, and ethyl acetate would be used in turn to extract most of the aglycones. Then, acetone, ethanol, methanol, and methanol-water (1:1) are used to extract polyhydroxy flavonoid, biflavonoid, chalcone, etc. components. Finally, diluted alcohol and boiling water are used to extract glycoside, and 1% HCl used to extract anthocyanidins.

2.1.4 Alkaline water or alkaline diluted alcohol extraction

Alkaline water (such as aqueous Na_2CO_3, NaOH and $Ca(OH)_2$ solutions) or alkaline diluted alcohol (such as 50% ethanol) can be used for extraction because most of the flavonoids contain a phenolic hydroxyl group. By acidifying the extract, flavonoids are precipitated. But this method is not satisfactory as many impurities co-exist. Furthermore, cautions have to be taken to avoid making the alkaline concentration too high to break down the basic structure of flavonoids during heating.

2.2 Isolation

The isolation of flavonoids includes two steps: the first is to isolate the flavonoids from other co-existing compounds and the second is to further isolate to get different flavonoid compounds. The common isolation methods used are solvent extraction, alkali extraction and acid precipitation, thin-layer chromatography, polyamide column chromatography, silica gel column chromatography, lead salt precipitation, boric acid complexation, pH gradient extraction, gel column chromatography, etc.[7,8]. With the development of modern technology, many new extraction and isolation methods have been introduced to TCM R&D and achieved good results.

2.3 New extraction and isolation methods

2.3.1 Ultrasonic extraction

Ultrasonic extraction (UE) technology is one of the new extraction and isolation methods applicable to TCM research. Taking advantage of the cavitation effect produced by ultrasonication, the effective ingredients of a plant can be extracted more quickly. Additionally, the secondary effects of ultrasonication (such as mechanical vibration, emulsification, diffusion, shredding and chemical effect, etc.) can also accelerate the diffusion and release of the target composition, which will make the extraction easier.

2.3.2   Ultrafiltration

Ultrafiltration (UF), a representative of membrane separation technology, can effectively remove the giant molecular substances in the extract by controlling the pore diameter size of the ultrafiltration membrane. The method is based on taking advantage of the porous semipermeable membrane and certain pressure to force small molecules to pass through the membrane to isolate, purify, and concentrate the extract.

2.3.3   Macroporous adsorption resin chromatography

Macroporous adsorption resin chromatography (MARC) has been used in TCM extraction and isolation since the late 1970s. It has the advantages of large adsorption capacity, easy regeneration and reliable effects, thus making it especially suitable for the isolation and purification of flavonoids and saponins at industrial scale[8,9].

2.3.4   Aqueous two-phase extraction

When combining two different water-soluble polymers' aqueous solutions with a concentration above critical point, two incompatible phases would naturally be displayed. This is known as an aqueous two-phase system. The principle of aqueous two-phase extraction (ATPE) is the selective partition for the substances in the aqueous two-phase system. Different substances have different partition coefficients in the two phases, thus when the substances go into the aqueous two-phase system, they would be selectively partitioned between the top and bottom phases. In this way, they are then able to be extracted and isolated. The common aqueous two-phase extraction systems are high polymer systems (e.g., PEG/dextran phase) and high polymer-inorganic salt phases (e.g., PEG/sulfate or phosphate phase)[10].

2.3.5   Supercritical fluid extraction

Supercritical fluid extraction (SFE) is a new extraction and isolation technology which uses supercritical fluids to extract and isolate the active ingredients of TCM. The principle of SFE is taking advantage of the strong solubility, good flow ability, and high transmissibility of a certain fluid when it is in the vicinity of the critical point to extract and isolate.

2.3.6   Enzymic extraction

Most of the impurities existing in TCM are starch, pectin, proteins, etc. To decompose and remove them, one can use enzymic extraction (EE). For instance, partially hydrolyzed starch, glucosidase or transglycosidase can be added into a root extract, which is rich in lipophilic, sparingly soluble or insoluble components, to convert these components into water-soluble glucosides[11]. Cellulase can hydrolyze the β-1,4 glucosidic bond to break up the plant cell wall, and thus can be used for the extraction of TCMs, which are rich in cellulose.

2.3.7   High-performance liquid chromatography

Sometimes, with paper chromatography, column chromatography, and thin-layer chromatography, the isolation result may not be ideal, and the use of a gas chromatogram is limited by its requirement of sample derivatization (e.g. silylation). But with high-performance liquid chromatography (HPLC), the isolation result is satisfactory. Considering the high cost, HPLC is mostly used for qualitative and quantitative analysis or small quantity sample preparation of flavonoids.

2.3.8   Micellar thin layer chromatography and microemulsion thin layer chromatography

Micellar thin layer chromatography (MTLC) uses a surfactant micelle solution as its mobile phase[12], while microemulsion thin layer chromatography (ME-TLC) uses a microemulsion as its mobile phase. A microemulsion is formed by a surfactant, oil and water in appropriate ratios and is a colorless, isotropic, and low viscosity thermodynamically stable system. Both microemulsion and micelle are association colloids, but have different basic characteristics[13]. Compared with micellar materials, microemulsions have a higher chromatographic efficiency due to larger solubilization capacity and ultralow interfacial tension.

2.3.9   Molecular imprinting technology

Molecular imprinting technology (MIT) is an extraction and isolation method of preparing high affinity polymer material, molecularly imprinted polymer (MIP), for targeting molecules. The polymer cavity of MIP can complement the spatial structures of the target molecules and interact with functional groups (like hydrogen bond, ionic bond or van der waals forces, etc.). This interaction is similar to that of enzyme and receptor. Jianchun Xie[14] et al. have reported the use of quercitrin as the target molecule to prepare MIP and the feasibility of applying MIP in the extraction of flavonoids directly from ginkgo leaves.

2.3.10   Other isolation techniques

Other isolation techniques, such as high speed countercurrent chromatography (HSCCC), molecular distillation (MD), microwave extraction (ME) and semi-bionics extraction (SBE) etc.[15,16], also have certain prospective applications in TCM extraction and isolation. But reports on these techniques are few and there is still a great need for further R&D.

Section 3   Identification and Structure Study of Flavonoids

3.1   Identification

The most common identification methods are physical and chemical identification, thin-layer chromatography (TLC), HPLC, etc.[6,7].

3.1.1   Physical and chemical identification

Physical and chemical identification is dependent on the physical and chemical properties of the flavonoid groups.

1. Color, optical rotation and solubility

Different flavonoid groups have some variabilities in color (e.g., flavone shows yellow, while flavanone is almost colorless), optical rotation (flavone has no optical rotation but flavanone does) and solubility (e.g., flavonoid aglycone is sparingly soluble in water, but flavonoid glycoside is often water-soluble). Based on these properties, the structure types of the flavonoids can be preliminarily determined. For details, see Section 1.2 covering "Physical and chemical properties of flavonoids".

2. Color reaction

Hydrochloric acid-Mg powder reaction: add a small amount of magnesium powder into the ethanol or methanol flavonoid solution, shake, then add a few drops of concentrated hydrochloric acid; color will appear in 1 to 2 minutes. Flavones, flavonols, flavanones and flavanonols show a positive reaction with a red to purple color and a few of them show blue or green (such as 7, 3′, 4′-flavones). When ring B has an –OH or –OCH$_3$ substituent group, the color will become deeper. When flavanoids are absent a 3–OH group or glycoside reacted at the 3-position, the color reaction will not be obvious. However, isoflavones, chalcones, aurones and catechins show a negative reaction. But some aurones, chalcones, and anthocyanidins can show red color upon the addition of concentrated hydrochloric acid without Mg powder. The sodium amalgam reduction reaction, aluminum trichloride color reaction, Gibbs reaction, sodium borohydride reduction reaction, zircon salt-citric acid reaction and ammine-strontium chloride reaction can also be used.

3.1.2   Thin-layer chromatography and spectral identification

Paper chromatography, polyamide thin film chromatography and silica gel thin-layer chromatography can be used to identify target compounds. According to the Rf value obtained, chemical staining, UV fluorescence detection, etc., can be used to identify or compare a compound with that of a known reference standard.

UV and IR spectra can also be used to make identifications according to the actual data obtained or by comparison with a standard spectrum. HPLC can be used by comparing with reference standards or a standard spectrum. Other techniques, like GC, NMR, MS, can also be used for identification. These methods and their results can be combined (especially TLC, TLC-Scan, HPLC and GC etc.) to obtain a fingerprint spectrum of flavonoids, which can make the identification of flavonoids quicker and more effective.

3.2   Structure study

For known flavonoids, one can generally use chemical methods (such as derivate preparation and color reaction etc.) and spectral analysis (such as IR, UV, NMR, and MS etc.) to determine their structure. If a reference standard is available, the structure will be more easily determined by contrasting TLC, IR spectra and mixing melting point. Even for the unknown flavonoids, with current modern technology (IR, UV, NMR, MS, etc.) and chemical methods, their structures can also be easily determined. For the flavonoids with special structures, complete synthesis would assist the structure determination. Numerous related researches have been performed, and for details, please see the related reference[1].

3.2.1   Spectral features of flavonoids

With the development of extraction and purification technology, flavonoids have already been systematically and thoroughly studied. The current research hot spots for flavonoids are the flavonoid glycosides, which contain more than one sugar unit. Because of the overlapped signals of the ^{13}C spectrum in many locations, it's hard to confirm the carbon signals, and thus difficult to use ^{13}C spectroscopy only to determine the number and type of the saccharides, or glycosidation shift to confirm the position of the glycosidic bond and the linkage sequence of the saccharides. In recent years, 2D-NMR technology has been rapidly developed and become a powerful tool for the structure determination of flavonoid glycosides and other compounds. It is easy to identify the position and linkage sequence of saccharides in flavonoid glycosides, even if the sample quantity is small, with DQPS-COPY and RCT combined with T1 value detection, 2D-NOE, and FAB-MS. Detailed information about

NMR can be found in the book "Structural Identification of Organic Compounds with Spectroscopic Techniques" (Yongcheng Ning. Beijing: Science Publishing House, 2000) and related journals.

MS analysis is also very important. EI-MS is an effective method and also commonly used to determine the structure of flavonoid aglycones, especially when the sample quantity is small. For the sample quantity, EI-MSR requires only 1 to 50 μg. FAB-MS, developed since the 1980s, is a solid sample analysis technology. Ionized glycoside compounds can be broken down by it to fragments which are similar to those of FD-MS. As it's equipped with an anion capture detector and the resulting anion mass spectrum can complement the cation mass spectrum, this makes the information obtained wider and more reliable. FD-MS and FAB-MS have their own advantages. FD-MS can provide detailed information in the high mass region, but cannot provide the aglycone related fragment information. FAB-MS can provide information about relative molecular mass and saccharide group fragments, and also the structures of aglycone fragments in the low mass region. Thus, FAB-MS makes up for the shortages of FD-MS. Actual operations generally use FAB-MS to analyze glycosides. Besides, electrospray ionization mass spectrometry (ESI-MS) and atmospheric pressure chemical ionization MS (APCI-MS) can be used. ESI and APCI are soft ionization techniques, that easily acquire the relative molecular mass information but with few or none of the fragmented ion peaks. To obtain further structure information, collision-induced dissociation (CID) will be needed. By adjusting the voltage of the collision cell, MS information at different dissociation degrees can be obtained. Furthermore, when the cascade MS technology (tandem MS or MS-MS) is combined with CID, simpler and more abundant MS information can be obtained.

There are two common MS fragmentation patterns of flavonoids: pattern I and pattern II.

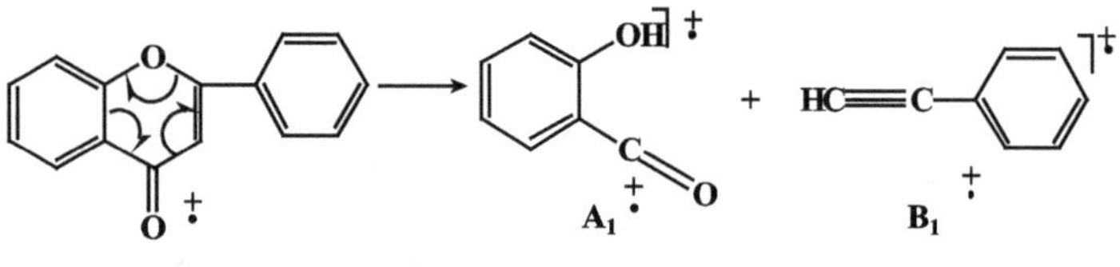

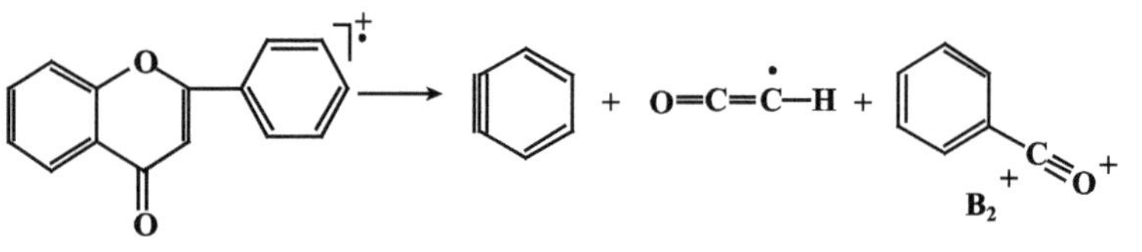

The absolute configuration of the chiral carbon in ring C of flavanones and flavanonols can be confirmed by circular dichroism (CD) and optical rotatory dispersion (ORD) spectroscopy.

With monocrystal x-ray diffraction, thousands of independent diffraction and valid observation points can be obtained. Thus one can know the coordinate of each atom, confirm the relative spatial position of each atom, and finally a direct and reliable stereograph can be obtained against which the molecular structure can be confirmed. For instance, in the reference article[16], use of monocrystal x-ray diffraction to confirm the linkage position and the configuration of the saccharides in semiaquilinoside has been reported.

The development of other hyphenated techniques, such as HPLC-NMR, HPLC-MS, LC-MS-MS GC-MS, etc., has facilitated the structure determination of flavonoids.

In summary, the technologies of structure determination have improved quickly and provided powerful tools for the structure determination of flavonoids.

3.2.2   Flavonoid aglycone structure study

For known flavonoids (including flavonoid aglycones and glycosides), generally one can use derivate preparation and spectral analysis to determine their structure. If a reference standard is available, the structure will be more easily determined by contrasting the IR spectrum and mixing melting point. Even for the unknown flavonoids, after initially determining their structure by derivate preparation and spectrum analysis, one can further validate and confirm the structure by spectroscopic determination or chemical degradation and conversion. For the flavonoids with special structures, complete synthesis would assist the structure determination. Details are shown below.

1. Derivate preparation

The most common derivates of flavonoids used are methide and acetylate. When preparing these derivates, on one hand, one can compare them with the chemical and physical data of the known compounds reported in the literatures. On the other hand, one can further compare their spectral data with those of the original compounds in order to determine their structure.

A. Preparation of acetylate derivate

Hydroxyl groups can be acetylated by acetic anhydride and acetyl chloride with a catalyst, such as pyridine, sodium acetate, concentrated sulfuric acid, 4-dimethylamino-pyridine or p-methyl benzenesulfonic acid. All the reagents used must be anhydrous. Acetic anhydride/pyridine is the most commonly used reagent group.

Some flavonoids are unstable or can be easily isomerized in alkaline solution. For such flavonoids and those flavonoid glycosides, which are difficult to be acetylated or the reaction products too complex during base catalysis, a trace amount of concentrated sulfuric acid or anhydrous P-methyl benzenesulfonic acid can be used as the catalyst.

B. Preparation of methide derivate

Diazomethane can be used to methylate the phenol group with a high yield of reaction product. If the 5-OH cannot be easily methylated due to the formation of the hydrogen bond, then dimethyl sulfate and other reagents can be used. Dimethyl sulfate can methylate all the hydroxyl groups in the flavonoids.

In addition, another full methylation method is rapid and simple, suitable for various glycosides and saccharides, and produces a high yield of reaction product. It's also very convenient for GC to isolate and identify the methide. The procedures are listed as follows: Dissolve 4 to 5 mg glycoside or saccharide in 0.3 to 0.5 mL of DMSO. Add 20 mg finely ground NaOH powder at room temperature. Stir for 6 min., then add 0.1 mL iodomethane and stir for another 6 min. to complete the reaction. Add 1 mL of water and 1 mL of chloroform. Take the chloroform layer, wash with water, then dry and further evaporate to dryness. Then, the final product can be purified using chromatography or crystallized with appropriate solvents. In actual operations, as excess alkali is added during the reaction, some acyl groups which are easily hydrolyzed may be removed, and make the newly formed hydroxyl groups become methylated. This situation would have to be taken into consideration when determining the structure.

2. Special spectroscopic determination

As the structure of flavonoid aglycones is relatively simple, regular spectra determination, derivate preparation and spectra comparison are enough to determine their structure. Only when the structure is special or further confirmation of the structure needs to be made, will special spectroscopic determination be needed.

A. UV spectrum determination for hydroxyl position diagnosis reagents

The common hydroxyl position diagnosis reagents used are CH_3ONa, CH_3COONa, $AlCl_3$, CH_3COONa/H_3BO_3, $AlCl_3/HCl$, etc. By comparing the UV spectra before and after reagent addition one can obtain the shift value and, based on this, can further decide the position of the hydroxyl. Details of the method and the relationship between the different UV spectrum changes incurred by different diagnosis reagents and flavones' hydroxyl positions can be found in the references[1,6]. This method is simple and the sample quantity required is very small (micrograms). Thus, it has been widely used.

B. ^{1}H-NMR-NOE and solvent shift value determination

Methoxyl position of a flavonoid can be determined by the NOE gain generated between the methoxyl proton and the ortho-aryl proton. The ^{1}H-NMR shift of methoxyl on different positions of flavonoid induced by benzene-d_6 has been reported, and this method has already been used for structure determination. These two methods are the most important used to identify methoxyl, especially the NOE method which is simple, reliable, and can be used to determine stereochemistry. For flavonoids containing a basic fused ring structure, the NOE relationship among protons can help determine the position of the fused ring skeleton and the proton stereochemistry. Furthermore, HMBC technology can easily determine the position of methoxyl groups according to the signal of the C atom that they are connected with. NOESY technology is very convenient for stereochemistry studies.

3. Chemical degradation

Chemical degradation uses a chemical method to break down the complicated structure of a compound into simpler structure compounds. Then structure study will be made for the degradation products, and based on this, to determine the original complicated structure. With the use of highly developed spectroscopic technologies, chemical degradation is seldom used in structure study today, but for compounds with odd or unusual structures, chemical degradation is still an important technique. The common chemical degradation methods used are oxidation, reduction, alkali degradation, etc.

4. Reported examples

[Example 9-1] Structure investigation of corylifolin and corylifolinin.

Corylifolin amd corylifolinin are flavonoids extracted from the TCM *Psoralea corylifolia* L. Corylifolinin can enhance coronary artery blood flow and inhibit Hela cells.

Corylifolin, molecular formula $C_{20}H_{20}O_4$, melting point 211 to 212°C; $[\alpha]$ is -22° (EtOH); UV absorption spectrum (EtOH):λ_{max}nm (lgε): 219 (4.49), 237 (4.45), 280 (4.24), 323 (3.94); Mg power/HCl reaction shows purple blue color. Corylifolin can be converted to chalcone in 10% EtOH at room temperature. It can be confirmed to be a flavanone based on its location and intensity in UV spectrum band II and band I, and its optical activity and ring opening reaction with alkali to form chalcone.

Besides the common flavanone fragments, the C_5H_9 residue is also present in its molecular formula, which can be hydrogenated to dihydride with a Pd/C catalyst, and further ozonized to acetone. This indicates that the C_5H_9 residue is γ,γ-dimethyl allyl. Corylifolin can form m-dihydroxyhenzene, ρ-hydroxybenzoic acid and acetophenone derivates bound with C_5H_9 residue by KOH liquation. This indicates the presence of the 4'-OH on the ring B and the γ,γ-dimethyl allyl substituted on ring A. That the ^{1}H-NMR singlet at δ7.27 ppm and δ6.22 ppm belong to the aromatic protons of acetophenone derivates indicates that the aromatic protons are in para-position. Thus, its partial structure is 2,4-dihydroxy-5-(γ,γ-dimethyl allyl)-acetophenone which has been confirmed by synthesis. Therefore, the structure of corylifolin is determined to be 7,4'-dihydroxy-6-(γ,γ-dimethyl allyl)-flavanone.

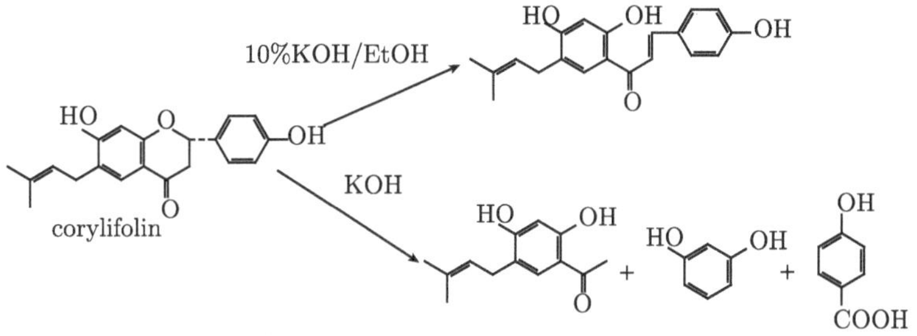

The absolute C-2 configuration of corylifolin is determined by comparing the optical rotation spectrum with that of (2S)-liquiritigenin. The two optical rotation spectrum curves are similar at 325 to 375 nm with positive Cotton effect. Thus, the configuration of corylifolin is 2S, similar with that of (2S)-liquiritigenin.

liquiritigenin   corylifolin

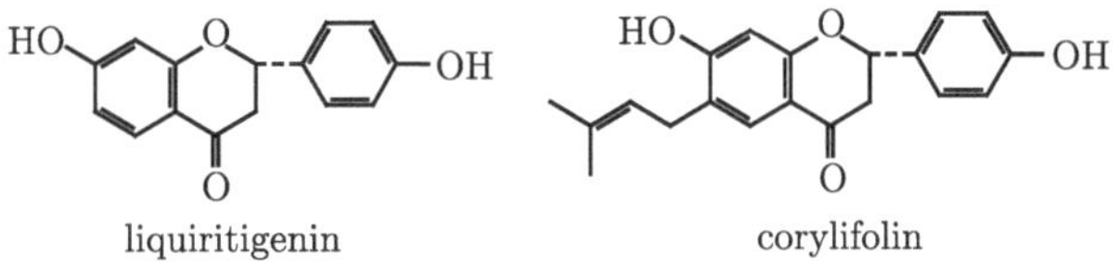

Corylifolinin

Corylifolinin, molecular formula $C_{20}H_{20}O_4$, melting point 166 to 167°C, no optical activity; UV absorption spectrum (EtOH) λ_{max} nm(lgε): 230 (sh, 4.16), 371 (4.16), shows the typical UV spectrum of chalcone. That corylifolinin can form a racemic flavonone in 5% HCl-EtOH solution via ring closure reaction also proves that corylifolinin is a chalcone. After ozonization and hydrogenation, acetone and dihydride are obtained. M-dihydroxyhenzene and p-hydroxybenzoic acid are formed by solid alkali liquation. p-Hydroxybenzoic acid and C_5H_9 substituted acetophenone are obtained by alkali degradation. The ^{1}H-NMR signals of this derivate at δ7.45 and δ6.37 ppm show a couple of AB type proton peaks (J = 9), and this indicates that the two aromatic protons are at an ortho position. Therefore, its partial structure is 2,4-dihydroxy-3-(γ, γ-dimethyl allyl)-acetophenone. Therefore, the structure of corylifolinin is determined to be 2',4',4-trihydroxy-3'-(γ,γ-dimethyl allyl)-chalcone.

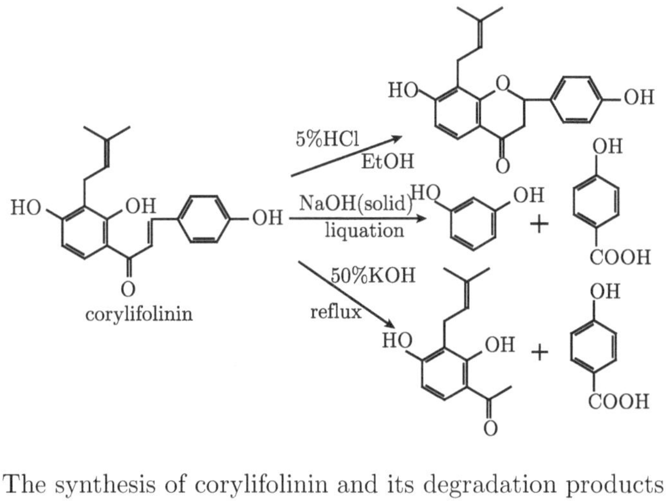

The synthesis of corylifolinin and its degradation products

The monosodium salt of 2,4-dihydroxy acetophenone suspended in anhydrous benzene is heated to reflux with 1-bromoisoprene for 19 h. Then, the alkali decomposition products of corylifolin and corylifolinin, – C_5H_9 substituted acetophenone derivates, are obtained. Under strong alkali conditions, the condensation reaction between the acetophenone derivatives of corylifolinin and p-hydroxy benzaldehyde would occur and form corylifolinin. Its ring closure compound is another proof for the corylifolinin structure.

3.2.3 Structure study for flavonoid glycosides

Flavonoids are widely present in the form of glycosides in the plant kingdom and most of them are O-glycosylated compounds linked with saccharides through an oxygen atom (such as baicalin). Others are C-glycosylated compounds linked with saccharides by a carbon atom (such as puerarin). If the glycosides were categorized according to the number of saccharide units that the aglycones were connected to, then the glycosides can be categorized as monoglycoside (baicalin and puerarin), bioside (hesperidin), triglycoside (robinin), etc. As most of the glycosides contain at least one saccharide unit, it's hard to obtain the molecular ion peak with EI-MS, but FD-MS or FAB-MS can be used. If the problem still cannot be solved, then full methylation can be used to enhance thermal stability and evaporation of the glycosides before determination. The saccharide protons are at $\delta 3.0$ to 6.0 ppm of the ^{1}H-NMR spectrum, while saccharide carbons are at $\delta 60$ to 105 ppm in the ^{13}C-NMR spectrum. From the number of saccharide protons and carbons shown in ^{1}H-NMR and ^{13}C-NMR spectra, one can know how many saccharides that the glycoside is composed of, and with further analysis, can even determine the structure of saccharides and their linkage with the aglycons.

Structural determination of flavonoid glycosides involves the determination of aglycones and saccharides, and the linkage of aglycons connected to saccharides. The structural determination of aglycones has already been discussed. Determination of the saccharide structure is generally performed for known saccharides, as most saccharides are already known and the varieties are limited. Thus, the structure determination for flavonoid glycosides is mainly used to study the linkage of aglycons to saccharides.

1. Hydrolysis of glycosides

Flavonoid glycosides are not hydrolyzed under common hydrolysis conditions. That C-glycoside would not be hydrolyzed by acid can be used in C-glycoside identification. However, O-glycoside can be hydrolyzed by acid and certain specific enzymes, and under appropriate hydrolysis conditions, O-glycoside can be partially or completely hydrolyzed. The structure of the flavonoids, the type of the saccharide, and their linkage to the basic structure of the flavonoids, has much more effect on the hydrolysis rate. Enzyme hydrolysis is a gentle, specific, and convenient method which will become more popular with the commercialization of enzymes.

2. Saccharide identification

Most of the saccharides from flavonoid glycosides are known. After acid hydrolysis, the water solution needs to be treated with polyamide chromatography or directly analyzed with paper chromatography for identification of the saccharide against the reference standard.

GC is also convenient for saccharide identification. Dry the water-soluble part of the acid hydrolysis solution under reduced pressure to prepare trimethyl siloxane derivatives (procedure: dissolve 5 mg sample in 0.5 mL anhydrous pyridine, then add 0.2 mL hexamethyl-disilazane and 0.1 mL trimethylchlorosilicane. Close tightly for 5 minutes, then dry under reduced pressure) or prepare full methylation derivatives. Identification can be performed by GC, using known saccharide derivatives as a comparison.

In recent years, with the development of ^{1}H-NMR and ^{13}C-NMR, the number of saccharides can be directly determined by the saccharide protons signals at $\delta3.0$ to 6.0 ppm and the saccharide carbons signals at $\delta60$ to 105 ppm. This method is simple and reliable. For deoxysaccharide, its deoxycarbons and protons are at a relative higher NMR field. For example, the methyl protons of rhamnose presents as a doublet in $\delta1.5$ ppm while methyl carbon appears in $\delta20$ ppm. These features can directly identify the properties and number of saccharides.

3. Glycosidation position

The glycosidation position of a flavonoid can be identified directly by ^{1}H-NMR and ^{13}C-NMR. From the chemical shift differences of a saccharide anomeric proton (H-1″) incurred by different glycosidation positions, one can obtain the information of the glycosidation position.

The glycoside carbon bond of the aglycone would always makes a 2 ppm glycosidation shift to high field in ^{13}C-NMR (compared with that of aglycone). However, the different glycosidation position would make the shift of the ortho- and para- carbons much different. The 3-O-glycoside of flavonol makes the C-2 signal shift about 9 ppm toward low field, and it has become the obvious feature of the 3-O-glycoside. 5-O-glycoside can break down the hydrogen bonding between the 5-OH and carbonyl group, and the ortho- and para- carbons of C-5 would shift about 3.0 to 4.5 ppm toward low field. The C-2 and C-4 on ring C would shift about 3 and 6ppm toward high field, respectively, while the C-3 would shift about 2.5 ppm towards low field. 7-O-glycoside and 4′-O-glycoside can make the carbons at the ortho- and para- positions shift about 0.5 to 1.0 ppm and 1.5 to 1.8 ppm toward low field, respectively.

4. Anomeric carbon

There are two kinds of configurations for the anomeric carbon; α- or β-configuration. Most of the glucoside, galactoside, glucuronide and xyloside compounds are β-configurations, while most of the arabinosides and rhamnosides are α-configurations.

Identification of the anomeric carbon configuration using ^{1}H-NMR and ^{13}C-NMR is more convenient, simpler, and more reliable than traditional methods.

5. C-glycoside

In recent years, with the development of isolation technology, the number of C-glycoside of flavonoids isolated from natural sources has increased dramatically. Besides the common flavonoid C-glycoside and flavonol C-glycoside, chalcone C-glycoside, flavanone C-glycoside and flavanonol C-glycoside have also been discovered.

^{13}C-NMR is a simple and reliable method to identify C-glycoside and O-glycoside. With this method, structure determination of C-glycoside can be made directly without removal of the saccharide from the aglycone. The anomeric carbon (C-1″) signal of C-glycoside is at δ71 to 78 ppm. Although the deoxycarbon of the deoxysaccharide is at high field, the saccharide carbon peak would always appear at δ60 to 82 ppm. At the same time, the glycosidated carbon signal of the aglycone would change from d peak to s peak and shift 6 to 11 ppm toward low field. The anomeric carbon (C-1″) signal of O-glycoside is at δ95 to 105 ppm, overlapped with that of the aromatic carbon. The signals of glycosidated carbon of the aglycone shift very little ($\pm$2 ppm) and always appear as s peaks.

CD spectroscopy can also be used to distinguish 6- and 8-C-flavonoid glycosides. In addition, the MS spectra of a fully methylated or fully deuterated-methylated C-glycoside can also assist in the C-glycoside structure determination.

Section 4   Pharmacological Study of Flavonoids

Flavonoids are widely present in nature and exhibit various bioactivities, which have attracted much attention. Many pharmacological studies of flavonoids have been reported. A brief overview appears below.

4.1   Cardiovascular system activity[6,18]

Many effective TCMs used to treat coronary heart disease contain flavonoids. Studies show that rutin, quercetin, puerarin and recordil (synthesized) have obvious coronary relaxation effects; quercetin, rutin, hyperin, puerarin, breviscapine, total flavones of pueraria (TFP) and total flavones of ginkgo (TFG) have shown protective effects against hypoxic ischemia brain damage; hyperin, silybin, luteolin and total flavones of seabuckthorn fruit have exhibited protective effects against heart muscle ischemia damage; TFG, puerarin and daidzein can not only lower cardio- and cerebral- vascular resistance, myocardial consumption of oxygen, and the production of lactic acid, but also have obvious protective effects against myocardial anoxia damage. Additionally, the total flavones of seabuckthorn fruit, puerarin, total flavones of *Lightyellow sophora* root and glycyrrhiza flavonoid (mainly the liquiritigenin and isoliquiritigenin) have shown anti-arrhythmia effects.

4.2   Anti-microbial and anti-virus activity[6,18]

Cyanidenon, baicalin and baicalein have, to a certain degree, anti-microbial effects; quercetin, distylin, morin and campherol demonstrate anti-viral effects; pure flavonoid isolated from *Chrysanthemum* and *Swertia bimaculata* show strong inhibitory effects on HIV virus; daidzein, genistein and biochanin A also have, to a certain degree, inhibitory effects on HIV virus.

4.3   Anti-tumor activity[19,20]

Flavonoids have various anti-tumor mechanisms. For instance, the anti-tumor activity of meletin is correlated with its antioxidant effect, inhibitory effect on related enzyme activity, reduction of tumor cell resistance, induction of tumor cell apoptosis, and estrogenic effect. The anti-tumor activity of silymarin is correlated with its antioxidant effect, inhibitory effect on related enzyme activity and inducing cell cycle arrest. Obviously, the effective and various anti-tumor activities of flavonoids, especially isoflavonoids, suggest that they have great application potential for tumor prevention and treatment.

4.4   Anti-oxyradical activity[18]

Most flavonoids have shown strong anti-oxyradical effects, with which their pharmacological activities are often correlated.

4.5   Anti-inflammatory and analgesia activity[6,18]

Rutin, hydroxyethylrutin, and distylin have shown visible inhibitory effects on arthritis and cotton ball granuloma caused by formaldehyde, rat paw edema induced by CAR, 5-HT and PGE. The dimeric procyanidin in wild buckwheat has shown anti-inflammatory, fever relieving and phlegm eliminating effects. Hyperin, rutin, meletin, and the total flavones of ginkgo (TFG) have shown positive analgesic effects.

4.6   Hepatoprotective activity[6,18]

It is reported that silybin can clearly inhibit the production of MDA in liver homogenate. Clinical research has shown that silybin has a positive therapeutic action against toxic hepatic lesions, acute and chronic hepatitis, and hepatic cirrhosis. Epimedium flavonoids, such as baicalein and baicalin, can protect the liver by inhibiting liver lipid peroxidation, increasing the SOD activity in mouse liver, and decreasing liver lipofuscin formation. The total flavones of *Hypericum japonicum* Thunb can lower the concentration of serum transaminase and improve liver function. Baicalein has shown protective effects against liver lipid peroxidation caused by adriamycin. Glycyrrhiza flavonoids have shown protective effects against liver cell ultrastructure injury caused by ethanol.

Furthermore, many studies have shown that flavonoids also have many other pharmacological activities, such as lowering blood pressure, lowering blood lipid, anti-aging, improving body immunity, purgation, cough relief, phlegm elimination, spasmolysis, and anti-allergic action. Because of their great varieties, wide presence in plants, and the low toxicity, flavonoids are important resources for new drug R&D and show great application potential.

Section 5   Flavonoid Assay Determination and Formulation

5.1   Flavonoid assay determination

Assay determination is very important for accurate identification of Chinese crude drugs, formulation process evaluation, and quality control of TCMs. The methods listed below are commonly used for the assay of flavonoids.

5.1.1   Ultraviolet spectrophotometry (UV)

The use of UV to determine the assay of flavonoids has already been stated in certain pharmacopeia. It's based on the two major absorption peaks of flavonoids under UV: band I (300 to 400 nm) and band II (240 to 280 nm). Sometimes, diagnostic reagents, such as $AlCl_3$, can be added as a chromogenic agent, and generally the complex formed by flavonoids and metal ion gives yellow or orange color. However, the UV method is not specific and can only be used to roughly estimate the total flavonoid content and cannot be used to quantitatively and qualitatively determine the specific pure flavonoid compound. However, it's still an easy and convenient method which can be used for initial quality control during production.

5.1.2   Thin-layer chromatography scanning (TLC scan)

TLC scan was once the most commonly used method to assay flavonoids, as it is easy to operate and shows good reproducibility. Examples include using fluorescence TLC scan to determine the puerarin content in kudzuvine root[21], using dual wavelength TLC scan to determine the icariside content[22], and using TLC scan to determine the daidzein content in red rice and its formulation Xuezhi Kang[23]. However, when compared with HPLC, the

TLC scan procedures are too tedious, and accuracy is also low. At present, TLC scan has already been replaced by HPLC.

5.1.3 HPLC

Since the late 1970s, HPLC has been more widely used to assay various chemical compounds. HPLC can quantitatively detect samples at nanogram or picogram levels. When determining the assay of flavonoids quantitatively, the common extraction methods used are methanol reflux extraction, chloroform shaking extraction, acetone reflux extraction, mobile phase extraction, and ultrasonic extraction. Many studies about using HPLC to determine the assay of flavonoids have been reported. HPLC has been regarded as a convenient, quick and efficient method for the quality evaluation of Chinese crude drugs, optimal collecting time determination, extraction process screening, and formulation quality control.

1. Determination of puerarin in kudzuvine root collected from different regions by HPLC[24]

HPLC has been used to determine the puerarin content in Kudzuvine root collected from Anhui, Shandong, Shanxi provinces, and also from *Puerariae thomsonii* Benth from Guangxi province, and *Pueraria lobata* Wild Ohwi from Hubei province, respectively. The results showed that the puerarin content in kudzuvine roots collected from Shangdong province is the highest and also the best quality.

2. Determination of baicalin in *Scutellaria baicalensis* (baical skullcap) roots by HPLC[25]

HPLC had been used to determine the baicalin content in *Scutellaria baicalensis* (baical skullcap) roots, which were collected (from South China botanic garden) on the 15th day of every month for a year. The results showed that the baicalin content in *Scutellaria baicalensis* roots collected from spring and autumn was the highest. This result is consistent with traditional experiences to harvest in spring and autumn, especially in autumn.

3. Determination of icariin in [*Epimedium sagittatum* (Sieb. et Zucc.) Maxim.] water extracts prepared with different purification processes[26]

HPLC has been used to determine the assay of icariin in [*Epimedium sagittatum* (Sieb et Zucc.) Maxim.] water extracts prepared with four different purification processes. Results showed that icariin content in [*Epimedium sagittatum* (Sieb. et Zucc.) Maxim.] water extract prepared with the alcohol sedimentation method is the highest; thus the alcohol sedimentation method has been regarded as the first choice for process purification.

4. Determination of flavonoid aglycone in three different *Ginkgo biloba* L. formulations[27]

HPLC has been used to determine the assay of flavonoid aglycon in three different *Ginkgo biloba* L. formulations (two homemade tablets and one imported tablet). Results showed no significant assay difference between the homemade tablets and the imported tablets.

5. Overseas research reports

The review by Merken, et al.[28] about using HPLC to determine the ingredients of flavonoids has comprehensively and systematically introduced the selection of chromatographic conditions, test facilities, sample preparation, and the features of different types of flavonoids using HPLC.

In summary, HPLC is a mature technology with high sensitivity, excellent reproducibility, high performance, and is easy to operate. It has become the first choice for assay determination, and is also a good method used to determine the flavonoids in Chinese crude drugs and formulations.

5.1.4   Capillary electrophoresis (CE)

Compared with UV and TLC methods, CE has many advantages, such as shorter analysis time, less solvent consumption, lower costs and less waste. It is suitable for the quick, quantitative analysis of certain flavonoids. For instance, high performance CE has been used to determine puerarin in kudzuvine root and its formulation Zhengxin Tai capsule[29]. Micellar electrokinetic capillary chromatography (MECC) has been used to determine *Ginkgo biloba* L. flavonoids, and capillary zone electrophoresis with electrochemical detection has been used to assay baicalin and baicalein in baical skullcap roots and their formulations[30].

Other assay determination methods, like GC, high-speed counter-current chromatography (HSCCC), HPLC-MS/ESI, GC-MS, solid phase micro extraction (SPME) and supercritical fluid extraction chromatography, have also been reported but the application of these methods is not common and further research is needed.

5.2   Flavonoid formulations

Because of the various pharmacological activities of flavonoids, preparation into suitable finished dosage forms is very important for R&D. Since the 1990s, with the improvement of process formulations and production techniques various *Ginkgo biloba* L. formulations have become available, including extracts, tablets, capsules, granules, oral liquids and injections, which can be widely used to treat cardio-cerebrovascular disease, liver, kidney and lung disorders, psychosis, diabetes, etc., with satisfactory effects[31]. Daidzein from the kudzuvine root has low toxicity, and is thus suitable for long-term use. It can be used to improve and eliminate dizziness, and stiff neck, etc. symptoms for patients with hypertension and coronary artery disease. It has a mild effect on blood pressure lowering and positive effect on sudden deafness and osteoporosis. It is of great value to develop this product. However, due to the isoflavone structure of the daidzein in kudzuvine root, its water solubility and liposolubility are unsatisfactory, which makes bioavailability poor.

When the daidzein of kudzuvine root and PVP are combined in proper ratios to make solid dispersion tablets, the bioavailability in vivo has increased significantly and the therapeutic effects also increase[32]. Icariin is the main active component in *Epimedium sagittatum* (Sieb. et Zucc.) Maxim. It has coronary and peripheral vascular relaxation effects and can improve blood circulation. When combined with hydroxypropyl-β-cyclodextrin (HP-β-CD) into ointment for external use on the skin, its release rate increased, therapeutic effect enhanced, and adverse effects were reduced[33]. Zhuanggu Shenbao capsules [containing *Epimedium sagittatum* (Sieb. et Zucc.) Maxim.] have been reported to have activity to prevent and treat corticosteroid osteoporosis and osteoporosis caused by long-term use of hydroxycarbamide or the adrenal cortex hormone prednisone acetate[34].

The development of modern technologies has provided an excellent environment for TCM formulations. Many superior isolation and purification techniques have been developed, including flocculate precipitation, SFE, UE, membrane separation, UF, ME, SBE and enzymatic chemical techniques. These advanced isolation and purification technologies have reduced impurities, improved product quality, and made smaller doses with stronger efficacy become possible. The growth of new excipients has pushed the development of TCM formulations to a new level. For instance, polyvidone and polyethylene glycol (PEG) are excellent carriers for solid dispersion, which have promoted the development of guttate pills. Acrylic resin can be used to prepare sustained-release arteannuin granules and has shown a delayed releasing effect, which can keep the blood concentration at a constant level. Chitosans have similar structure to that of dextran amine, with great advantage when used as a drug carrier. Chitosans, as excipients, have many functions, including use as a suspending agent, adhesive, delayed releasing agent, controlled releasing agent, microsphere carrier, membrane for medicinal use, etc. The use of chitosans in TCM delayed release microspheres has opened a new page for TCM sustained-release formulation development[35,36]. Other delivery systems used for TCM formulations, such as liposome, nanocapsule, microsphere, magnetic microsphere, microcapsule, solid dispersion, saturation technique, etc., also have great potential in TCM development and further investigation for their use as flavonoid formulations needs to be made[37,38].

With the increasing demands for health and medical care, pharmaceutical formulations will focus more on "three effects" (more efficient, long acting and quick) and "three smalls" (small dosage, small toxicity and small side effects). The ideal TCM formulation should have a stable and definite active marker compound as well as reproducible pharmacological and clinical data. However, currently, most TCM formulations still have much further to go to reach this target. Modern formulation technology and advanced materials will greatly promote the R&D of flavonoid formulation.

Section 6   Conclusion

In summary, flavonoids are of great variety and widely present and exhibit various pharmacological activities. There are already many flavonoid products in the market. Further studies and development of flavonoids, with technologies like new isolation methods, new formulation techniques, new analysis techniques, modern biotechnologies, information technologies, etc., will make it possible to develop more new anti-cancer, anti-aging and anti-cardiovascular disease drugs with great market potential and contribution to society.

References

[1] Department of Natural Products Chemistry, Shanghai Institute of Materia Medica of Chinese Academy of Sciences, Handbook of Identification of Flavonoids. Beijing: Science Press, 1981.

[2] Harborne J B et al. The Flavonoids: Advances in Research. London, New York: Chapman & Hall, 1982.

[3] Franks L et al. Flavonoids and Bioflavonoids: Studies in Organic Chemistry. New York: Elsevier, 1985.

[4] Harborne J B et al. The Flavonoids: Advances in Research Since 1986. London, New York: Chapman & Hall, 1986; 1994.

[5] Agrawal P K et al. Carbon-13 of Flavonoids. New York, Oxford: Elsevier.1989.

[6] Xu Rensheng. Chemistry of Natural Products (Second Edition), Beijing: Science Press, 2004, 526-571.

[7] Xiao Conghou. Chemistry of Chinese Herbal Medicine. Shanghai: Shanghai Sicentific and Technology Press, 1994, 265-274, 274-279, 265, 331.

[8] Li Yuan, Zhang Ming. Guangdong Pharmaceutical Journal, 1999, 9(2): 4-6.

[9] Ma Zhengshan. Chinese Traditional Patent Medicine, 1997, 19(12): 40.

[10] Mishima K et al. Solvent Extraction Free Paper Session of Japan. II. Fukuoka, 199.

[11] Xu Lianying, Tao Jiansheng, Feng Yi etc. Chinese Traditional Patent Medicine, 2000, 22(1): 6-19.

[12] Lin Huigai, Ge Zaochuan, Li Zhiliang etc. Acta Pharmaceutica Sinica, 1991, 26(6): 471-474.

[13] Kang Chun, Wen Liyu, Ding Zhongbo. Chinese Journal of Pharmaceutical Analysis, 2000, 202: 121-124.

[14] Xie Jianchun, Luo Hongpeng, Zhulili et al. Acta Physico-Chimica Sinica, 2001,17(7): 582-585.

[15] Feng Nianpin, Fan Guangpin, Han Zhaoyang. Chinese Journal of Information on Traditional Chinese Medicine, 2000,7(10):15.

[16] Qu Guilong, Guo Haiming. Chinese Traditional and Herbal Drugs, 2000, 31(4): 310.

[17] Cui Chengbing, Chen Yingjie, Yao Xingsheng et al. Journal of Shenyang Pharmaceutical University, 1991, 8(1): 36-38.

[18] Huang Hesheng, Ma Chuangen, Chen Zhiwu. China Journal of Chinese Materia Medica, 2000, 25(10): 589-592.

[19] Shi Min. Foreign Medical Sciences (Section of Hygiene), 2001, 28(2): 96-99.

[20] Han Bing, Yang Junshan. Chinese Traditional and Herbal Drugs, 2000, 31(11): 873.

[21] Shang Xiaohong, Dong Chuan, Feng Yucai et al. Chinese Journal of Analytical Chemistry, 2001, 29(1): 115.

[22] Yu Aipin, Lu Jinqin, Yu Nancai. Hubei Journal of Traditional Chinese Medicine, 2001, 23(1): 49.

[23] Xu Yanchun, Wei Luxue, He Dalin et al. China Journal of Chinese Materia Medica, 2001, 26(1): 33-34.

[24] Xiao Xuefeng, Gao Lan. Chinese Traditional and Herbal Drugs, 2001, 32(3): 220.

[25] Xu Jiwen, Luo Suxiang. Journal of Guangdong College of Pharmacy, 2000, 16(4): 314-315.

[26] Zhao Xiaomei, Nie Qixia, Zhang Baoxian et al. Chinese Journal of Experimental Traditional Medical Formulae, 2000, 6(6): 3-5.

[27] Li Shuxia, Tang Lei, Wu Jiexiong et al. Guangdong Pharmaceutical Journal, 2001, 11(1): 20-21.

[28] Merken H M, Beecher G R. Journal of Agriculture and Food Chemistry, 2000, 48(3): 577-599.

[29] Li Hongbin, Sun Lihua, Hu Yongming et al. Chinese Journal of Modern Applied Pharmacy, 2001, 18(3): 213-214.

[30] Chen Gang, Li Lijun, Ye Jiannong. Journal of Instrumental Analysis, 2001, 20(1): 27-29.

[31] Guo Haipin, Li Huizhen, Guan Jun. Tianjin Pharmacy, 2001, 13(3): 62-64.

[32] Huang Qihong, Hu Longrong. Journal of Guangdong College of Pharmacy, 2001, 17(2): 87-88.

[33] Lou Zhihua, Liu Yanli. Chinese Traditional Patent Medicine, 2001, 23(5): 321-323.

[34] Liu Jiangqin, Liang Tong, Zhuang Haiqi, et al. Chinese Journal of Modern Applied Pharmacy, 2001, 18(3): 192-194.

[35] Zhe Peiying, Zhang Yejing, Li Jian, et al. Chinese Traditional and Herbal Drugs, 2001, 32(3): 273-275.

[36] Zhang Zhirong, Huang Yuan. Chinese Traditional and Herbal Drugs, 2000, 31(spl): 1-5.

[37] Wang Xueli, Zhu Chunyan, Yang Shiling. Chinese Traditional and Herbal Drugs, 2001, 32(2): 176-178.

[38] Zhang Jinqing, Zhang Zhirong. Chinese Traditional and Herbal Drugs, 2000, 31(spl): 23-25.

De-Yun KONG

CHAPTER 10

Anthraquinones

Section 1   Introduction

Anthraquinones are the derivatives of anthraquinone[1]. The C-1, C-4, C-5, C-8 of anthraquinone (9,10-anthracenedione) are also known as α position, the C-2, C-3, C-6, C-7 known as β position, and C-9, C-10 known as *meso* position. The substituents of anthraquinones could be methyl, hydroxymethyl, carboxyl, or formyl as well as hydroxyl or methoxyl. Hydroxyl groups are present in most naturally occurring anthraquinones.

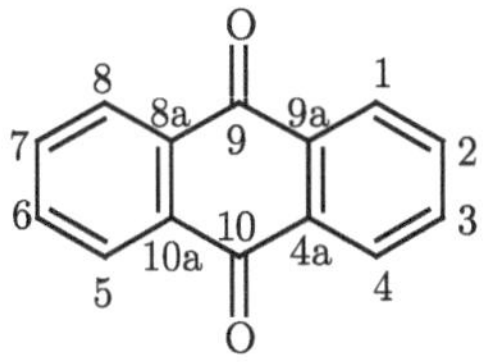

Anthraquinone

Generally the anthraquinones also include dithranols, oxidized dithranols, anthrones and the dimers of anthrones.

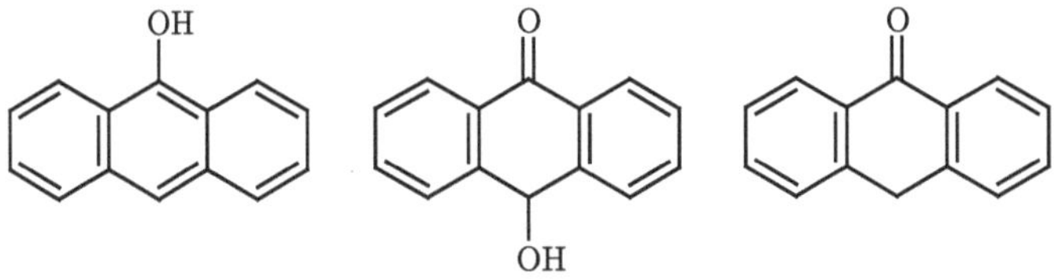

Dithranol   Oxidized dithranol   Anthrone

The hydroxyl dithranols and anthrones are in generally found in fresh plants, and could be gradually oxidized. *Meso*-hydroxyl dithranols are easily oxidized, but their glucosides are stable. Anthraquinones can be found in plants in free form or their glycoside form. The most common sugar found in anthraquinone glycosides is glucose, and others include rhamnose, arabinose, xylose, and so on.

Most naturally occurring anthraquinones are found in higher plants, such as fungi and lichens. Those found in higher plants occur mostly in Rubiaceae, and partly in Rutaceae, Rhamnaceae, *Cassia* (Leguminosae), *Rheum* and *Rumex* (Polygonaceae), Bignoniaceae, Verbenaceae, *Digitalis* (Scrophulariaceae), and Liliaceae. Those found in fungi occur in *Aspergillus* and *Penicillium*.

1.1   Structure

1.1.1   Monosubtituted anthraquinones

The groups of monosubtituted anthraquinones are generally located at β position. Tectoquinone (β-methyl anthraquinone) is from *Tectona grandlis*. fil (Verbenaceae), *Acatypha india* (Euphorbiaceae), *Morinda umbellata* L. (Rubiaceae), and so on. It is one of a few anthraquinones without a hydroxyl. The heartwood of teak (*T. grandis*) is extremely resistant to attack by insects or fungi. Tectoquinone present in teak heartwood may be responsible for this resistance[2]. 2-Hydroxymethyl-anthraquinone is found in *Morinda parvifolia* (Rubiaceae) and exhibits anti-leukemia effects in vivo.

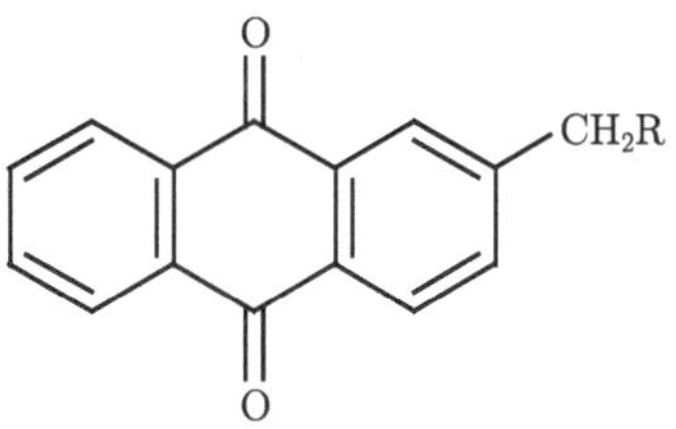

	R
Tectoquinone	H
2-Hydroxymethylanthraquinone	OH

1.1.2   Disubstituted anthraquinones

Only a few disubstituted anthraquinones exist in nature. In most of them, both substituted groups are located at the same ring. Alizarin, mp 290°C, is the representative compound in *Rubia cordifolia* L., *R. tinctorum*, *Morinda umbellata* Linn., and *M. citrifolia*. Its glucoside exists in *Libertia coerulescens*. Alizarin-2-methyl ether generally co-exists with alizarin in the above plants. *R. tinctorum* was widely cultivated in Western Europe when its roots were the main source of alizarin.

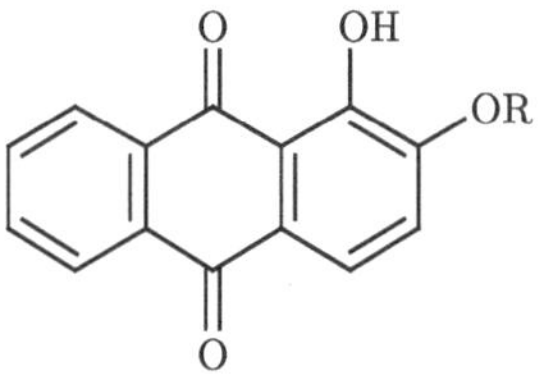

	R
Alizarin	H
Alizarin-2-methyl ether	CH$_3$

1.1.3   Trisubtituted anthraquinones

Trisubtituted anthraquinones are widely distributed, and emodin type anthraquinones and dencichine type anthraquinones are the most important ones.

1.1.3.1   Emodin type anthraquinones

Most of these types of compounds are yellow, and their hydroxyl groups are located in two separate rings. The principal anthraquinones in *Rheum* species are emodin type anthraquinones. Chrysophanol (chrysophanic acid) is widely distributed in both higher and lower plants. Its 1-*O*-D-glucopyranoside (pulmatin), 8-*O*-D-glucopyranoside (chryso-phanein) and diglucopyranoside also occur in *R. Palmatum*.

Aloe-emodin exists in the rhizomes of *Rheum palmatum* and *R. officinale*, roots of *Rumex patientia* and *Cassia mimosoides*, leaves of *R. gelinae*, seeds of *C. obtusifolia*[3] and *C. occi-dentalis*, and the fruits of *Rhamnus officinalis*. Aloe-emodin and aloe-emodin-8-*O*-glucoside found in the rhizomes of *R. Palmatum* are the antimicrobial ingredients of rhubarb. Cathartic action induced by aloe results from aloin, the C-glucoside of aloe-emodin, mp 146 to 148°C. Rhein occurs naturally, both in free form and its glycoside form.

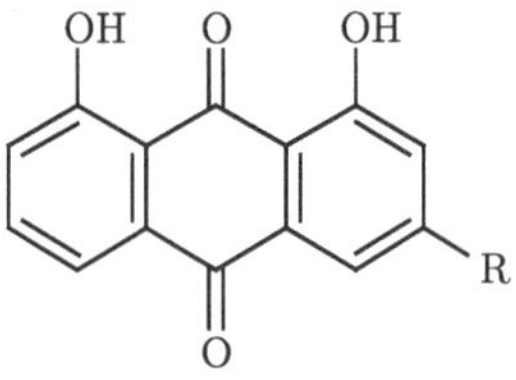

	R
Chrysophanol	CH$_3$
Aloe-emodin	CH$_2$OH
Rhein	COOH

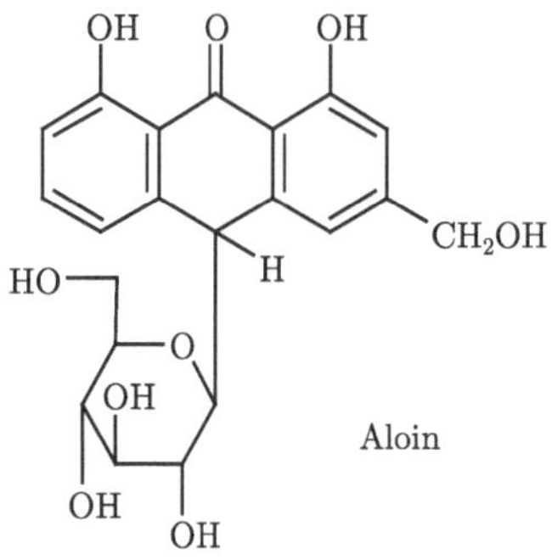

Aloin

Sennosides A and B are two isomers of sennosides, the dimers of rhein, isolated from rhubarb and *Cassia angustifolia*. The hydrolysis of sennoside A resulted in the aglycone (sennidin A) and two molecules of glucose. Sennidin A is the dimer of anthrones connected covalently between C_{10} and $C_{10'}$ in *trans*, which is easily broken to form stable anthrones. The cathartic action induced by rhubarb and *C. angustifolia* resulted in the transformation of anthrone glucosides from sennoside A in the intestines.

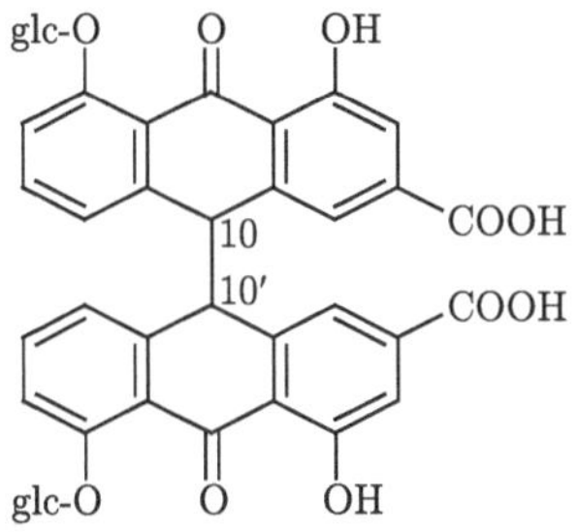

Sennoside A mp 200~240°C(dec.), $[\alpha]_D$−147°
Sennoside B mp 180~186°C(dec.), $[\alpha]_D$−100°

1.1.3.2   Madder type anthraquinones

Most anthraquinones in Madder plants are madder type anthraquinones, in which hydroxyls are located in the same ring. The majority of them are orange. Pupurin, mp 263C°, an important anthraquinone, occurs in *Rubia tinctorum*, *R. cordifolia* and *Galium* and is another important pigment in *R. cordifolia*.

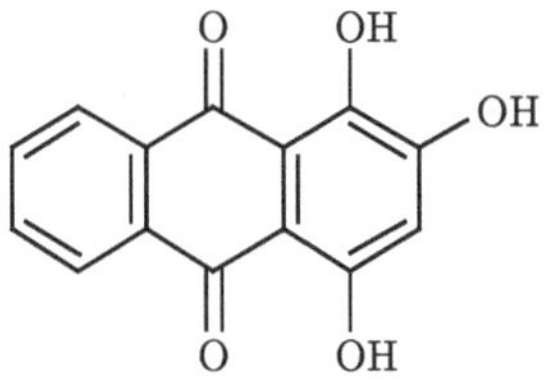

Purpurin

1.1.4   Tetrasubstituted anthraquinones

Emodin is one of the most widely distributed naturally occurring anthraquinones and is a representative of the family of tetrasubstituted anthraquinones. The glycosides of emodin occur in higher plants and fungi and are the principal active ingredients of rhubarb. Emodin-8-O-glucoside can be found in the bark of *Polygonum cuspidatum* (Polygonaceae), rhizomes of *Rheum sublanceolatum*[4] and *Ligularia kanaitizensis* (Compositae)[5]. Physcion (rheochrysidin) usually co-occurs with emodin and chrysophanol. Pseudopurpurin, mp 222 to 224°C, is one of the madder anthraquinones. Its glucosides occur in roots of certain madder plants, such as *Rubia tinctorum* (Rubiaceae).

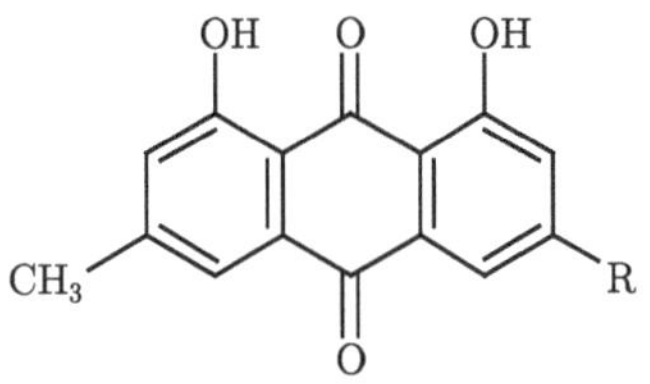

	R
Emodin	OH
Physcion	OCH$_3$

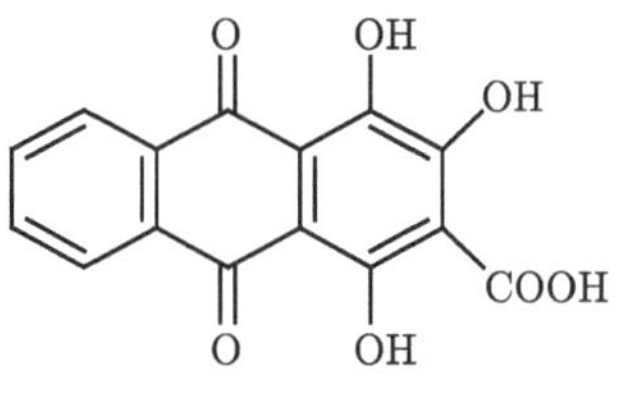

Pseudopurpurin

1.1.5   Pentasubstituted anthraquinones

A number of naturally occurring pentasubstituted anthraquinones exist in certain insects, molluscs, and molds. 1,4,6,8-Tetrahydroxy-3-methyl-anthraquinone (catenarin) was isolated from *H. catenarium* mycelium (15% dry weight).

Aspergillus versicolor (Streptomyces), as its name indicates, is a multi-colored organism from which many colored compounds structurally characterized with a six-carbon side chain at position 2 are obtained. Examples include averythrin and averantin.

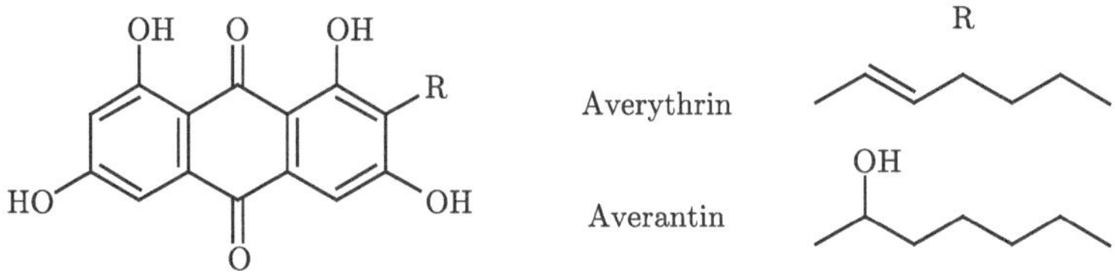

1.1.6   Hexasubstituted and heptasubstituted anthraquinones

Few hexasubstituted and heptasubstituted anthraquinones are found in nature. In addition to chrysophanol and its analogs, cassia seeds also contain some hexasubstituted anthraquinones such as aurantio-obtusin, mp 265 to 266°C, obtusin, mp 242 to 243°C, chryso-obtusin and chryso-obtusin-β-O-glucoside. Chryso-obtusin-β-O-glucoside could strongly inhibit platelet aggregation. 2,5,7-Trihydroxyemodin, a heptasubstituted anthraquinone present in the lichen *Mycoblastus sanguinarius*, may be the most hydroxy–substituted anthraquinone and its extraction with acetone was difficult by heating for 2 weeks due to its strong polarity.

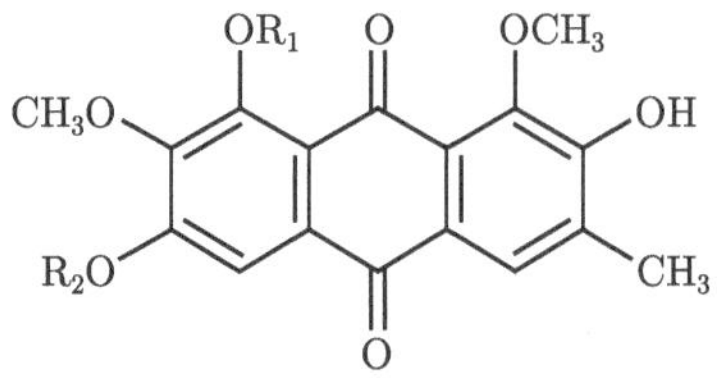

	R_1	R_2
Aurantio-obtusin	H	H
Obtusin	H	CH_3
Chryso-obtusin	CH_3	CH_3

1.1.7   Anthracyclinones

Anthracyclinones come from *streptavidin*. Many of them, such as topoisomerase II inhibitor, show anti-tumor and anti-bacterial biological activities. Doxorubicin (adriamycin) and daunorubicin from *Strepomyces speucetins* are the representative compounds. Doxorubicin, red-orange needle, was easily soluble in water, mp 229 to 231°C, hydrochloride mp 204 to 205°C, $[\alpha]_D^{20}248$ (c = 0.1, methanol). Daunorubicin is slightly different, with mp 208 to 209°C, hydrochloride mp 188 to 190°C, $[\alpha]_D^{20}248 \pm 5$ (c = 0.05 to 0.1, methanol).

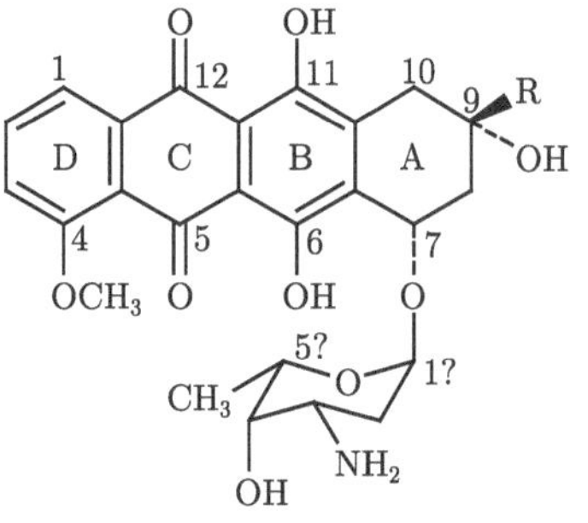

Daunorubicin R = COCH$_3$
Doxorubicin R = COCH$_2$OH

1.2   Biological Activity

Many naturally occurring anthraquinones exhibit cathartic action which does not interfere with the absorption of nutrients from the small intestine. SAR of cathartic action with anthraquinones is as follows: 1) The anthraquinone glycosides show stronger cathartic action than aglycones, and dithranols stronger than corresponding anthraquinones. The esterification of the phenolic hydroxyl in anthraquinones led to the disappearance of cathartic action. 2) The anthraquinone glycosides with carboxyl groups show stronger cathartic action than the corresponding glycosides without carboxyl groups, and their dimers (such as sennoside) also present strong cathartic action. Only anthraquinone glycosides could induce significant cathartic action in the large intestine, but anthraquinones couldn't because the majority of free anthraquinones have been absorbed or damaged before arriving in the large intestine, suggesting the sugars in the glycosides play an important role for cathartic action.

Certain anthraquinones (such as rhein, emodin, and aloe-emodin) show stronger antibacterial effects than their glycosides. Anthracyclinones possess good antibiotic effects for Gram-positive bacteria, mycoplasma, and certain yeasts, but they are inactive toward Gram-negative bacteria. Because of their high toxicity, they are not used as clinical antibacterials.

Rhubarb significantly inhibits inflammation for a variety of experimental animals, which is due to the processing method. Emodin from rhubarb has anti-inflammatory effects and can significantly inhibit the carrageenian-induced paw swelling and increased abdominal capillary permeability caused by acetic acid, as well as the carrageenan-induced acute pleurisy. Emodin can be easily absorbed into the body, showing fast action, efficacy toward early inflammatory exudation, increased capillary permeability, as well as for acute inflammation[6]. Emodin could also enhance normal immune functions[7,8].

Some anthraquinones (such as emodin, aloe-emodin) can eliminate hydroxyl radicals, and are remarkable antioxidants. Their antioxidant activity is dependent on the location of the hydroxyl(s)[9]. Through chemiluminescence detection, it has been proved that rhubarb can remove O_2, H_2O_2 and other reactive oxygen species, inhibit lipid peroxidation, and is an effective antioxidant[10].

Doxorubicins isolated from natural sources are antineoplastic antibiotics. Their common properties include interaction with DNA in a variety of different ways, including intercalation (squeezing between the base pairs), DNA strand breakage and inhibition with the enzyme topoisomerase II. However, they lack the specificity of the antimicrobial activity and thus produce significant toxicity. Anthracyclines are among the most important antitumor drugs available. Both are widely used for the treatment of several solid tumors while daunorubicin is used exclusively for the treatment of leukemia. Doxorubicin may also inhibit polymerase activity, affect the regulation of gene expression, and produce free radical damage to DNA. Doxorubicin possesses an antitumor effect against a wide spectrum of tumors, either grafted or spontaneous. The anthracyclines are cell cycle-nonspecific.

Animal experiments have demonstrated that emodin from *Polygonum cuspidatum* can inhibit 1-nitro-pyrene (1-NP) mutagenesis and confront 1-NP gene toxicity. Both water extracts of *P. cuspidatum* and emodin can significantly reduce the 1-NP-induced mutation in a dose-dependent way, as the experiments with ^{32}P-labeled *Salmonella* TA98 confirmed the the direct inhibitory effect of emodin on 1-NP-induced mutation.

Emodin showed a strong inhibitory effect on protein tyrosine kinase (PTKs). The kinetic analysis of PTK inhibition demonstrated that emodin was a competitive inhibitor of ATP and also a non-competitive inhibitor of peptide substrates containing tyrosine, implying that the inhibitory effects of emodin on p56lck were induced by blocking ATP-binding. Tests showed that the inhibitory effects of emodin on diverse tumor cell lines were not directly related to its cytotoxicity, and therefore, emodin was a good tumor cell growth inhibitor[11].

Section 2   Physical and Chemical Properties

2.1   Physical properties

Anthraquinone glycosides are easily soluble in hot water, alkali solution, methanol, ethanol, acetic acid and pyridine, slightly soluble in cold water, and difficult to dissolve in ether, toluene, chloroform, and so on. Anthraquinones in free form are almost insoluble in cold water, slightly soluble in cold ethanol, soluble in hot ethanol, toluene, chloroform, ethyl ether, pyridine, acetone, acetic acid and alkali aqueous solutions. The free anthraquinones in aqueous alkali solutions are cherry red in color and can't be extracted by general organic solvents. After acidification, they can be extracted into an organic phase.

Most anthraquinones or their glycosides, yellow or orange-red crystals and may show fluorescence dependent upon the pH. Some anthraquinones could be sublimated. Anthraquinone glycosides can be hydrolyzed by heating in aqueous acid.

2.2   Coloration

Natural drugs containing anthraquinones are usually colored (e.g., the rhizomes of rhubarb are yellow), which could be significantly enhanced by ammonia or base. Hydroxyanthraquinones and their glycosides with free phenolic hydroxyl(s) show significant red-purple color in the presence of base. The color is related to the conjugate structure between phenolic hydroxyl and carbonyl. Dithranols, anthrones and anthrone dimers can color after oxidation to form anthraquinones. The color reagents of anthraquinones and their glycosides are as follows: ammonia, 10% KOH methanol solution, 3% NaOH solution or sodium carbonate solution, 50% piperidine solution in toluene, and saturated lithium carbonate solution. Anthraquinones also show red to purple red color in the presence of sulfuric acid.

Anthraquinones could be colored with 0.25% to 0.5% magnesium acetate in methanol. The solution could also serve as a chromatograpic color reagent. The α-hydroxyl in anthraquinones is required for this coloration, and the colors depend on the number and location of hydroxyls in the anthraquinones.

2.3   Acidity

The majority of natural anthraquinones are hydroxyanthraquinones and show some acidity, depending on the number and the location of the phenolic hydroxyl and carboxyl groups. The anthraquinones with carboxyl (e.g., rhein) are strongly acidic and can be dissolved in sodium bicarbonate solution. β–Hydroxyl anthraquinones are acidic because of the influence of electronic attraction of the carbonyl on hydroxyl, and can be dissolved in aqueous Na_2CO_3 solution. β Hydroxyl anthraquinones are weakly acidic because the hydrogen bond between hydroxyl and adjacent carbonyl limits the dissociation of the hydroxyl proton. Thus, they dissolve only in the sodium hydroxide solution.

The acid strengths of free anthraquinones are listed as: -COOH > containing more than two β-hydroxyl > containing one β-hydroxyl > containing two α-hydroxyl > containing one α-hydroxyl.

Section 3   Extraction and Isolation

3.1   Extraction

Anthraquinones in natural drugs often occur in free and glycoside form. The extraction methods used depend on which form of anthraquinone is desired.

3.1.1   Extraction with organic solvent

The herbal powder is heated with 90% ethanol. The ethanolic solution is concentrated in vacuo, the residue is partitioned between H_2O and $CHCl_3$. The free anthraquinones are available in $CHCl_3$ phase. The aqueous phase is partitioned successively with some polar organic solvents such as n-butanol to obtain anthraquinone glycosides. As most anthraquinones contain a hydroxyl or carboxyl, they are often present in plants in the form of sodium, potassium, calcium or magnesium salts. Therefore, before extraction with organic solvents, natural herbs are often treated with acid to make anthraquinones completely free.

3.1.2   Treatment with lead salt

Addition of lead acetate solution into an extraction solution with anthraquinone glycosides results in the precipitation of anthraquinone glycosides with lead salt, which then is filtered, washed, suspended in water, treated with H_2S gas and filtered. The filtrate is then adjusted to a neutral pH and dried by evaporation. The residue is subjected to silica gel column chromatography and the pure anthraquinone glycosides are available through crystallization or refinement.

3.1.3   Acido-basic treatment

The plant powder is treated with ammonia or 0.1% to 0.5% NaOH solution at room temperature. The alkali solution is then extracted using ether to remove lipophilic substances, and the aqueous solution is acidified and extracted with ether. Residue of the ether phase is treated with toluene, methanol, and ethanol, respectively, to obtain free anthraquinones. In an ethanolic extract without free anthraquinones, the addition of KOH will lead to the precipitation of anthraquinone glycoside salts, which are treated with acetic acid to give total anthraquinone glycosides.

3.2   Isolation

If the acidic hydrolysis is applied to herbs, anthraquinones and their glycosides will be obtained in free form. The herbs are refluxed in a mixture of organic solvent (such as chloroform and toluene) and 20% sulfuric acid, and the free anthraquinones are transferred into the organic layer which is then extracted successively with various alkali solutions (5% sodium bicarbonate, 5% sodium carbonate, and 1 or 5% sodium hydroxide). The alkali solutions are neutralized with acid and then extracted with organic solvent. Total anthraquinones with different acidity can be obtained from the corresponding organic solvent extracts. Total anthraquinone glycosides could be isolated with a similar process.

Chromatography is the common separation method for anthraquinones in the laboratory. Chromatography with DCP, silica, polyamide, and dextran gel is available for the separation of both anthraquinones and anthraquinone glycosides. In general, aluminum oxide is not used as chromatographic adsorbent for the isolation of anthraquinones, as the adsorption of anthraquinones to aluminum oxide is too strong to be eluted.

When using silica gel column chromatography eluted with a mixture of toluene, ethyl acetate, and methanol for the separation of anthraquinones, small amounts of oxalic acid or citric acid should be added to the silica in order to prevent the dissociation of the anthraquinones. A mixture of ethyl acetate, methanol, and water is used as the eluent for silica gel column chromatography of anthraquinone glycosides.

Hydroxyanthraquinones can be isolated by using polyamide chromatography based on the strength difference of hydrogen bonds between polyamide with methanol as eluent. Sephadex (e.g., Sephadex LH-20) chromatography separation is based on the size difference of anthraquinones. The eluent is often methanol or a mixture of ethanol and water.

Drop countercurrent chromatography (DCC) can be also used for the isolation of anthraquinones. For example, it is difficult to apply silica gel chromatography for the isolation of the lipophilic extracts from root bark of *Psorospermum febrifugum*, since the irreversible adsorption on silica will result in the loss of effective anthraquinones. However, anthraquinone and two other compounds have been isolated quantitatively in only one step by DCC[12,13].

By using DCC and HPLC, chrysophanol-1-*O*-glucoside, chrysophanol-8-*O*-glucoside and physcion-8-*O*-glucoside, were isolated from roots of *R. palmatum* (Polygonaceae). Fifty grams of *R. palmatum* were refluxed in a mixture of benzene and 20% sulfuric acid (5:l) for 4 hours, after which the supernatant was discarded. The residue was washed two times with benzene. The combined benzene layer was washed three times with water, then with 5% sodium bicarbonate three times. The combined alkaline extracts were acidified with hydrochloric acid and a yellow precipitate was generated and refined by sublimation to give rhein in yellow needles. The benzene layer was then extracted three times with 5% sodium carbonate solution. After acidification of the alkaline extracts and sublimation of precipitate under vacuum, emodin was obtained as orange needles. The benzene layer was then extracted with 5% potassium hydroxide three times and the combined alkaline extracts were acidified. The orange precipitate was formed and washed repeatedly with chloroform. Refinement of the orange precipitate with toluene gave aloe-emodin as orange needles. The chloroform solution from the precipitate was extracted with 5% KOH two times, the acidification of the extracts resulted in precipitate washing, and was then subjected to cellulose chromatography and eluted with water-saturated petroleum ether to give chrysophanol and physcion[14].

chrysophanol-1-*O*-glucoside $R_1 = $ Glc, $R_2 = R_3 = H$
chrysophanol-8-*O*-glucoside $R_2 = $ Glc, $R_1 = R_3 = H$
physcion-8-*O*-glucoside $R_1 = H$, $R_2 = $ Glc, $R_3 = OCH_3$

Rhein yellow needles (sublimation), mp 321 to 322°C, dissolve slightly in water, soluble in alkaline and pyridine, slightly soluble in ethanol, benzene, chloroform, ethyl ether and petroleum ether. UV λ (ε) nm: 229 (36800), 258 (20100), 435 (11100). IR KBr (cm^{-1}): 1710(C), 1637(chelating C = O). $C_{15}H_8O_6$.

Emodin orange needles (ethanol), mp 256 to 257°C, difficult to dissolve in water, soluble in ethanol and alkali solution. ^{13}C-NMR (DMSO-d6) 123.88 (C-7), 164.32 (C-1), 107.71 (C-2), 165.40 (C-3), 108.88 (C-4), 120.22 (C-5), 148.04 (C-6), 161.19 (C-8), 189.44 (C-9), 180.92 (C-10), 132.51 (C-10a), 112.99 (C-8a), 108.68 (C-9a), 134.77 (C4a), 21.44 (CH$_3$). $C_{15}H_{10}O_5$.

Aloe-emodin orange needles (toluene), mp 223 to 224°C. Soluble in hot ethanol. Aloe-emodin shows yellow color in ethanol and benzene, and crimson in ammonia and sulfuric acid. $C_{15}H_{10}O_5$.

Chrysophanol monoclinics (ethanol), mp 196°C (sublimation). Almost insoluble in water, slightly soluble in cold ethanol and petroleum ether, soluble in boiling ethanol and benzene, chloroform, ethyl ether, acetic acid and acetone. IR KBr (cm^{-1}): 1680, 1630, 1607. UV λ (ε) nm: 224 (4.73), 257 (4.48), 277 (4.18), 287 (4.18), 429 (4.14). MS m/z (%): 254 (100). $C_{15}H_{10}O_4$.

Physcion red brick crystals, mp 203 to 207°C. Soluble in benzene, chloroform, toluene and pyridine, slightly soluble in acid and ethyl acetate, insoluble in methanol, ethanol, acetone and ether. UV λ (ε) nm: 266 (4.45), 255 (4.22), 267 (4.25), 288 (4.22), 440 (4.02). IR KBr (cm^{-1}): 1570 (C = C), 1625 (chelating C = O), 1668 (free C = O), 3400 (OH). $C_{16}H_{12}O_5$.

Section 4   Spectral Characteristics

4.1   UV

UV-visible spectra of anthraquinones show characteristic maxima related with the substituents and their location, which is useful for deducing structure. Four significant maxima at 252, 272, 325 and 405 nm, related with the benzene-like structure and quinone-like structure, are generally found in the UV spectrum.

benzene-like structure   quinone-like structure

A majority of natural anthraquinones are substituted with the hydroxyl group. Five major UV maxima could be found: maximum I around 230 nm; maximum II at 240 to 260 nm (caused by benzene-like structure); maximum III at 262 to 295 nm (quinone-like structure); maximum IV 305 to 389 nm (benzene-like structure); maximum V > 400 nm (quinone-like structure).

The maxima and intensity of hydroxyanthraquinones depend on substituents as well as their number and arrangement. Anthraquinones with one phenolic hydroxyl show maximum I at 222.5 nm, and when the number of phenolic hydroxyl increases, the maximum is shifted to a longer wavelength. Logε value of maximum III greater than 4.1 suggests the existence of an α-phenolic hydroxyl, otherwise there is no α-phenolic hydroxyl. When anthraquinones are substituted by -CH$_3$, -OH or -OCH$_3$ at α position, a red shift of maximum IV will be observed, but its intensity is lower. When these groups are located at β position, the intensity of the maximum will be enhanced. In general, the number of α-hydroxyls considerably impacts maxima in visible region. For example, anthraquinones containing one α-hydroxyl have maximum at 400 to 420 nm and those containing two or more α-hydroxyls have maximum over 430 nm.

4.2   IR

Characteristic IR peaks of hydroxyanthraquinones are the carbonyl stretching vibration near 1670 cm^{-1}, three hydroxyl stretching vibrations at 3600 to 3150 cm^{-1}, and benzene skeleton vibration at 1600 to 1480 cm^{-1}.

The chemical environments for two carbonyls in a non-substituted anthraquinone are identical, so only one $v_{C=O}$ peak could be found at 1675 cm^{-1} (paraffin paste). The $v_{C=O}$ peaks for anthraquinones with or without α-OH are generally found at 1679 to 1653 cm^{-1}, or at 1660 to 1630 cm^{-1}, respectively. Two $v_{C=O}$ peaks of the anthraquinones with one α-OH resulted from the fact that only one carbonyl chelated with α-OH and the other did not. Two $v_{C=O}$ peaks for 1,8-dihydroxyl anthraquinones are respectively located at 1678 to 1661 cm^{-1} for non-chelated carbonyl and at 1626 to 1616 cm^{-1} for chelated. Only one $v_{C=O}$ peak for 1,4- or 1,5-dihydroxy anthraquinones could be observed, since two carbonyls in the molecule are respectively chelated with different α-OH.

4.3   MS

The molecular ion peak of anthraquinones is often the base peak. The mass spectrum of anthraquinones is characterized by the successive loss of two carbon monoxide molecules, resulting in peaks of m/z 180 (M-CO) and 152 (M-2CO). Strong peaks of (M-3CO) and (M-4CO) could be observed in the mass spectrum of anthraquinones with one or two hydroxyls, in addition to the peaks of (M-OH). The strong MS peak of (M-CHO) will appear in anthraquinones with α-OCH$_3$. In EIMS, the aglycone ion is usually the base peak for anthraquinone glycosides.

4.4   NMR

The ^{1}H-NMR signals of aromatic hydrogens in anthraquinone can be divided into two categories: α-hydrogens located at the C = O negative shielding show downfield shifts to δ 8.07, and those of β-hydrogen are upfield to δ 6.67.

These chemical shifts and the shape of aromatic hydrogen are dependent on the substituents, their number, and location. For example, the signal of an isolated aromatic hydrogen is singlet. The signals of *ortho* hydrogen are doublet (J = 6.0 to 9.4 Hz), those of *meta* hydrogen are double doublet (J = 0.8 to 3.1 Hz). In anthraquinones, the methyl will enable the chemical shifts of *ortho*- and *meta*-hydrogen upfield 0.15 and 0.10 ppm, and hydroxyl and methoxyl will enable them upfield 0.45. Carboxyl will make those of *ortho*-H downfield 0.8.

Because of the hydrogen bonding between hydroxyl and carbonyl groups, the signals of α-hydroxyl appear at δ 12.25 for anthraquinones with one α-hydroxyl, at δ 11.6 to 12.1 for 1,8-dihydroxyl anthraquinones, and at δ <10.9 for those with an α-hydroxyl and a group substituted at *ortho* position of α-hydroxyl.

^{13}C-NMR is important in the identification of anthraquinones. The carbon atoms of anthraquinones could be divided into four categories. The chemical shifts of these carbons for anthraquinone without substituents are as follows: α-carbon 126.6, β-carbon 134.3, carbonyl carbon 182.5, quarter carbon 132.9.

Anthraquinones with a substituent in one ring only show little modification of the chemical shifts of the carbons of the non-substituted rings; this means that the impact of a substituent will only affect the substituted ring. It is a useful rule for the structural identification of anthraquinones[15]. The impact of diverse groups substituted at α- and β-positions on the carbon chemical shifts of anthraquinones are listed in Table 10-1.

For ^{13}C-NMR of monosubstituted anthraquinones, the differentiation value of chemical shifts between the observed and deduced, as shown in Table 10-1, will be less than 0.5. For disubstituted anthraquinones with two groups at different rings, a relatively important differentiation will be expected, and the data in Table 10-1 should be modified. For example, when the substituent is -OH or -OCH$_3$, in 1,3-disubstituted anthraquinones a chemical shift of C-2 should be corrected by -3; C-4 by +4; in 1,4-disubstituted anthraquinones, C-1, C-3 by +3; C-2 by +2; C-4 by +2.5; C-13, C-14 by -1.6.

The rule of ^{13}C-NMR chemical shift change of anthraquinones indicated in Table 10-1 can be applied to deduce the structures of various substituted anthraquinones[16].

For example, UV, IR, and ^{1}H-NMR spectra of 1,3,6-trihydroxy-2-methyl-anthraquinone and those of 1,3,7-trihydroxy-2-methyl-anthraquinone were almost identical. Wu Lijun et al. deduced the structure of the former using ^{1}H- and ^{13}C-NMR spectra[17]. In ^{1}H-NMR of the compound, the signal δ 7.12 (1H, s) was assigned to H-4. In order to determine the location of hydroxyl at position 6 or 7 (i.e., the assignment of 7.14 (1H, d, J = 2.5 Hz), 7.37 (1H, d, J = 2.5 Hz), 7.95 (1H, d, J = 8.5Hz) in ABX system), HMQC was studied carefully. Thus, signals at δ 182.0 and 185.6 were assigned to C-9 and C10, and those at δ 7.37 and 7.95 to H5 and H8, respectively. In conclusion, a hydroxyl is substituted at position 6 rather than 7.

Table 10-1 ^{13}C-NMR Chemical Shifts and Modifications of Anthraquinones($\Delta\delta$)

C	C1–OH	C2–OH	C1–OCH$_3$	C2–OCH$_3$	C1–CH$_3$	C2–CH$_3$	C1–OAc	C2–OAc
C-1	34.7	−14.4	33.2	−17.1	14	−0.1	23.6	−6.5
C-2	−10.6	28.8	−16.1	30.3	4.1	10.1	−4.8	20.6
C-3	2.5	−12.8	0.8	−12.9	−1.0	−1.5	0.3	−6.9
C-4	−7.8	3.2	−7.4	2.5	−0.6	−0.1	−1.1	1.8
C-5	0	−0.1	−0.7	−0.1	0.5	−0.3	0.3	0.5
C-6	0.5	0.0	−0.9	−0.6	−0.3	−1.2	0.7	−0.3
C-7	−0.1	−0.5	0.1	−0.1	0.2	−0.3	−0.3	−0.5
C-8	−0.3	−0.1	0	−0.1	0	−0.1	0.4	0.6
C-9	5.4	0	−0.7	0	2.0	−0.7	0.9	−0.8
C-10	−1.0	−1.5	0.3	−1.3	0	−0.3	−0.4	−1.1
C-10a	0	0	−1.1	0.3	0	−0.1	−0.3	−0.3
C-8a	1.0	0.2	2.2	0.2	0	−0.1	2.0	0.5
C-9a	−17.1	2.2	−12.0	2.1	2.0	−0.2	−7.9	5.4
C-4a	−0.3	−7.8	1.4	−6.2	−2.0	−2.3	1.6	−1.6

Section 5   Examples

5.1   Anthraquinones from *Rhynchotechum vestitum*[18−20]

Rhynchotechum vestitum Hook. f. et Thoms (Gesnericeae), Chinese name Mao-Xian-Zhu-Ju-Tai, is used as a traditional Chinese folk medicine for the treatment of hepatitis A and B in the South Yunnan Province of China.

R. vestitum was first percolated with 95% ethanol. The combined extracts were concentrated in vacuo to yield syrup, which was partitioned between CHCl$_3$ and H$_2$O. The CHCl$_3$ layer was treated with 5% sodium carbonate solution. β-Sitosterol and lupeol were isolated from the chloroform solution. Three anthraquinones, rubiadin, rubiadin-1-methylether, and a new compound rhynchotechol, were obtained from the sodium carbonate solution. The above basic aqueous layer was subjected to porous resin and eluted with a gradient of H$_2$O-EtOH. The fraction eluted with 30% EtOH was chromatographed on silica gel eluted with CHCl$_3$-MeOH (4:1) to give a new anthraquinone, munjistin-1-methyl ether, and five anthraquinone glycosides, rubiadin-3-O-β-glucoside, lucidin-3-O-β-glucoside, lucidin-1-methyl ether-11-O-β-glucoside, lucidin-3-O-β-primeveroside, and rubiadin-1-methyl ether-3-O-β-primeveroside. The last one is a new compound. The fraction eluted with 50% ethanol was further chromatographed on silica gel to give rubiadin-3-O-β-primeveroside. Part of the 95% ethanol extracts of *R. vestitum* were successively partitioned between H$_2$O and petroleum ether, CH$_2$Cl$_2$, acetic acetate and n-butanol. The water layer was subjected to porous resin and eluted with a gradient of H$_2$O-EtOH. 6-OH-rubiadin was isolated from the eluate. The acetic acetate fraction was subjected to column chromatography on silica gel with a chloroform/methanol gradient to obtain 1,3-dihydroxy-2-carboethoxy-anthraquinone.

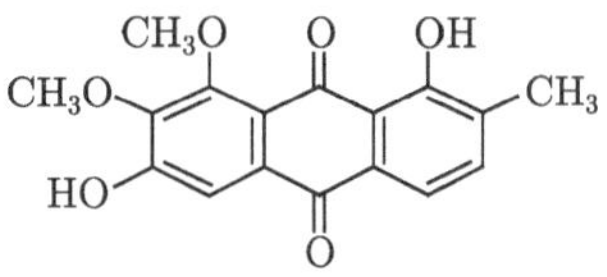

rhynchotechol 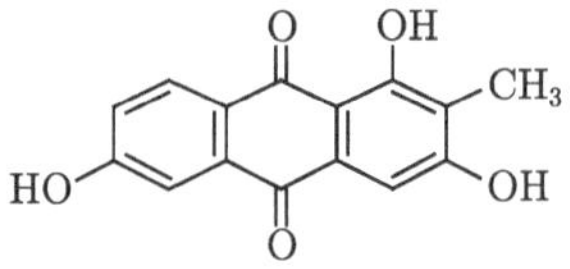   6-OH-rubiadin

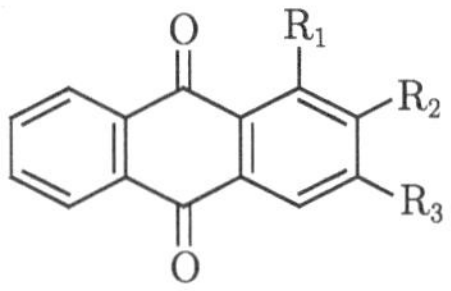

Name	R1	R2	R3
rubiadin	OH	CH$_3$	OH
rubiadin-1-methylether	OCH$_3$	CH$_3$	OH
munjistin-1-methylether	OCH$_3$	COOH	OH
1,3-dihydroxy-2-carboethoxy-anthraquinone	OH	COOEt	OH
rubiadin-3-O-β-glucoside	OH	CH$_3$	O-β-glc
lucidin-3-O-β-glucoside	OH	CH$_2$OH	O-β-glc
lucidin-1-methyl ether-11-O-β-glucoside	OCH$_3$	CH$_2$O-β-glc	OH
rubiadin-1-methyl ether-3-O-β-primeveroside	OCH$_3$	H	O-β-prim
lucidin-3-O-β-primeveroside	OH	OH	O-β-prim
rubiadin-3-O-β-primeveroside	OH	H	O-β-prim

Note: prim=-D-xylopyranosyl-D-glucopyranosyl; Glc=D-glucopyranosyl.

Rhynchotechol, orange crystals, mp 236.5 to 238°C (ethanol), was assigned the molecular formula C$_{17}$H$_{14}$O$_6$ (HRMS). UV maxima of 222 nm, 281 nm, 413 nm suggested a hydroxy anthraquinone. The free hydroxyl (3353 cm^{-1}), non-chelated carbonyl (1656 cm^{-1}) and chelated carbonyl (1630 cm^{-1}) were observed in IR. The ^{1}H-NMR signals indicated five substituents, one at δ 2.8 for β-methyl (α-methyl at δ 2.4), two at δ 13.37 and 6.45 for α- and β-hydroxyls (disappearance in the presence of D$_2$O) and two at δ 4.1 (3H, s) and 4.0 (3H, s) for two methoxyls. In ^{1}H-NMR, one α-H at δ 7.68 (s), two *ortho* aromatic hydrogens at δ 7.68 and 7.45 (d, J = 8 Hz), the same W$_{1/2}$ for signals at δ 7.68 and β-methyl indicated the presence of a propylene-like structure. According to the fragmentation rule of anthraquinone and fragments at m/z 181, 135, 153, 107, the five substitutents should be identified as follows.

To determine the location of three substituents, the reduction of rhynchotechol was performed as follows:

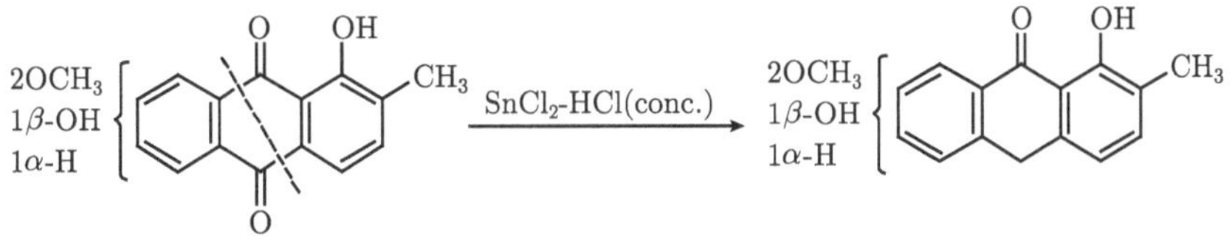

Rhynchotechol

Comparing the ^{1}H-NMR of two compounds, it was found that the signals of 4-H, 5-H in the reduced compound were up field about 1 ppm, respectively, indicating that a non-substituent was located in both sides of a non-associated carbonyl, i.e., the isolated hydrogen located at C-5. The location of the remaining two -OCH$_3$ and one β-OH were determined through NOE technology. The structure of rhynchotechol was deduced as 1,6-dihydroxy-2-methyl-7,8-dimethoxy 9,10-anthraquinone, which was further verified by HMQC and HMBC.

Munjistin-1-methyl ether, yellow crystals, mp 270°C (decomposition), was assigned the molecular formula C$_{16}$H$_{10}$O$_6$([M]$^+$ at m/z 298 and ^{13}C-NMR data). A positive coloration with FeCl$_3$ and KOH and UV maxima of 243 nm and 283 nm indicated a hydroxyanthraquinone. The ion peak fragments at m/z 281 [M-OH]$^+$ and 254 [M-CO$_2$]$^+$ in the EIMS indicate the existence of OH and COOH. The peak at m/z 280 [M-H$_2$O]$^+$ suggested that OH was *ortho* to carboxyl, since the formation of a six-member transition state with the two substituents can facilitate H$_2$O loss. The NMR signal at δ 3.80 (CH$_3$O) showed the presence of the third substituent on the anthraquinone, and δ_H 7.16 (1H, s, isolated aromatic proton) and δ_H 7.79 to 7.87, 8.07 to 8.13 (4H, symmetrical AA'BB' type of aromatic protons) indicated that one aromatic ring of the anthraquinone was unsubstituted and the other substituted by -OH, -COOH and –OCH$_3$. The signals at ν 1675.9 and 1654.7 cm^{-1} and δ_C 179.0 and 183.2 indicated that the two quinone carbonyls were not chelated with the phenol group. Thus, the OH had to occupy at β position. The isolated proton at δ 7.16 could be assigned the α position (C-4), which was shifted up field about 0.9 ppm relative to that of a characteristic α-H (δ 8.07), because of the electron releasing effects of the OH and OCH$_3$ groups at C3 and C1. As no NOE was observed between OCH$_3$ and H-4, the methoxyl must be at C1. Thus, the structure of this compound was determined to be 2-carboxy-3-hydroxy-l-methoxy-9,10-anthraquinone. It is known as munjistin-1-*O*-methyl ether. The proposed structure was elucidated by HMQC and HMBC.

Lucidin-1-methyl ether-11-*O*-β-glucoside pale yellow needles, mp 218 to 220°C was assigned the molecular formula C$_{22}$H$_{22}$O$_{10}$ (EIMS, [M]$^+$ at m/z 335 and ^{1}H and ^{13}C-NMR). UV maxima at 241.3, 274.5 and 335.0 nm and IR bands at 3591, 1670 and 1583 cm^{-1} suggested a hydroxyanthraquinone glycoside. An isolated signal at δ_H 3.88 (3H, s), an AB spin system at d$_H$ 4.60 and 4.89 (each 1H, d, J = 10.1Hz) and the signals at d$_C$ 162.0 and 162.5 (two oxygenated aromatic carbons) and δ_C 59.2 (one oxygenated secondary aliphatic carbon) indicated an anthraquinone aglycone possessing three substituents, methoxyl, aphenolic hydroxyl and hydroxymethyl. Furthermore, the carbon signals of the glycosyl group suggested a glucose residue attached to one of the hydroxyls. Four aromatic protons in a symmetrical AA'BB' type pattern and one in an isolated α position indicated that one aromatic ring in the anthraquinone was unsubstituted and that the other was substituted at C-1, 2 and 3. Most anthraquinones isolated from this plant have no substituent in one aromatic ring and always have a carbon substituent at position 2 in the other ring. Therefore, the hydroxymethyl was placed at C-2 based on comparative and biogenetic grounds. The methoxyl was placed at C-1, since the possibility of its placement at C-3 was ruled out by the absence of a NOE between CH$_3$O and H-4 and the δ_C of CH$_3$O (62.6, i.e., over 60) was in agreement with an *ortho, ortho*-disubstituted arrangement. Thus, the third substituent, the phenolic hydroxyl, had to be at C-3, which was in agreement with the lack of chelating between the phenol group and the quinone carbonyl (^{1}H and ^{13}C-NMR). In order to locate the glucosylated hydroxy group (OH-3 or OH-11), the ^{13}C-NMR data of this compound were compared with those of the reference compound, lucidin-1-methyl ether 3-*O*-β-primeveroside. Several significant differences involving the δ_C values of this compound were noted: (1) C-11 was shifted to a lower field (-7.2 ppm) and C-2 to a higher field; (2) The downfield shift of C-3 and upfield shift of C-9a were observed; (3) C-1' was shifted to lower field. This change was due to the decreased shielding of the anomeric carbon from the tertiary pyranoside in C-3 to the secondary pyranoside in C-11. Furthermore, the δ of the anomeric proton was shifted to δ 4.31 from δ 5.10 in the reference compound, which also strongly suggested that C-1' was linked to alcoholic oxygen rather than a phenolic one. The above data confirmed that OH-11 was glucosylated and that OH-3 was free. Based on the J value (7.8 Hz) of the doublet of H-1′, the anomeric configuration was β. Thus, this compound was identified as lucidin-1-methyl ether 11-O-β-glucoside. The proposed structure was further confirmed with HMBC.

5.2   Anthraquinones from *Rheum hotaoense*[21]

The property and taste of the dry root and rhizome of *Rheum hotaoense* C. Y. Cheng et C. T. Kao are described as bitter and cold. *R. hotaoense* used as Rheum shows antibacterial, anti-inflammatory, anticoagulant, but no obvious cathartic activity. *R. hotaoense* has been used to obtain anthraquinone glycosides as raw materials for reducing blood lipid.

Ten anthraquinones were isolated from the methanol extract of *R. hotaoense* by silica gel and polyamide column chromatography. With chemical and spectroscopic methods, these anthraquinones were identified as chrysophanol, physcion, emodin, aloe, rhein, chrysophanol-1-O-β-D-glucopyranoside, chrysophanol-8-O-β-D- glucopyranoside, physcion-8-O-D-β-glucopyranoside, aloe-8-O-β-glucopyranoside, and chrysophanol-8-O-β-D-(6′-O-galloyl)-glucopyranoside.

Chrysophanol-8-O-β-D-(6′-O-galloyl)–glucopyranoside, orange needles (acetone), mp 213 to 216°C. The characteristic MS peaks suggest a quinone structure in the molecule. ^{13}C-NMR shows the signals of anthraquinone, hexose, and galloyl. Chrysophanol, glucose and gallic acid were detected from its hydrolysis solution by TLC. UV and IR data suggest the existence of a free hydroxyl. Compared with the ^{1}H-NMR spectra of chrysophanol, the H-4 is downfield of H-7; ^{13}C-NMR spectra of C-8 were more upfield that those of C-1, indicating that glycosides are linked at position 8. From ^{13}C-NMR and ^{1}H-NMR data, sugar was identified as D-glucose. That C-6 and H-6 of sugar are downfield shifts suggests galloyl at C-6 of glucose. The structure is deduced as chrysophanol-8-O-β-D-(6′-O-galloyl)-glucopyranoside, a new compound (see the figure below).

5.3   Anthraquinones in *Morinda elliptica*[22]

Ismail et al. isolated 2-formyl 1-hydroxy anthraquinone and 10 other known anthraquinones from *Morinda elliptica*.

2-Formyl 1-hydroxy anthraquinone, Orange crystals; mp 183 to 185°C (CHCl₃), gives a molecular ion peak at m/z 252.0414 (HR-MS) corresponding to the molecular formula $C_{15}H_8O_4$. The UV-vis spectrum showed absorption maxima at 229, 278, 331 and 407 nm. These were suggestive of a 9,10-anthraquinone structure. A hathochromic shift (407 to 531 nm) upon addition of NaOH suggested the presence of OH at C-1, which was further confirmed by the band for a chelated carbonyl at 1638 cm⁻¹. In addition, the ^{1}H-NMR spectrum showed the characteristic downfield signal for the chelated hydroxyl group at δ 13.26. The ^{1}H-NMR spectrum also revealed the presence of an aldehyde proton at δ 10.63 and six aromatic protons. The ^{13}C-NMR showed 15 carbon atoms, which include three carbonyl carbons (downfield signals at δ 181.8, 188.0 and 188.9). The exact location of the substituents was established after careful examination of the NMR pattern of the aromatic protons using HMQC and HMBC. There were five signals in the aromatic region and the pattern suggested that one ring of the anthraquinone moiety was unsubstituted. Two sets of multiplets centered at δ 8.35 and 8.32, integrating for one proton each, were attributed to H-8 and H-5, respectively. Another set of multiplets at δ 7.88, integrating for two protons, was assigned to H-6 and H-7. The remaining two aromatic protons, a hydroxyl and a formyl group, were to be accounted for as substituents on another ring. A set of *ortho* coupled doublets at δ 8.23 and 7.89 (both 1 H, J = 8.0 Hz) indicated that the two aromatic protons were next to each other. This ruled out the possibility of placing the formyl group at C-3, since the hydroxyl had already been established at C-1. Its structure and ^{13}C-NMR signal were attributed as follows, and the deduced structure was further verified by HMQC and HMBC.

References

[1] Thomson R H. Anthraquinones in: Naturally Occurring Quinones, 2nd Ed. New York: Academic Press Inc 1971.

[2] Rudman, P.; da Costa, E. W. B. Relationship of Tectoquinone to Durability in *Tectona grandis*, Nature, 1958, 181(4610): 721-722.

[3] Hao Y J, Sang Y L, Zhao Y Q. Chinese Traditional and Herbal Drugs, 2003, 34(1): 18.

[4] Xiang L, Zheng J H, Guo D A, Kou J P, Fan G Q, Duan Y P, Qin C.Chinese Traditional and Herbal Drugs, 2001, 32(5): 395.

[5] Li Y S, Wang Z T, Luo S D, Chen J J, Zhang M. Journal of China Pharmaceutical University, 2001, 32 (5): 342.

[6] Qin H. Chinese Traditional and Herbal Drugs, 1999, 30(7): 522.

[7] Zhang J, Weng F H, Li H Q, Yao Z. Chinese Traditional and Herbal Drugs, 2001, 32(8): 718.

[8] Cai S X. Chinese Journal of Information on Traditional Chinese Medicine, 1997, 4(3): 23.

[9] Yeng C, Duh P D, Chuang D Y. Food Chemistry, 2000, 70: 437.

[10] Chen J W, Hu T X. Chinese Pharmaceutical Journal, 1996, 31(8): 461.

[11] Shang M Y, Ma X J, Cai S Q. Chinese Traditional and Herbal Drugs, 1997, 28(7): 433.

[12] Marston A, Potterat O, J Chromatogr, 1988, 450(1): 3450.

[13] Marston A, Chapuis J C, Sordat B, Msonthi J D, Hostettmenn K. Planta Med, 1986, (3): 207.

[14] Kan Y M. Chinese Herbal Medicine Chemistry Experiment Operation Technique. Peking: China Medicine Science and Technology Press, 1988, 149.

[15] Liu G M, Ding W G, Chen Z N. World Pharmacys, 1989, 4(3): 98.

[16] Itokawa H,Mihara K,Takeya K. Chem Pharm Bull. 1983, 31: 2353.

[17] Wu L J, Zhang W, Wang S X, Li X, Hua H M, Hou B L, Hou J Zhang Z F. Journal of Shenyang College of Pharmacy, 1994, 11(2): 138.

[18] Liu G M, Chen Z N, Yao T R, Ding W G.Acta Pharmaceutica Sinica, 1990, 25(9): 699.

[19] Lu Y, Xu P, Chen Z N, Liu G M. Phytochemistry, 1998, 47(2): 315.

[20] Lu Y, Xu P, Chen Z N, Liu G M. Phytochemistry, 1998, 49(4):1135.

[21] Li J L, Wang A Q, Li J S, He W Y, Kong M. Chinese Traditional and Herbal Drugs, 2000, 31(5): 321.

[22] Ismail N H, Ali A M, Aimi N, Kitajima M, Takayama H, Lajis N. Phytochemistry, 1997, 45: 1723.

Yang LU

CHAPTER 11

Coumarins

Section 1 Introduction

Coumarins, a class of natural products with a characteristic fragrant odor, are common to numerous plants. Coumarin is the lactone derived from *cis-ortho*-hydroxycinnamic acid, known as benzo-α-pyrone or benzo-1, 2-pyrone.

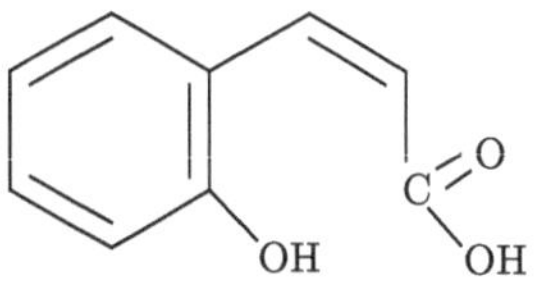

cis-ortho-hydroxycinnamic acid 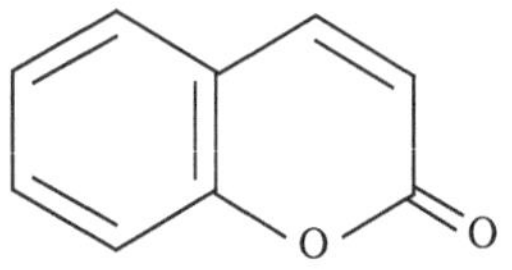   coumarin

Naturally occurring coumarins can exist as aglycones or as glycosides, and are widely distributed in more than 30 families, including about 150 species of plants. They are abundantly found in the families of Umbelliferae, Rutaceae, and Moraceae of dicotyledonous plants, and also exist in Leguminosae, Oleaceae, Solanacea, Compositae, Orchidaceae, and other families.

Most natural coumarins have been assigned trivial names which were derived from the genus or species name, or a combination of both, of the plant in which they were found.

Natural coumarin is characterized by its large number and various structural types[1−5]. About 2000 different coumarins were reported in literature from 1920 to 2003. Indeed, in the field of natural products, coumarins exhibit the best example of a class of compound with the greatest number of simple structure biogenetic modifications.

In recent years, with the development and application of modern chromatographic and spectroscopic techniques, many new types of coumarins have been discovered, such as prenyl-furocoumarin type sesquiterpenoid derivatives[6], a novel ethylene-dioxo derivative of coumarin[7], and some coumarins possessing unexpected structural features, e.g., coumarin sulfate, 3,4,7,-trimethylcoumarin without any oxygenated substituents, and tetraoxygenated coumarin, etc.

A variety of methods and technology are available for the study of coumarins, including supercritical fluid extraction (SFE), liquid chromatography under different pressures (LPLC, MPLC, and HPLC), vacuum liquid chromatography (VLC), and capillary electrophoresis (CE), and have been used advantageously for separation. As for structure determination, two-dimensional nuclear magnetic resonance spectroscopy (^{1}H-^{1}HCOSY, NOESY, HMQC, HMBC, etc.) is able to determine the structure with a small amount of sample and without any chemical reaction.

Several simple approaches with higher yield have been developed[8] for chemical synthesis, including a single-step synthesis of coumarins[9]. Interesting synthetic results have been reported for the first asymmetric synthesis of optically pure bicoumarin, (+) and (-) isokotanin A[10], synthesis of thirty new compounds of 6- or 7-styryl coumarin derivatives[11], and a series of novel derivatives of 3-bromo-4-methyl-7-methoxy-8-aminocoumarin[12] used for antitumor screening.

The diversity of biological activities is another characteristic of coumarins. For example, the antibiotic activity of novobiocin, the photosensitizing effects of certain linear furanocoumarins in the treatment of psoriasis and mycosis fungoides, the anticoagulant action of dicoumarol, and the acute hepatotoxicity and carcinogenicity of certain aflatoxins. Recently, calanolides, a series of dipyranocoumarins, have been isolated from the genus *Calophyllum* with the activity of inhibiting HIV-1 reverse transcriptase[13]. Khellactones, a type of angular pyranocoumarins, showed activity toward coronary vasodilatation[14]. Some coumarins with inhibition of NO synthesis[15−17] and having estrogen-like activity[18] were discovered. A bioassay based on the *Artemia salina* (brine shrimp) method of detecting potential phototoxic linear furanocoumarins was also developed[19].

A new review (1999–2003) for the study of coumarins, including 131 new compounds and the development of separation, structural determination, stereochemistry, synthesis and bioactivity, has been reported recently[20]. The advances in pharmacological research[21] and biological activities[22] in the last ten years are also reviewed.

Section 2   Structural Types

The classification of coumarins is based on several basic skeletons formed by the biosynthetic pathway. Generally, coumarins can be divided into four classes: simple coumarins, furanocoumarins, pyranocoumarins, and other coumarins.

2.1   Simple Coumarins

As the name implies, the simple coumarines have the substituents only at the benzenoid positions. A vast majority of natural coumarins carry an oxygen substituent at C-7, with only a few exceptions. 7-Hydroxycoumarin, commonly known as umbelliferone, is often regarded as the precursor, both structurally and biogenetically, of the more complex coumarins.

In plants, coumarin can be derived from phenylalanine or tyrosine. One of the biogenetic pathways of umbelliferone may be illustrated as follows (Figure 11-1):

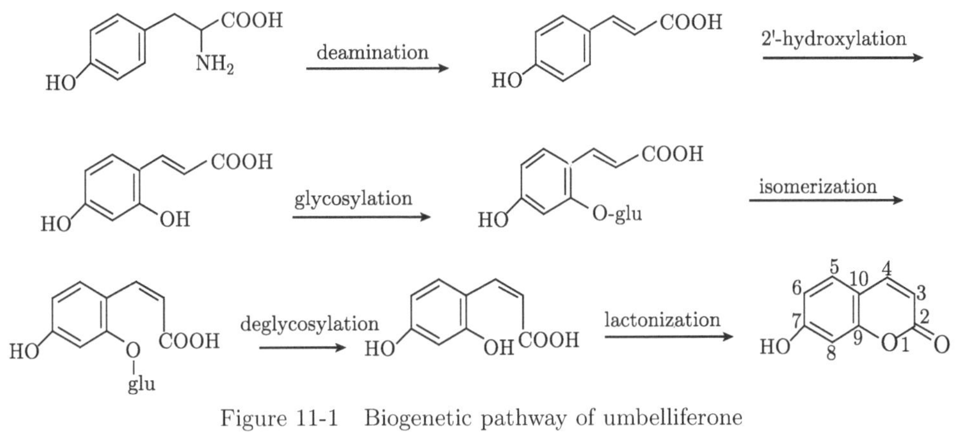

Figure 11-1   Biogenetic pathway of umbelliferone

For umbelliferone, the C-5, C-6, C-8 of the benzenoid ring may have to be substituted by oxygenated groups, such as hydroxy, methoxy, ethylenedioxy, isopentenyloxy, etc. A common feature of coumarins is the presence of isoprenoid chains attached to the carbon atom of the ring directly or through an oxygen atom.

Common examples of simple coumarins:

7-oxygenated:

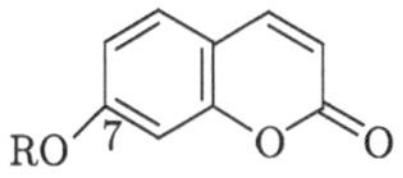

R: H   umbelliferone

       glu   skimmin

5, 7-dioxygenated:

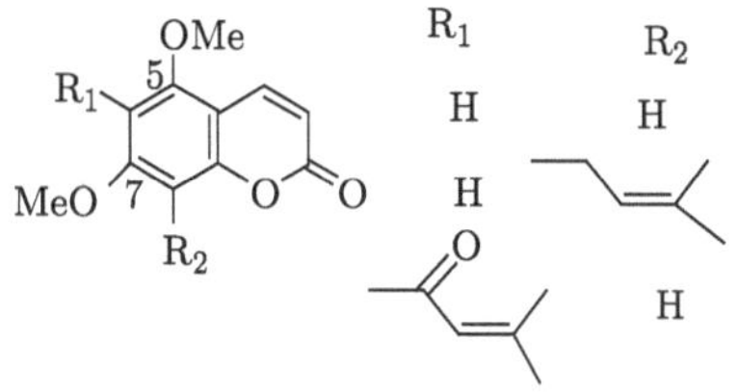

	R_1	R_2	
	H	H	limettin
	H	(prenyl)	coumarayin
	(acyl)	H	angelicone

6, 7-dioxygenated: 7, 8-dioxygenated:

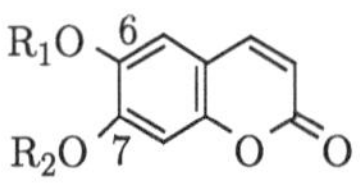

R_1	R_2	
H	H	esculetin
Me	H	scopoletin
Me	β-glu	scopolin
Me	Me	scoparone

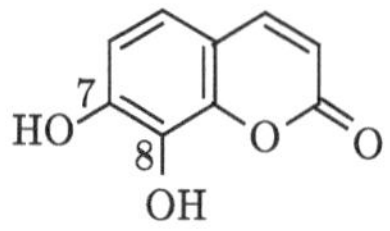

daphnetin

2.2 Furanocoumarins

Biogenetically, an additional furan ring in furanocoumarin can be formed when the prenyl group of coumarin interacts with an *ortho*-phenolic group. Oxidative interaction can lead to the hydroxyisopropyldihydrofuran moiety, which, by dehydrogenative loss of the hydroxyisopropyl group, produces the unsubstituted furan ring. When the prenylation of umbelliferone occurs at C-6, this interaction leads to demethylsuberosin, via (+) marmesin, and finally to the linear furanocoumarin psoralen. On the other hand, prenylation at C-8 leads to the angular furanocoumarin angelicin, via osthenol and (-) columbianetin. An alternative oxidative cyclization may form hydroxydihydropyrans, which upon dehydration, produce the linear and angular pyranocoumarins xanthyletin and seselin, respectively. Thus, the coumarin in each furano- and pyrano- class can be further divided into two types: the "linear" and the "angular".

The biogenetic pathways of furano- and pyrano- coumarin are shown in Figure 11-2[1a].

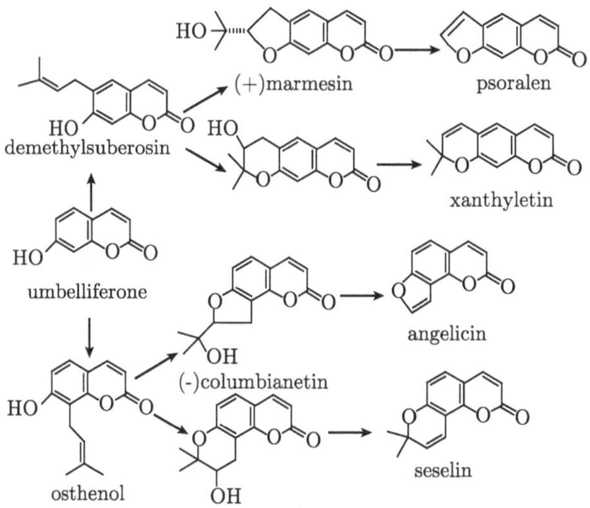

Figure 11-2 Biogenetic pathways of furano- and pyrano-coumarins

2.2.1   6,7-Furanocoumarins

Psoralen is typical of a 6,7- or linear furanocoumarin. Therefore, coumarins with similar structures are also known as psoralen coumarins. Common substituents such as methoxy, isopentenyloxy are at the C-5 or/and C-8 positions. For example:

R_1	R_2		R_1	R_2	
H	H	psoralen	OMe	OMe	isopimpinellin
OMe	H	bergapten	H	—O⌐⌐	imperatorin
H	OMe	xanthotoxin			

The linear furan group is also present in the dihydrofuran form.

2.2.2   7,8-Furanocoumarins

Angelicin (also known as isopsoralen) is typical of a 7,8- or angular furanocoumarin, and thus, also known as isopsoralen coumarins. It is most common to see a methoxy group substituted at the C-5 or both the C-5 and C-6 positions in this group. For example:

R_1	R_2	
H	H	angelicin
OMe	H	isobergapten
OMe	OMe	pimpinellin

The angular furan group also presents in the dihydrofuran form.

2.3   Pyranocoumarins

2.3.1   6,7-Pyranocoumarins

Xanthyletin is a representative of the linear type of pyranocoumarins. The methoxy and isopentenyl are the common groups substituted at C-5 or C-8 position. For example:

R_1	R_2	
H	H	xanthoxyletin
OMe	H	xanthoxyletin
H	OMe	luvangetin
OMe	✕⌐	poncitrin

2.3.2   7,8-Pyranocoumarins

Seselin is an example of an angular pyranocoumarin. Oxygenated substituents at C-5 or C-6 are commonly found in its derivatives. For example:

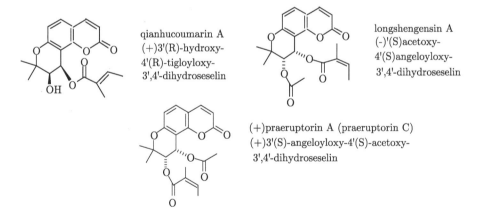

In recent years, khellactones, a series of monoacyl or diacyl derivatives of angular dihydropyranocoumarins with coronary vasodilatation activity isolated from the genus *Peucedanum* L. Khellactone ($R_1 = R_2 = H$) can be present in many stereo configurations. Their activity is characterized by the *cis*-form of C-3′ and C-4′[14].

For example, qianhucoumarin A to E[23−25] and (+) praeruptorin A (praeruptorin C)[26] were isolated from the traditional Chinese medicine Bia-Hua-Qian-Hu, roots of *Peucedanum praeruptorum*, and longshengensin A[27] was isolated from Nan-Ling-Qian-Hu, roots of *Peucedanum longshengense*. The common acyloxy groups in the esters are acetoxy, angeloyloxy, tigloyloxy, 2-methylbutyryloxy, isovaleryloxy and senecioxy. The absolute configurations of qianhucoumarin A, (+) praeruptorin A (praeruptorin C) and longshengensin A are shown as follows:

qianhucoumarin A
(+)3′(R)-hydroxy-
4′(R)-tigloyloxy-
3′,4′-dihydroseselin

longshengensin A
(-)′(S)acetoxy-
4′(S)angeloyloxy-
3′,4′-dihydroseselin

(+)praeruptorin A (praeruptorin C)
(+)3′(S)-angeloyloxy-4′(S)-acetoxy-
3′,4′-dihydroseselin

The declining effects of praeruptorin C on spontaneous contractile behavior and action potential were observed in cultured myocardial cells of neonatal rats. Experimental results indicate that the reduction of cardiac oxygen demand is likely to be one of the mechanisms of the anti-anginal effect of Qian-Hu[28].

2.4   Other Coumarins

The other coumarin group mainly consists of coumarins with substituents on the pyrone ring, dipyranocoumarins and bicoumarins.

2.4.1   3- or 4-Phenylcoumarins and 3,4-Benzocoumarins

For example:

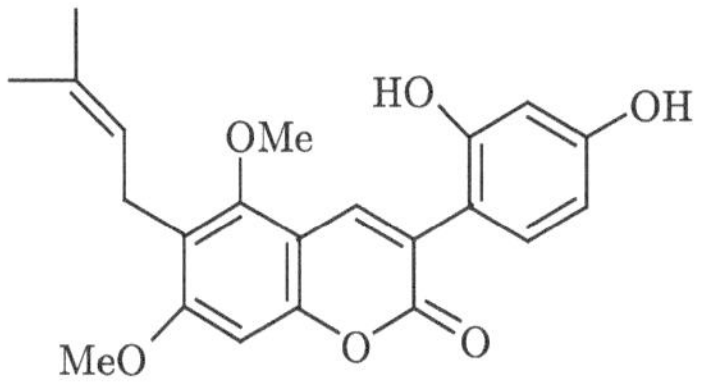

isoglycycoumarin 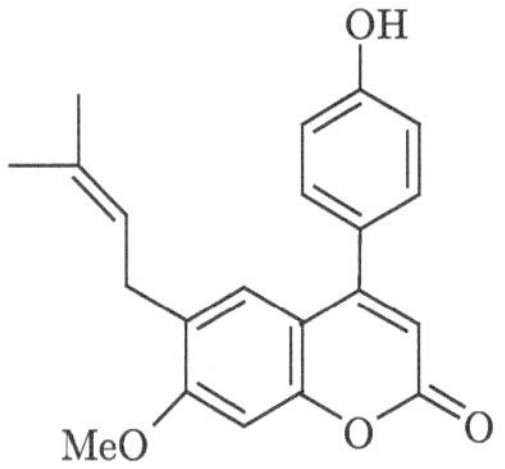   inflacoumarin A 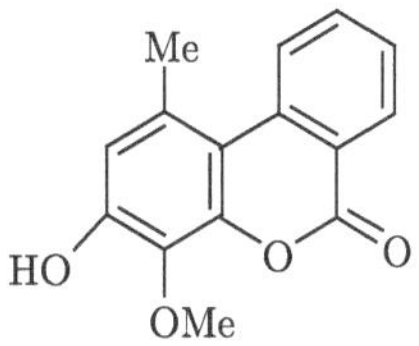   autumnariniol

2.4.2   4-Hydroxycoumarin derivatives

In addition to the common hydroxyl and methoxyl group substituted at C-4, a series of compounds known as coumestans are formed by connection of 4- hydroxyl and 3- phenyl group, such as wedelolactone, demethylwedelolactone, and a novel isodemethylwedelolactone isolated from the Chinese traditional medicine *Eclipta prostrata* L.[29]. The research progress of wedelolactone[30] and its anti-inflammatory-immune activity[31] has been reported recently.

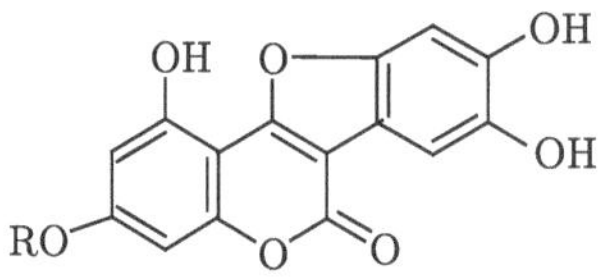

R = CH₃ wedelolactone
R = H demethylwedelolactone

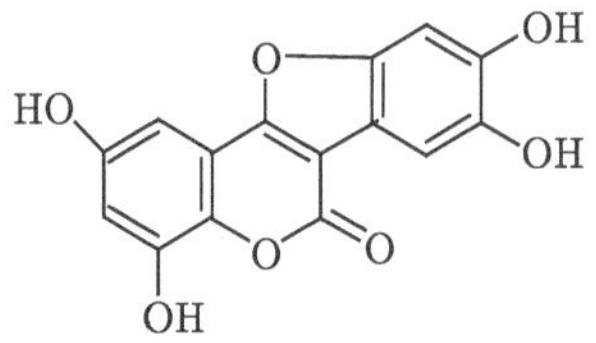

isodemethylwedelolactone

Novobiocin is a 4,7-dihydroxycoumarin-7-O-glycoside with a nitrogen containing moiety at C-3. It is known as an antibiotic metabolite of *Streptomyces*.

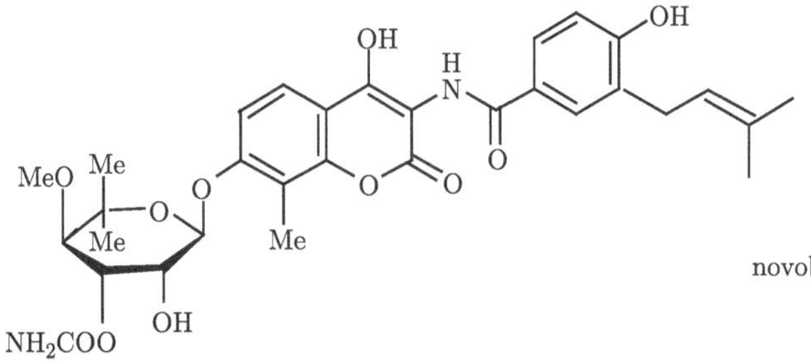

novobiocin

Recently, a number of prenyl-furocoumarin type sesquiterpenoid derivatives have been isolated from *Ferula ferulaeoides*, in which the 4-hydroxyl group is linked to a 3-prenyl sidechain, forming a furocoumarin moiety[6]. The 4-hydroxyl group may also form a chromocoumarin at the 3,4- position, such as that found in frutinone A, which has antibacterial activity and is isolated from *Polygala fruticosa*.

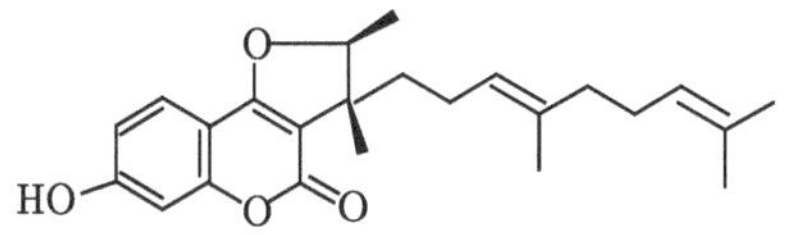

prenyl-furocoumarin type sesquiterpenoid derivative 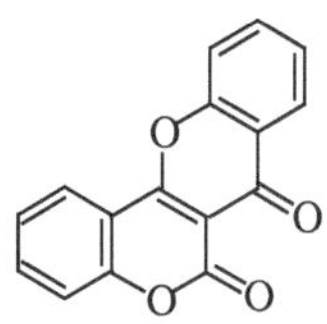   frutinone A

2.4.3   Calanolides

Calanolides, novel dipyranocoumarins, have recently been isolated from different species of *Calophyllum* L. This is a new class of non-nucleoside reverse transcriptase inhibitors (NNRTIs) for HIV-1. They fall into three basic structural types[13]: (a) tetracyclc dipyranocoumarins in which the C rings have a *gem*-dimethyl group, such as (+)-calanolide A; (b) tetracyclic dipyranocoumarins with reversed C and D pyran rings, that is, the *gem*-dimethyl is present in ring D, as in the (+)-pseudocordatolide C; and (c) tricyclic pyranocoumarins, such as calophyllolide, which contains a noncyclized equivalent of D ring of the calanolide tetracyclic structure. Individual members of the groups vary with respect to the C-4 substituents on the lactone ring, where methyl, n-propyl, or phenyl groups may be encountered.

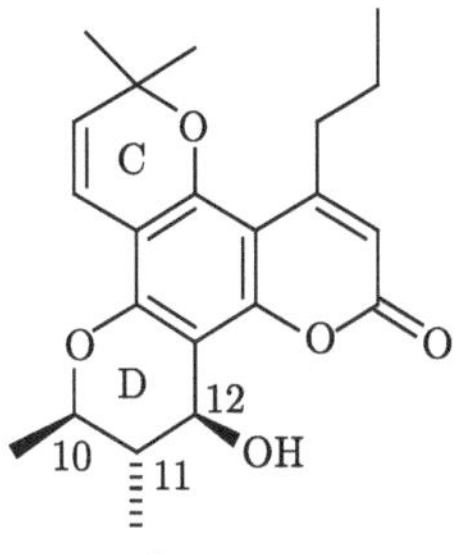

(+) calanolide A 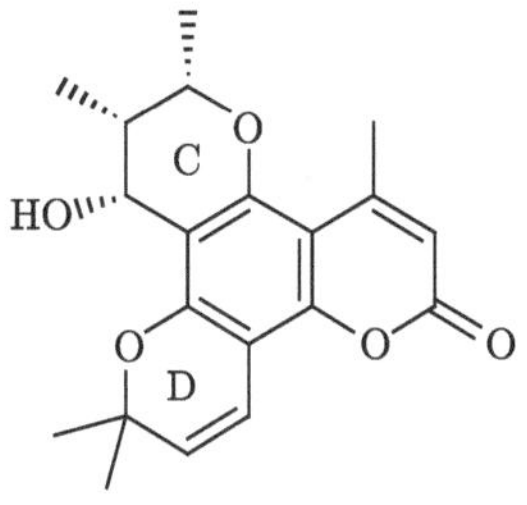   (+) pseudocordatolide C 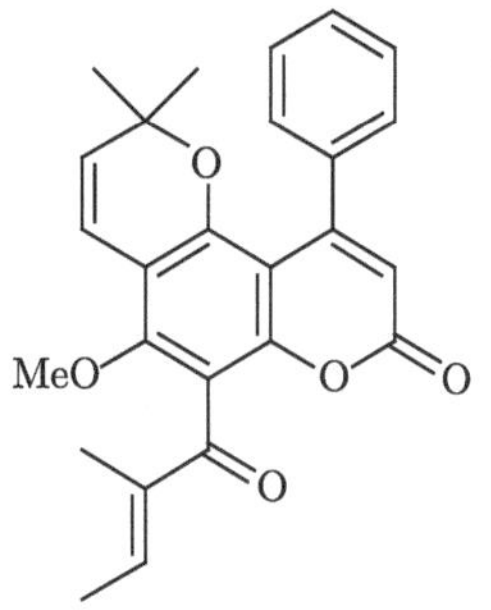   calophyllolide

(+)-Calanolide A, isolated from *Calophyllum lanigerum* var. *austrocoriaceum* is the representative of the structure having HIV-1-RT inhibitory activity[32]. Research indicated that the trans orientation of the C-10 and C-11 methyl groups and the 12β-OH of the saturated D ring are of considerable importance for the activity[33]. (+)-Calanolide A is the only naturally occurring anti-HIV agent at phase II clinical trial. However, its quantity is relatively low from nature. Total synthesis of racemic calanolide A was completed[34]. The enantioselective synthesis of optically active (+)-calanolide has also been reported[35].

Structural modifications of (+)-calanolide A have been carried out. However, no compound has showed anti-HIV-1 activity superior to the parent compound. Hopefully, calanolide A will be a promising anti-HIV drug in the near future. A review of *Calophyllum* coumarins as potent anti-HIV agents has been published recently[36].

Bicoumarins

Dicoumarol is a bicoumarin, a dimer of 4-hydroxycoumarin, known to have anti-platelet aggregation activity.

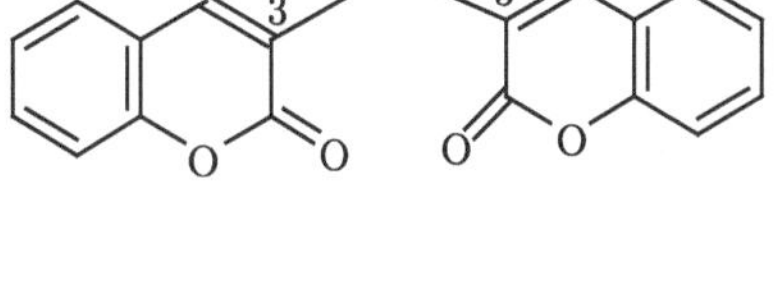

Dicoumarol

The coumarin moieties of the dimer can be connected either by simple coumarins or pyranocoumarins in linear-linear or linear-angular form. The nature of the linkage can be either by two coumarins, or through oxygen, methylene, or another group. As for dicoumarol, the linkage is C-3-CH_2-C-3′. However, many bicoumarins are linked by the C-8 of one coumarin unit with the C-3, C-5, C-6, or C-8 of another coumarin unit.

In recent years, some new bicoumarins have been discovered. For example, bicoumastechamin, a novel 8-6′ directly linked dimer, was isolated from the roots of *Stellera chamaejasme* or "Lang Du" in Chinese traditional medicine[37]. Cnidimonal and cnidimarin, two novel bicoumarins with 8-8′ or 8-5′ indirect linkages, respectively, were isolated from a traditional Chinese medicinal herb, the fruit of *Cnidium monnieri*, for the treatment of erectile dysfunction and frigidity[38]. From the twigs of *Pistacia chinensis*, two new ingredients of 3, 3′dimers of 4-aryldihydrocoumarin have been isolated. Their structures were determined as the only difference in the stereochemical linkage between the two moieties. These compounds are found to have noteworthy inhibitory activity against [^{3}H] estradiol binding to estrogen receptors and are the first examples of bis-4-arylcoumarins, which have shown estrogen-like activity[18] as new types of phytoestrogen.

Section 3   Physical and Chemical Properties

Because of the powerful spectroscopic techniques that have been widely employed, the earlier approach based on chemical investigation for structural elucidation is no longer practical. Thus, the physical and chemical properties of coumarin are now mainly used to confirm structural features based on their spectroscopic analysis and to distinguish isomers.

3.1   Fluorescence

Most coumarins are solid, normally crystalline, and exhibit fairly sharp melting points, apart from a number which decompose on heating. Coumarin can be evaporated or distillated by steam. It is insoluble or sparingly soluble in water, but soluble in petroleum ether, benzene, ether, chloroform, ethyl alcohol and other solvents.

The distinguishable physical property of most natural coumarins is the fluorescence they display in UV light (365 nm), many 7-oxygenated coumarins even fluoresce under visible light. This feature has been employed widely for their detection on paper and thin-layer chromatograms as the spots can be easily located. It is often possible to make tentative assignments of the structural class of a coumarin based on the color it displays[1c].

7-Alkoxycoumarins generally have a purple fluorescence whereas 7-hydroxy-coumarins and 5,7-dioxygenated coumarins tend to have blue fluorescence. In general, furanocoumarins possess a dull yellow or ocher fluorescence, with the blue-fluorescing psoralen, 6-methoxyangelicin, purple-fluorescing angelicin and 8-hydroxy-5-methoxypsorelen being the notable exceptions.

Intensified fluorescence upon treatment with ammonia vapor is an indication of a phenolic group in the coumarin. A blue fluorescence changing to an intense green fluorescence in alkaline solution, which may then disappear, is an indication of a 7-hydroxycoumarin.

The study of the relationship of the structure and the fluorescence of methoxycoumarin derivatives has been reported recently[39].

3.2   Reactions with Alkali

When treated with hot basic solution, the lactone ring of the coumarin is hydrolyzed slowly to become the coumarinate salt, in which the *cis* configuration is retained and dissolved to give a yellow solution. Acid, even carbon dioxide, leads to prompt relactonization and regeneration of coumarin, because the intermediate *ortho*-hydroxy-*cis*-cinnamic acid (coumarinic acid) is unstable. Prolonged treatment of coumarin with hot basic solution results in a salt in *cis* to *trans* form. If acidification occurs again, it can produce the stable *ortho*-hydroxy-*trans*-cinnamic acid (*O*-coumaric acid) without lactonization. Although the mechanism is not absolutely certain, it is likely that an addition-elimination pathway prevails with hydroxide ion adding to C-4 of the coumarinate, thereby permitting rotation about the single bond from C-3 to C-4[1d](Figure 11-3).

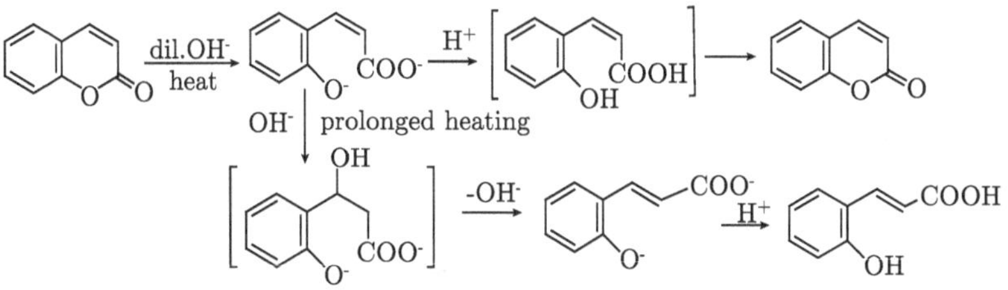

Figure 11-3　Opening and closing of the coumarin lactone-ring

3.3　Reactions with Acid

Coumarin can undergo a number of reactions with acid, for example, cleavage of allyl and related ethers.

Many coumarins contain prenyl, geranyl, or farnesyl ether residues. Such grouping in which the allyl ether system is present, characteristically undergo facile hydrolytic cleavage under mild acidic conditions. Gently warming an ethanol solution containing a trace of hydrogen chloride or a few drops of concentrated hydrochloric acid or sulfuric acid for a short time is generally sufficient to produce high yields of the phenol derivatives.

3.3.1　Alkene hydration

The side-chain double bond of coumarin can be converted to a hydroxyl group by acid hydration. The formation of such hydroxylated derivatives has potential practical importance. The virtually non-toxic aflatoxin B_{2a} is produced by the acid-catalyzed addition of water to the highly toxic aflatoxin B_1, and this has led to the suggestion of acid treatment as a possible means of detoxification of contaminated foods[1d].

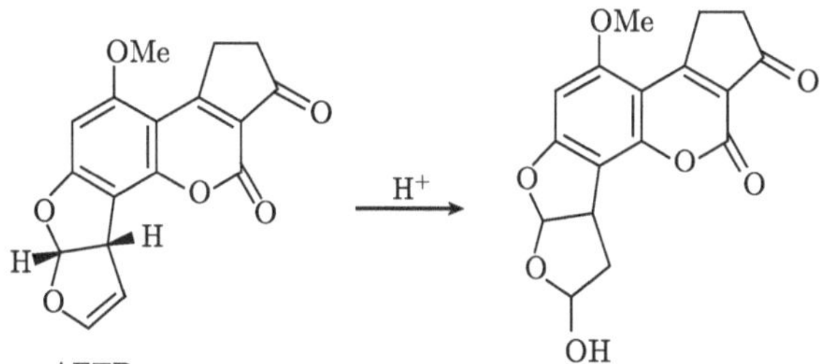

Section 4　Isolation

4.1　Extraction

In general, the isolation procedure initially depends on successive extractions of fresh or dried plant material with solvents of increasing polarity. Light petroleum, ether, acetone, and methanol are commonly used. Although most oxygenated coumarins are sparingly soluble in light petroleum, this is often used for the initial extraction since oils extracted with the coumarins tend to exert a solubilizing effect. Consequently, coumarins frequently crystallize directly from the concentrated extract, either during Soxhlet extraction or during cooling of the extract.

When the plant leaf is used as a raw material, the separation of ether-soluble coumarins from chlorophylls, waxes, and other constituents of similar solubility usually gives rise to additional problems which were solved effectively by the following example[1b]. During the extraction from the fresh leaves of *Angelica archangelica*, a number of furanocoumarins and chlorophylls were dissolved in boiling methanol. The filtrate, after adjustment to 60% aqueous methanol, was washed twice with hexane to remove chlorophylls and other fatty materials. The hexane layer was back-extracted once with 60% aqueous methanol, which was then combined with the main methanolic solution. Virtually all of the coumarins were retained in the aqueous methanolic part, while over 70% of the total weight of the original methanol extract was removed. Concentration of the aqueous methanolic solution gave an aqueous suspension which was then extracted with ether to recover the coumarins.

Another convenient approach was found to extract the plant material initially with acetone. The extract was concentrated to about one-third of its original volume and the dark heavy tar which separated was removed by filtration through cellulose. After evaporation of acetone, the residue was dissolved in chloroform and dried. This, however, did not previously remove the large amounts of fats and waxes which were then removed by hexane using column chromatography[1b].

It is important to note that each author often uses a specific extraction method, sometimes very different from techniques of others. A paper was reported to compare seven different extraction procedures and the resulting quantities[40].

4.2   Lactone Separation

Lactone separation is an early procedure for separation of coumarin and still is useful. At alkali conditions, the lactone ring of coumarin opens to become water-soluble coumarinate salts and other components can be removed by ether extraction. Acidification of the alkaline solution results, in most cases, in spontaneous lactonization and regeneration of coumarin. However, caution should be used as many coumarins are alkali- and/or acid-sensitive and can lead to the isolation of artifacts rather than the native coumarins.

4.3   Fractional Crystallization

Prior to the emergence of chromatography, fractional crystallization was widely used, either alone or in combination with fractional precipitation. The latter method was based on the low solubility of most oxygenated coumarins in light petroleum. It involves the precipitation of coumarin fractions by gradual addition of light petroleum to the ether extract.

4.4   Vacuum Sublimation and Steam Distillation

High-vacuum sublimation is a very simple way to purify coumarin that is thermally stable. However, thermally induced rearrangement or degradation may occur during the process. For example, the isoprene group of coumarins may be lost by heat to form a phenol group.

Steam distillation was used for the isolation of relatively unstable phenolic coumarins, such as separation of xanthyletin from other coumarins.

4.5   Chromatographic Methods

Chromatographic methods are the most common and effective for the separation and purification of coumarin at present. Column chromatography and thin-layer chromatography are the most commonly used. Best separation and purification results can be achieved with repeats or combinations of different chromatographic techniques.

4.5.1   Column chromatography

Because coumarin structures can be changed easily, the choice of the column absorbent should be very selective.

Degradation and loss of coumarins can occur with basic alumina. Acid-washed and neutral alumina has been found to give satisfactory separation results. However, sometimes the hydroxylic coumarins can be strongly absorbed on acidic alumina and are difficult to wash off.

Silica gel is the most commonly used absorbent with the common solvents, such as hexane-ether, hexane-ethylacetate, petroleum ether-ethylacetate, and petroleum ether-acetone. It has been recognized recently, however, that aldehyde artifacts may be produced from a pinacol-pinacolone rearrangement induced by the acidity of the silica gel used in the isolationof the 2,3-dihydroxyprenyl coumarins[1b].

Other absorbents, including polyamide, activated carbon, 18-alkylated silica gel, Sephadex LH-20, and macroporous resin, are also frequently used in conjunction with silica gel columns. Sephadex LH-20, as an absorbent in size exclusion chromatography, is commonly used in the purification of the final product. For example, the earliest separation difficulties of wedelolactone and demethylwedelolactone were overcome successfully by the use of Sephadex LH-20[41].

4.5.2   Other chromatographic techniques

Preparative thin-layer chromatography (P-TLC) has been used advantageously for the further separation of partially purified mixtures of coumarins. Solvents such as dichloromethane were found to give good separation of several linear furanocoumarins. Other combinations of solvents which have been found to produce good separation are chloro-form-ethylacetate, dichloromethane-ether, hexane-ethylacetate-methanol, hexane-acetone, and hexane-chloroform-ether. Coumarins are easily detected from their fluorescence under UV irradiation. Individual bands from the P-TLC are scraped off, washed with chloroform or acetone, and then crystallized to give coumarins.

4.5.3   High-performance liquid chromatography

High-performance liquid chromatography (HPLC) has the advantages of high-speed separation and automated operation over other chromatographic methods. Analytical HPLC is a very effective means to find traces of coumarin. Preparative liquid chromatography under different pressure, such as flash chromatography, LPLC, MPLC, and HPLC, played a major role in the separation and purification of coumarins with similar structures.

Recently, the development of capillary electrophoresis for the analysis of coumarins has also been reported[42].

Section 5   Spectroscopic Identification

A variety of spectroscopic methods have been developed for the identification of known coumarins and for the structural elucidation of novel coumarins[1c].

5.1   Ultraviolet Spectroscopy

Coumarin shows absorption bands at 274 (log ε 4.03) and 311 nm (log ε 3.72) which can be attributed to the benzene and pyrone rings, respectively. UV spectra are useful for distinguishing coumarins from chromones. Chromones normally have a strong absorption at 240 to 250 nm (log ε 3.8), whereas coumarins usually have minimums at this wavelength.

5.1.1   Alkylcoumarins

The introduction of a methyl group into the coumarin structure results in very small shifts of the UV λ_{max}. The 3-Me group leads to a small hypsochromic shift in the λ_{max} of the pyrone ring, while 5-Me, 7-Me, and 8-Me lead to a bathochromic shift of the λ_{max} of the benzene ring. No significant effect on the λ_{max} can be observed in the substitution of longer alkyl groups with minimal or no chromophoric properties of their own.

5.1.2   Oxygenated coumarins

The spectra of 7-oxygenated coumarins, such as 7-hydroxycoumarin (umbelliferone) show strong absorption bands at ~217 and 315 to 330 nm (log ε ~4.2) with weak peaks or shoulders at 240 and 255 nm (log ε ~3.5). The UV spectra of 7-hydroxycoumarin, 7-methoxycoumar-in, and 7-β-D-glucosyloxycoumarin are virtually identical. 5,7- and 7,8-dioxygenated coumarins have very similar spectra and resemble those of 7-oxygenated coumarins, except that the λ_{max} between 250 and 270 nm are slightly more intense (log ε 3.8 to 3.9). 6,7-dioxygenated coumarin can be readily differentiated from 5,7- or 7,8-dioxygenated coumarin as the two strongest bands are found at ~230 and 340 to 350 nm while two other bands of almost the same intensity (log ε 3.7 to 3.8) appear at ~260 and ~300 nm. The spectra of 6,7,8-trioxygenated coumarins, with their λ_{max} at 335 to 350 nm, more closely resemble those of 6,7- than those of 7,8-dioxygenated coumarins. On the other hand, the spectra of 5,6,7-trioxygenated coumarins, with their λ_{max} at 325 to 330 nm, are more similar to 5,7- than to 6,7-dioxygenated coumarins.

5.1.3   Spectral shifts

Electron delocalization of phenoxide ion with the pyrone-carbonyl group is possible for salts of 4-,5-, and 7-hydroxycoumarins, but not for those of 6- and 8-hydroxycoumarins. Consequently, when UV spectra are recorded in an alkaline medium, the intensities of the λ_{max} of the first group increase with 5- and 7-hydroxycoumarins showing noticeable bathochromic shifts; for 7-hydroxycoumarin, the long-wavelength band shifts from 325 nm (log ε 4.15) to 372 nm (log ε 4.23), while in the second group, bathochromic shifts also occur but with a simultaneous fall in log ε of the long-wavelength band. Thus, the difference of 6-hydroxycoumarin from a 7-hydroxycoumarin in base is obvious.

Sodium acetate is a weak base and ionizes only the most acidic hydroxyl groups. Consequently a bathochromic shift in the UV λ_{max} on addition of sodium acetate with no further shift on addition of sodium methoxide suggests the presence of a free hydroxyl group at C-5 or C-7. This leads to the difference of 7-hydroxy-6-methoxycoumarin (scopoletin), shifts from 344 nm to 391 nm with increased intensity, and 6-hydroxy-7-methoxycoumarin (isoscopoletin), shifts from 347 nm only to 350 nm with noticeably dropped intensity.

Aluminum chloride is also of diagnostic value for some coumarins. 5,7-Dihydroxycoumarins are easily distinguished from 6,7- and 7,8-dihydroxycoumarins because only the *ortho*-dihydroxycoumarins form complexes with aluminium chloride to give bathochromic shifts. These shifts can be as little as 7 nm for 7,8-dihydroxycoumarin (daphnetin) or as much as 28 nm for 6,7-dihydroxycoumarin (esculetin).

5.1.4   Furanocoumarins

Linear furanocoumarins (psoralens) show four zones of absorption at λ_{max} nm (log ε) of 205 to 225 (4.0), 240 to 255 (4.06 to 4.45), 260 to 270 (4.18 to 4.26), and 298 to 316 (3.85

to 4.13). Angular furanocoumarins (angelicins) can readily be distinguished from psoralens by absence of the λ_{max} at 242 to 245 and 260 to 270 nm which are characteristic of linear curmarins.

Psoralens mono-oxygenated at C-5 or C-8 can also be readily differentiated from their UV spectra. The former shows maxima at ~268 nm which is absent in the latter. Moreover, the former shows λ_{max} at ~310 nm which appears at ~300 nm in the latter.

5.2 Infrared Spectroscopy

IR spectroscopy is generally used in the determination of the functional groups and for the identification of an unknown sample with a known structure by superimposing their spectra.

5.2.1 C—H stretching frequencies

Two or three bands of weak to medium intensity have been observed at the region of 3025 to 3175 cm^{-1} in the spectra of furanocoumarins. These absorptions are due to the C-H stretching vibrations of the pyrone, benzene, and furan rings.

5.2.2 C=O stretching frequencies

The C=O stretching frequency of coumarins (α-pyrones) is usually observed at the region 1700 to 1750 cm^{-1}, whereas in chromones (γ-pyrones) it is found at ~1650 cm^{-1}. 6-O- and 7-O-coumarin glycosides have their C=O frequencies below 1700 cm^{-1}.

A study of psoralens in paraffin mulls has shown that the C=O band shifts to a frequency higher than 1720 cm^{-1} when an alkoxyl group is attached to C-5 with C-8 unsubstituted, but when the alkoxyl group is at C-8, a frequency lower than 1720 cm^{-1} is obtained.

The IR spectra of pyranocoumarins show a strong C=O absorption band at 1717 to 1730 cm^{-1}, which shifts to 1735 to 1750 cm^{-1} in dihydropyranocoumarins.

3-Aryl coumarins with a free OH group at C-2$'$ show the C=O absorption at 1660 to 1680 cm^{-1}. This is presumably due to intramolecular hydrogen bonding between the 2$'$-hydroxyl and pyrone-carbonyl groups. The low C=O frequency (1660 cm^{-1}) of bicoumarin dicoumarol is also revealed to have strong intramolecular hydrogen bonding between the two halves of the molecule.

5.2.3 C=C skeletal vibrations

Normally there are three strong absorption bands at the region 1600 to 1660 cm^{-1} in the IR spectra of coumarins. This pattern of absorption provides a quick method of differentiation from isomeric chromones, the absorption of which is generally much simpler.

In the spectra of furanocoumarins, in addition to the aromatic bands at ~1500 and ~1600 cm^{-1}, a strong sharp band appears at 1613 to 1639 cm^{-1} which is attributed to the C=C stretching of the furan ring.

5.2.4 Other absorptions

Two bands found at the regions 1088 to 1109 and 1253 to 1274 cm^{-1} in the spectra of furanocoumarins are considered to be characteristic C-O stretching of the furan group. The bands at the 740 to 760 and 870 to 885 cm^{-1} region are due to the in-plane and out-of-plane deformations, respectively, of the furan C—H bonds.

5.3   Nuclear Magnetic Resonance Spectroscopy

A broad range of ^{1}H and ^{13}C NMR techniques have been widely applied to the structural elucidation of naturally occurring coumarins. Some of the spectra-structure correlation rules summarized by systematic examination are available and generally lead to the unambiguous structure of virtually any unknown or novel natural coumarin[1c].

5.3.1 ^{1}H-NMR

5.3.1.1   Ring Protons

H-3 and H-4: A pair of doublets, $J = 9.5$ Hz, at δ 6.1 to 6.4 and 7.5 to 8.3 (in CDCl$_3$) from H-3 and H-4, respectively, is a strong indication of a coumarin without substitution on the pyrone ring. Compared to other aryl protons, in general, the H-3 signal is highest field, while H-4 is at the lowest field.

The majority of natural coumarins have an oxygen functional group at C-7 which leads to an increase in the electron density at C-3, thereby causing the resonance of H-3 to move to higher field by ~0.17 ppm.

The H-4 resonance is found in the region δ 7.5 to 7.9 in coumarins lacking a C-5 oxygen functional group. An oxygen or alkyl substituent at C-5, however, characteristically shifts the resonance of H-4 downfield by ~0.3 ppm (the *peri* effect), H-4 then found at δ 7.9 to 8.2.

H-5, H-6, and H-8: In theory, the H-5 and H-6 of a 7-oxygenated coumarin should appear as a pair of *ortho*-coupled doublets with the H-6 signal also *meta*-coupled to an H-8 signal of similar chemical shift. In reality, however, H-5 is found as a doublet, $J = 9$ Hz, at δ 7.38, downfield from the H-6 and H-8 signals which appear as a two-proton multiplet at δ 6.87. The signals for these three benzenoid protons are always flanked by the doublets from H-3 and H-4.

When an oxygen functional group is substituted at C-5 in a 7-oxygenated coumarin, the protons at C-6 and C-8 are now *meta* related and give rise to a pair of doublets, $J \sim 2$ Hz. In general it is not possible to assign the signals from H-6 or H-8 on the basis of their chemical shifts alone. However, careful examination of the spectrum may reveal long-range coupling, $J = 0.6$ to 1.0 Hz, of H-8, when present, with H-4.

Many 7-oxygenated coumarins are known to contain alkyl or alkoxyl groups at C-8. The presence of two *ortho*-related protons of H-5 at δ ~7.3 and H-6 at δ ~6.8 can be recognized instantly since they give rise to another pair of doublets having a coupling constant of 9 Hz.

5.3.1.2   Ring Substituents

(a) Ar-Me: δ 2.45 to 2.75; Ar-OMe: δ 3.8 to 4.4

(b) prenyl (3-methylbut-2-enyl) group, δ

2Me (non-equivalent) 1.6~1.9 (3H×2, s), one or both of which may show allylic coupling, J~1Hz.
= CH   5.1~5.7 (1H, t, br, J = 7Hz)
Ar-CH2   3.3~3.8 (2H, d, J = 7Hz)
Ar-O-CH2   4.3~5.0 (2H, d, J = 7Hz)

(c) 1,1-dimethylallyl group, δ

2Me 1.5 (6H, s)
= CH2 5.1 (2H, m)
= CH~6.25 (1H, dd, J = 18, 10Hz)

5.3.1.3 Furanocoumarins and Pyranocoumarins

The presence of an unsubstituted furan ring is easily recognizable from the pair of doublets, $J \sim 2.5$ Hz, which arise from H-2' and H-3'. The signals from the former resonate at $\delta \sim 7.5$ to 7.7, while the latter are found at $\delta \sim 6.7$ in the linear series and at $\delta \sim 7.0$ in the angular furanocoumarins.

Natural pyranocoumarins are invariably *gem*-substituted at C-2' with methyl groups which resonate as a six-proton singlet at δ 1.45. The two olefinic protons show as a pair of doublets, $J = 10$ Hz, centered at δ 5.3 to 5.8 (H-3') and at δ 6.3 to 6.9 (H-4').

5.3.1.4 Nuclear Overhauser Effects

Measurements of NOE have been especially useful in structural elucidation of a number of coumarins in which all four positions on the benzenoid ring are substituted. For example:

On saturation of methoxyl signal at d 3.82 of poncitrin, the integrated intensities of the doublets arising from H-4 of coumarin and H-4' of pyran ring were increased by 9 and 13%, respectively.

From the above experiment, it was concluded that the methyoxy group must be close to H-4, and hence must be located at C-5 and also proximate to the pyran ring which, therefore, had to be linearly fused.

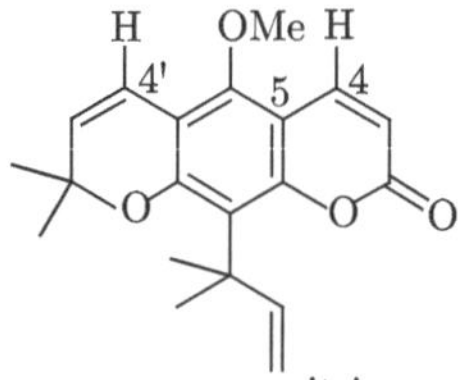

poncitrin

poncitrin

5.3.2 ^{13}C-NMR

5.3.2.1 Coumarin Nucleus and Its Simple Derivatives

Chemical shifts of nine sp^2 carbon atoms in coumarin are found in the 100 to 160 ppm region; both the signals of C-2 and C-9 can be easily recognizable at lower field (Table 11-1).

Table 11-1 δ Value (CDCl$_3$) of Carbon Atoms in Coumarin Nucleus[1c,43]

C	2	3	4	5	6	7	8	9	10
δ	160.4	116.4	143.6	128.1	124.4	131.8	116.4	153.9	118.8

The δ value of the carbonyl-carbon atoms (C-2) of most coumarins has been found to be approximately the same at about 160 ppm. The effect of OH and OMe groups on the benzenoid ring is quite characteristic in that the signal from the newly formed quaternary carbon atom is found approximately 30 ppm downfield from the value observed in coumarin while the carbons *ortho* and *para* to the substituent move upfield by ~ 13 and ~ 8 ppm, respectively. The shift effect to *meta* carbon is generally small; particularly the effects of Me and COOH to *meta* position are nearly negligible, for example, 7-hydroxycoumarin (Table 11-2):

Table 11-2 δ Value (DMSO) of Carbon Atoms in 7-Hydroxycoumarin[44]

C	2	3	4	5	6	7	8	9	10
δ	160.7	111.5	144.3	129.6	113.3	161.6	102.5	155.7	111.5

Comparing the ^{13}C-NMR data of 7-hydroxycoumarin with that of coumarin, the shift effects of OH group are consistent with the general rules mentioned above.

5.3.2.2   Furanocoumarins and Pyranocoumarins

Based on the carbon spectrum data of psoralen and scoparone (Tables 11-3 and 11-4), the difference between the linear and angular types is mainly found in the δ values of C-6 and C-8 atoms of which can be identified easily.

Table 11-3 δ Value of Carbon Atoms in Psoralen (Linear Furanocoumarin)[45]

C	2	3	4	5	6	7
δ	161.1	114.7	144.2	120.0	125.0	156.6
C	8	9	10	2′	3′	
δ	99.6	152.2	115.6	147.0	106.6	

Table 11-4 δ Value of Carbon Atoms in Scoparone(Angular Pyranocoumarin)[46,44]

C	2	3	4	5	6	7
δ	160.4	112.2	143.5	127.5	114.6	155.9
C	8	9	10	2′	3′	4′
δ	108.8	149.8	112.2	77.2	130.4	113.1

5.3.3   2D-NMR

2D-NMR such as ^{1}H-^{1}H COSY, HMQC, HMBC, NOESY etc. is now widely applied in the structural elucidation of more complex or novel coumarins in nature. For example, a new pyranocoumarin, qianhucoumarin F with a rare structure was isolated from the Chinese traditional medicine *Peucedanum praeruptorum*[47]; a novel 6,7-disubstituted coumarin was isolated from the Chinese medicinal plant *Zanthoxylum dimorphophyllum* var. *spinifolium*[48]; two new coumarin glycosides were found in the *Pleurospermum rivulorum* distributed in Yunnan Province of China[49]. All of these novel structures were elucidated by using 2D-NMR as a powerful tool in the assignments of the chemical shifts of ^{1}H and ^{13}C.

5.4   Mass Spectrometry

Fragmentation patterns resulting from EI-MS of many natural coumarins have been determined, rationalized and proved to be of a great assistance in structural studies. Soft ionization methods such as FAB-MS and ESI-MS also have become increasingly used for the measurement of the molecular ion of coumarins.

5.4.1   Simple coumarins

Coumarin shows a strong molecular ion (M$^+$, m/z 146, 76%) upon electron impact, and a benzofuran ion (m/z 118, 100%) as the base peak by loss of carbon monoxide from the pyrone ring. The loss of CO from the coumarin-carbonyl group is a characteristic feature of the mass spectra of most coumarins. The benzofuran ion is broken down further by consecutive loss of CO and an H atom.

$$\text{m/z146(76\%)} \xrightarrow[*]{-CO} \text{m/z118(100\%)} \xrightarrow[*]{-CO} C_7H_6 \cdot^+ \xrightarrow{-H\cdot} C_7H_5^+$$

m/z146(76%)   m/z118(100%)   m/z90(43%) m/z89(35%)

An asterisk (*) under an arrow denotes that the transition has been substantiated by the observation of the appropriate metastable ion.

7-Hydroxycoumarin also shows a strong molecular ion (m/z 162, 80%) with the base peak (m/z134, 100%) formed from it by loss of CO from the pyrone-carbonyl group. The remaining oxygen atoms are subsequently removed as CO (m/z 106, 11%) followed by CO (m/z 78, 32%) or as CHO (m/z 105, 25%).

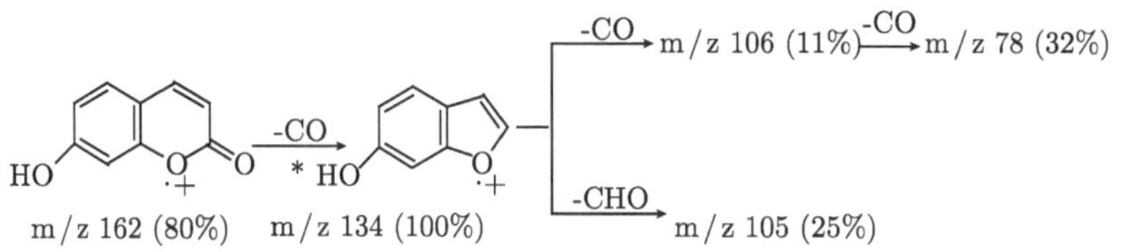

7-Methoxycoumarin forms the molecular ion as the base peak and a strong [M - CO]$^+$ ion which becomes a stable quinone from conjugated oxonium ion (m/z 133), after losing the methyl radical from the methoxyl group.

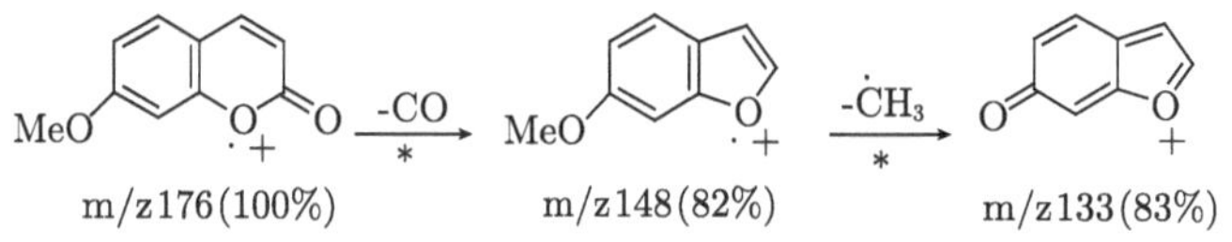

5.4.2   Furanocoumarins and pyranocoumarins

The presence of a furan ring in a furanocoumarin does not alter the fundamental fragmentation process observed for structurally simple coumarins, namely the ready elimination of CO from the pyrone ring. However, in methoxyfuranocoumarins such as xanthotoxin, loss of a methyl radical can give rise to a conjugated oxonium ion, followed by the loss of CO.

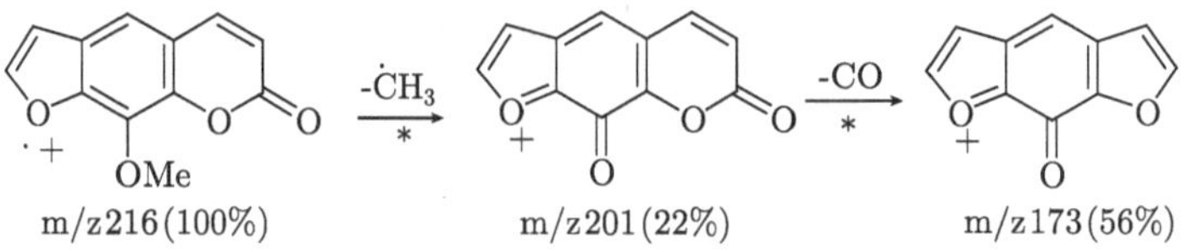

The mass spectrum of a 2′, 2′-dimethylpyranocoumarin such as sesilin is dominated by the loss of a methyl radical and generation of a stable benzopyrylium ion which is frequently the base peak.

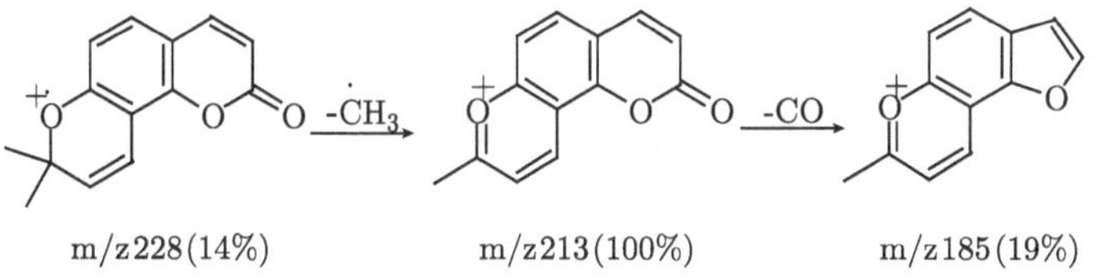

Section 6   Research Example[37]

6.1   Plant Material

The roots of *Stellera chamaejasme* L. (Thymelaeaceae), known as " Lang Du" in Chinese traditional medicine, have been used for the clinical treatment of mange, stubborn skin ulcers, malignant tumors, chronic tracheitis, and tuberculosis for many years in China.

A novel bicoumarin (**1**) as well as twelve other constituents were isolated from the plant material.

6.2   Extraction and Isolation

The air-dried ground plant roots (20 Kg) were extracted with 95% EtOH and a portion of the concentrated extracts was successively partitioned with petroleum ether and ether. The ether extract (1750 g) was subjected to silica gel column chromatography, eluted with a hexane and acetone mixture of increasing polarity. The hexane-acetone (1:2) fractions were collected and the solvent was evaporated to give a fraction, of which the water insoluble portion was further separated by RP-8 chromatography and eluted with 40% EtOH to yield **1** (9 mg).

6.3   Structural Determination

Compound 1, yellow square crystal, mp 264 to 265°C, was assigned the molecular formula $C_{19}H_{12}O_6$ (HRMS, $[M^+] = m/z$ 336.0612, calculated 336.0630). The IR absorption at 3420, 1725, 1600, and 1500 cm^{-1} suggested the presence of hydroxyl, carboxyl (lactone) and benzyl groups. UV absorption at 212 nm (log ε 4.44), 275 nm (w), and 325 nm (log ε 4.37) showed the characteristics of a 7-oxygen-substituted coumarin. The ^{1}H NMR (in DMSO-d_6) indicated one methoxyl signal (δ 3.79), eight aromatic proton signals (δ 6.24 to 8.05), and a hydroxyl proton signal (δ 10.62). The two pairs of typical H-3 and H-4 signals at δ 6.24 (1H, d, J = 9.7 Hz), 7.93 (1H, d, J = 9.7 Hz), and δ 6.26 (1H, d, J = 9.5 Hz), 8.04 (1H, d, J = 9.5 Hz) predicted that the molecule was composed of two C-3 and C-4 unsubstituted coumarin moieties. This also matched the fourteen degrees of unsaturation required by its molecular formula. The positions of the two correlated protons (δ 7.75, 1H, d, J = 8.7 Hz, and δ 7.18, 1H, d, J = 8.8 Hz), two singlet signals (δ 6.89, 1H, s, and δ 7.46, 1H, s), together with the connection of the two moieties were determined by NOESY. In its NOESY spectrum, the correlation between H-4' (δ 7.93) and H-5' (δ 7.46), H-8' (δ 6.89) and OH (δ 10.62), H-4 (δ 8.04) and H-5 (δ 7.75), together with H-6 (δ 7.18) and OMe (δ 3.79) were observed. It is obvious that the compound consisted of two coumarin moieties connected through C-8 and C-6'. The structure is also confirmed by its ^{13}C-NMR data and EI-MS decomposition fragments shown in Figure 11-4. Therefore, the structure of compound **1** was determined as 7-methoxy-7'-hydroxy-[8,6']-bichromenyl-2, 2'-dione named bicoumastechamin.

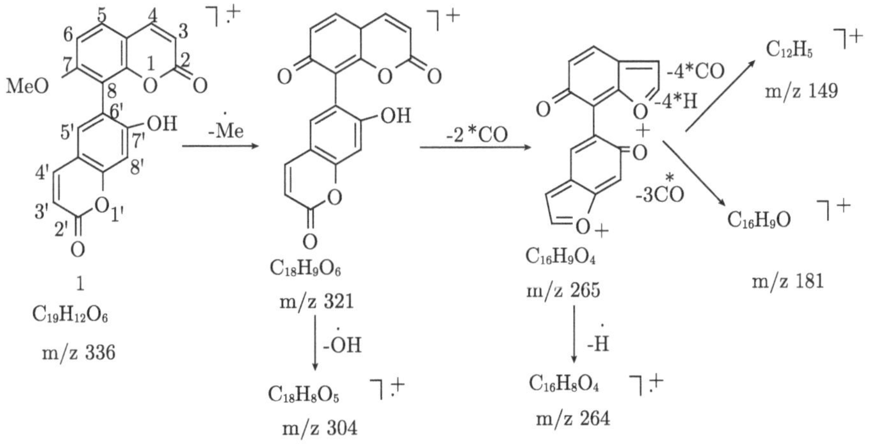

Figure 11-4   Structure and EI-MS fragments of compound **1**

References

[1]   Murray R D H, et al. The Natural Coumarins, Occurrence, Chemistry and Biochemistry. New York: John Wiley & Sons Ltd; 1982, a: 1-12; b: 13-20; c: 22-53; d: 55-95.

[2]   Murray R D H. *Nat Prod Rep*, 1989, 6: 591.

[3]   Murray R D H. *Nat Prod Rep*, 1995, 12: 477.

[4]   Estevez-Braun A, et al. *Nat Prod Rep*, 1997, 14: 465.

[5]   Chen Z N. Coumarins. In: Xu R S, et al. eds. Chemistry of Natural Products. 2nd ed. Beijing: Scientific Press; 2004, 589.

[5]   Kojima K, et al. *Chem Pharm Bull,* 2000, 48(3): 353.

[6]   Quadri-Spinelli T, et al. *Planta Med*, 2000, 66: 728.

[7]   Ishii H, et al. *Chem Pharm Bull*, 1991, 39(11): 3100.

[8]   Paknikar S K, et al. *J Indian Inst Sci*, 1994, 74: 277.

[10] Lin G Q, et al. *Tetrahedron Lett*, 1996, 37(17): 3015.

[11] Xu S, et al. *Acta Pharm Sinica*, 2000, 35(2): 103.

[12] Nofal Z M, et al. *Molecules*, 2000, 5: 99.

[13] Mckee T C, et al. *J Nat Prod*, 1996, 59: 754.

[14] Kong L Y, et al. *Nat Prod Res Develop*, 1994, 6(1): 50.

[15] Kang T H, et al. *Planta Med*, 1999, 65: 400.

[16] Kim N Y, et al. *Planta Med*, 1999, 65: 656.

[17] Wang C C, et al. *Bioorg Med Chem*, 2000, 8: 2701.

[18] Nishimura S, et al. *Chem Pharm Bull*, 2000, 48(4): 505.

[19] Ojala T, et al. *Planta Med*, 1999, 65: 715.

[20] Kong L Y, et al. Progress in chemical study of coumarins. In: Yu D Q, et al. eds. Advances in Natural Product Chemistry. Beijing (China): Chemical Engineering Press; 2005, 168.

[21] Li Y Y, et al. *J Chin Med Mater*, 2004, 27(3): 218.

[22] Zhang S Y, et al. *China J Chin Mat Med*, 2005, 30(6): 410.

[23] Kong L Y, et al. *Acta Pharm Sinica*, 1993, 28(6): 432.

[24] Kong L Y, et al. *Acta Pharm Sinica*, 1993, 28(10): 772.

[25] Kong L Y, et al. *Acta Pharm Sinica*, 1994, 29(1): 49.

[26] Chen Z X, et al. *Acta Pharm Sinica*, 1979, 14(8): 486.

[27] Huang P, et al. *Acta Pharm Sinica*, 1997, 32(1): 62.

[28] Wang H X, et al. *Acta Pharm Sinica*, 1995, 30(11): 812.

[29] Zhang J S, et al. *Acta Pharm Sinica*, 2001, 36(1): 34.

[30] Xu R M, et al. Progress in the research of wedelolactone. *Chinese Scientific Paper Online*, 2006 Oct 19, P. 1-7. Available from: http: // www. paper. edu.cn.

[31] Xu R M, Studies on the hepatoprotective effect of *Eclipta* and the anti-inflammatory-immune activity of wedelolactone [Doctoral Dissertation]. Shanghai (China): Shanghai Jiaotong University; Nov. 2007.

[32] MckeeT C, et al. *J Nat Prod*, 1998, 61: 1252.

[33] Matthee G, et al. *Planta Med*, 1999, 65: 493.

[34] Kucherenko A, et al. *Tetrahetron Lett*, 1995, 36(31): 5475.

[35] Deshpande P P, et al. *J Org Chem*, 1995, 60: 2964.

[36] Wang L, et al. Recent progress in *Calophyllum* coumarins as potent anti-HIV agents. In: Liang X T, et al. eds Medicinal Chemistry of Bioactive Natural Products. New York: John Wiley & Sons, Inc.; 2006, 325.

[37] Xu Z H, et al. *J Asia Nat Prod Res* 2001, 3: 335.

[38] Cai J N, et al. *J Nat Prod*, 2000, 63: 485.

[39] Takadate A, et al. *Chem Pharm Bull*, 2000, 48(2): 256.

[40] Bourgaud F, et al. *Phytochemical Analysis*, 1994, 5: 127.

[41] Yu P Z. Phenylpropanoids. In: Tan R X, et al. eds. Plant Constituents Analysis. Beijing: Scientific Press; 2000. 460.

[42] Tegtmeier M, et al. *Pharmazie*, 2000, 55(2): 94.

[43] Ji X S, et al. *Chin J Magn Reson,* 1997, 14(2): 185.

[44] Gong Y H, et al. Chemical Shifts of[13]C NMR in Natural Organic Compound. Kunming: Yunnan Scientific Technology Press; 1986. 189.

[45] Macias F A, et al. *Magn Reson Chem*, 1989, 27: 705.

[46] Macias F A, et al. *Magn Reson Chem*, 1990, 28: 732.

[47] Kong L Y, et al. *Chin J Magn Reson*, 1994, 11(3): 245.

[48] Tao Z Y, et al. *Acta Pharm Sinica*, 2001, 36(7): 511.

[49] Xiao Y Q, et al. *Acta Pharm Sinica*, 2001, 36(7): 519.

Ze-Nai CHEN

CHAPTER 12

Lignans[1,2]

Section 1   Introduction

The term lignan was used to describe a group of natural products that are primarily derived from the oxidative coupling of phenylpropanoid units. Lignans are usually phenylpropanoid dimers. In 1961, Freudenberg, et al. numbered the phenylpropanoid C6-C3 carbon skeleton as 1-9, and the C6-C3 unit as 1′-9′. Lignans are usually referred to as dimers of phenyl-propanoids, linked by β-β' (8-8′) (**1**)[3]. Gottieb coined the term neolignan for compounds composed of two phenylpropanoid units that are linked other than by the C8-C8′ (**2**)[4]. Later, he also coined the term oxyneolignans for such lignans whose two C6-C3 units were linked by oxygen (**3**).

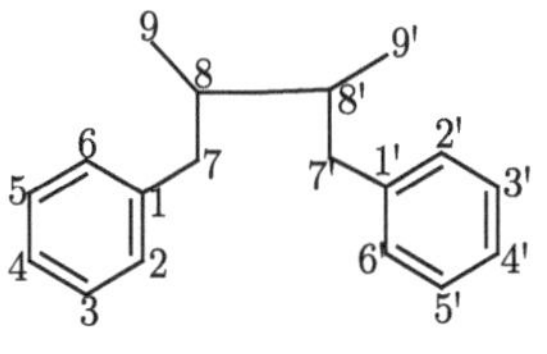

1 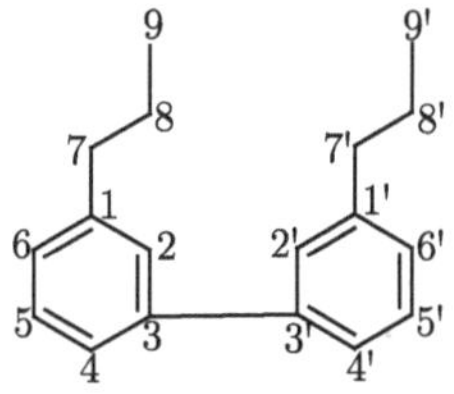   2 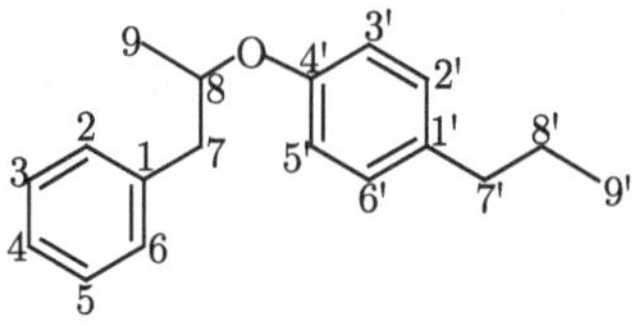   3

Phenylpropanoid units in lignans are of various types, such as cinnamyl alcohol, cinnamic acid, propenyl phenol, and allyl phenol. There are often oxygen substitutions on the two phenyls groups, particularly at C_3, C_4, C_5 or C_3', C_4', C_5'. Oxygen groups in the side chain are varied, especially when they are the terminal oxygen groups, and can be hydroxyl, carboxyl, carbonyl, etc. When the oxygen groups are condensed through dehydration, additional rings, such as tetrahydrofurans and lactones will occur. This may partly account for the complex structures of lignans.

1.1 Nomenclature

At first, lignans were usually given trivial names or semi-systematic names. With the increase of the number of lignans, similar lignans may be named differently, e.g., there are various names for the following carbon skeletons.

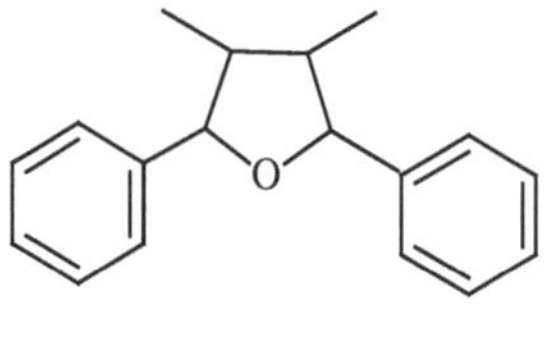

7, 7′-epoxylignane
2, 5-diaryltetrahydrofuran
7, 7′-epoxylignane
7-O-7′, 8, 8′-lignan
2, 5-diaryl-3, 4-dimethyltetrahydrofuran

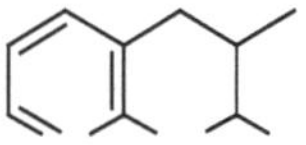

2, 7′-cyclolignane
4-aryl-1, 2, 3, 4-tetrahydronaphthalene
8, 4′-oxyneolignane 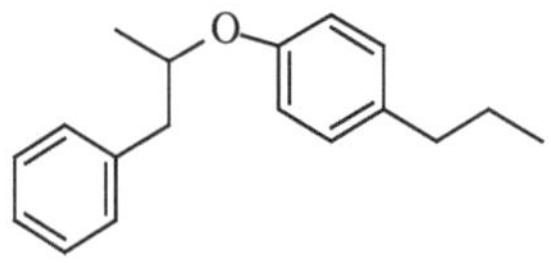
8-O-4′-lignan
(8-O-4′)-neolignan

In order to resolve the problem, Chemical Abstracts (CA) invented its own systematic nomenclature. However, sometimes it made the issue more complicated. Under these circumstances, IUPAC recommended the nomenclature of natural products in 1976 and 1979. Once again in 2000, it recommended the nomenclature of lignans and neolignans[5]. The main approach is to select C6-C3 as the basic unit. The benzyl carbon linked to the side chain is numbered as 1, the phenyl ring is numbered from 1 to 6, and the side chain is numbered from 7 to 9, sequentially. Another unit is numbered in the same way except adding a prime symbol (′) to the numbers. If the two C6-C3 units are attached by C8-C8′, they are called lignans, otherwise they are called neolignans. If two C6-C3 units are attached by an ether oxygen bond other than a carbon-carbon bond, they are called oxyneolignans. The additional rings in the structures, the substitutions, and the stereoisomers are still denoted by the traditional organic chemistry nomenclature. Some illustrations are shown below[6].

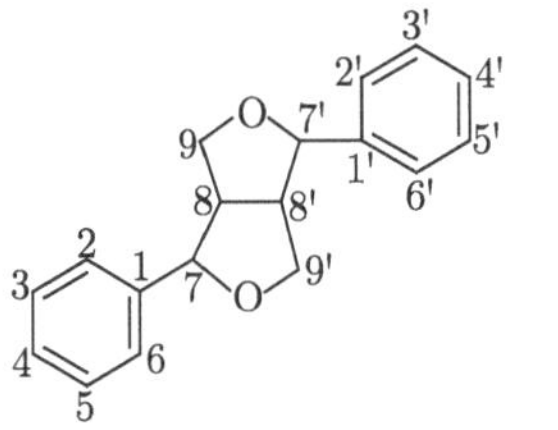

7,9′,7′,9-diepoxylignan

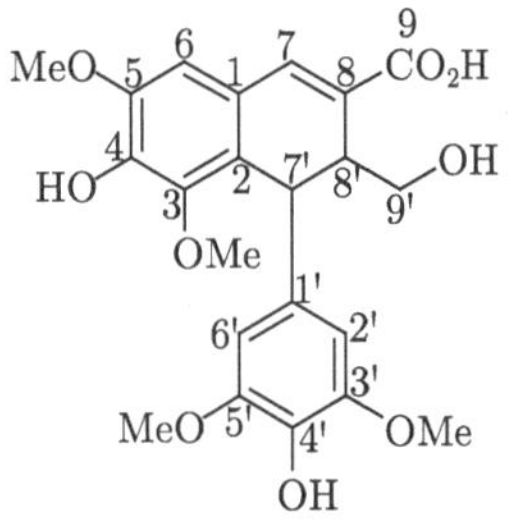

4,4′,9′-trihydroxy-3,3′,5,5′-tetramethoxy
-2,7′-cycloligan-7-en-9-oic acid

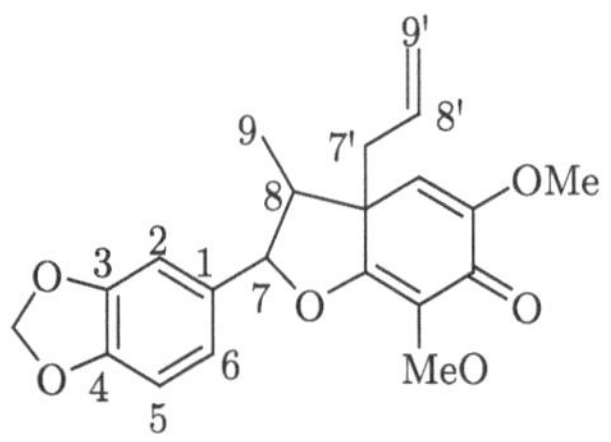

3′,5′-dimethoxy-3,4-methylenedioxy
-2′,7-epoxy-4′H-8,1′-neolignan--8′-en-4′-one

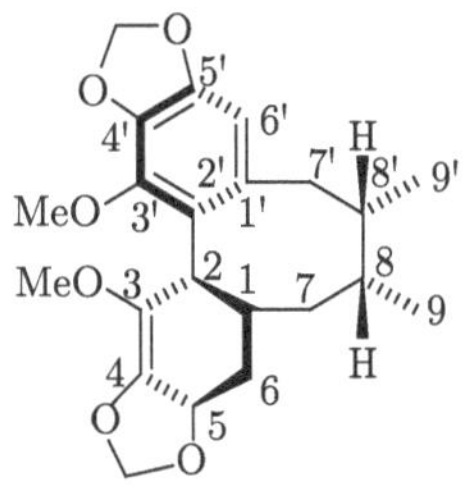

rac-(8α,8′α)-3,3′-dimethoxy-4,5:4′,5′
-bis(methylene-dioxy)-2,2′-cyclolignan

It is noteworthy that there are obvious differences between IUPAC and CA nomenclature. Podophyllotoxin is illustrated as an example.

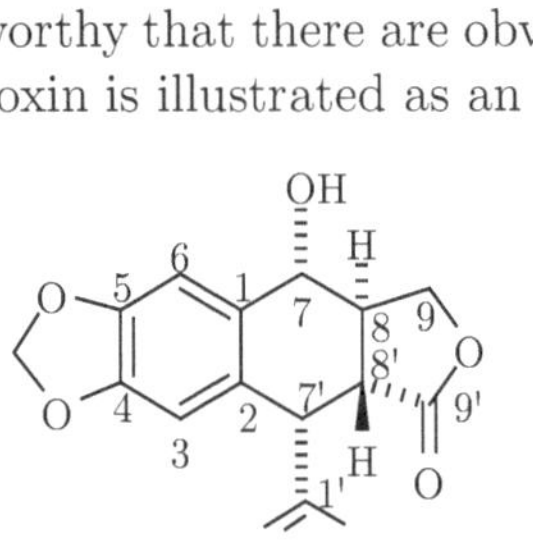

(7α, 7?α, 8α, 8?β)-7-hydroxy-3?,4?,5?-

trimethoxy-4,5-methylenedioxy-2,7?-

cyclolignano-9?, 9-lactone

(IUPAC nomenclature)

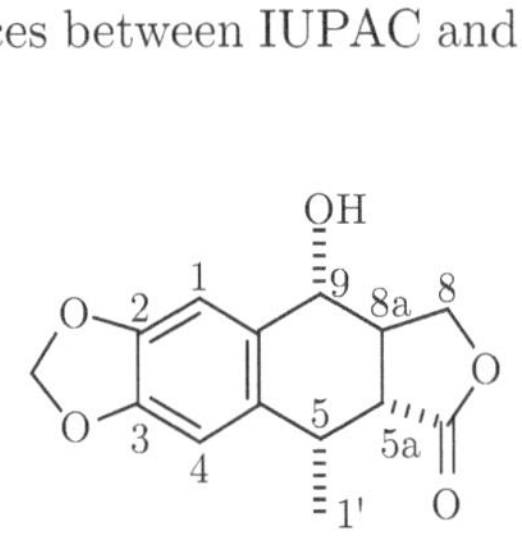

Furo [3?, 4?=6, 7] naphtho [2, 3 -d] -1, 3-

dioxol-6-(5αH)-one, 5, 8, 8a, 9-

tetrahydro-9-hydroxy-5-(3', 4',

5'-trimethoxy-phenyl)-, (5R, 5aR, 8aR, 9R)-

(CA nomenclature)

1.2   Biosynthesis of Lignans

Some researchers estimated that lignans and lignins were formed by the oxidative coupling of two coniferyl alcohols. Cell-free enzyme preparations from *Forsythia suspensia* reduce ferulic acid to coniferyl alcohol. The enzyme requires magnesium ions for full activity. The reduction process is *via* the phosphorylation to allyl or propenyl side chain. Coniferyl alcohol gives phenolic radicals by phenolic oxidase. In regards to the resonance formation mechanism, coniferyl alcohol can produce various phenolic radicals which give rise to diverse lignans and lignins.

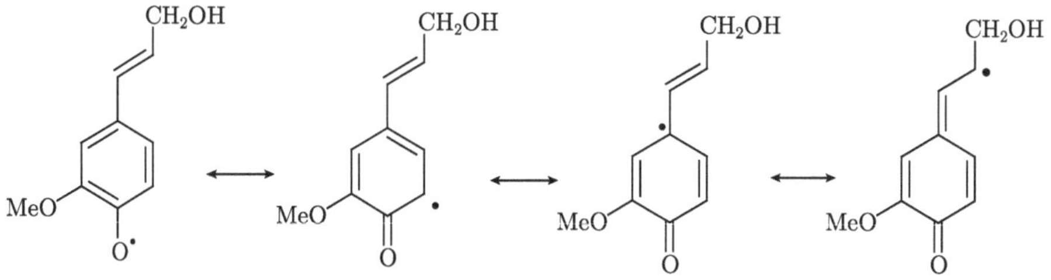

The resonance intermediates of coniferyl alcohol

In *Forsythia suspensia*, it was demonstrated that (+)-pinoresinol was formed by the stereo-selective oxidatize coupling of coniferyl alcohol. Subsequently, (+)-pinoresesinol was sequentially reduced to (+)-lariciresinol and (-)-secoisolariciresinol. Two isofunctional forms of the reductase were responsible for this process. Both catalyze the sequential, NADPH-

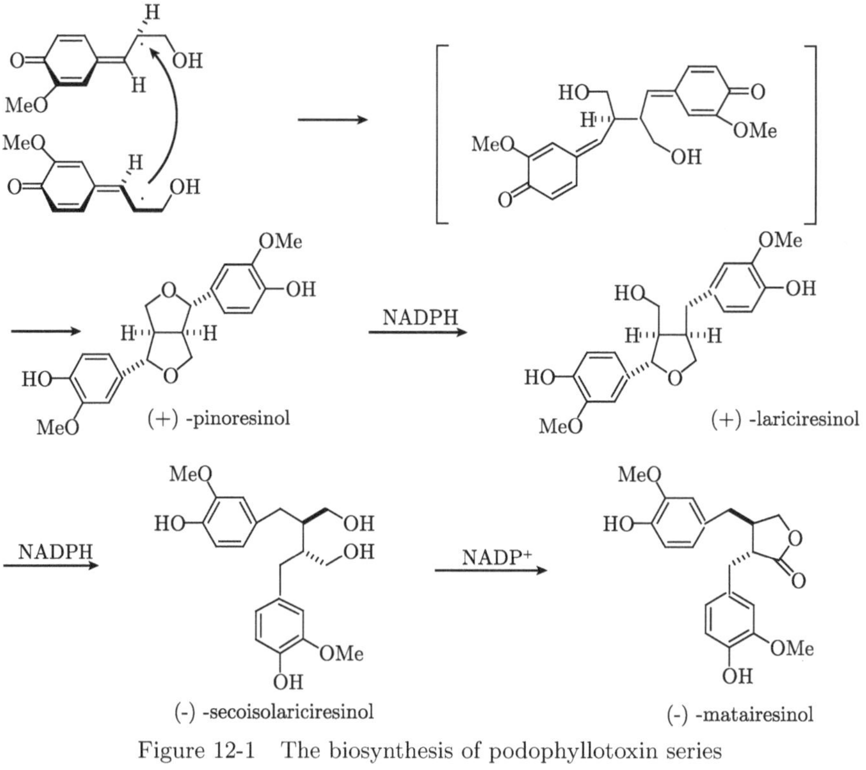

Figure 12-1   The biosynthesis of podophyllotoxin series

dependent, stereoselective reduction of (+)-pinoresesinol and (+)-lariciresinol and have similar kinetic parameters and molecular weight. The stereoselectivity of this process results in inversion of the configuration at C-2 and C-5 of pinoresesinol, a process which is envisaged to occur either by a concerted S_N2 mechanism or though reduction of an intermediate quinomethane. Stereoselective dehydrogenation of (-)-secoisolariciresinol then occurs to give (-)–matairesinol which is considered to be the branch point leading to other important groups of lignans, such as the podophyllotoxin series[7] (Figure 12-1).

The same biosynthetic pathway also operates in *Zanthoxylum ailanthoides*. In contrast, (+)-lariciresinol (**6**) acted as an intermediate in the biosynthesis of arctigenin[8] (Figure 12-2). Compound **6** was formed by caffeic alcohol (**4**) through the cleavage of a (+)-pinoresinol (**5**) benzylether bond. Given that pinoresinol can be reduced to lariciresinol in cell-free extract of *Zanthoxylum ailanthoides*, NADPH was proven to be present.

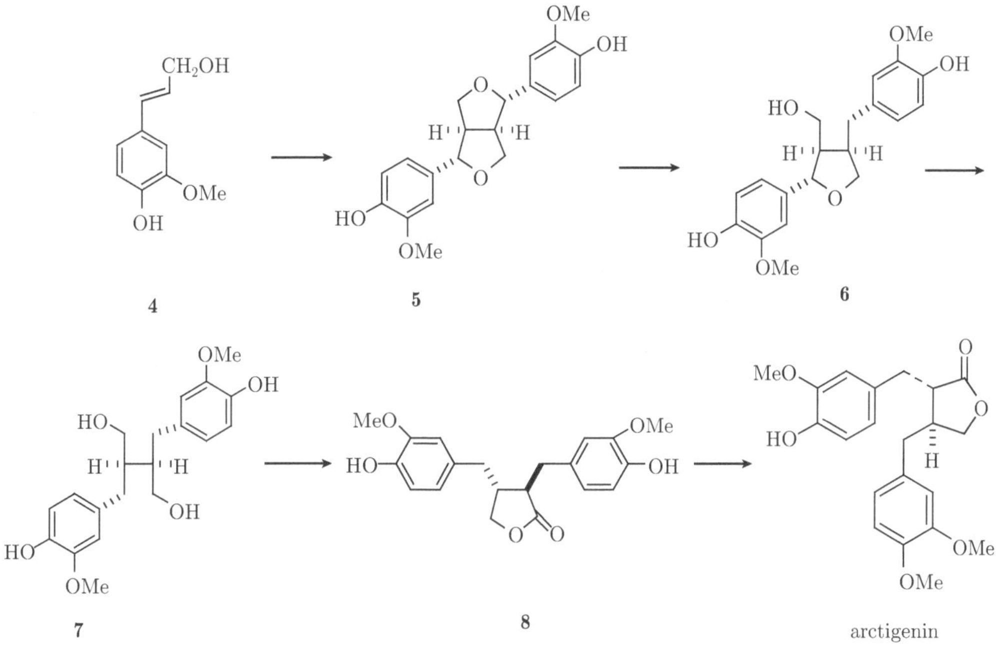

Figure 12-2   The biosynthetic pathway of arctigenin

1.3   Isolation and Identification

In the past 20 years, studies on lignans have made great progress. This is not only because several lignans demonstrated important bioactivity, but also because they usually have complex structures. Novel techniques and methods have promoted the advance of this field.

Regarding chemical structures, lignans are devoid of common characteristic reactions, comprising more than ten molecular carbon skeletons. In this point, lignans are not one kind of natural product. They mainly exist in lipiphilic fractions and can be purified over silica gel column chromatography. Lignans bearing a phenol hydroxyl can be isolated in the same way as hydroxybenzene substances. Using TLC, lignans can be detected by spraying 5% phosphomolybdic acid in ethyl alcohol solvent or 30% H_2SO_4 in ethyl alcohol solvent. Different kinds of lignans show different colors after the plate is heated at 100°C for several minutes. This experiment can help determine skeleton types. SFE (super fluid extraction) has also been efficiently applied in the isolation of lignans[9].

Lignans can be qualitatively and quantitatively analyzed by normal-phase or reverse-phase HPLC. HPLC coupled with FAB-MS or ESI-MS affords a fast analytical method. Examples include the direct analysis of secoisolariciresinol and its glucosides from flaxseed without isolation[10] and quantitative analysis of schizandrin from human blood plasma[11].

1.3.1   UV spectrum

Chemical structures, even stereo structures of most lignans, can be determined by analysis of their spectral data. UV spectroscopy is used to determine the structures of their phenyl rings. Most lignans have two independent phenyl rings which display similar absorption in the UV spectrum and the absorption potency is the sum of the two independent absorptions. Lignans usually exhibit phenyl derivatives' absorption type, e.g., dibenzylbutanes and benzofurans exhibit absorption maxima at ca 230 nm (lgε >4.0) and 280 nm (lgε 3.5). If there is a double bond in their side chain which is conjugated with the benzyl ring, they show characteristic absorption at ca 235 nm, 295 nm, and 335 nm, such as in dibenzylbutyrolactones. If their side chains form a phenyl ring, they show naphthalene derivatives' characteristic absorption at about 220 nm, 260 nm, 295 nm, 310 nm, and 350 nm, such as in arylnaphthalenes. Benzylfuran lignans exhibit absorption at about 320 nm. In regards to dibenzocyclooctadiene lignans, characteristic absorptions corresponding to oxygen substituted dibenzyls at 220 nm, 250 nm, and 285 nm will be observed.

Podophyllotoxin and its derivatives are used here as an example to introduce UV spectral characteristics in lignans. Podophyllotoxin (**9**), deoxypodophyllotoxin (**10**), and podophyllotoxin glycoside (**11**) demonstrate absorption maxima at 290 to 294 nm (ε 4400 to 4800), which is the sum of two chromophores (methylenedioxy phenyl, 283 nm (ε 3300); trimethoxy phenyl, 270 nm, (ε 650). The absorption peaks show minor red shift due to the alkyl substitution of the phenyl rings, just like safrole (**14**). The UV absorption peak (270 nm, ε 650) occurs red shift compared with methylenedioxy phenyl. α-peltatin (**12**) and β-peltatin (**13**) have similar structures with that of podophyllotoxin The minor difference between them is that the former is 6-hydroxy while the latter is 7-hydroxy, which makes the UV absorption peaks of the former slightly more blue shift than the latter, just like myristicin (**15**). The UV absorption peak is blueshift compared to safrole.

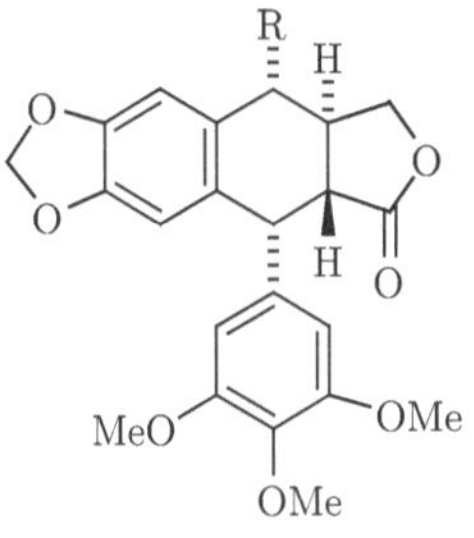

	R	λmax (log ε)		R	λmax (log ε)
9	OH	292 (3.65)	**12**	H	274 (3.41)
10	H	293.5 (3.68)	**13**	Me	273 (3.26)
11	O-glc	291 (3.62)			

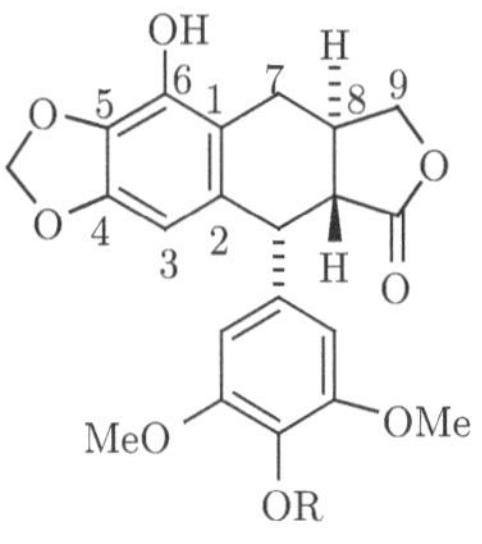

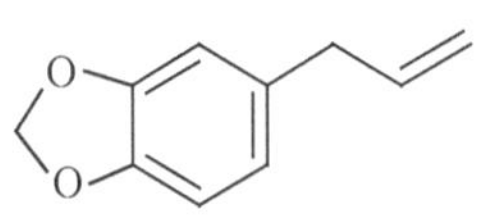

14	**15**

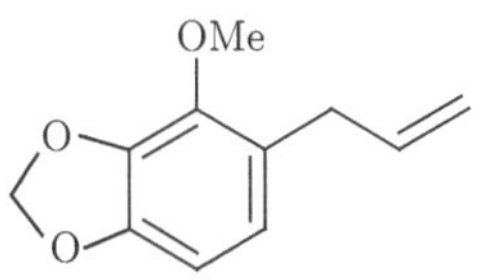

UV spectroscopy can help determine the state of the conjugation system, e.g., α-, β-, γ-, etc. Apopicropodophyllin (**16, 17, 18**) and dehydrogenpodophyllotoxin (**19**) are podophyllotoxin derivatives. Compound **17** has a similar UV spectrum as that of podophyllotoxin because its ethylenic linkage is not conjugated with phenyl rings. Compound **16**'s ethylenic linkage is conjugated with one phenyl ring, which results in a minor red shift of its absorption peak. Compound **18**'s ethylenic linkage connects a phenyl ring and carbonyl, thus, its UV spectra red shifts a little more than **16**. Dehydrogenpodophyllotoxin has similar UV spectra as naphthalene derivatives.

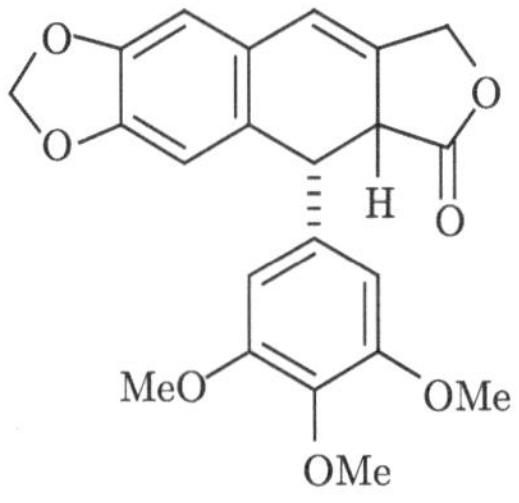

311nm (3.88)

16

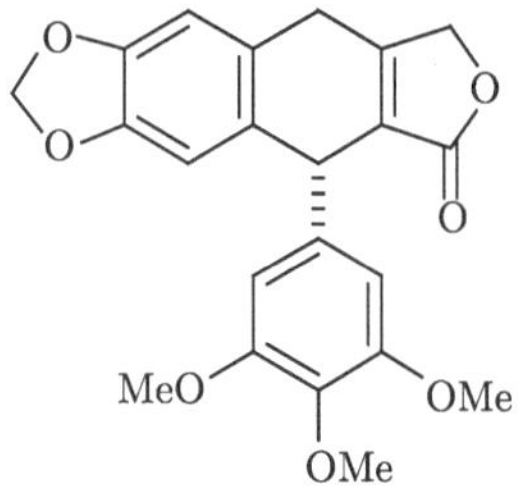

290 nm (3.66)

17

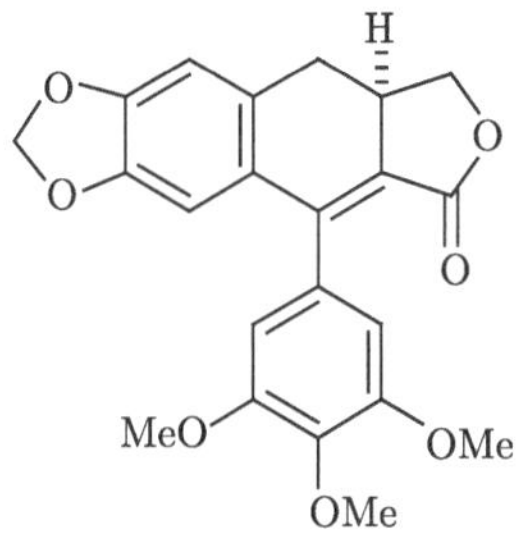

350 nm (4.10)

18

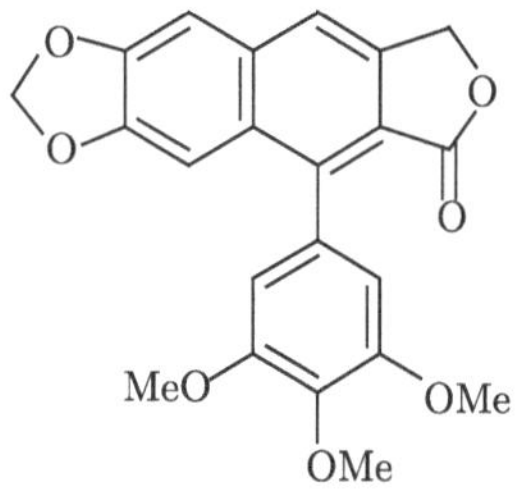

323 nm (4.02)
356 nm (3.72)

19

1.3.2   IR spectrum

Infrared spectra can be used to identify the cyclooctadiene ring oxidation and substitution pattern. In the IR spectra, most lignans exhibit a characteristic phenyl ring frequency at 1500 to 1600cm^{-1}. Above that, 1760 to 1780cm^{-1} shows isolated five-member lactones, 1740 to 1760cm^{-1} shows conjugated five-member lactones, 1725cm^{-1} shows unsaturated five-member lactones, 1625cm^{-1} corresponds to the double bond of side chains, 1670 cm^{-1} corresponds to ketones, and 1640 cm^{-1} corresponds to the diketene.

1.3.3   Nuclear magnetic resonance (NMR) spectroscopy

^{1}H-NMR and ^{13}C-NMR, as well as 2D-NMR, are principal means to determine lignan chemical structures. We will introduce their usages in the examples below. There were some scholars who tried to summarize NMR spectroscopy of various types of lignans, but it was difficult to put into practice because of structural variations. In fact, ^{13}C-NMR spectra are relatively useful for the determination of lignan molecular skeletons and the substitution of the side chain[12].

Section 2   Structural Types and Characteristics of Lignans

2.1   Dibenzylbutanes

Dibenzylbutanes are simple kinds of lignans linked by C8-C8'. Their side chains can be oxidized to hydroxyl, carboxyl, carbonyl or form an ethylenic link by dehydration. Examples include dibenzylbutylene, dibenzylbutanol, and dibenzylbutyric acid. Another example includes nordihydroguaiaretic acid (**20**), which is a racemic compound isolated from *Larrea divaricata* and can be used as an antioxidant in foodstuffs. Phyllanthin (**21**) from *Phyllanthus niruri*[13] and compound **22** from *Schisandra rubriflora*[14], are both dibenzylbutanediols.

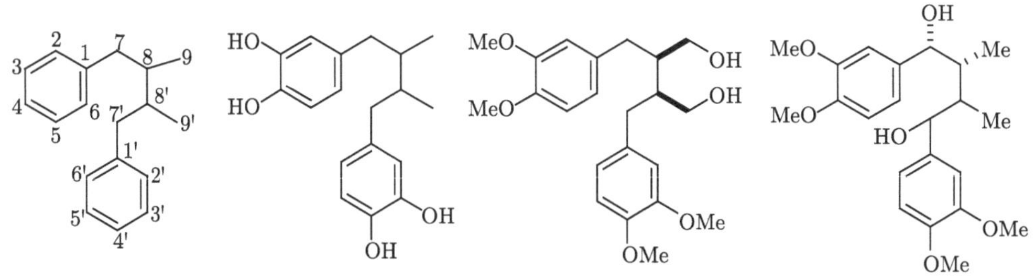

20   21   22

In the biosynthetic pathway, dibenzylbutanes are putative processors of other types of lignans. Dibenzylbutanes can be transformed into other types by chemical methods, e.g., compound **23** can be oxidized to compounds **24**, **25** and **26** by DDQ in TFA[15].

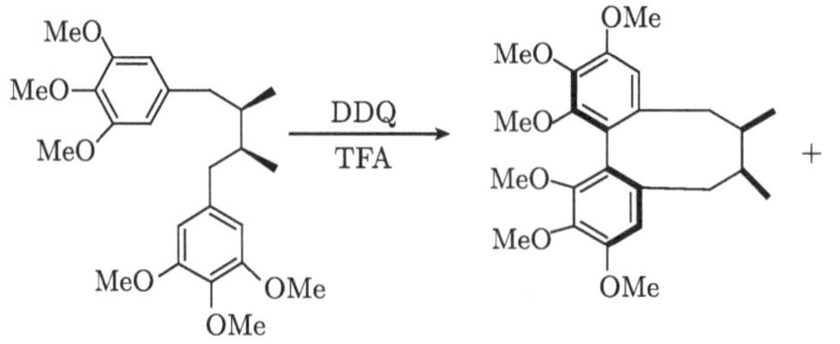

23   24

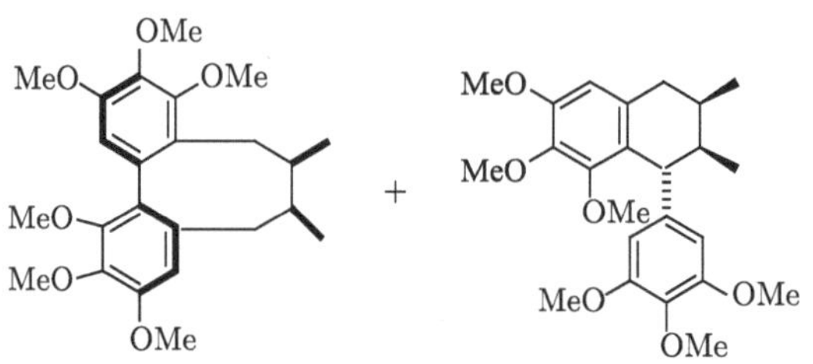

25   26

2.2   Dibenzylbutyrolactones

Dibenzylbutyrolactones, also called lignanolides, are formed through esterization by C9 carboxyl and C9' hydroxyl. Matairesinol (**27**) is one typical lignan isolated from *Podocarpus spicatus*. There are dehydrogen and dedihydrogen derivatives of dibenzylbutyrolactones, according to the unsaturated degree of the lactone rings.

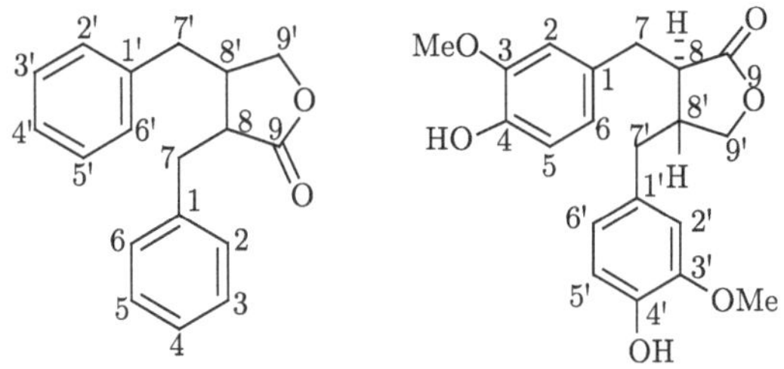

3,4,8-trihydroxy-3′,4′-dimethoxylignano-9,9′-lactone (**28**)

Phenax angustifolius a plant found in Central America, is used as a traditional pesticide. Its leaves are soaked in 70% EtOH and the extract partitioned between *n*-butanol and water. The *n*-butanol fraction was chromatographed over LH-20, eluted with methanol, and then subjected to C18 bondapack reverse phase silica gel. MeOH:H_2O (4:6) was used as an eluent.

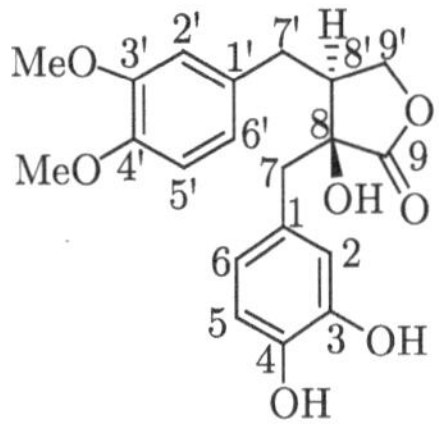

Finally, an oil-like substance (**28**) was obtained. $[\alpha]_D^{25}$ =−39.1°; EI MS gave molecular ion peak at m/z 374. ^{1}H-NMR (600 MHz) showed two methylenes, one other methylene combined with oxygen, and two methoxys. There were six phenyl protons corresponding to two phenyl rings. It showed a lactone ring and twelve phenyl carbons in the ^{13}C-NMR spectrum. Its ^{1}H and ^{13}C signals were assigned on the basis of its 2D-NMR. Based on the ^{1}H-DQF-COSY spectrum, the coupling relationships between 9′-H_2 and 8′-H, 7′-H_2 were verified; 9′-H_2 was assumed to be a methylene of 2,3-trans-γ-butyrolactones[16].

The chemical shift of the 8′-C (δ 77.4 ppm) in compound **28** was consistent with that in benchequiol, whose butyrolactone was *trans*, but was inconsistent with that in guayadequiol (δ 75.9 ppm), whose butyrolactone was *cis*.

In the HMBC spectrum, correlations between 9′-H_2(δ 4.00 ppm) and 7′-C (δ 32.2 ppm), 8′-C (δ 44.6 ppm), 8-C (δ 77.4 ppm), 9-C (δ 180.6 ppm); 7-H_2(δ 3.14 ppm, δ 2.98) and 8′-C, 8-C, 9-C were observed.

The ^{1}H and ^{13}C-NMR data were assigned on the basis of the HMQC and HMBC spectra.

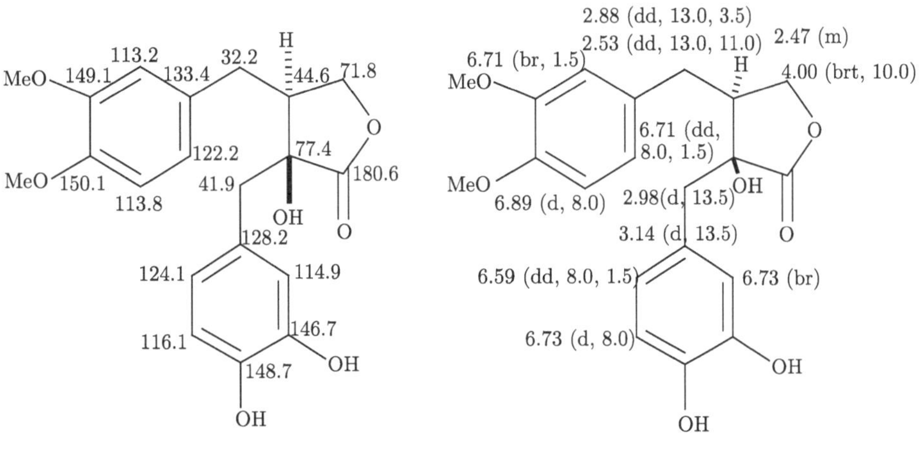

The compound is chemically named 2-hydroxy-2-(3′,4′-dihydroxyphenyl) methyl-3-(3″,4″-dimethoxyphenyl) methyl-γ-butyrolactone in the original paper[17].

2.3   Arylnaphthalenes

Arylnaphthalenes include arylnaphthalene, dihydroarylphthalene, and tetrahydroarylphthalene. They are also called cyclolignolides because C-9 and C-9′ often constitute one γ-lactone and C-7, C-8, C-7′, C-8′ comprise one naphthalene ring with one phenyl ring. Arylnaphthalenes are one kind of the largest lignans, which are widely distributed in nature. Such lignans have been carefully studied, especially podophyllotoxin and its derivatives, for their clinical values. In the references, if the lactone carbonyl is at the top, then it is normal or retro. Please note the numberings for podophyllotoxin, as in some references they differ from the present paper.

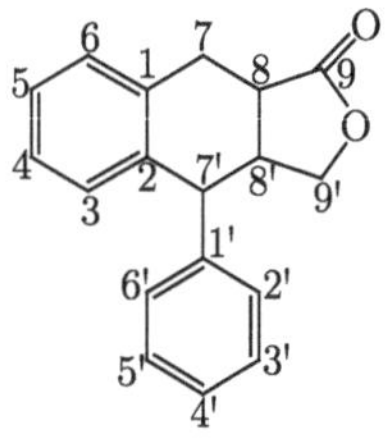

normal 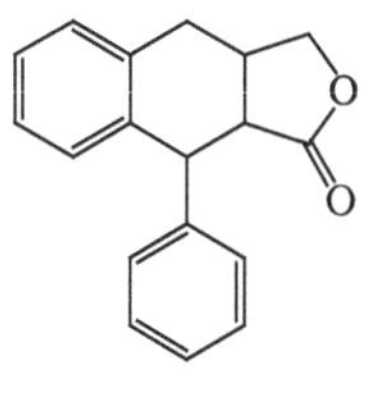   retro 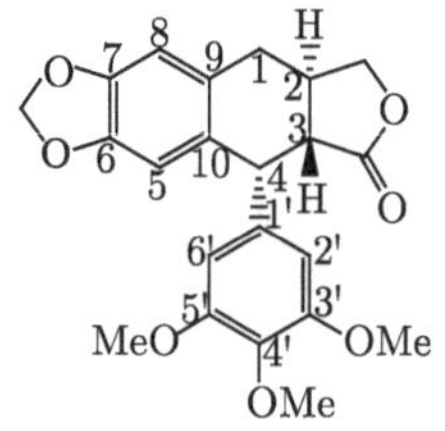   29

Podophyllotoxin can be transformed to picropodophyllotoxin in alkali solvent. C8-C8′ is changed from *trans* configuration to *cis*. In acid, epipodophyllotoxin transformed to podophylotoxin, with its 7β-OH changed to 7α-OH. The transformation process can be illustrated as follows:

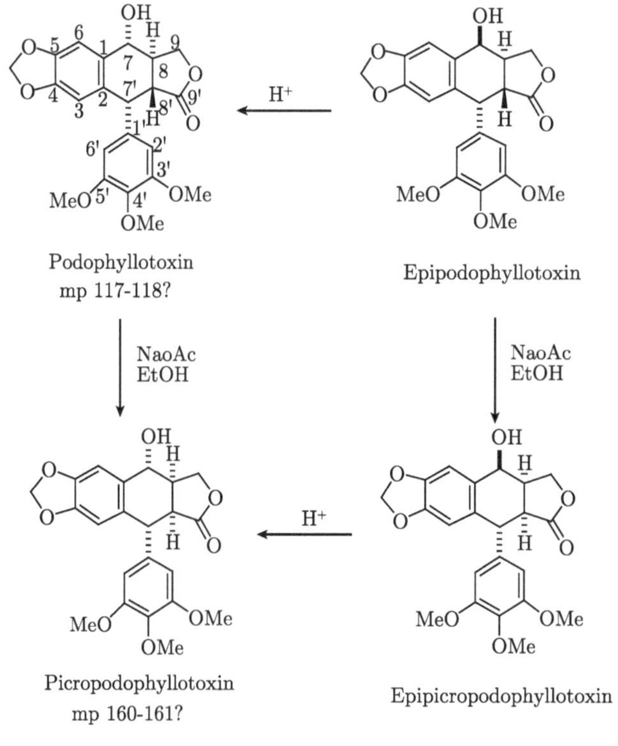

Podophyllotoxin
mp 117-118?

Epipodophyllotoxin

Picropodophyllotoxin
mp 160-161?

Epipicropodophyllotoxin

Both phyllamyricolide B (**30**) and phyllamyricin C (**31**), from *Phyllanthus myrtifolius*, are arylnaphthalenes[18].

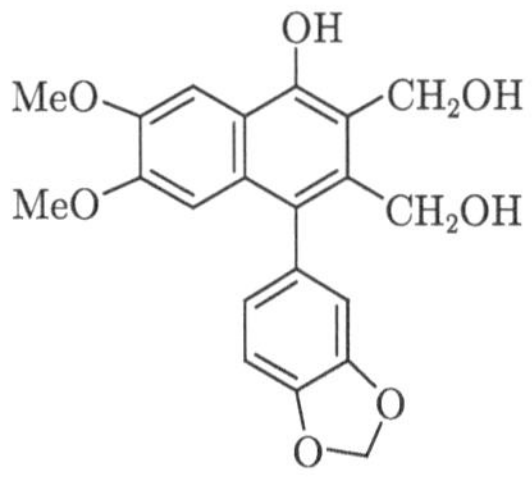

30 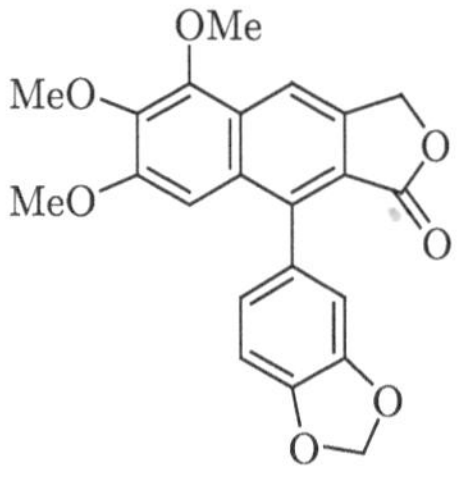   31

2.4   Tetrahydrofurans

Tetrahydrofurans are some of the largest lignans. New tetrahydrofurans are continually discovered. They can be further divided into three subtypes, i.e., 7-O-7′, 7-O-9′, 9-O-9′. There may be four chiral carbons in their structures, so their stereochemistry is also very interesting.

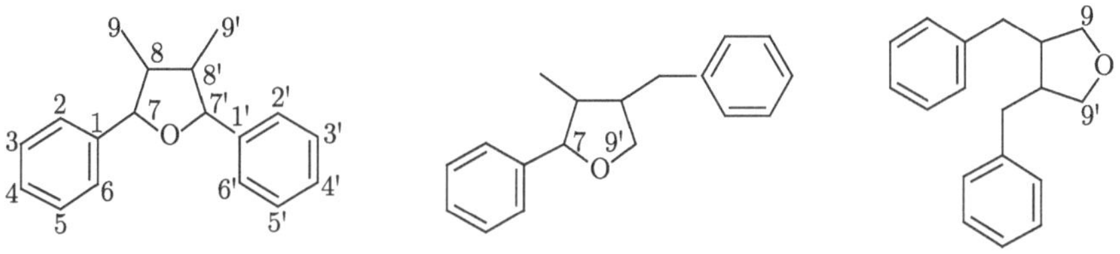

7-O-7′   7-O-9′   9-O-9′

Furoguaiacin (**32**) is a constituent from the resin of *Guaiacum officinale*. It belongs to the 7-O-7′ subtype. This compound can turn into guaiac blue (**33**) when it combines with alkali or oxygenase.

Lancea tibetica is used to cure coughs and colds. Tibeticoside (**34**), one kind of tetrahydrofuran glycoside, is isolated from the whole plant[19].

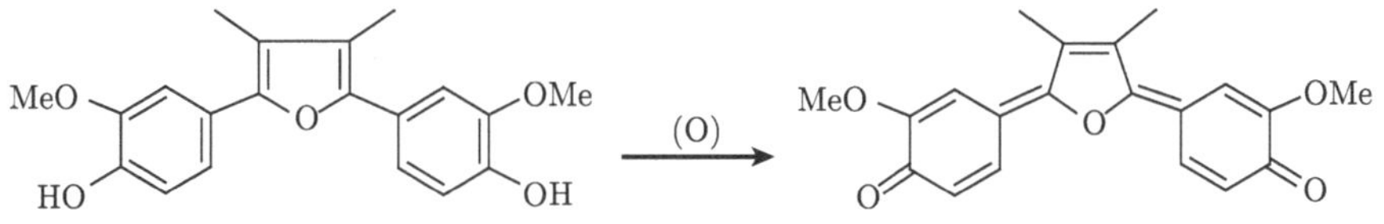

32   33

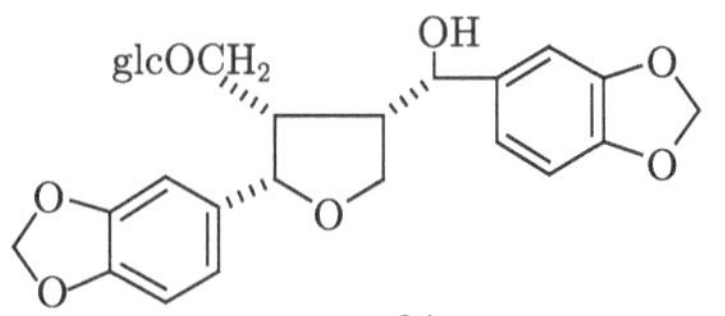

34

Such compounds are not stable in acid, e.g., in the 2% $HClO_4$ solvent of HOAC, veraguensin (**35**) can change into **36** by rearrangement and dehydration.

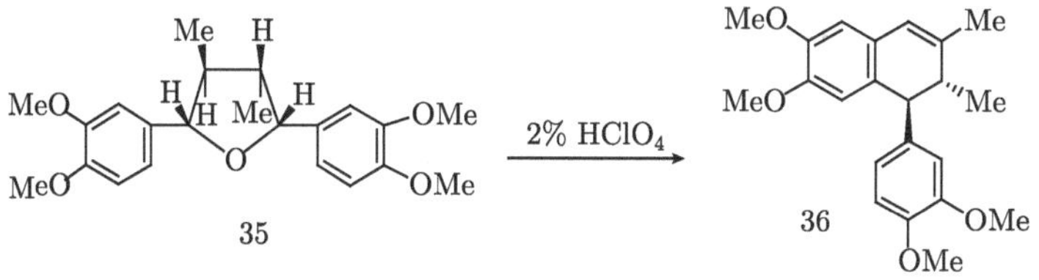

$$35 \xrightarrow{2\% \text{ HClO}_4} 36$$

2.5   2,6-Diaryl-3,7-dioxabicyclo[3,3,0]octanes

These kinds of lignans are also called bisepoxylignans because there are four chiral carbons in the structures, and because two tetrahydrofuran rings of such lignans always *cis*-fuse,

the maximum numbers of their stereoisomers correspond so that they only contain three chiral carbons. If two aryls at 7- and 7'- are *trans* equatorial or *cis* axial conformation in referrerence to the ditetrahydrofuran rings, they are called symmetric lignans, such as (+)sesamin (**37**) and diasesartenin (**38**). On the contrary, if two aryls at 7- and 7'- are *cis* equatorial or *trans* axial conformation in reference to the ditetrahydrofuran rings, then they are asymmetric lignans.

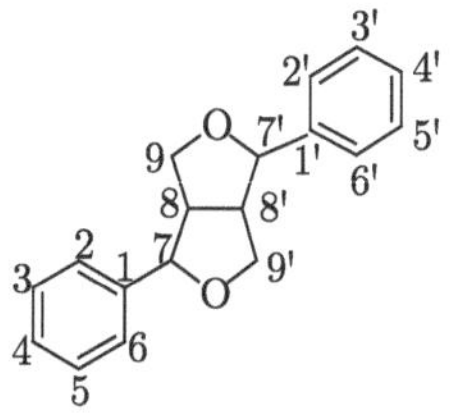

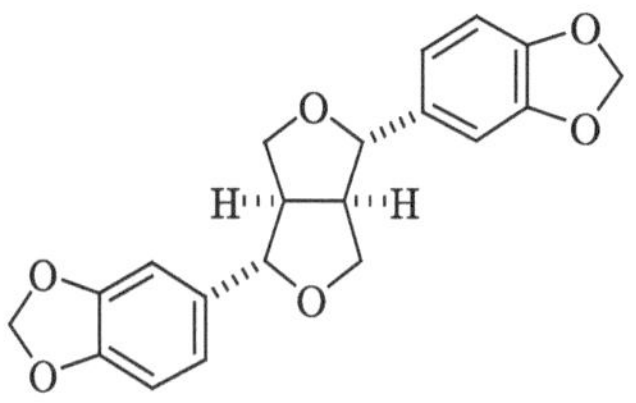

37

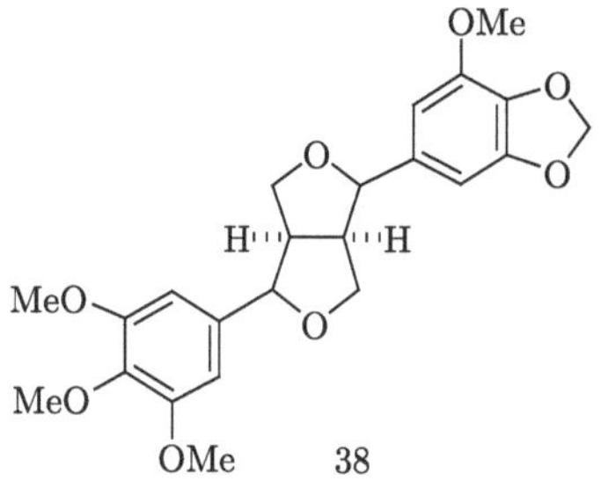

38

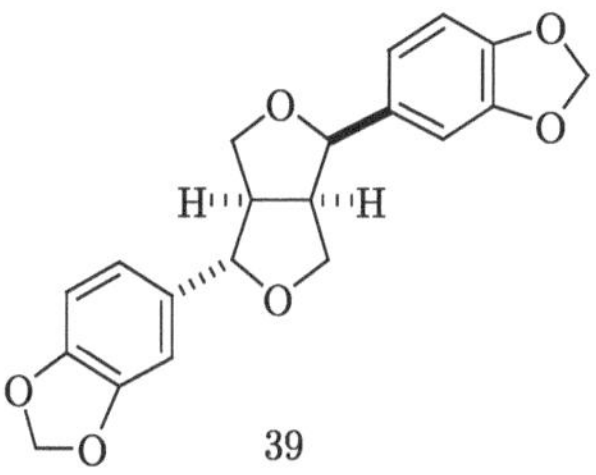

39

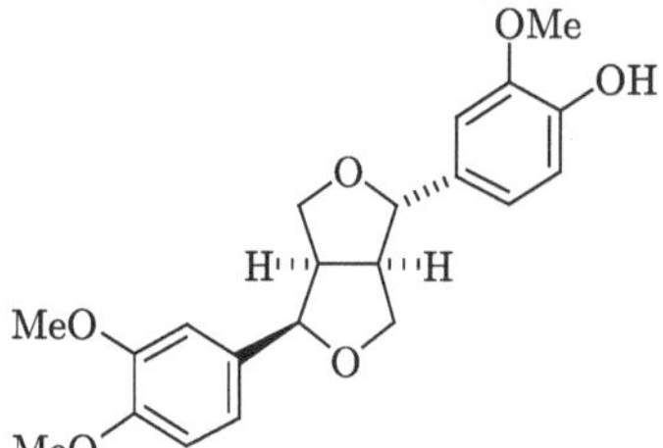

40

Such lignans are easy to isomerize because benzyl ether is readily cleaved and also easily recycled. Thus, symmetric lignans can isomerize to various asymmetric compounds in acid solvent. For example, (+)sesamin (**37**) can isomerize to (+)asarinin (**39**) and (+)pinoresinol (**41**) can isomerize to (+)epipinoresinol (**42**). Considering this problem, one should avoid using acid in the separation process. In addition, such lignans are readily oxidized by HNO_3 (e.g., compound **41** changes to **43** through methylation and bromization, and **44** is obtained through further oxidation by HNO_3[20]).

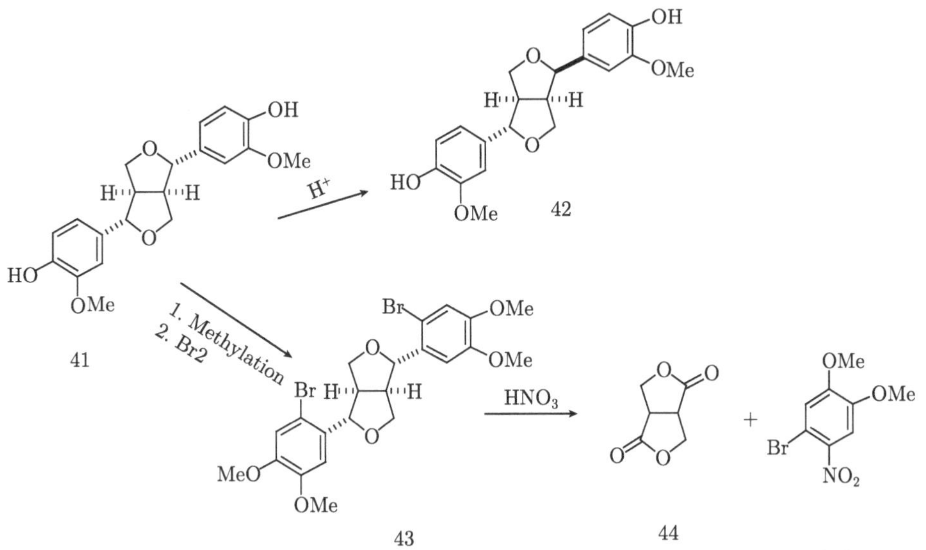

Structural determination of clemaphenol A (45) from *Clematis chinesis*[21]

The stem of *Clematis chinese* was extracted with boiling 95% EtOH and the EtOAC soluble fraction of the extract was subjected to silica gel chromatography to give one lignin (**45**). HR MS provided its molecular formula as $C_{20}H_{22}O_6$, $[\alpha]_D^{25}$ +72° (CHCl₃). IR: 3400 cm^{-1} (br, hydroxyl), 1610 to 1460 cm^{-1} (phenyl), 860 cm^{-1}, 830 cm^{-1}(1,2,4- or 1,3,4-trisubstitutedphenyl). In the ^{1}H-NMR spectrum: δ 5.60 ppm (OH), δ 3.90 ppm (-OCH₃), δ 3.10 ppm (8, 8′-H), δ 4.74 ppm (7,7′-H); δ 3.87 ppm(dd, 8.8, 3.3) and δ 4.26 ppm (dd, 8.8, 6.6) corresponding to 9-H₂ and 9′-H₂ in a bisepoxylignan were observed. It was presumed to be a symmetric lignan given that 9-H and 9′-H showed dd signals instead of multiple signals. There were only three groups of phenyl protons signals: δ 6.90 ppm (d, 1.1), δ 6.82 ppm (dd, 8.2, 1.1), δ 6.89 ppm (d, 8.2), which implied two phenyl rings bearing the same substitution type (1,2,4- or 1,3,4-trisubstituted phenyl). δ3.90 ppm was assigned to CH₃O- and δ 5.60 ppm 1(br-) was assigned to the hydroxyl proton. Combined with its^{13}C-NMR data, this compound was presumed to have two possible structures: either the same structure as (+) pinoresinol (**41**) or a new structure, illustrated as **45**.

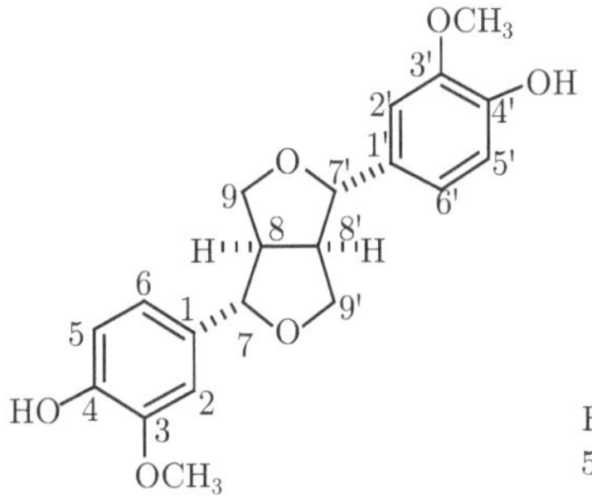

41

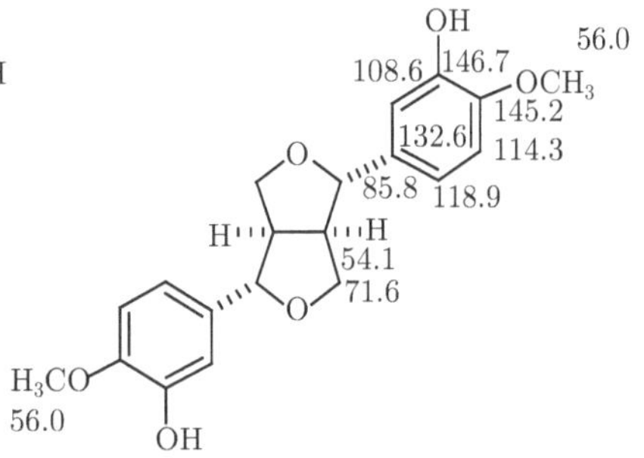

45

The key problem is to decide the position of the phenyl hydroxyl. By the classical approach, compound **45**, ferulic acid and isoferulic acid were simultaneously developed on the same polyamide layer using 90% MeOH as eluent. Gibbs reagent was used as a color developing reagent. The substance will turn blue color which last a long time if there is a dissociative proton in the *para* position of the phenyl hydroxyl. In this experiment, both compound **45** and isoferulic acid turned blue and the color lasted a long time. Though ferulic acid also turned the compound blue, the color faded in air and changed to yellow. Based on this experiment, compound **45** was established to have the same substitution type as isoferulic acid.

2.6   Dibenzocyclooctenes

Dibenzocyclooctenes are mostly discovered from *Schisandra*. Because some dibenzocyclooctenes demonstrate critical bioactivities, these lignans have attracted great attention. For example, steganacin showed antileukemic activities, and schisantherin A and its derivatives showed hepatoprotective effects and can lower SGPT levels.

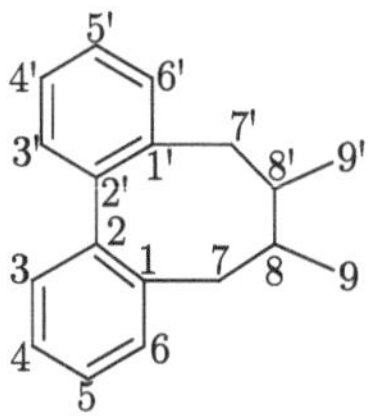

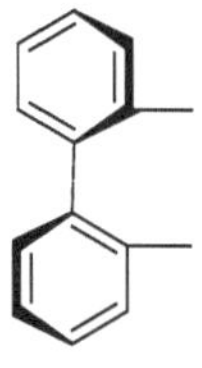

Diphenyl R 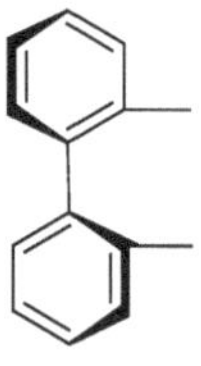   Diphenyl S 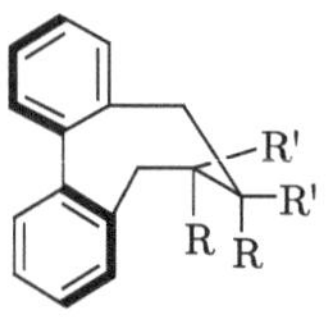   TBC 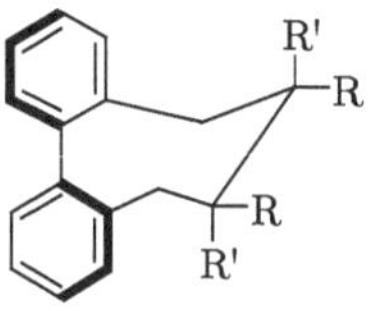   TB

Such lignans have more isomers because they contain many chiral units in their structures. They are divided into R and S according to the configuration of the dibenzyl. The characteristic compounds gomisin D, which is dibenzyl(S), and gomisin A, which is dibenzyl(R), are elucidated by x-ray diffraction analysis. The cyclooctenes of such lignans are mostly winded boat-chair configuration, and a few are twined boat configuration.

The absolute stereochemistry of the biphenyl chromophore can be deduced by circular dichroism (CD) spectroscopy. If the CD spectrum of a lignan derivative shows both (-) cotton effect at 250 nm and a (+) cotton effect at 220 nm, the biphenyl unit has the S-configuration. Conversely, a lignan with an R-configuration yields a CD spectrum with a (+) cotton effect at 250 nm and a (-) cotton effect at 220 nm. Regarding specific rotation, though most S-configuration dibenzyls are levorotatory, R configuration dibenzyls are dextral, although there are some S configuration lignans reported to be dextral. Therefore, it is not always exact to determine their absolute stereochemistry by specific rotation.

According to the biphenyl configurations and cyclooctene structures, dibenzocyclooctenes can be categorized into various series which can be characterized by one typical compound, such as gomsin, kadsurin, schisantherin, kadsulignan, stegane (**46**), schizandrin-wuweizisu, and heterolitin.

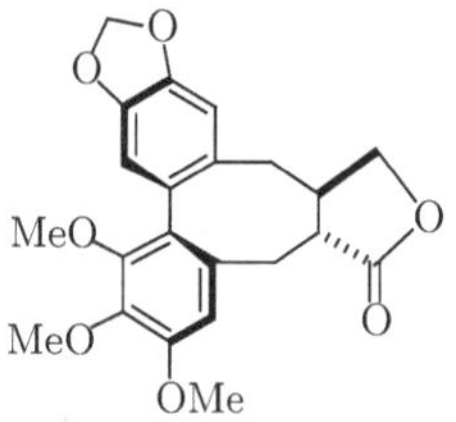

Such lignans often have many names. Isokaduranin (**47**) is also known as deoxygomisin O or gomisin N, and deoxygomsin A (**48**) can also be called γ-schizandrin or wuweizisu B.

46

As a result, it is necessary to know lignan chemical names and how to use them correctly. For example, (-)steganone (**49**) is chemically named $(2R\alpha)$-$(8\beta, 8'\alpha)$-3,4,5, -trimethoxy-4', 5'-methylenedioxy-7'-oxo-2,2'-cyclolignano-9,9'-lactone.

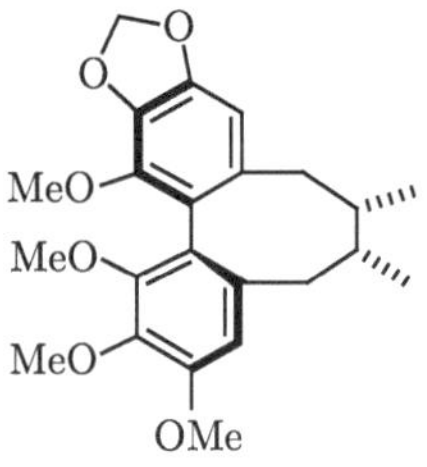

47 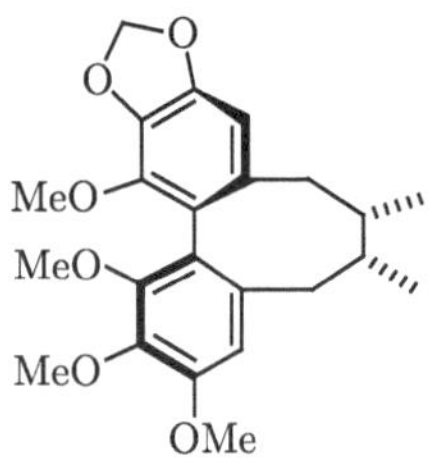   48 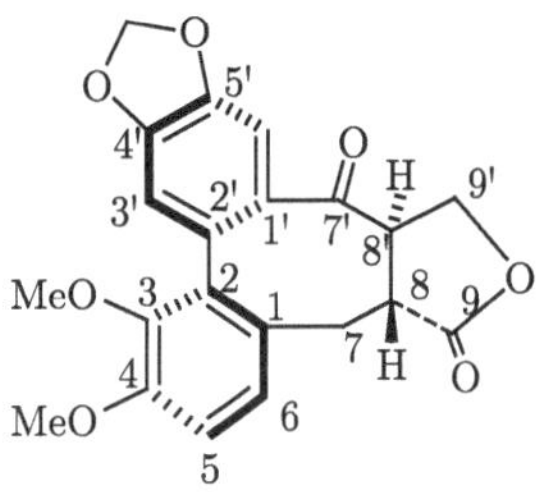   49

2.7 Benzofurans

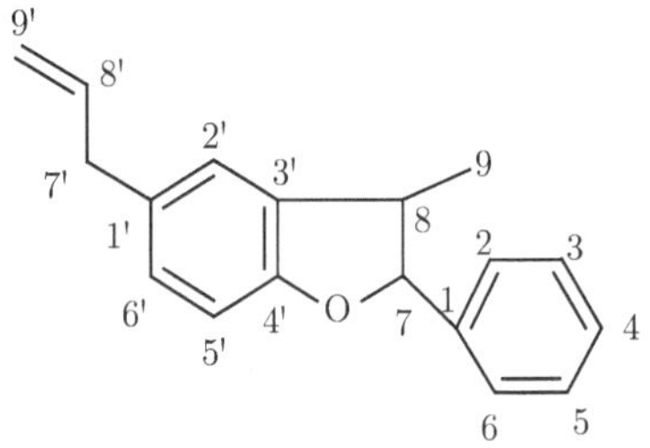

Benzofurans are neolignans. One side chain of the C6-C3 unit constitutes one furan ring with another benzyl ring. The furan ring and benzyl ring attached to the furan ring sometimes are hydrogenated; thus, there are dihydrofuran, tetrahydrofuran, and hexahydrofuran derivatives. These types can often be found in Lauraceae and Piperaceae.

Kadsurenone (**50**), hancinone (**51**) and kadsurin A (**52**), isolated from *Piper futokadsura* and *Piper hancei*, show potent inhibitory effect against PAF.

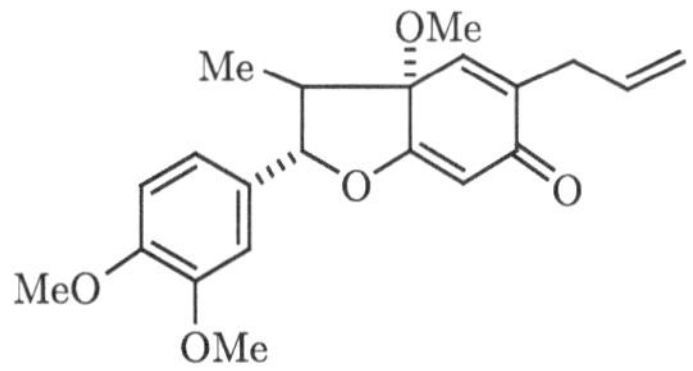

50 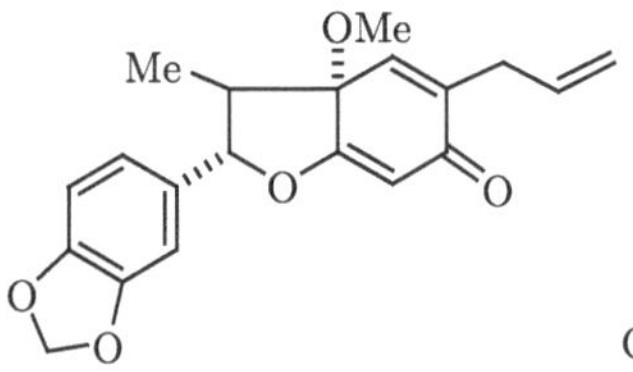   51 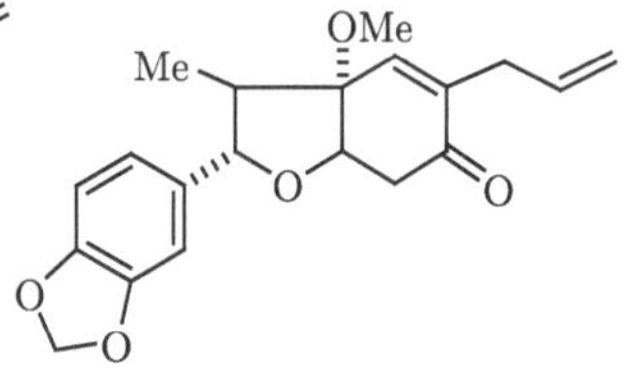   52

Isolation and identification of quinquenin L$_1$(**53**) from *Panax quinque foliuins*[22]

The compound was initially extracted from the leaves of *Panax quinque foliins* and then mainly separated by repeat gel chromatography and reverse-phase silica gel chromatography from the *n*-butanol fraction.

Compound **53**, white powder, mp 224 to 226°C, D-glucose can be detected when hydrolyzed in 10% H$_2$SO$_4$ solvent. Its ^{1}H-NMR (C$_5$D$_5$N) showed five phenyl protons: δ 7.48 ppm(1H, br, d, J = 8.5 Hz), δ 7.32 ppm (1H, brs), δ 7.20 ppm (1H, d, J = 8.5), δ 7.04 ppm (1H, brs), δ 6.91 ppm (1H, brs), which correspond to one 1,3,4,-trisubstituted phenyl ring and one 1,3,4,5,-tetrasubstituted

53

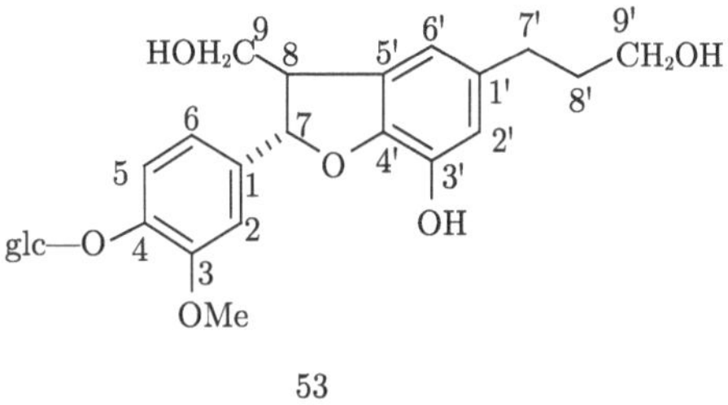

phenyl ring. In addition, there are two CH$_3$O signals at δ 3.67 ppm and 3.59 ppm, and a group of glucose signals at δ 5.66 ppm (d, J = 7.0 Hz) and 4.0 to 4.8 ppm (6H). In the ^{13}C-NMR spectra, there are six carbon signals besides phenyl signals and CH$_3$O signals at δ 87.9 ppm, 64.4 ppm, 62.3 ppm, 56.2 ppm, 36.1 ppm, and 32.7 ppm. From the above data, compound **53** was determined as a lignan containing one furan ring. By comparing with reported data, it was approved as a known compound first isolated from *Larix leptolepis*. Its chemical name is 3′,9,9′,- hydroxyl-3-methoxy-4′,7-epoxy-neolignane-4-O-β-D-glucopyranoside.

2.8   Bicyclic (3,2,1)-Octanes

These kinds of lignans contain two aliphatic rings. There are eight carbons in the aliphatic rings. One C6-C3 unit is partly hydrogenated while the side chain of another C6-C3 unit is only attached to the hydrogenated phenyl ring. To date, these compounds are mostly discovered in tropical plants, such as Piperaceae, Ocotea, and Aniba. Both puberulin A (**54**) and C (**55**) are examples of these kinds of lignans, which can be isolated from *Piper puberulum*[23].

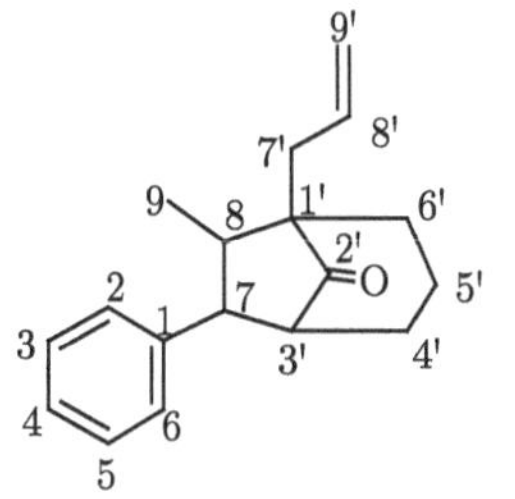

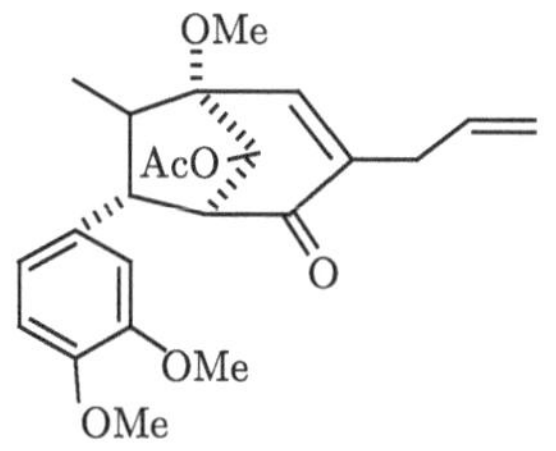

54

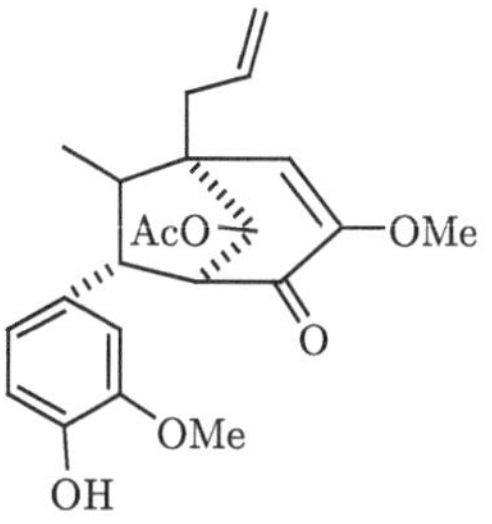

55

2.9   Benzodioxanes

When two phenyl hydroxys at the 3-,4-position form a dioxane with the side chain of another C6-C3 unit, they are called benzodioxanes. For example, eusiderin A (**56**) from *Eusideroxylon zwageri*, americanol A (**57**) and isoamericanol A (**58**) from *Phytolacca americana* are benzodioxanes[24]. Eusiderin A is a typical compound of this kind[25].

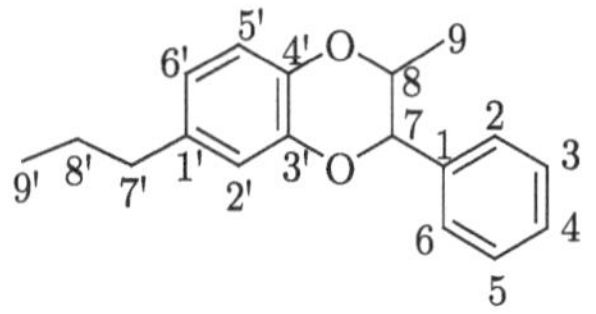

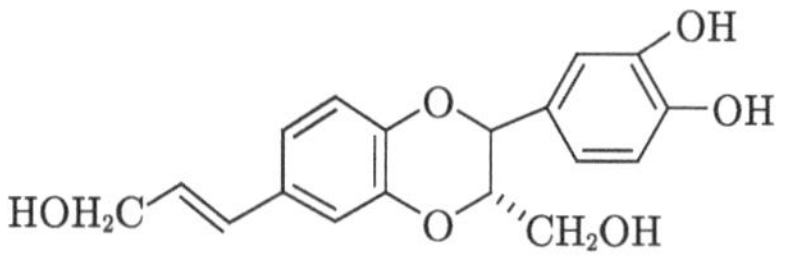

57

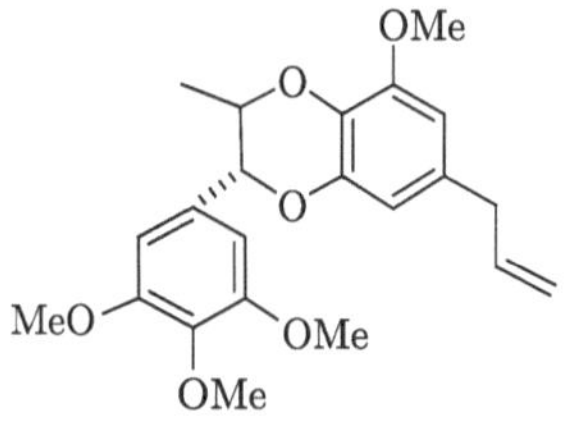

56

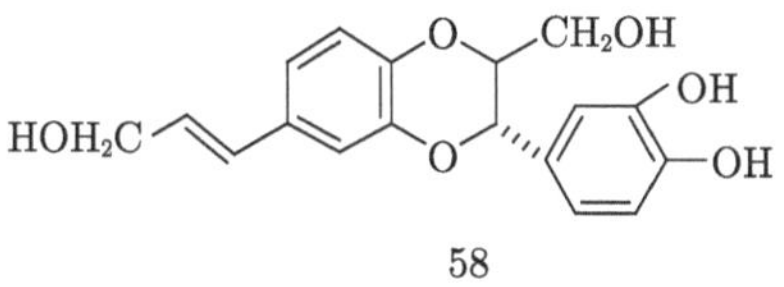

58

2.10   Biphenyl Derivatives

Biphenyl derivatives are directly connected by the two phenyls, mainly by C3-C3′. Magnolol (**59**) was obtained from the bark of *Magnolia officinalis*. Its isomer, honokiol (**60**) was obtained from the bark of *M. obovata*. Compound **61** from *Irvigia malayana* and **62** from *Beta vulgaris* contained more oxygen substituents. Their UV spectra were very characteristic[26,27].

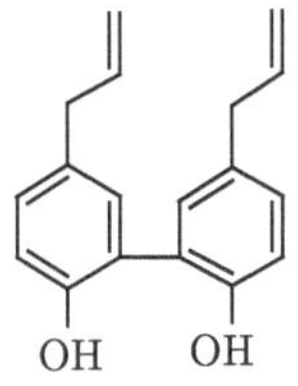

59 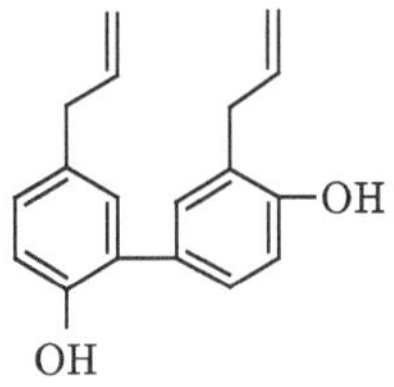   60

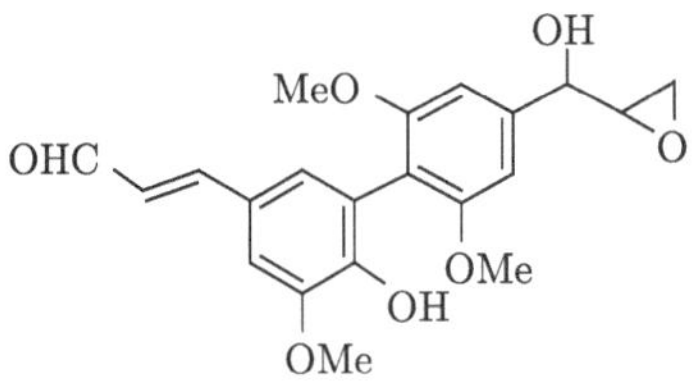

61 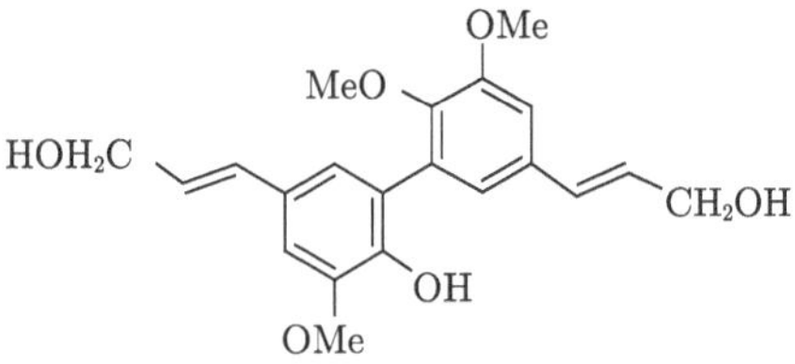   62

2.11 Oligomeric Lignans

Oligomeric lignans contain more than two C6-C3 propanoid units, including sesquilignan, dimeric lignan, trilignan, tetralignan, etc. In recent years, more and more new oligomeric lignans were discovered from nature, and various oligomeric lignans usually are found in the same plant.

Simonsinol (**63**) can be obtained from the bark of *Illicium simonsii*[28]; compound **64**, isolated from *Illicium dunnianum*[29]; compound (**65**) purified from the root *Salvia Yunnanensis*[30] and cyclohexa-2,5-dienone (**66**) from the stem of *Coptis chinese franch.*[31] All are known as sesquilignans.

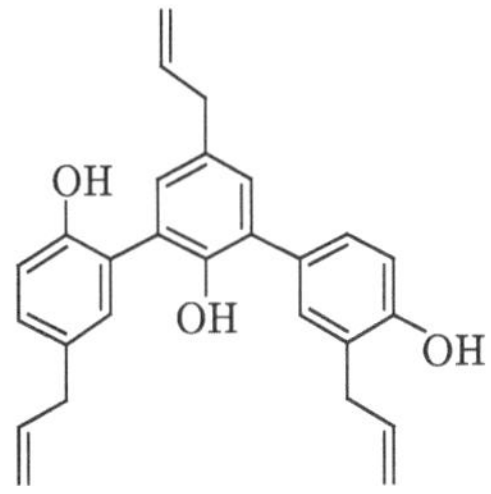

63 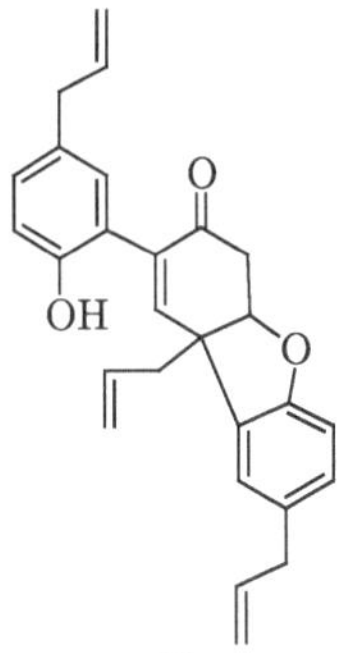   64

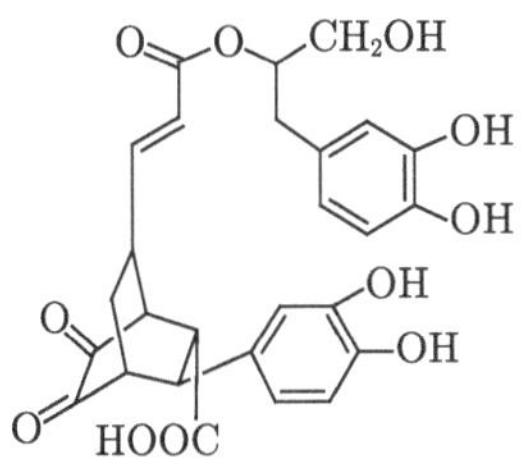

65 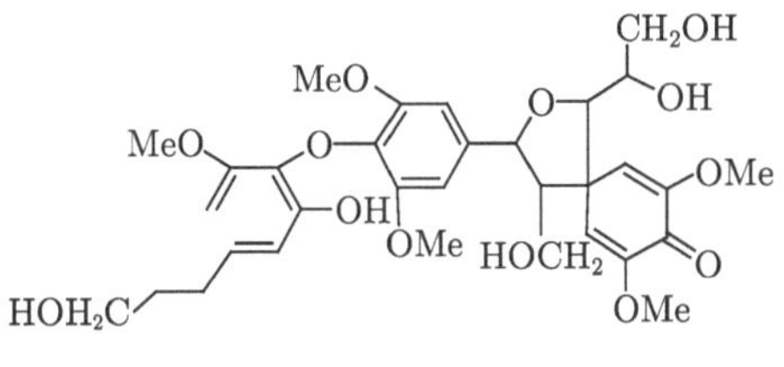   66

2.12 Miscellaneous

Besides the lignans we discussed above, there are other types of lignans (e.g., biaryl ethers, norlignans, spirodienones, flavonolignans, coumarinolignans, lignan-iridoid complexes, etc.).

Isomagnolone (**67**), obtained from *Illicium simonsii*, is a binaryl ether. Two C6-C3 units are connected by oxygen and thus, it can also be called an oxyneolignan[32]. Compound **68**, from *Ehretia ovalifolia*[33], and cannabisins E (**69**) and F (**70**), from *Cannabis sativa*[34], are also all oxyneolignans.

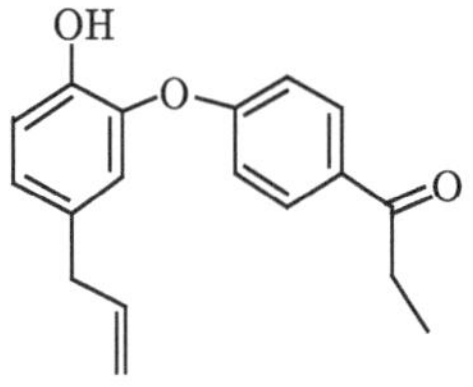

67

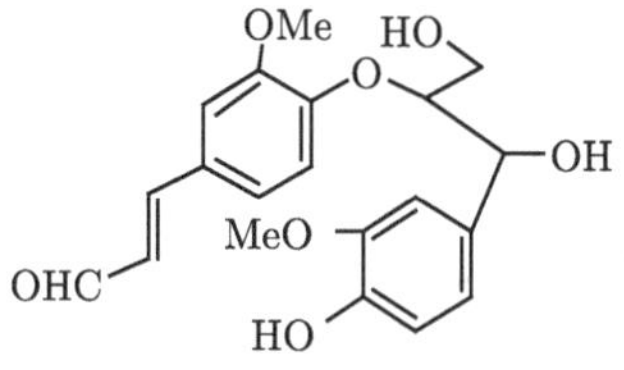

68

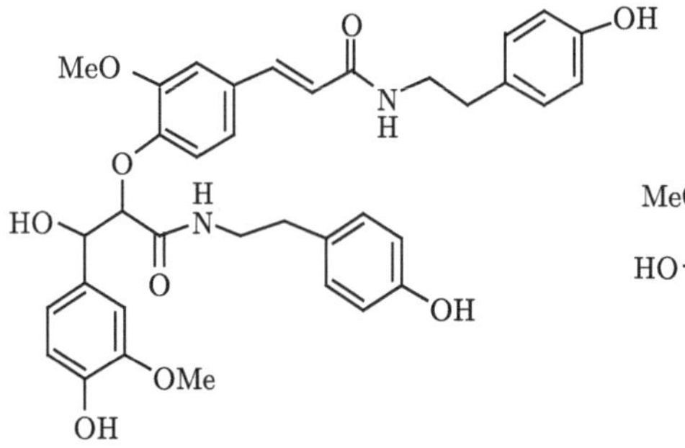

69

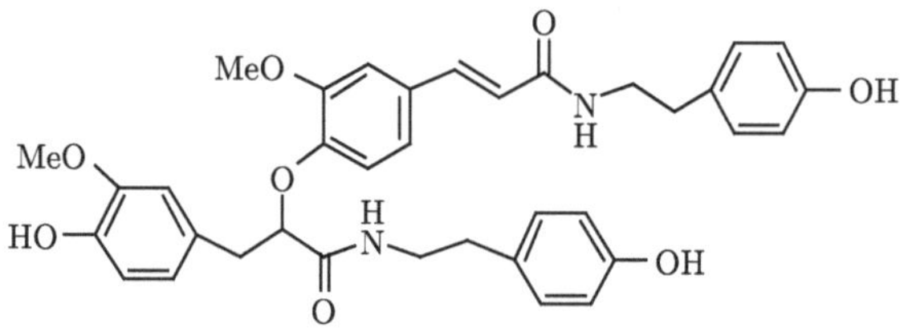

70

Some lignans lose one or two carbons of their side chains and are called norlignans. Compound **71**, from *Curculigo capitulate*[35], and ailanthondol (**72**), from *Zanthoxylum ailanthoides*[36], are such compounds.

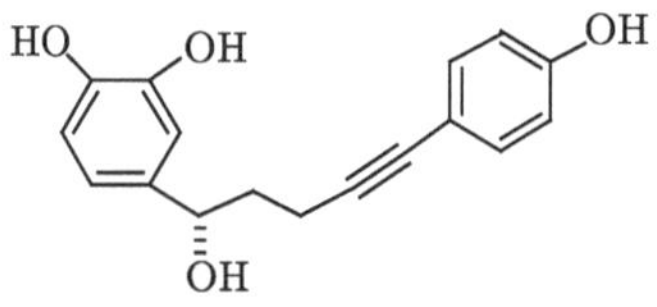

71

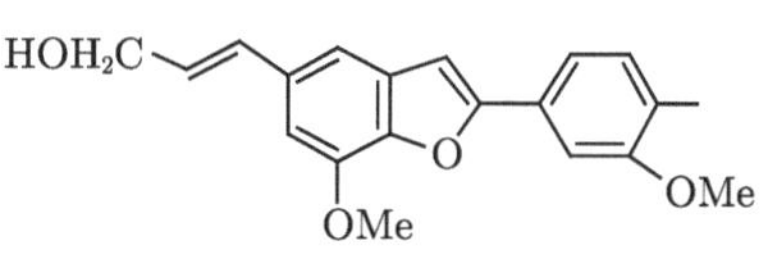

72

Futoenone (**73**), isolated from *Piper futokadsura*, and compound **74**, isolated from the rhizome of *Coptis japonica*[37], are spirodienones.

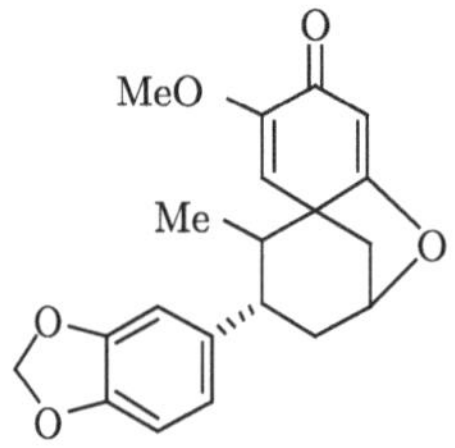

73

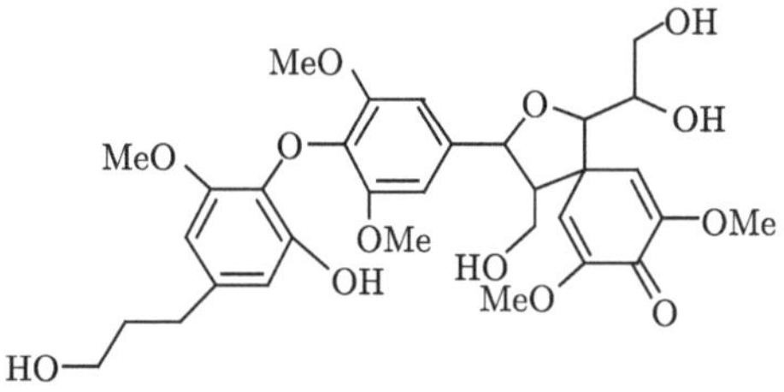

74

Sinaiticin (**75**) is a flavonolignan isolated from the leaves of *Verbascum sinaiticun*[38]. 5′-Demethoxycadensin G (**76**) is a xanthonolignan isolated from *Cratoxylum cochinchinese*[39]. Compound **77** is one lignaniridoid found in the root of *Buddleia davidii*[40].

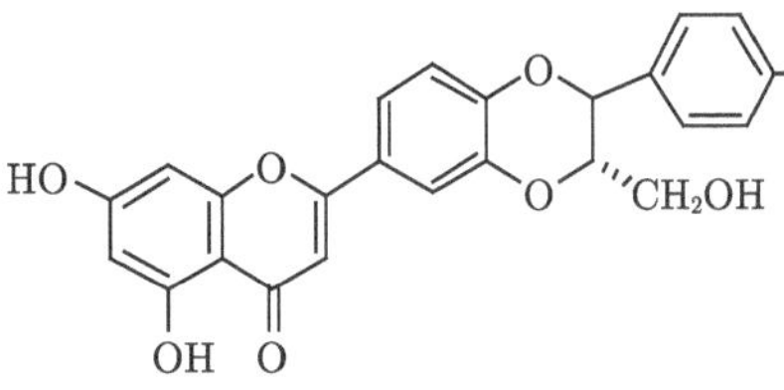

75 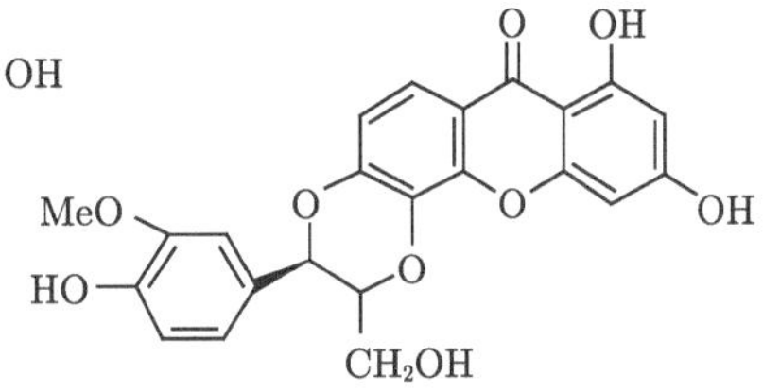   76

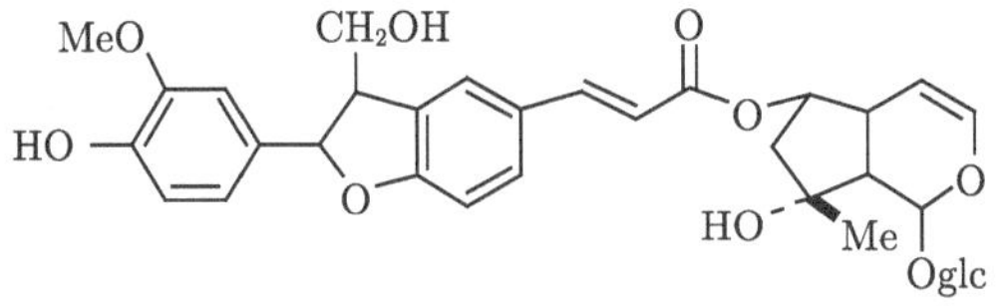

77

Brevitaxin (**78**) is a diterpenelignan isolated from *Taxus brevifolia*[41]. Compound (**79**) is a coumarinlignan isolated from *Hemidesmus indicus*[42].

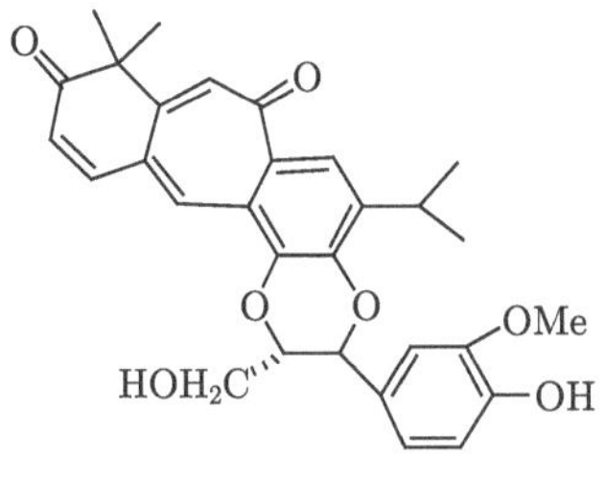

78 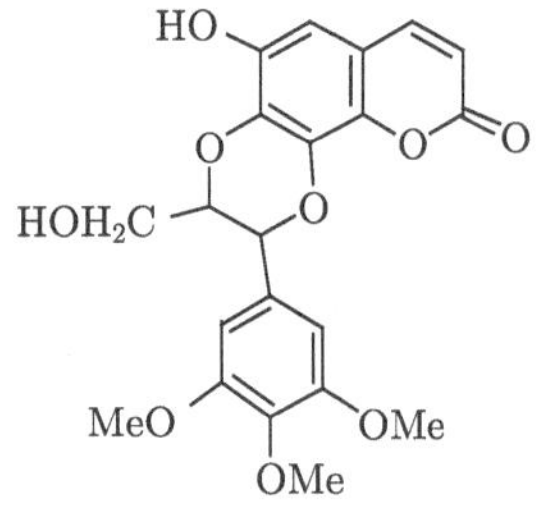   79

Section 3   Bioactivities of Lignans

Natural lignans have various types of structures and demonstrate a broad spectrum of bioactivities. The main bioactivities of lignans are presented below.

3.1   Anti-tumor

Some lignans can inhibit the growth of tumor cells due to their cytotoxic effects. For example, podophyllotoxin, is the the main active component of podophyllin, an herbal exract of *Podophyllum peltatum* used for the treatment of genital warts. Podophyllotoxin was also found in other podophyllum plants, such as *P. pleianthum* and *Sinopodophyllum emodi*. The resin of these plants was used to remove warts or used as an evacuant. Further research revealed that podophyllotoxin was the principal active constituent in the plant. Therefore, it was used to treat skin cancer and warts. However, because of its strong toxicity, podophyllotoxin was difficult to use in hospitals. A series of compounds were subsequently made from podophyllotoxin by chemical modification or semisynthesis. Now, etoposide (VP-16) and teniposide (VM-26) are commonly used as anticancer drugs. Further anti-tumor mechanism research showed podophyllotoxin analogues were inhibitors of topoisomerase II, one key enzyme in the replication of DNA. A number of peculiar antitumor agents were further discovered from podophyllotoxin derivatives by screening this key enzyme[43].

Some other lignans, such as machilin, matairesinol, magnolol, and honokiol, also exhibit cytoxicity[44].

3.2   Anti-virus

Arnebia euchroma has been used to treat measles. A caffeic acid tetramer, a K^+ or Na^+ salt in the plant, was isolated from it and has exhibited HIV inhibiton[45]. Gomisin G, isolated from *Kadsura interior*, was found to inhibit HIV in one experiment[46].

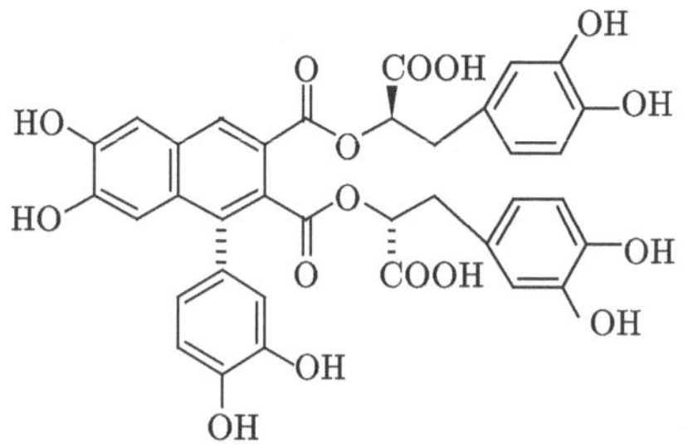

Rerojusticidin B and phyllamyricin, from *Phylianthus myrtipolinus*, showed inhibitory effects against reverse transcriptase and DNA polymerase[47]. Podophyllotoxin analogues demonstrated inhibitory effects against measles and HSV-I.

3.3   Cardioprotective Effects

Lignans from *Piper futokadsura* resist the activity of PAF. Kadsurenone is the most potent, since it can inhibit 95% cell growth of at the concentration of 3.0 μmol/L. (+)-Gomisin M_1, one dibenzocyclooctene isolated from *Kadsura heteroclite*, can inhibit PAF with an EC_{50} value of 3.0 μmol/L. Pregomisin, from *Schisandra sphenanthera*, is also reported to be an active agent against PAF[49].

Some arylnaphthalenes and their derivatives can lower lipid levels. Not only can they reduce serum cholesterol levels, but also improve high-density lipoprotein levels in rats[50]. Many lignans show antioxidative activity[51]. The antioxidative properties of sesame seeds, which have long been used as a health food to prevent aging, appear to be related to their lignan components.

3.4   Hepatoprotective Effects

Dibenzocyclooctenes from *Schisandra chinensis* and *S.sphenanthera* can lower SGPT levels, for example, scisantherin A and wuweizisu C. It was believed that the methylenedioxy group is the active group. Bifendate is an intermediate in the synthesis of wuweizisu C, and can inhibit SGPT and improve the symptoms of hepatitis. It has been used to treat persistent liver hepatitis[52].

3.5   Miscellaneous

Schizandrin demonstrates CNS tranquilization activity and was believed to be the principal active compound in *Schisandra chinensis*. Tranquilization and muscle relaxation activity of

Magnoliol officinalis is related to its content of magnolol and honokiol. Lignans also show anti-allergic, anti-inflamation, anti-bacterial, and pesticidal activities.

References

[1]   Xu R.S., Chemistry of Natural Products, Science Press, Beijing, 1993, 710.

[2]   Yao X.S. Chemistry of Natural Medicines, Peoples Medical Publishing House, Beijing, 1994, 143.

[3]   Freudenberg K. et al. Tetrahedron, 1961, 15, 115.

[4]   Gottlieb O.R. Fortschr. Chem. Org. Naturstoff., 1978, 35, 1.

[5]   IUPAC Commission, Pure Appl. Chem., 1999, 71, 587.

[6]   Moss G.P., Pure Appl. Chem., 2000, 72, 1493.

[7]   Davin L.B. et al. Science, 1997, 275, 362.

[8]   Katayama T. et al. Mokuzai, Gakkaishi, 1997, 43, 580.

[9]   Slanina J. et al. Planta Med., 1997, 63, 277.

[10] Harris R.K. et al. Cereal Foods World, 1993, 38, 147.

[11] One H. et al. J. Chromatogr. B., 1995, 674, 293.

[12] Agrawal P.K. et al. Magn. Reson. Chem., 1994, 32, 753.

[13] Satyanarayana P. et al. Tetrahedron 1991, 47, 8931.

[14] Wang H. et al. Chin. Chem. Lett., 1993, 4, 31.

[15] Carroll A.R. et al. Aust. J. Chem., 1994, 47, 937.

[16] Inagaki I. et al. Chem. Pharm. Bull., 1972, 20, 2710.

[17] Rastrelli L. et al. J. Nat. Prod., 2001, 64, 79.

[18] Lee S.S. et al. J. Nat. Prod., 1996, 59, 1061.

[19] Su B.N. et al. Planta Med., 1999, 65, 558.

[20] Gireger H. et al. Tetrahedron, 1980, 36, 3551.

[21] He M. et al. Acta Pharmaceutica Sinica. 2001, 36(4), 278.

[22] Wang J.H. et al. Chinese Traditional and Herbal Drugs, 2001, 32, 15.

[23] Zhang S.X. et al. J. Nat. Prod., 1995, 58, 540.

[24] Fukuyama Y. et al. Chem. Pharm. Bull, 1992, 40, 252.

[25] Dias S.M.C. et al. Phytochem., 1986, 25, 213.

[26] Mitsunaga K. et al. Nat. Medicines, 1996, 50, 325.

[27] Micard V. et al. Phytochem., 1997, 44, 1365.

[28] Kouno I. et al. Chem. Pharm. Bull, 1994, 42, 112.

[29] Sy L.K. et al. Phytochem., 1996, 43, 1417; 1997, 45, 211.

[30] (a)Tanaka T. et al. J. Nat. Prod., 1996, 59, 843.
      (b)Chem. Pharm. Bull, 1997, 45, 1596.

[31] Yoshikawa K. et al. J. Nat. Prod., 1997, 60, 511.

[32] Kouno I. et al. Chem. Pharm. Bull, 1994, 42, 112.

[33] Yoshikawa K. et al. Phytochem., 1995, 39, 659.

[34] Sakakibara I. et al. Phytochem., 1995, 38, 1003.

[35] Chang W.L. et al. J. Nat. Prod., 1997, 60, 76.

[36] Sheen W.S. et al. Phytochem., 1994, 36, 213.

[37] Yoshikawa K. et al. J. Nat. Prod., 1997, 60, 511.

[38] Afiti M.S.A. et al. Phytochem., 1993, 34, 839.

[39] Venkatraman G. et al. Tetrahedron Lett., 1996, 37, 2643.

[40] Yamamoto A. et al. Phytochem., 1993, 32, 421.

[41] Arslanian R.L. et al. J. Nat. Prod., 1995, 58, 583.

[42] Das P.C. et al. Indian J. Chem. Section B, 1992, 31, 342.

[43] Cho S.J. et al. J. Med. Chem., 1996, 39, 1396.

[44] Hirano T. et al. Life Sci. 1994, 55, 1061.

[45] Kashiwada Y. et al. J. Nat. Prod., 1995, 58, 292.

[46] Chen D.F. et al. Bioorg. Med. Chem., 1997, 5, 1715.

[47] Liu KCSC et al. Med. Chem. Res., 1997, 7, 168.

[48] Han G.Q. et al. J. Chin. Pharm. Sci., 1992, 1(1), 20.

[49] Lee I.S. et al. Arch. Pharm. Res., 1997, 20, 633.

[50] Kuroda T. et al. Chem. Pharm. Bull., 1997, 45, 678.

[51] Nitao J.K. et al. J. Chem. Ecol., 1992, 18, 1661.

[52] Liu G.T. et al. Acta Pharmaceutica Sinica, 1983, 714.

Chang-Qi HU

CHAPTER 13

Other Natural Bioactive Compounds

This chapter includes some compounds with biological significance but not discussed in other chapters.

Section 1   Sulfur Compounds

Plants containing sulfur compound(s) have a distinct obnoxious odor. Examples are garlic, onions, leeks, and scallions. Some plants in the Cruciferarae family, such as Song-lan (*Isatis tinctoria*), Ting-li-zi (*lepidium apetalum*), Luo-bo or radish(*Raphanus sativus*), and Ji-cai (*Capsella bursa-pastoris*), smell of sulfur due to the sulfur compounds they contain. Most of these sulfur compounds have bioactivity and should be given attention.

1.1   Allicin and Diallyltrisulfide

Garlic (*Allium sativum*) is a well-known health food product. It is a Chinese herb and has been used historically as an antibacterial, antifungal, anti-inflammatory, and hypolipidemic agent and for cancer prevention. In the process of crushing allicin, diallyldisulfide oxide **6**, is produced. Allicin is the main antibacterial ingredient of garlic. When crushing garlic, unstable alliin **1**, or (+)-*S*-allylsysteine-*S*-oxide, is first produced, which is then decomposed to allyl-sulfenic acid **2** and aminopropenoic acid **3** by allicinase. Compound **2** further decomposes to allylsulfinic acid **4** and allylmercaptan **5**. Allicin **6** is formed when **4** and **5** further lose water[1].

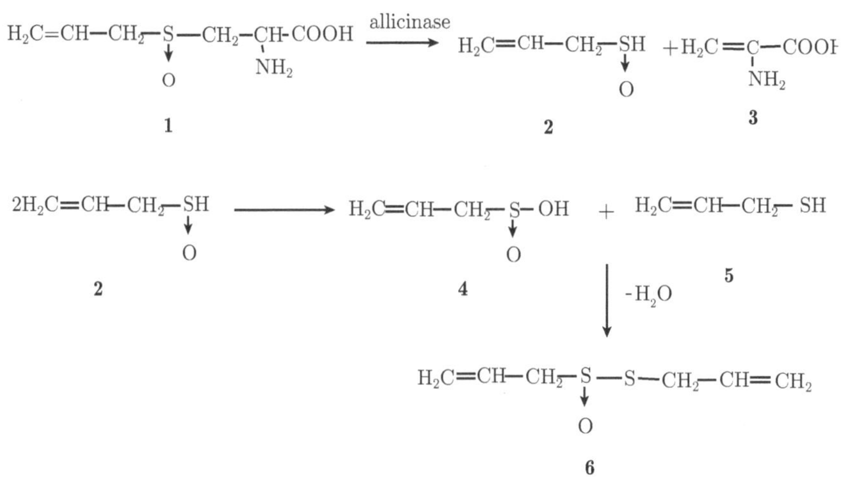

Alliin is a needle-like crystal with mp 164 to 166°C, $[\alpha]_D^{20} + 63.5°$, (c 2, H_2O). Allicin is an oil with d_4^{20} 1.112; n_D^{20} 1.561 and pH 6.5.

In 1975, Treeman et al. proved that alliin consisted of 85% alliin **1**, 13% of (+)-*S*-allylsysteine-*S*-oxide **7**, and 2% of *S*-allylsystein **8**. Treated with sallicinase, the mixture further produced methyl allylsulfide oxide **9**, dimethyl-sulfide oxide **10**, and another sulfide oxide R-SO-S-R' (R, R' represent methyl, propyl, propylene etc.)[2]. Allicin can be thermally decomposed into allyl-sulfenic acid **2** and methyl mercaptal **11**, which upon Diels-Alder reaction at room temperature dipolymerized to 3-vinyl-4H-[1,2]-dithiin **12** and 2-veyl-4H-[1,3]-dithiin **13**[3].

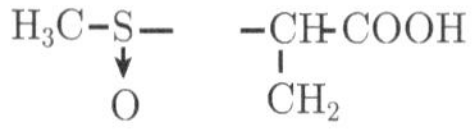

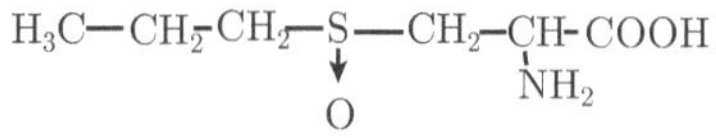

In 1996 Zhang XJ et al. indicated alliin should not be considered as a stable sulfoxide, as it may be easily oxidized to sulfide (*S*-allyl-L-cysteine) with the combination of allicinase. If combined with oxygen in the air, alliin will immediately oxidize to sulfoxide[4]. There are various reports about the sulfides of garlic, which are mainly due to the instability of sulfides and produce many degradants and polymers.

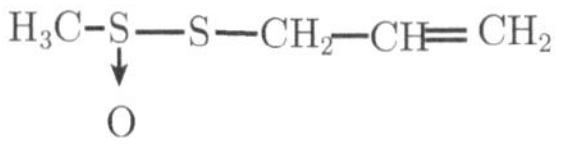

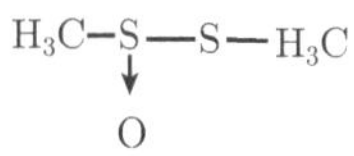

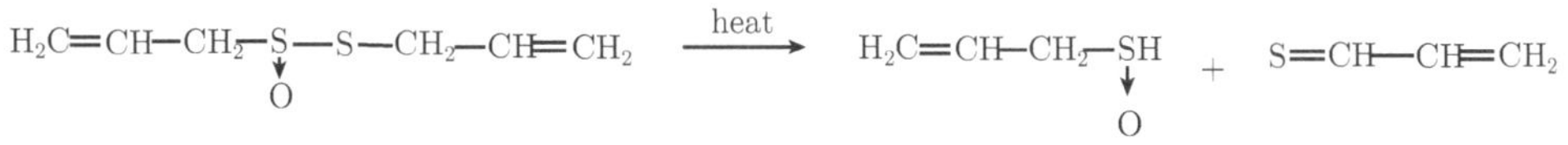

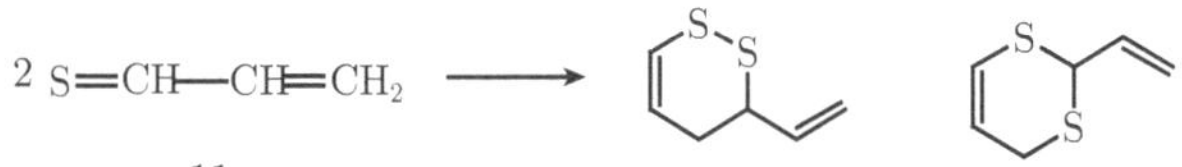

Lang YJ[5] found the antifungal component in garlic is diallyl-trisulfide or diallyl-trisin **17**, a relatively stable compound. It can be obtained by the following procedure: 50 kg of the crushed garlic was mixed with an equivalent amount of water and steam-distilled to collect the product by saline ice. The obtained garlic oil was separated (170 g) and further distilled to collect the fraction at 85 to 100°C/1 mm Hg (120 g). The result was mainly diallyl-trisulfide (40%, n_D^{20} 1.572 to 1.582). The structure was identified chemically: allyl-chloride reacted with sodium thiosulfate to give sodium allyl-thiosulfate **15**, which reacted with sodium sulfide to get sodium allyl disulfide **16**. When **15** reacted with **16**, diallyl-trisin **17** was produced.

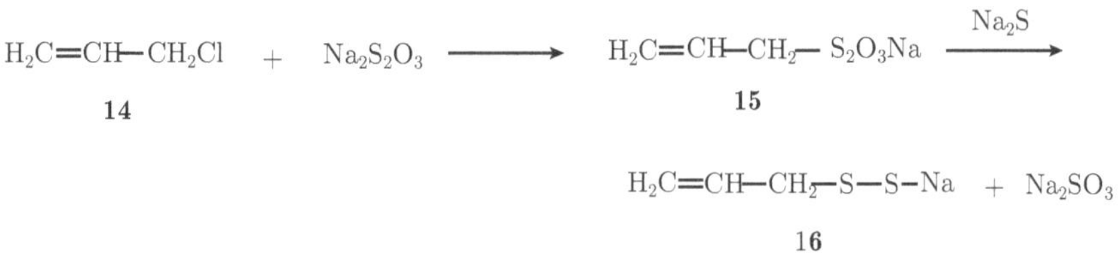

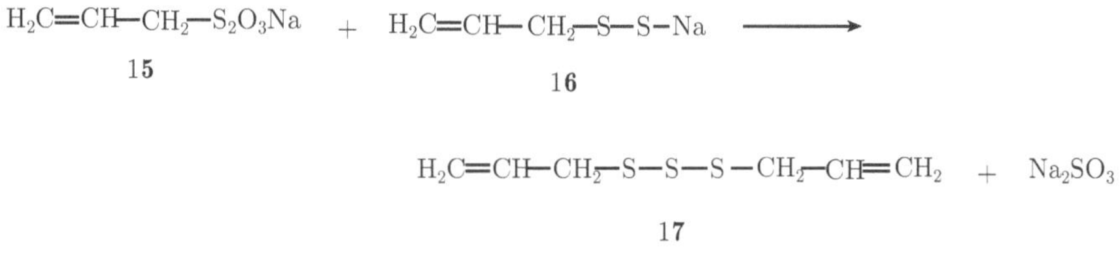

MS of diallyltrisin: m/z 178 (M^+), 146, 118, 105 (CH_2=CH–S_2^+), 73 (CH_2=CH-CH_2–S^+), 41 (CH_2=CH-CH_2^+). ^{1}H-NMR (δgpm): 3.48 (4H, C$\underline{H}_2$–S-), 5.22 (4H, -CH=C$\underline{H}_2$), 5.82 (2H, -C$\underline{H}$=CH$_2$). IR (ν_{max} cm^{-1}): 1830(-CH=CH$_2$), 1000 (=CH-).

Mass spectroscopy is a major tool for analysis of garlic components. In addition to molecular ion peaks, the fragment peaks are useful for structural determination. GC-MS can be used for both separation and structure identification[6]. However, as GC-MS uses high temperature operation, it is not appropriate in case of thermally unstable components. Alternatively, LC-MS provides a more convenient means for analyses of garlic components to a new level. The application of reverse phase HPLC is useful to examine garlic components, such as allyl-systeine-S-sulfoxide, γ-L-glutamyl-S-alk(en)yl-L-systeine, 5-alk(en)yl-L-systeine sulfoxide, and others[7,8].

Injection of diallyl-tristin is used clinically in China for the treatment of *Candida albicans*, *Streptococcus*, and other fungal infections. It is also used for the treatment of respiratory, digestive tract, meningitis, and other bacterial infections.

Deodorized garlic has been gradually developed for industrial production. The common methodology is microwave treatment with pH adjustment using citric acid. The crushed garlic is soaked in a 0.03% solution of alum and adjusted to pH 4 by citric acid. After 2 hours, the mixture is treated with an 850 W microwave for 3 to 4 minutes and then cooled down quickly. This garlic was considered completely deodorized with high retention of active ingredients[9].

Recently a new anticancer seleno-amino acid in garlic, γ-glutamyl-Se-methyl-seleno–systein (GGMSC **18**), was reported by the Clement group. After oral administration, it was enzymolyzed to the known anticancer compound Se-methyl-selenocysteine, MSC **19**, and both **18** and **19** showed the same anti-tumor effect in vivo. GGMSC shows good absorption and selenium distribution to organs after oral administration, and is excreted through urine[10,11].

$$NH\text{–}CO\text{-}CH_2CH_2\text{-}CH\text{-}COOH$$
$$H_3C\text{-}Se\text{–}CH_2\text{-}CH\text{-}COOH \qquad NH_2$$
$$NH_2$$
$$H_3C\text{-}Se\text{-}CH_2\text{-}CH\text{-}COOH$$

18   **19**

The selenium content in natural garlic is about 1×10^{-4} to 3×10^{-4}. However, the selenium content may be up to 1×10^{-3} or more by using selenium based fertilizer during the growth of garlic.

1.2   Bioactive Components in Scallions and Onions

Scallion (*Allium fistulous*) and onion (*Allium sepal*) are common flavoring foods. Both of them contain unstable (+)-S-prop(-1-en)yl-L-sustain-S-oxide **20**, which can be enzymolyzed to lachrymatory prop-1-enylsulfennic acid **21**. Under alkaline conditions, compound **20** becomes an alliin isomer (cycloalliin **22**) which, in acidic conditions (6 M HCl), gave thiazine **23**, 2-methyltaurine **24**, and cysteic acid **25**[12,13]. Compound **21** can be further decomposed to propenol, propanal, and methylpentenal.

In addition, Chinese chives (*Allium fuberosum*) contain alliin and similar sulfur compounds[14].

1.3   Rorifone

Rorifone **26** is a bioactive compound isolated from the Chinese herb *Roripa montana*, Cruciferae[15]. It has antitussive and anticough effectiveness. The structure of **26** was proven by spectral analysis and total synthesis. It is a sulfonyl and cyano-long chain alkyl compound with the molecular formula $C_{11}H_{21}O_2NS$. It is a needle-like crystal, mp 45 to 46°C. Compound **26** is a neutral compound, insoluble in acid and alkali but soluble in alcohol. An orange-red color appeared when it was sprayed with Dragendorff reagent.

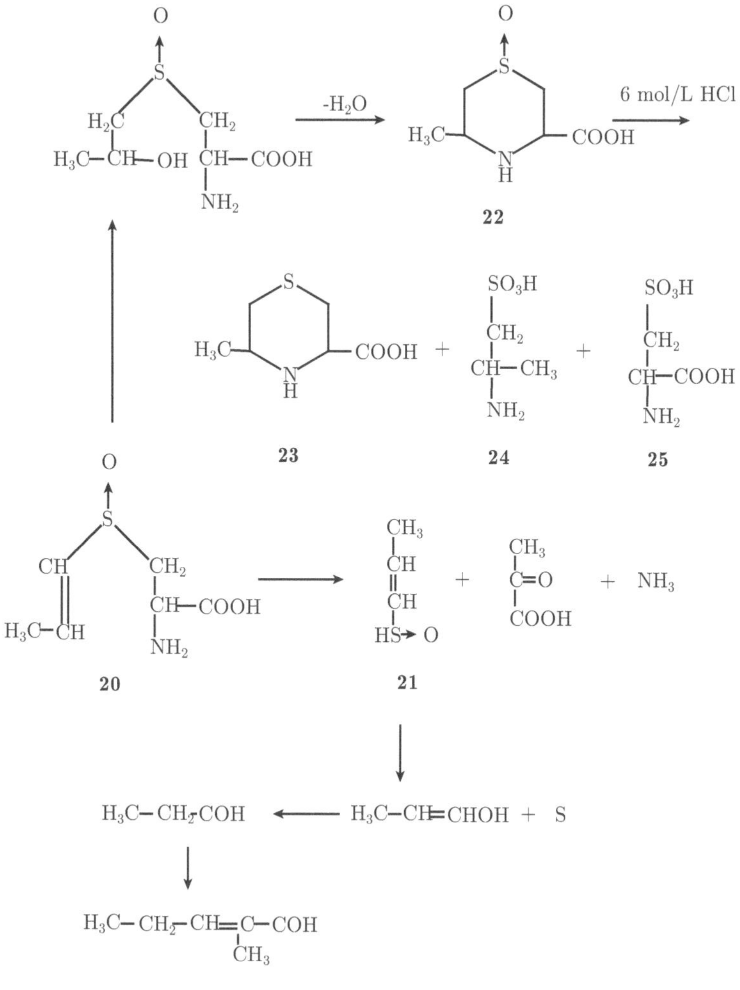

Isolation of rorifone: The whole plant powder was soaked with water, boiled for 1.5 hour, the filtered water extract was collected and the residue was boiled with fresh water for another hour. Water extracts were combined and concentrated under vacuum. After filtration of the concentrated, ethanol was added to reach about 75% and then further concentrated to 1/10 the later volume. An equal volume of water was added and allowed to stand at low temperature. The precipitated material was washed with 0.1 M NaOH twice and then water until neutral, crude rorifone was obtained. It was dissolved in chloroform, filtered, concentrated to dryness, and then crystallized with alcohol. Yield of pure rorifone was about 0.2%. The crude rorifone can be also purified by column chromatography eluted with ethyl acetate. The fraction that exhibited an orange-red color by Dragendorff reagent, was collected, concentrated, and finally crystallized from alcohol.

Spectral data: IR ν_{max} (cm^{-1}): 2240 (-CN), 1312, 1290, 1132 (-SO$_2$). ^{1}H-NMR (δ ppm): 2.90 (3H, s, S-CH$_3$), 3.03 (2H, t, -CH$_2$-CH$_3$), 1.30 to 1.90 (14H, m, -CH$_2$ x 7), 2.34 (2H, t, CH$_2$CN). EIMS (m/z): 232 (M$^+$), 216 (M - CH$_3$)$^+$, 79 (CH$_3$SO$_2^+$), fragments 152, 138, 124, 110, 96, 82, 68, 54 corresponded to 40 (CH$_2$CN) + n x 14 (CH$_2$), but no fragment of

$n = 9 + 40$. This means only 8 CH_2 at the left side of CH_2CN.

In addition to rorifone, rorifamide **27** was also separated from the plant. The molecular formula of rorifamide is $C_{11}H_{23}O_3NS$ and mp is 138°C, but it does not have antitussive and anti-cough activity.

$$H_3C- SO_2- CH_2—(CH_2)_7- CH_2CN \qquad\qquad H_3C- SO_2- CH_2—(CH_2)_7- CH_2- CONH_2$$

26   27

1.4   Other Natural Sulfur Products

Sulfonate and isothiosulfocyanate are reported as anti-inflammatory and analgesic agents. Sinalbin **28** is present in seeds of white mustard, *Brassica* alba[16]. Sinigin **29** is contained in seeds of mustard, *Brassia juncea* and the whole plant of Ji-cai (*Capsella bursa-pastoris*)[17]. The leaves (known as Da-qing-ye) and roots (known as Ban-lan-gen) of Song-lan (*Isatis tinctoria*) are famous Chinese medicine herbs. The major sulfur compounds from this plant are glucobrassicin **30**, neoglucobrasisicin, 1-methoxyl-3-indolylmethyl-glucosinolate **31** and 1-sulfo-3-indolylmethyl-glucosinolate **32**[18]. The plant is used in TCM to treat colds or as antibacterial, anti-inflammatory and antifungal agents. However, the mechanisms and the roles of these sulfur compounds, are unclear.

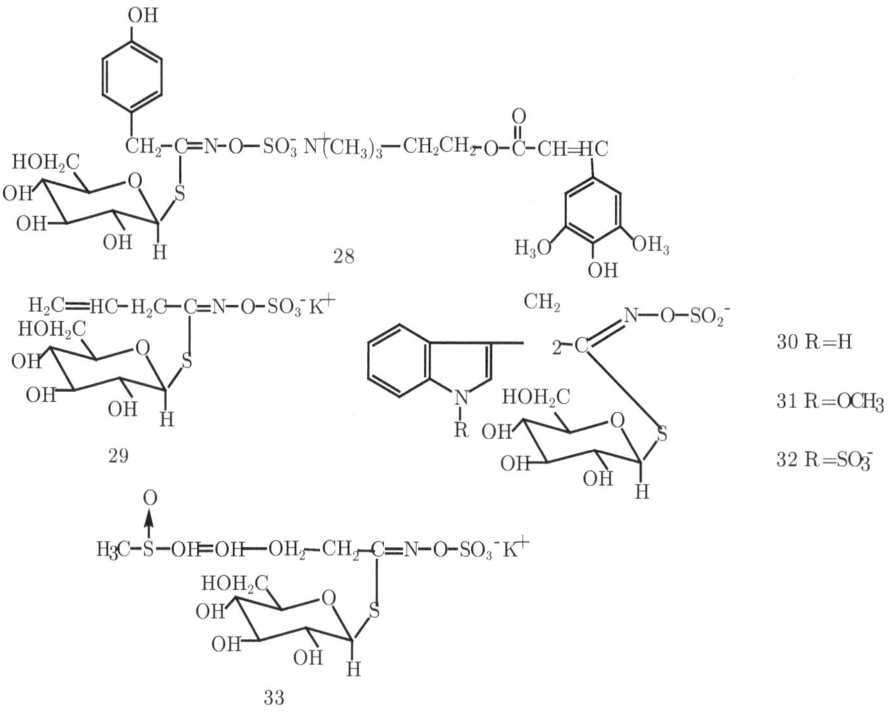

Another sulfur compound, glucoraphenin **33**, was isolated from the roots of radish or Luo-bo (*Raphanus sativus*)[19]. Compound **33** was reported to be effective in boosting cytochrome P-450 (CYP)-associated monooxygenases and the postoxidative metabolism[20].

Section 2   Cyanide Compounds

There are several cyanic natural compounds, many of them conjugated with sugar to form an α- or β-cyanoglycoside. The natural amygdaline or lactril of bitter almond seeds and phyllanthin of *pyllanthus niruri* belong to α–cyanoglycoside, and sarmentosin belongs to β-cyanoglycoside. Most α-cyanoglycosides easily release hydrocyanic acid when treated with acid or enzyme and have toxicity. β-cyanoglycosides are not toxic as they do not usually release hydrocyanic acid. Cyanoglycosides are easily dissolved in water to form hydrates with water, but are only slightly soluble in alcohol and not soluble in chloroform, ethyl ether and other non-polar solvents. The extraction methods of cyanoglycosides are generally the same as that for glycosides and normally involves alcohol to avoid enzymolysis.

Spectral characteristics of cyanoglycoside include a cyan group with a displayed $UVg\lambda_{max}$ at 119 nm, ν_{max} 2250 cm^{-1} in IR, and δ 120 ppm signal in ^{13}C-NMR. The sugar-based signals are similar to those of most glycosides.

2.1   Amygdalin

Amygdalin **34** can be isolated from bitter almond, (*Prunus armenica*), peaches, plums, cherries and apples. It contains two sugars as α-cyanoglycoside, formed from aglycone, mandelnitrile **36** and gentiobiose. It appears as a crystal, with the molecular formula $C_{20}H_{27}NO_{11}3H_2O$, mp 200°C, $[\alpha]_D$ -42° (anhydrous). One glucose was lost via hydrolysis by emulsin to give prunasin **35**, then further enzymolysis yielded mandelnitrile **36** and glucose. Thermal degradation of amygdaline gave benzaldehyde and hydrocyanic acid. Under acidic conditions, amygdaline is decomposed into hydrocyanic acid, benzaldehyde, and glucose. Thus, amygdalin has some toxicity.

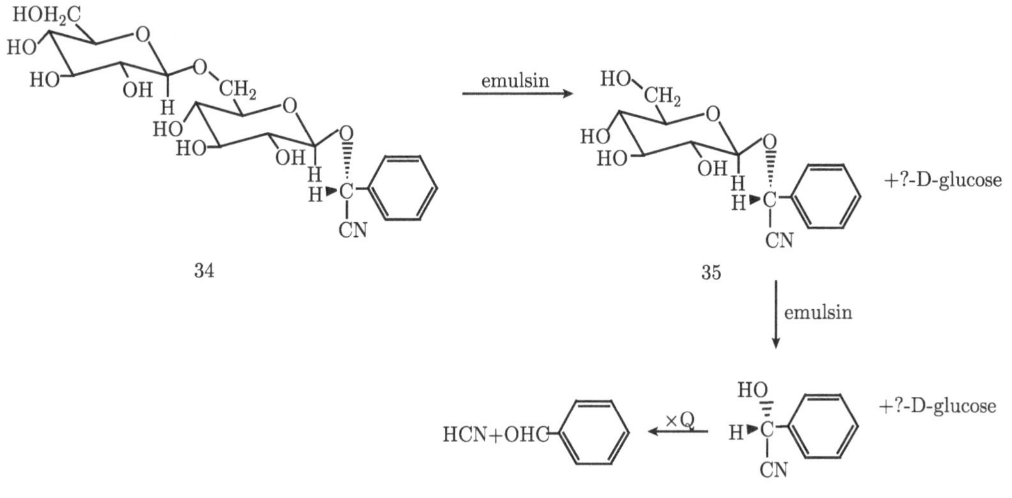

Extraction: After the oil was squeezed out of the bitter almonds, it was extracted with hot alcohol by reflux. The extract was filtered and evaporated to obtain crude amygdaline crystal. The pure compound was obtained by crystallization with ethyl ether. Yield of amygdaline was as high as 3%.

In the early 1950s, amygdaline was declared to have anticancer effects and known as vitamin B$_{17}$ for years. In the 1980s, a study from the American National Cancer Institute (NCI) showed that it had no anticancer effect. Then, the U.S. FDA announced that the anticancer activity of vitamin B$_{17}$ was invalid[21−23].

2.2 Sarmentosin and Isosarmentosin

Sarmentosin **37** is an antihepatitis compound isolated from the Chinese herb *Sedum sarmentosum* (family Crassulaceae). It is an α-cyanoglycoside, colorless, transparent, gummy compound, easily dissolved in water, with the molecular formula $C_{11}H_{17}O_7N$, $[\alpha]_D$ -17.4° (c 0.62, H_2O)[24]. Clinical trials showed it lowers the serum glutamic-pyruvic transaminase (SGPT, or ALT) level. The fact that only glucose was found after acid hydrolysis or enzymolysis indicates that its aglycone is unstable. Sarmentosin is not toxic because no hydrocyanic acid is released due to a double bond connection of the cyanogroup.

Extraction: The ground plants were refluxed with 95% ethanol, the extract was concentrated, dissolved in water, and chloroform was added to remove chlorophyll. The water layer was then concentrated and put on a granular charcoal column, eluted with water, and then 70% ethanol. The ethanol elute was concentrated and put on a silica gel column for separation, eluted with water saturated ethyl acetate and then ethyl acetate-methanol (8:1). The sarmentosin part was collected (positive reaction with α-phenol). The crude sarmentosin was further purified through chromatography on a silica gel column and eluted with ethyl acetate-acetone (3:2). The yield of sarmentosin was about 0.1%.

Spectral data: UV (MeOH) λ_{max} (log ε): 212 nm (4.02). IR ν_{max} (cm^{-1}): 3540 (-OH), 2238 (-CN), 1640(C=C), 110 (-O-). MS (m/z): 276 (M$^+$+ 1), 257 (M$^+$- H_2O). ^{1}H-NMR (C_5D_5N) δ ppm: 6.70 (1H, t, $J = 6$ Hz, = CH), 4.30 (1H, d, $J = 12$ Hz, anomeric proton of glucose), 3.82 to 4.20 (7H, m, -CH$_2$ x 3, C$_{5'}$-H), 4.48 to 4.76 (3H, m, C$_{2'-4'}$-H of glucose), 5.20 to 7.40 (5H, -OH x 5). ^{13}C-NMR (D_2O) δ_c ppm: 103 (d, C$_{1'}$), 74.3(d, C$_{2'}$), 77.2 (d, C$_{3'}$), 70.8 (d, C$_{4'}$), 77.0 (d, C$_{5'}$), 62.0 (t, C$_{6'}$), aglycone part: 63 (t, - OCH$_2$), 145 (d, = CH), 117.4 (s, = C), 68.4 (t, -CH$_2$OH), 117.7 (s, -CN).

Penta-acetyl sarmentosine was produced by acetylation with acetylanhydride and sulfuric acid. The result was a colorless, crystalline needle with mp of 80°C.

Spectral data: MS (m/z): 485 (M$^+$), 443 (M$^+$- COCH$_3$), 426 (M$^+$- OCOCH$_3$). ^{1}H-NMR (CCl$_4$) δ ppm: 1.91, 1.96, 2.01, 2.07 (15H, s, 1:2:1:1, CH$_3$CO x 5), 3.70 (1H, m, C5'-H), 4.13 (2H, d, $J = 5$ Hz, C$_{6'}$-2H), 4.57 (1H, βtC$_{1'}$-H), 4.96 (3H, m, C 2′,3′,4′-H), aglycone part: 4.48 (2H, d, $J = 7$ Hz, -CH$_2$OAc), 4.60 (2H, s, -O-CH$_2$-), 6.54 (1H, $J = 7$ Hz, = CH).

According to the spectral data above, the structure of sarmentisin was determined as **37**, which has been later confirmed by x-ray diffraction analysis of its penta-acetate.

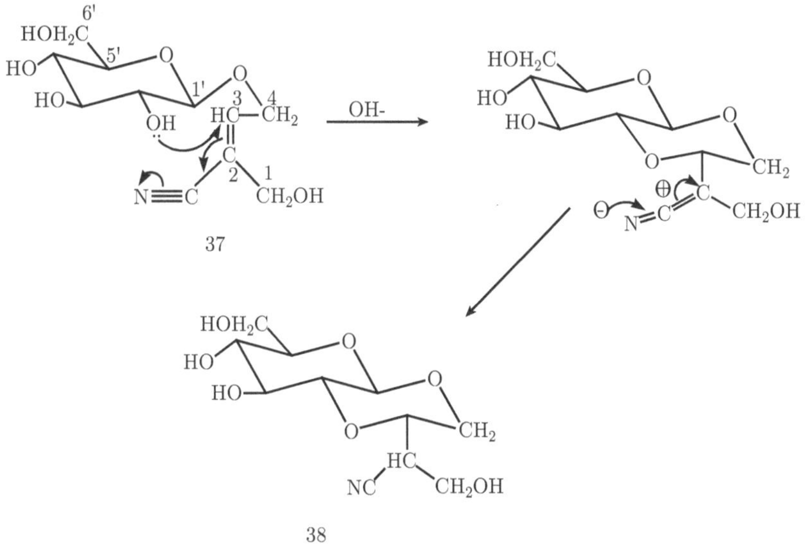

In dilute alkali solution, sarmentosin was isomerized to isosarmentosin **38** at room temperature [$C_{11}H_{17}O_7N$, ethanol crystallization, mp 212°C, $[\alpha]_D$ +51.4° (c, 0.1, H_2O)]. However, there was no bioactivity found.

Isosarmentosin is easily dissolved in water but difficult in organic solvent, stable in acid and decomposed under basic conditions by heating. It has no UV absorption. MS showed molecular ion m/z 275, IR (KBr) ν_{max} (cm^{-1}): 3460, 3270, (-OH), 2230 (-CN), no double bond absorption. ^{1}H-NMR (C_5D_5N, δ ppm): 3.20, 3.48 (1H, m, -O-C<u>H</u>-), no alkene proton signal.

These results indicated that sarmentosin in basic conditions may undergo inter-molecular rearrangement through Michael reaction to form bicyclical isosarmentosin. The proton NMR signal δ 3.20 corresponds to -C<u>H</u> (CN)-CH$_2$OH), and δ 3.48 to –<u>H</u>-CH (CN)-CH$_2$OH. Table 13-1 shows ^{13}C-NMR comparison data.

Table 13-1 ^{13}**C-NMR Data of Sarmentosin and Isosarmentosin**

	Sarmentosin	Isosarmentosin		Sarmentosin	Isosarmentosin
C-1	68.4 t	69.3 t	C-2″	74.3 d	80.0 d
C-2	117.4 s	37.1 d	C-3″	77.2 d	78.8 d
C-3	145.0 d	60.1 d	C-4″	70.8 d	71.0 d
C-4	63.0 t	72.0 t	C-5″	77.0 d	73.8 d
C≡N	117.7 s	120.0 s	C-6″	62.0 t	61.7 t
C-1′	103.0 d	98.2 d			

Section 3   Coixenolide

Coixenolide is a yellow oil isolated from the seeds of the Chinese herb Yi-Ren (*Coix lachrymajobi*, family Graminaceae). It showed anti-cancer activity in animal tests.

Extraction and isolation[25]: 2 kg of the seed powder was extracted three times with 5 L acetone at room temperature. The extracts were combined and evaporated to get 86.6 g reddish brown syrup which was then dissolved in petroleum ether, filtered and concentrated to get 85 g, yield 4.25%; 20 g of the latter syrup was dissolved in petroleum ether and put on a silica gel column for chromatography and successively eluted with 1 L petroleum ether, 500 mL of ethyl ether, acetone, ethyl acetate, and methanol. Evaporation of the solvents gave 3.75 g, 13g, 1.5 g, 0.6 g and 0.1 g extracts, respectively. Ehrlich ascites sarcoma in mice tests showed that only the petroleum ether fraction was active. Fifteen grams of the active part was dissolved in petroleum ether and washed twice with 2 M KOH solution to get an acidic portion (11.7 g) and an alkali-insoluble neutral portion (3 g). Tests on mice showed that the neutral part was active but the acidic part was weak in activity. The neutral active part was then dissolved in 10 mL chloroform and put on a 1% bromothymol blue solution-soaked aluminum oxide column and eluted with chloroform. The elution was continued until the yellow band moved to the end of column, and then eluted with acetone until the color turned blue. The chloroform elute was neutral (2.4 g) and the acetone elute (0.52 g) was acidic. Both are oils. The neutral oil (250 mg) was dissolved in 6 mL petroleum ether, put on a silica gel column for chromatography, and followed by step-wise elution using ethyl ether-petroleum ether as solvent, 4:96 (100 mL), 10:90 (100 mL), 50:50 (200 mL) and separated to three parts. The first two parts were combined to get A and the last part is B. Ehrlich ascites carcinoma tests in mice showed B was an active ingredient known as coix-ester **39**.

Structural elucidation[26]: Coix-ester **39**, molecular formula $C_{38}H_{70}O_4$ determined by elemental analysis, n_D 1.4705, $[\alpha]_D^{20}$0°. Palladium carbon hydrogenation of compound **39** gave a tetrahydro-compound **40** by absorption of two molecules of hydrogen. Compound **40** is a colorless crystal with mp 56 to 61°C. Hydrolysis of **40** in 0.1 M NaOH/methanol under heat gave succinic acid **41** and crude crystals of mixed organic acids (mp 56 to 61°C). Crude crystals were further separated by RP Kieselguhr chromatography to give equivalent molecules of palmitic acid **42** (mp 63°C) and steric acid **43**. When compound **39** was directly hydrolyzed by heating with ethanol KOH without hydrogenation, an oil mixture of acids and 2,3–butanediol was obtained. Fractionation of the mixture yielded acid A (bp 107°C, $C_{16}H_{30}O_2$) and acid B (bp 125°C, mp 37°C, $C_{18}H_{34}O_2$). Both A and B were identified by preparation of their β-bromophenacyl esters, compared with standards, and found to be *cis*-palmitoleic acid **44** and *trans*-vaccenic acid **45.**

IR ν_{max} (cm^{-1}): Except ester absorption at 1730, a peak at 960 (trans-vinyl vibration absorption) showed up in **39**. The peak at 960 cm^{-1} in compound **45** appeared even stronger, but was not observed from the spectra of compounds **40** and **44**.

The TCM medicine known as Kanglaite injection is used clinically in China for treatment of cancer and was prepared with coix-oil. However, it is not clear whether the effective component is the coix-ester or other ingredient.

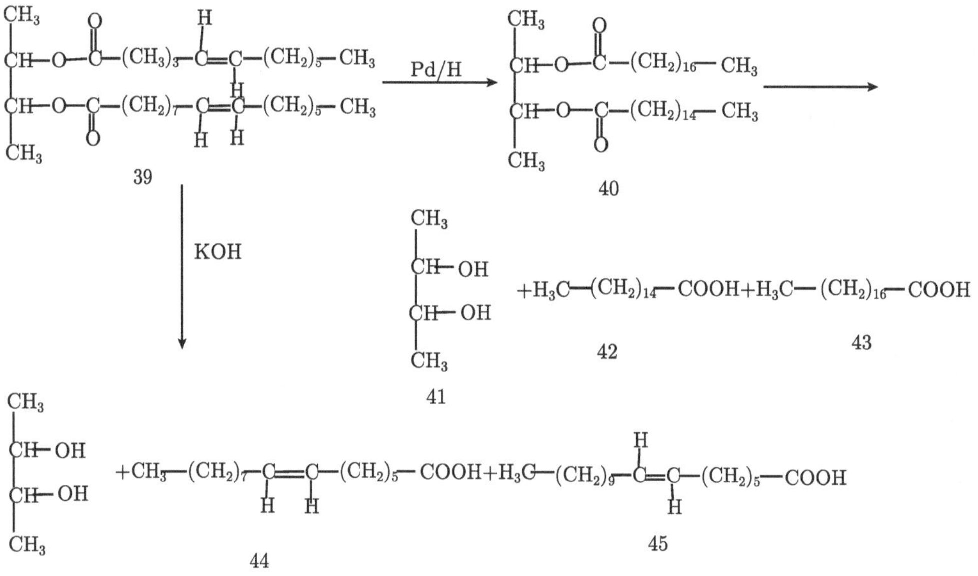

Section 4   Resveratrol

Resveratrol **46** is a stilbene compound, isolated from grape seeds or pericarp and its structure was identified as 3′,4,5′-trihydroxyl-stilbene, $C_{14}H_{12}O_3$, light yellow crystal, mp 265 to 267°C[27].

IR ν_{max} (KBr, cm^{-1}): 3200 to 3300 (OH), 1630 (>C=C<), 1600, 1595, 1510 (aromatic), 965 (trans-CH=CH-).

^{1}H–NMR (d_6-acetone, δ ppm): 6.28 (1H, t, J = 2 Hz, C4′-H),), 6.55 (2H, d, J = 2 Hz, C2′, 6′-H), 6.76 (1H, d, J = 17 Hz) and 7.08 (1H, d, J = 17 Hz, *trans*-CH=CH-), 6.82 (2H, d, J = 9 Hz) and 7.40 (2H, d, J = 9 Hz, aromatic Hs, A_2B_2 type, q), 8.23 (3H, brs, 3 × OH).

^{1}H-NMR (d_6-acetone, δ ppm) of triacetyl resveratrol: 2.23 (9H, S, 3 × OAc), 6.38 (1H, t, J = 2 Hz, C4′-H), 7.20 to 7. 40 (4H, aromatic Hs and —CH = CH—), 7.08 (2H, d, J = 9 Hz) and 7.60 (2H, d, J = 9 Hz, aromatic Hs, A_2B_2 type, q).

Research showed compound **46** has anti-cancer and anti-oxidant effects. Its content in red wine is more than in white wines and, thus, moderate red wine consumption is helpful to prevent aging and cancer. In addition to grape seed, TCM herbs such as Hu-zhang (*Polygonum cuspidatum*) and Da-huang (*Rheum palmatum*) also contain resveratrol.

Extraction: Ground powder of *Scirpus fluviatiles* roots (4.8 kg) was refluxed with n-hexane four times, and then with methanol five times. The methanol extract was concentrated to yield a brown-black gel (550 g), which was dissolved in 2 L of water and repeatedly extracted with butanol (total 8 L). The butanol extract was concentrated (26 g) and then purified by silica gel column chromatography eluting with chloroform. Around 0.13 g of betulinic acid was obtained. The chloroform insoluble part (10.3 g) was separated by silica gel column chromatography and eluted with a mixture of benzene–ethyl ether. Resveratrol (0.81 g) and 3,3′,4,4′ four hydroxyl stilbens were separated after elution with benzene-ethyl ether (6:4).

There are many synthetic methods to obtain resveratrol and its analogs. The following is an example.[28,29].

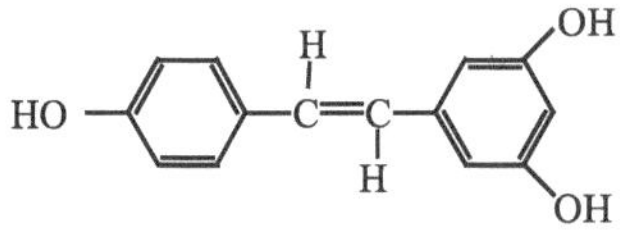

46

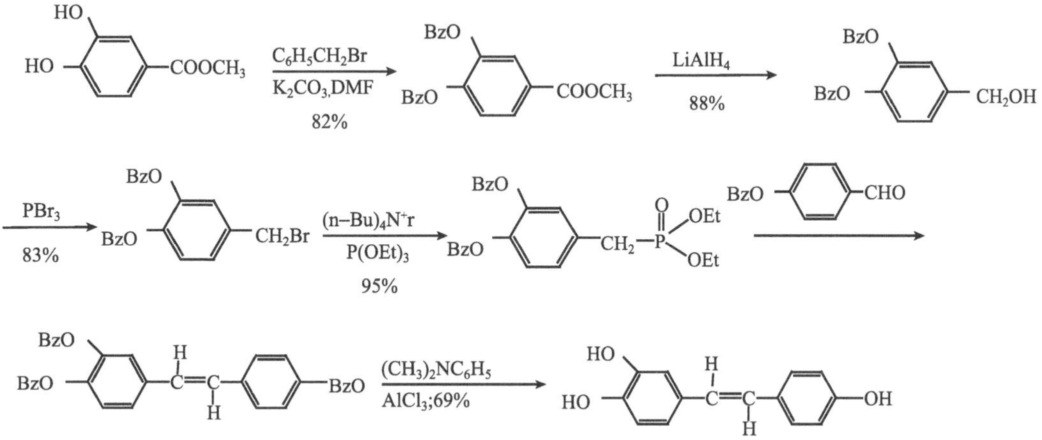

Section 5   Muscone

Muscone **62**, 3-methyl-cyclopentadecanone, is the major and active ingredient in an important Chinese medicine She-xiang with content of about 1.5 to 2%. She-xiang is obtained from the dried secretions of mature male musk (*Mouschcus berezovskii*). It has a strong aromatic smell. In TCM it is used for treatment of coronary heart disease, angina, and hepatic encephalopathy. It is a main component of a famous TCM formula Liushen pill (six-miracle pill), used for treatment of diphtheria, tonsilitis, laryngitis and other infectious diseases. Pharmacological experiments showed muscone has central nervous excitement (low dose) and inhibition effects (high dose), as well as anti-inflammatory, cardiotonic, anti-cancer and anti-pregnancy effects. In addition to **62**, She-xiang also contains protein, amino acids, lecithin, androstanol, cholesterol, urea, ammonium carbonate and other components.

Although muscone is the main component of She-xiang, it cannot be substituted for the latter. Recently, several research institutions in China have accomplished studies on the production of artificial She-xiang. It contains muscone and several other components, and appears that it can be substituted for natural She-xiang in clinical uses.

Muscone is a ring structured 3-methyl-cyclopentadecanone with the formula $C_{16}H_{30}O$. Natural muscone has a negative optical rotation $[\alpha]_D$ -11.3° (c 0.91, MeOH), bp 328°C, bp$_{0.5}$ 130°C, d_4^{17} 0.9221, n_{17}^D 1.4802, strong smell, does not dissolve in water and can be dissolved in alcohol.

Musk sources are limited in nature and the structure of muscone is not complicated. There are several synthetic methods for muscone which have been reported in literature. Stoll et al. reported starting from 15-hexadeca-dione, cyclolized to dehydromuscone, then hydrogenated to muscone[30]. There is also a report of starting from cyclododecanone through expansion into pentadecanone and then to muscone[31]. Wang et al.[32] reported a six-step synthesis of (±) muscone and nine-step synthesis of (R)-(-)-muscone by using cyclodecanone, the readily available petrochemical product, as starting material. The key step is azobisisobutyronitrile (AIBN)- and tri-n-butyltin hydride (Bu$_3$SnH)- induced free radical expension of α-ethoxycarbonyl-αt(2″-methyl-3″-phenyl-selenopropyl) cyclododecanon.

The synthesis of racemic muscone DL-**52** and optically active L-**62** is described below.

Preparation of 2-ethoxycarbonyl-one (**48**) from cyclododecanone (**47**) was performed first, then reacted with 2-methyl-1-benzo-selenium-3-iodo-propane **49** in a solution of benzene and NaH to obtain compound **50**, which was refluxed with AIBN and Bu$_3$SnH in benzene solution. The free radical was expanded to 3-methyl-5-ethoxycarbonyl-cyclo-decapentanone **51**. Next, it was hydrolyzed in concentrated hydrochloric acid, catalyzed by cupric acetate and finally oxidized with lead acetate for decarboxylation. After hydrogenation, racemic muscone **52** was obtained with total yield of 51%.

Preparation of compound **49**: 2-methyl-1,3-propanediol **53** was reacted with tosylsulfonyl chloride to give p-tosylsulfonate, reacted with sodium iodide to get iodide **54** and then reacted with diphenyl selenium and sodium borohydride to give **49**.

Racemic muscone **52**, colorless oil, C$_{16}$H$_{30}$O, bp$_{0.01}$ 90°C, n_D^{26} 1.4767.

MS: m/z 238 (M $^+$, 35%), 223 (M - CH$_3$)$^+$, 209 (M - C$_2$H$_5$)$^+$, 195 (M - C$_3$H$_7$)$^+$, 180 (M - C$_2$H$_5$CO)$^+$, 125 (C$_7$H$_{13}$CO)$^+$, 97 (C$_5$H$_9$CO)$^+$, 86 ([CH$_2$-C (OH) C$_3$H$_7$]$^+$, 100%), 69 (C$_3$H$_5$CO)$^+$, 55 (C$_2$H$_3$CO)$^+$, 41 (C$_3$H$_5$)$^+$. ^{1}H–NMR(δ ppm): 0.93 (3H, d, $J = 6.5$ Hz, 3-CH$_3$), 1.45 [23H, m, (4 -14) CH$_2$-, 3-H], 2.30 (4H, m, 2-CH$_2$, 15-CH$_2$). IR (ν_{max} cm^{-1}): 1700 (-CO), 1110 (-C-CO-C). UV: λ_{max} nm (log ε): 285 (1.54).

Preparation of optically active muscone **62**: Reaction of optically active (S)-2-methyl-3-bromine propan-1-ol **55** with dihydropyran gave (S)-1-(2-methyl-3–bromopropyl) 2-tetrahydro pyranyl ether **56**. The next step is preparation of compound **60**. Reaction of compound **48** with **56** gave compound **57**, which was reacted with pyridinium p-toluene sulfonate (PPTS) to remove the tetrahydropyran protection group to give compound **58**. Then it was tosylsulfonated and iodo-exchanged to give iodine-containing compound **59**. Finally, substitution of iodine by benzene selenium yielded the benzene selenium compound **60**. Treatment of **60** with AIBN and Bu$_3$SnH generated the expanded cyclic compound, 15-cyclo-ketone ester **61**. The latter, through hydrolysis, decarboxylation, and hydrogenation, gave optically active muscone **62**, with total yield 32%.

58 X = OH
59 X = I
60 X = SePh

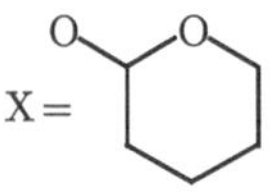

57 X =

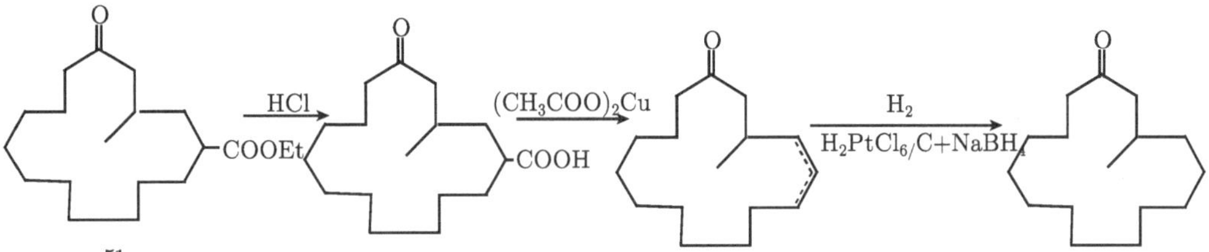

51   52

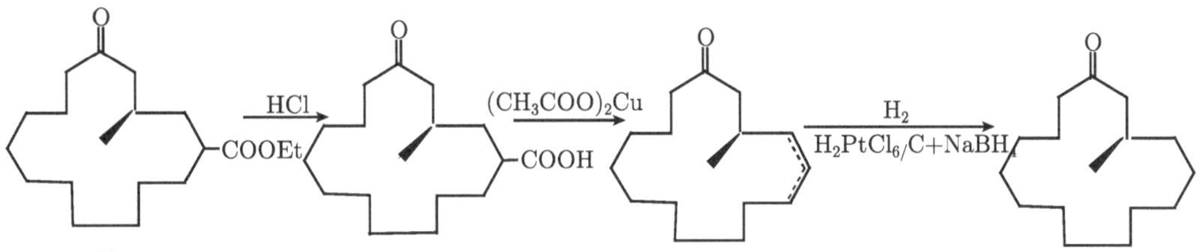

61   62

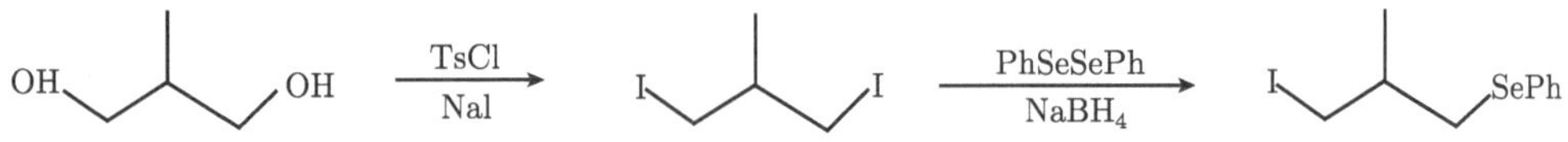

53   54   49

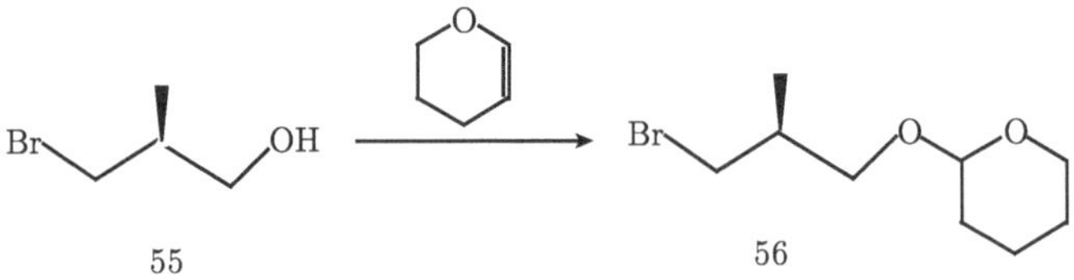

55   56

MS and IR spectra of (R)-muscone were the same as with the racemate. ^{1}H-NMR (δ ppm): 0.93 (3H, d, J = 6.5 Hz, 3-CH$_3$), 1.5 [23H, m, (4 -14)-CH$_2$-, 3-H], 2.24 (4H, m, 2-CH$_2$, 15-CH$_2$).

Section 6   Cardiac Glycosides

6.1   Introduction

Cardiac glycosides are steroidal glycosides that are able to enhance the contraction of heart muscle. Their molecules have the common steroid skeleton with the C-17 unsaturated five-or six-member lactone ring, and C-3 linked with various six-carbon sugars.

In 1869, Nativelle first isolated the cardiac glycoside digitalin from purple digitalis (*Digitalis purpurea*). In 1935, Stoll found that the earlier isolated cardiac glycoside was not the original, but hydrolyzed by an enzyme to form secondary glycosides. Their cardiotonic activities are much weaker than original glycosides[33]. By suppressing the enzymatic action, they first isolated the original purpurea glycosides A and B from *Digitalis purpurea*, and then successively isolated many original cardiac glycosides from *Digitalis lanata*, *Strophanthus kombe*, sea green onions (*Scilla maritima*) and *Nerium indicum*.

63 $R_1 = R_2 = R_3 = H$, $R_4 = CH_3$

64 $R_1 = OH$, $R_2 = R_3 = H$, $R_4 = CH_3$

65 $R_1 = R_3 = H$, $R_2 = OH$, $R_4 = CH_3$

66 $R_1 = R_2 = H$, $R_3 = OH$, $R_4 = CHO$

67 $R_1 = OCOCH_3$, $R_2 = R_3 = H$, $R_4 = CH_3$

68   69   70   71

Most cardiac glycosides are distributed in the seeds, roots, stems, and leaves of plants from the Scrophulariaceae, Apocynaceae, Asclepiadaceae, and Liliaceae families. Generally, the highest levels of cardiac glycosides are contained in seeds. Cardiac glycosides are also contained in TCM herbs, such as roots of *Rohdea japonica* (Liliaceae) and seeds of *Strophanthus divaricatus* (Asclepiadaceae).

The most commonly used cardiac glycosides in clinics are digoxin, cedilanid (or deacetyl-lanatoside C), K-strophanthin, and convallatoxin for treatment of acute and chronic congestive heart failure and rhythm disorders. Cedilanid and digoxin are the most useful. The physical data of common cardiac glycosides are listed in Table 13-2.

Table 13-2   Physical Data of Common Cardiac Glycosides

Cardiac glycoside				Aglycone				Sugar component	Source
Name	M.F.	M.P. (^{o}C)	$[\alpha]_D$	Name	M.F.	M.P. (^{o}C)	$[\alpha]_D$		
purpurea glycoside A	C_{47} H_{74} O_{18}	280	10.8^o (75% EtOH)	digitoxigenin **63**	C_{23} H_{34} O_4	250	19.1^o (MeOH)	3D-digitoxose −D-glucose	*D. purpurea*
lanatoside A	C_{49} H_{76} O_{19}	248	31.1^o (95% EtOH)	digitoxigenin **63**				2D-digitoxose -D-3-acetyl -digitoxose -D-glucose	*D. lanata*
purpurea glycoside B	C_{47} H_{74} O_{19}	242	15.6^o (75% EtOH)	gitoxigenin **64**	C_{23} H_{34} O_5	235	38.6^o (MeOH)	3D-digitoxose) -D-glucose	*D. purpurea*
lanatoside B	C_{49} H_{76} O_{20}	248	36.7^o (95% EtOH)	gitoxigenin **64**				2D-digitoxose -D-3-acetyl -digitoxose— D-glucose	*D. lanata*
lanatoside C	C_{49} H_{76} O_{20}	248	33.4^o (95% EtOH)	digoxigenin **65**	C_{23} H_{34} O_5	222	27.0^o (MeOH)	2D-digitoxose -D-3-acetyl- digitoxose-D -glucose	*D lanata*
deacetylanato- side C cedilanide	C_{47} H_{47} O_{19}	268	12.0^o (75% EtOH)	digoxigenin **65**				3D-digitoxose -D-glucose	*D. lanata*
dogoxin	C_{41} $H_{64}O_{14}$	265	13.3^o (py.)	digoxigenin **65**				3D − digitoxose **69**	*D. lanata*
K- strophanthoside	C_{42} H_{64} O_{19}	220	13.9^o (MeOH)	strophanthidin **66**	C_{23} H_{32} O_6	232	43.1^o (MeOH)	D-cymarose **70**-D-glucose	*strophanthus kombe*
K- strophathin	C_{36} H_{54} O_{14}	195	32.6^o (H_2O)	strophanthidin **66**				D-cymarose -D-glucose	
convallatoxin	C_{29} H_{42} O_{10}	242	-15° (py.)	strophanthidin **66**				L-rhamnose	*Convallaria majalis*
oleandrin	C_{32} H_{48} O_9	250	-52.1^o (MeOH)	oleandrigenin **67**	C_{23} $H_{36}O_5$	225	-8.5^o (MeOH)	L-oleandrose **71**	*Nerium indicum*
scillaren A	C_{36} H_{52} O_{13}	270	-71.9^o (MeOH)	scillarenin **68**	C_{24} H_{32} O_4	238	-16.8^o (MeOH)	L-rhamnose– D-glucose	*Scilla maritinuma*

6.2   Characteristics and detection of cardiac glycosides

Cardiac glycosides are neutral, colorless, crystalline, bitter, and optically active. Most cardiac glycosides are dissolved in ethanol or methanol, and some are water soluble, but difficult to dissolve in ether, chloroform and other non-polar solvents. The more hydroxyl groups contained in the molecule, the more hydrophilic the compound is. Likewise, the more sugars contained in the molecule, the more soluble it is in water. At room temperature, cardiac glycosides are relatively stable. However, in acid or enzyme conditions, they can be hydrolyzed or isomerized. According to the strength of the acid, they are hydrolyzed into glycoside with one or more sugars or aglycone. Characteristics of cardiac glycosides are related to their structures. The β-unsaturated lactone ring is the main bioactive functional group. The linking of sugar increased their solubility, absorption and excretion. Under normal circumstances, the more sugar in the molecule the more effective and less toxic the compound is.

Spectral analyses are the major tools used to identify their structures. Additionally, there are some commonly used reagents which can be used to determine the steroid structure, α, β-unsaturated ketone, and deoxysugars.

6.2.1   Lieberman-Burchard reaction

The sample is dissolved in a small amount of acetic anhydride and put in a test tube. If the sample is not quite soluble in acetic anhydride, a small amount of ethanol can be used to dissolve the sample, and then the solvent is evaporated and acetic anhydride is added. The addition of concentrated sulfuric acid along the wall will cause a purple ring to appear between the two layers, and a blue color in the acetic anhydride layer. This is a proof that the sample contains a steroidal moiety.

6.2.2   Kedde reaction

The sample solution is dropped onto a filter paper, then the Kedde reagent is added (1 g 3,5-dinitrobenzoic acid dissolved in 50 ml methanol, followed by the addition of 50 mL 1 M KOH). Purplish-red spots indicate that the sample contains an α,β-unsaturated lactone.

6.2.3   Keller-Kiliani reaction

The sample is dissolved in a test tube with 0.5% $FeCl_3$/acetic acid solution. Then, by adding concentrated sulfuric acid along the wall, a brown or other color will appear between the two liquid layers and a blue color in the acetic acid layer. This indicates that the sample contains a 2-deoxy sugar.

6.3   Extraction and Separation of Cardiac Glycosides

Extraction and separation of cardiac glycosides is complex and difficult, because their contents in plants are usually low (below 1%) and often mixed with similar compounds, such as saponins. For example, digitalis leaf contains digitalis saponins (digitonin). Additionally, a plant always contains dozens, of similar cardiac glycosides and every glycoside may have original and secondary glycosides (partially produced in the process of extraction by hydrolysis or enzymolysis) and aglycone. All of these factors increase the difficulty of separation and purification work. Therefore, repeated chromatographic separation may be necessary to obtain a pure compound. At the end of this chapter we will give an example of our own commercial method of isolation and separation of cardiac glycosides from *Digitalis lanata*.

6.4   The Structures of Cardiac Glycosides

Cardiac glycosides consist of aglycones and sugars. Aglycones can sometimes be isomerized and sugar moiety can be hydrolyzed by acid or enzyme. As many plants already contain enzymatic or isomeric enzymes, the literature in the past often gave conflicting reports on the isolated cardiac glycosides. Now these issues have been resolved. Below, we will discuss the structures of cardiac glycoside aglycones and sugars.

6.4.1   Sugar moiety of cardiac glycosides

Most cardiac glycosides can use the following general formula:

(1) R-O-(D-O-) $_{1-3}$ -(end-glucose-O-) $_{1-2}$-H

(2) R-O-(D-O-) $_{1-3}$-H

(3) R-O-(end-glucose-O-)-H

R: aglycones, D: deoxysugar

Cardiac glycosides are different from other glycosides, and in addition to containing common glucose and rhamnose, they often contain many unique deoxysugars. From the above formula, a deoxysugar is always connected directly with the aglycone, then glucose. In some cases, the aglycone is directly linked with only one deoxysugar or one glucose. So far, it has been found that all deoxysugars in cardiac glycosides are hexomethylose (6-deoxysugar or methyl-pentose), either C-3-hydroxymethylated sugars, or C-2-reduced sugars (2-deoxy-methyl-hexose, 2,6-dideoxysugar, 2-deoxymethyl pentose) or both. Besides, some sugar moieties contain acetyl groups. In addition to common deoxysugars, such as digitoxose **69**, cymarose **70** and eoleandrose **71** (shown in Table13-2), other examples include thevetose **72**, sarmentose **73** and diginose **74**.

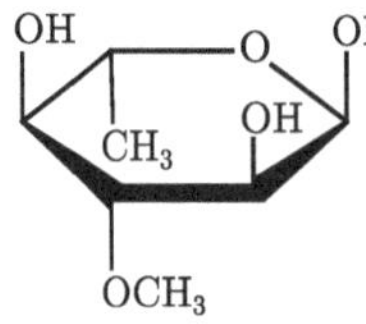

72 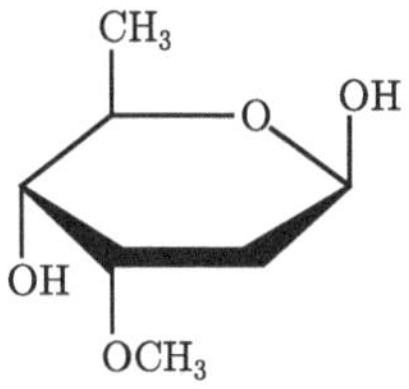   73 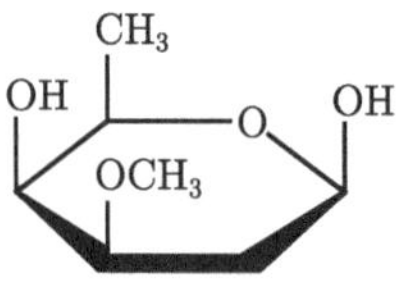   74

6.4.2   Aglycones of cardiac glycosides

Aglycones of cardiac glycosides have a steroid skeleton with C-17 linked with a five- or six-unsaturated lactone ring in β-position, the β-OH on C-3 linked with hexose, and C-10 mostly occupied by a β-methyl or CHO or CH$_2$OH group. Additionally, there are one to three hydroxyl groups in each molecule, generally at the C-3, C-14, and C-16 positions. A small number of compounds also have a hydroxyl group at C-5. In general, the A/B ring is *cis*, B/C ring is *trans*, and C/D is *cis* fused.

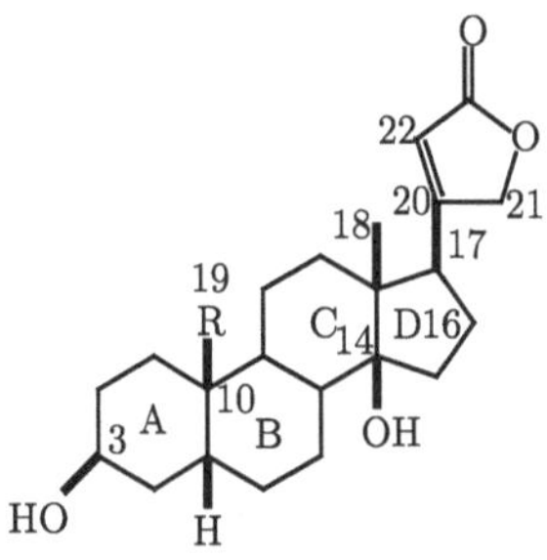

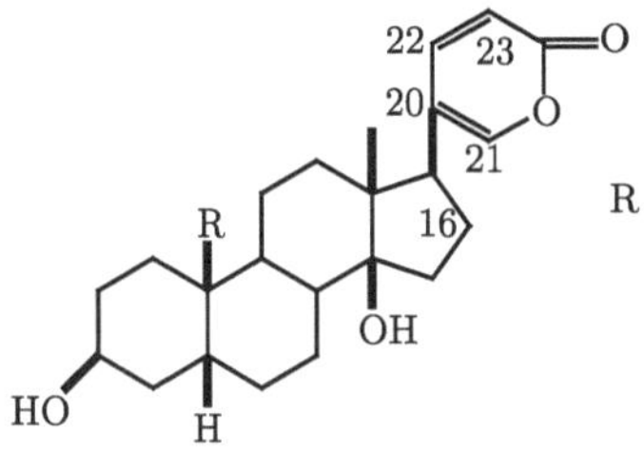

R = CH$_3$, CH$_2$OH, CHO

The lactone ring is opened under alkali conditions and closed with acid, but when opened by alcoholic alkali, it will not be closed by acid but rearranged to an isomer. The double bond moves from position α, β to β, γ and is then linked with C-14 OH to form an ether ring **75**.

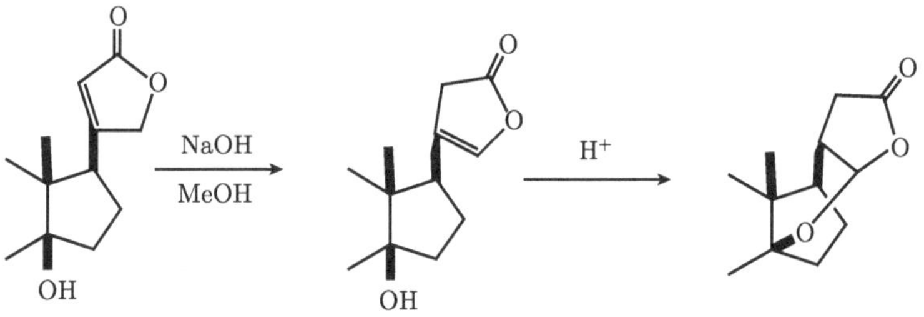

Sometimes, in addition to the enzymes that hydrolize the glycosidic bonds, the plants also contain enzymes which isomerize the glycoside to an allo form with no cardiotonic activity. In this case, the lactone ring is in α position and cannot form the above ether ring in the alcoholic base. This is also circumstantial evidence that the C-17 lactone ring and C-14-hydroxyl group are *cis*-linked and both are in β positions.

The same phenomenon can be seen for double unsaturated bonds of hexalactone: ring opened under base and closed with acid. However, in the alcohol alkali condition it will form an ester (in case of methanol it will form methyl ester). The free enol loses water and is then cyclolized to an isomer of an oxygen ring compound **76** with a C-14 hydroxyl group.

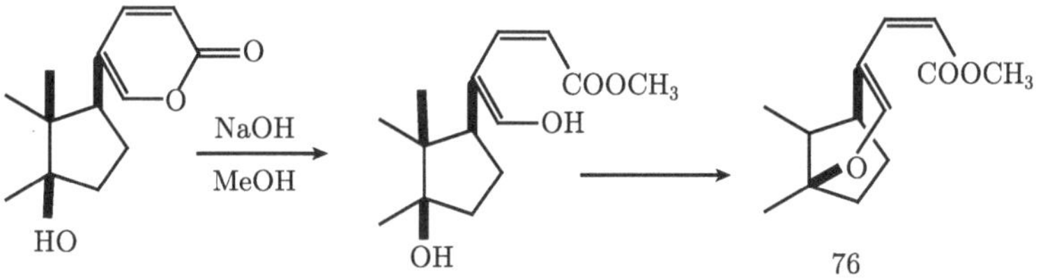

6.5 Spectroscopy of Cardiac Glycosides

UV (λ_{max} nm): a strong absorption peak at 220 indicates α, β-unsaturated γ-lactone ring and an α, β, γ, δ unsaturated δ-lactone ring shows a strong absorption peak at 300.

IR ν_{max} (KBr cm^{-1}): in addition to the general hydroxyl absorption peak near 3500, an unsaturated γ-lactone ring showed two strong absorption peaks at 1720 and 1780. Additionally, two strong absorption peaks were observed at 1718 and 1740 for the double unsaturated δ-lactone ring. Acetyl-containing compounds, such as oleandrigenin **67** and acetyl-digitoxose in lanatosides B and C, have one more acetyl absorption peak near 1760. Aldehyde compounds, like K-strophanthin or its aglycone **66**, have one more aldehyde absorption peak near 1725.

^{1}H-NMR (C$_5$D$_5$N, δ ppm): two tertiary methyl group signals are found near 1.00, with the C-18 methyl signal usually lower than C-19-methyl. The C-3 proton signal is multiple at 3.50, and when it links with sugar, the proton signal will be lower than 0.2. If the C-19 methyl was replaced by an aldehyde, such as **66**, its aldehyde proton signal (single peak) will be at 10.00. For aglycones, the C-17 proton signal is near 2.80 (multiple) and coupled with two C-16 protons (near 2.50). Alkene proton signals of a γ-unsaturated lactone ring are at

5.80 and its two methylene signals are at 4.8 and 5.1 (both are double with $J = 12$ Hz). Three proton signals of the δ-unsaturated lactone ring are at the range of 5.00 to 6.00. The sugar C-1' proton is a doublet, and their J value depends on the C-2' proton circumstances. When the two protons are vertical coupling (such as glucose), $J = 6$ to 8 Hz, and if the two equatorials are coupling (rhamnose), $J = 1$ to 2 Hz. The rhamnose methyl signal is in the vicinity of δ 1.50, doublet, $J = 6$ Hz. The number of glycoside sugars can be predicted by the number of anomeric proton signals. ^{13}C-NMR (δ ppm): The data of most aglycones steroidal molecules and common sugars can be found from reference books. The following tables list data of cardiac glycosides **70** through **74**, their sugar moieties and aglycones[34].

Table 13-3 ^{13}C-NMR Data of 70 through 74 Methyl Sugar Compounds

	70	71	72	73	74
C-1'	97.6	95.9	98.9	97.3	98.2
C-2'	36.4	35.8	73.8	33.6	33.1
C-3'	78.7	79.3	84.8	80.3	79.1
C-4'	74	77.1	76.6	67.9	67
C-5'	71.1	69.1	68.9	69.9	71.2
C-6'	18.9	18.6	18.5	17.5	17.6
OCH$_3$	58.1	56.9	60	56.7	55.1

Table-13-4 ^{13}C-NMR Data of Aglycones 66 to 69

C	66	67	68	69	C	66	67	68	69
1	30	30	30	24.8	13	50.3	50.4	56.4	50.1
2	28	28	27.9	27.4	14	85.6	85.2	85.8	85.3
3	66.8	66.8	66.6	67.2	15	33	42.6	33	32.2
4	35.5	33.5	33.3	38.1	16	27.3	72.8	27.5	27.5
5	35.9	36.4	36.4	75.3	17	51.5	58.8	46.1	51.4
6	27.1	27	26.9	37	18	16.1	16.9	9.4	16.2
7	21.6	21.4	21.9	18.1	19	23.9	23.9	23.8	195.7
8	41.9	41.8	41.3	42.2	20	177.1	171.8	177.1	177.2
9	35.8	35.8	32.6	40.2	21	74.5	76.7	74.6	74.8
10	35.8	35.8	35.5	55.8	22	117.4	119.6	117	117.8
11	21.7	21.9	30	22.8	23	176.3	175.3	176.3	176.6
12	40.4	41.2	74.8	40.2					

MS: mass spectra of cardiac glycosides are more complex, as their molecular ion peaks are weaker and less characteristic. In addition to the general molecular ion peaks, there are peaks of M-18, M-36 which indicate the presence of hydroxyl groups.

6.6   Cardiac Glycosides of *Digitalislanata*

As described before, scientists have isolated cardiotonic purpurea glycosides A and B and more than 30 other cardiac glycosides from the leaves of *Digitalis purpurea*. Later it was found that the cardiotonic activity from the leaves of *Digitalis lanata* is four times stronger than that of the leaves of *Digitalis purpurea*. Further studies showed the difference is that lanatoside A and B have one more acetyl group on the C-3' position of the third digitoxose than in purpurea glycosides A and B. In addition, they contain the unique digitalis glycoside C (lanatoside C, **77**) as the main component, which can act quickly and be excreted rapidly and, thus, *Digitalis lanata* replaces purple digitalis in application.

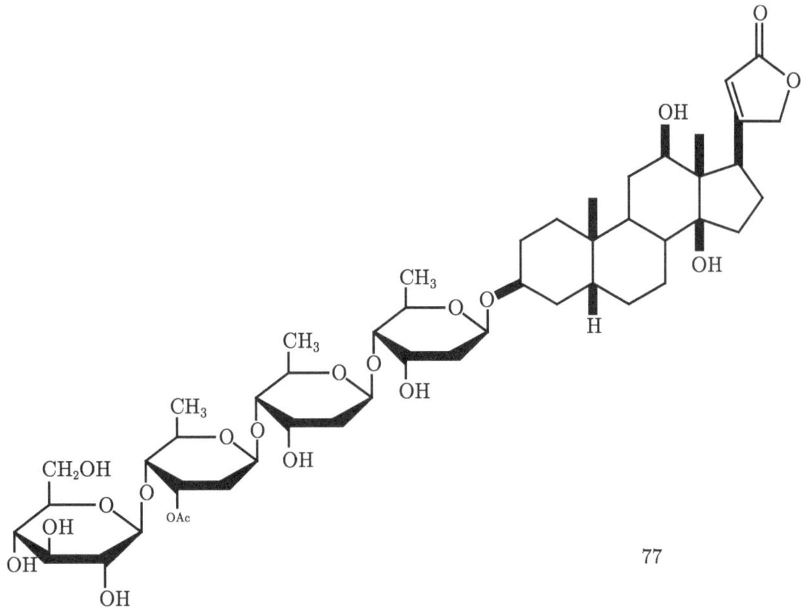

77

The sugar moieties of *Digitalis lanata* glycosides (lanatoside) A, B and C are the same. Their difference is that different hydroxyls are replaced in the aglycone. Compound **63** has no hydroxyl group at C-12 but **64** has one hydroxyl group at C-16. The cardiotonic role of A and B is much weaker than that of lanatoside C[36,37].

One characteristic of lanatoside C is it has one βtOH at C-12. Lanatoside A, B, and C all have the same sugar moiety, three digitoxoses (one acetoxyl group at C-3$'$) linked with C-3 in the β position of aglycone, and the last sugar is glucose. All the sugars are 1-4 linkage. Deacetylation of digitoxose is easily achieved via alkaline hydrolysis. The market cardiotonic medicine, deslanatoside C or cedilanide, has no acetyl group. It is stable, has higher solubility, and is made into an injectable form for emergency use. It usually works within 20 to 45 min. If the end of the glucose is hydrolyzed by the plant's own enzyme, then it is known as another market product, digoxin. Its bioactivity is slower (starting at 1 to 2 hr), long-lasting, and often made into a tablet for oral use in patients with heart disease. Both are used for the treatment of acute and chronic congestive heart failure and atrial fibrillation.

Cedilanid is soluble in methanol, slightly soluble in ethanol and water, and not soluble in chloroform and ether. Digoxin has no end glucose and no acetyl in digitoxose. It has higher solubility in dilute alcohol, pyridine and the mixture of chloroform and alcohol, but not in ether or chloroform.

There are several methods reported in the literature for the extraction and separation of cedilanid and digoxin. Here we will introduce our own method, which was developed through collaboration with the manufacturer in the 1970s. It is relatively simple and economic.

6.6.1   Extraction of cedilanid

To the ground leaves of *Digitalis lanata*, five times the volume of 70% ethanol was added. The mixture was then heated and stirred at 60°C for 2 hr. After filtration, the residue was extracted with 70% ethanol twice. All extracts were combined and concentrated to an alcohol concentration of about 20%, cooled, filtered by use of a cloth bag, and concentrated to the equal amount of the original leaves. The concentrate was washed with 1/3 the volume of chloroform, ethanol was added to the alcohol concentration of about 22%, then extracted with chloroform (1/3 volume) three more times. The chloroform extract was concentrated under vacuum to dryness to get crude glycoside. After adding a small amount of methanol and water (e.g., for 10 g crude glycosides, one should add 3 to 5 mL methanol and 0.25 mL water) to get a crystal mixture of lanatoside A, B, C, the product was then washed with acetone-ether (1:1), collected, and dried.

Purification of lanatoside C: ratio of crude glycosides-methanol-chloroform-water is 1:100: 500:500. First, the crude glycosides are dissolved in methanol, filtered, then chloroform and water are added and mixed well and let stand. The chloroform layer mainly consists of lanatoside A and B, and lanatoside B and C in the water layer. The water layer was concentrated and cooled to obtain lanatoside B, C powder mixture. The latter was filtered and the partition with the solvents above was repeated. Lanatoside C was obtained from the water layer by concentration. The chloroform layer, containing lanatoside A and B, was treated separately.

The process is summarized below:

| Ground leaves | $\xrightarrow[60°C]{70\% \text{ EtOH}}$ | ethanol extract | $\xrightarrow{\text{concentrated to}} \atop 20\% \text{ EtOH}$ | concentration |

| $\xrightarrow[\text{gel elimination}]{\text{cooling, filtration}}$ | filtration | $\xrightarrow{\text{concentration}}$ | concentration | $\xrightarrow{\text{CHCl}_3 \text{ washing}}$ |

| water layer | $\xrightarrow[\text{to } 20\%]{\text{add EtOH}}$ | dil. alcohol | $\xrightarrow{\text{CHCl}_3}$ | CHCl_3 extract | $\xrightarrow{\text{evaporation}}$ | residue |

| $\xrightarrow{\text{MeOH+H}_2\text{O}}$ | crystalline crude glycoside | $\xrightarrow[\text{partition}]{\text{MeOH, CHCl}_3, \text{H}_2\text{O}}$ | CHCl_3 (lanatoside A, B) |
| | | | H_2O layer $\xrightarrow{\text{concentration}}$ |

| lanatoside B, C | $\xrightarrow[\text{partition}]{\text{MeOH, CHCl}_3, \text{H}_2\text{O}}$ | H_2O layer | $\xrightarrow{\text{conc.}}$ | lanatoside C | $\xrightarrow{\text{Ca(OH)}_2}$ | cedilanid |

Deacetylation of lanatoside C: lanatoside C (3 g) was dissolved in five times the volume of methanol. Then, an equivalent amount of freshly prepared 0.15% aqueous solution of calcium hydroxide was added and the solution was stirred at room temperature overnight. The next day, if the pH was 7.0 (otherwise it was adjusted using 1% HCL) it was concentrated to precipitate cedelanide crystal and the crude crystal was recrystallized with methanol (mp 265°C, decomposed).

6.6.2   Extraction of digoxin

Equal amounts of ground leaf powder of *Digitalis lanata* and water were mixed and maintained at 40 to 50°C for enzyme fermentation for 20 hr, stirring the mixture once every 2 to 3 hr. Then the mixture was refluxed twice by four times and three times volume of 80% ethanol under stirring, each 2 hr, cooled and filtered. The filtrate was concentrated until containing about 20% ethanol, after cooling chlorophyll gel was precipitated. The upper layer was obtained by filtration with a cloth bag, and then the filtrate was extracted with chloroform three times (about 1/5 of the water amount each time). The combined chloroform extract was concentrated to 1/5 volume of the original leaf powder, and then washed with 10% sodium hydroxide water solution to hydrolyze the acetyl group and eliminate chlorophylls until the alkali solution turned colorless. Finally, it was washed with 1% sodium hydroxide solution once and then water to neutral pH. The chloroform liquid was evaporated to dryness and crystallized by adding a small amount of acetone and left to stand overnight. Digoxin was precipitated and recrystallyzed by 80% ethanol (a small amount of activated carbon was added if necessary) to obtain pure digoxin. The process is summarized as follows:

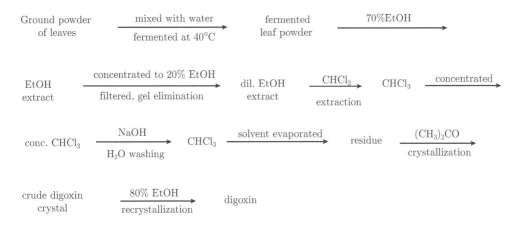

The acetyl group was hydrolyzed during the alkali solution washing. The crude digoxin mixture can be purified by repeated crystallization with 80% ethanol. This is based on the secondary glycoside solubility difference in the solvent: digitoxigenin glycoside is the most soluble, next is digoxin and then digitoxigenin glycoside.

In the extraction methods of cedilanid and digoxin above, ethanol was used as extraction solvent. Ethanol is more economical and suitable for the Chinese industry. If ethyl acetate is used as the extraction solvent, the procedure and purification could be simpler and the yield higher, but the process would be more expensive.

Bibliography

[1] Xu RS; Ye Y; Zhao WM. Natural Products Chemistry. 2nd ed. Beijing: Science Press, 2004

[2] Lin QS. Chemistry of Composition of Chinese Herbs. Beijing: Science Press, 1977

[3] Xu RS and Chen ZL. Extraction and Separation of Effective Components of Chinese Herbal Medicines. 2nd ed. Shanghai: Shanghai Science and Technology Publishing, 1982

[4] Stoll A and Jucker E. Herzglykoside in Modern Methoden der Pflanzenanalyse. Ed III. Berlin: Springer-Verlag, 1955 and 1961

[6] M. Dean F M. Naturally Occurring Oxygen Ring Compounds. London, New York: Butterworth, 1963

[7] Gong Y.H. ^{13}C-NMR Analysis of Natural Products. Kunming: Yunnan Science and Technology Press, 2005

[8] Cong PZ. MS Application in Natural Organic Chemistry. Beijing: Science Press, 1987

References

[1] Stoll A. et al. *Helv Chem Acta*, 1949, 32197

[2] Freeman G et al. *J Sci Food Agr*, 1975, 261869

[3] Calvey EM. *J Chromatography Sci*, 1994, 3293

[4]   Zhang XJ et al. *Chinese Condiment*, 1996, 1024

[5]   Lang YJ et al. *Chinese Herbal Medicine*, 1981, 124

[6]   Zhang CH et al. *Nat Prod R & D*, 1997, 953

[7]   Brion M. *J Chem Soc*, 196114

[8]   Mutsh-Eckner M et al. *J Chrom*, 1992, 625183

[9]   Yang PY et al. *Sci Tech Food Industry*, 2000, 2163

[10] Dong Y et al. *J Cancer Res*, 2001, 612923

[11] Clement IP, *J Nutr*, 1998, 1281845

[12] Spare CG et al. *Acta Chem Scand*, 1961, 151280

[13] Virtanen AI, *Angew Chem Internat*, 1962, Edit 1, (6)299

[14] Hegnauer R, *Chemotaxonome der Pflanzen II*, 1963320

[15] Tang ZJ et al . *Sci Sinica*, 1974, 116

[16] Ettlinger MG et al. *J Am Chem Soc*, 1956, 784172

[17] Waser J et al. *Nature*, 1963, 1981297

[18] Elliot MC, *Phytochemistry*, 1970, 91629

[19] Friis P, *Acta Chem Scand*, 1996, 20698

[20] Barillari J et al., *J Agric Food Chem*, 2007, 555505

[21] Schwartzmaier U, *Chem Ber*, 1976, 1093250

[22] Moertel CG et al. *New Engl J Med*, 1982, 306201

[23] Herbert V, *Am J Clin Nutr*, 1979, 321121

[24] Fang SD et al. *Acta Chim Sinica*, 1982, 40273

[25] Ukita T et al. *Chem Pharm Bull*, 1961, 943

[26] Tanimura A, *Chem Pharm Bull*, 1961, 947

[27] Nakajima K et al. *Chem Pharm Bull*, 1978, 263050

[28] Thakkar K et al. *J Med Chem*, 1993, 362950

[29] Chen YP et al. *Chin J Pharm*, 2000, 31334

[30] Stoll M et al. *Helv Chem Acta*, 1947, 302019

[31] Chen WZ et al. *Chemistry*, 1998, 244

[32] Wang DS et al. *Acta Chemica Sinica*, 1995, 53909

[33] Stoll A. *Helv Chim Acta*, 1951, 341460

[34] Stoll A et al. *Helv Chim Acta*, 1933, 16: 1049

Ren-Sheng XU

CHAPTER 14

Marine Natural Products

Section 1   Introduction

Three fourths of our planet is covered with water, which is not only the origin of life but also a rich source of living creatures. It turns out that around 15% of biological species (about 280,000) live in the oceans, which includes more than 2,000,000 species of lower grade organisms (sponges, corals, molluscs, etc.) except for the ordinary creatures such as fish, shrimp, and seashell, etc. These marine organisms play a crucial role in the marine ecological system, even though they are not familiar to people. Moreover, this distinctive marine ecological system (high pressure, high salt, oxygen and light deficiency) resulted in a fierce struggle for existence among different kinds of organisms. In order to survive under such a rigorous ecological environment, many marine organisms have to produce a series of small chemical molecules with unusual structures and significant bioactivities, namely secondary metabolites. Theses chemical substances are produced to defend the animals and their larvae against predators, and against the settlement of microorganisms like fungi or bacteria, and act as pheromones as well. Recent research revealed that the differences in living environments of marine organisms and terrestrial creatures led to a tremendous difference in their biosynthetic approaches for secondary metabolites. Secondary metabolites from marine sources display greater chemical diversity compared with those from terrestrial organisms. Many marine origin compounds have proven by modern pharmacology to be effective therapies against various diseases, which draws great interest from chemists, biologists, and pharmacologists.

The systematic studies on the chemistry and bio-properties of marine secondary metabo-lites started in the 1960s. Initiated by America scientists, researchers from Europe, Japan, and other countries devoted themselves to basic studies of marine bio-chemistry, biology, ecology, and pharmacology related to fatal diseases of human beings, such as tumors and cardiovascular disease. Marine natural product chemistry has experienced an explosive growth over the past 40 years, and remains prominent as a research field. Nearly 20,000 metabolites have been isolated from marine sources. Among these metabolites, several hundred molecules proved to have potency of various bio-activities and hold promise for development.

As a newly emerging course of study, marine natural product chemistry examines lower grade marine organisms (animals, plants, and micro-organisms, etc.). It mainly focuses on marine toxins, marine small molecules of secondary metabolites and related biochemical pharmaceuticals, and marine chemical ecology.

Marine natural product chemistry is closely related to other emerging natural science subjects, and at the same time, remains inter-penetrating, interacting, and correlating with different branches of science. Early studies on marine natural product chemistry followed and mimicked the methodology and research direction of phytochemistry and entomic-ecological chemistry, with the structural elucidation mainly focused on the basis of the traditional chemical approach. Research activity was progressing very slowly at the early stages because of the great difficulty in structural elucidation owing to the complexity, novelty and oddity of marine natural products. Developments in chromatographic techniques and in the collection, storage, and identification of bio-samples, as well as the widespread use of high resolution mass spectrometers and high-field NMR spectrometers, have enabled complete structural elucidation of small amounts of material. The purification and structural elucidation of new compounds is no longer an obstacle for the chemical investigation of marine natural sources, and even a compound with a tiny amount of 1 mg could be structurally elucidated. Meanwhile, the establishment of molecular bio-screening models and high-throughput bio-screening technique ensures a basis of bio-assay guided investigation of small molecules and the subsequent discovery of related bio-properties. These developments have made it possible for marine natural product research to be highly efficient, elaborated, ecological, and target-guided, and as a consequence, resulted in a boom of bioassay-guided chemical investigations on marine organisms.

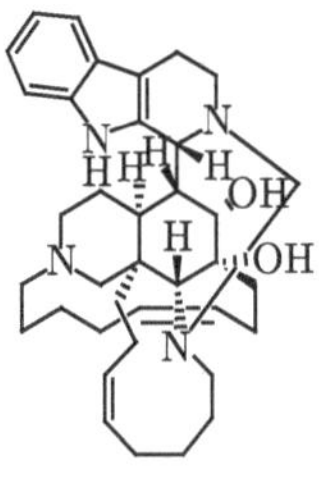

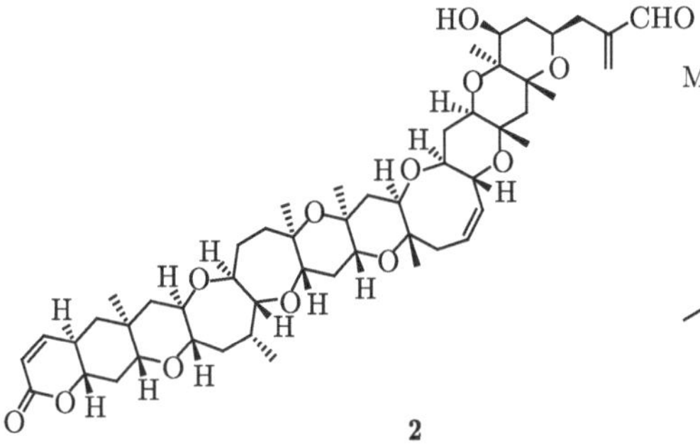

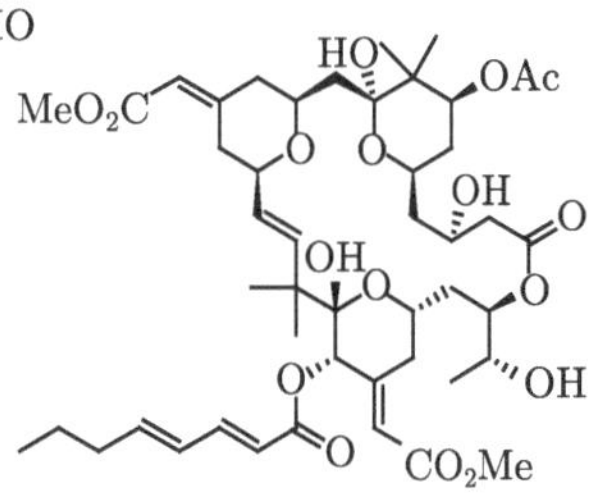

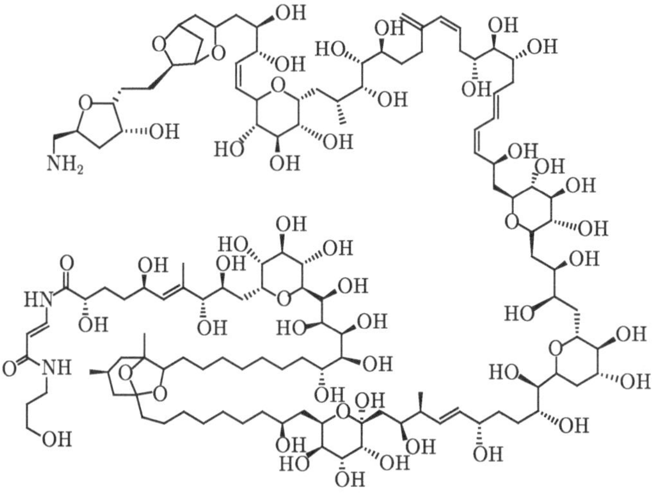

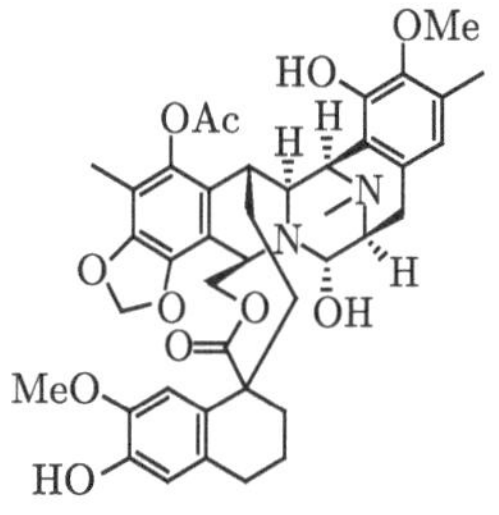

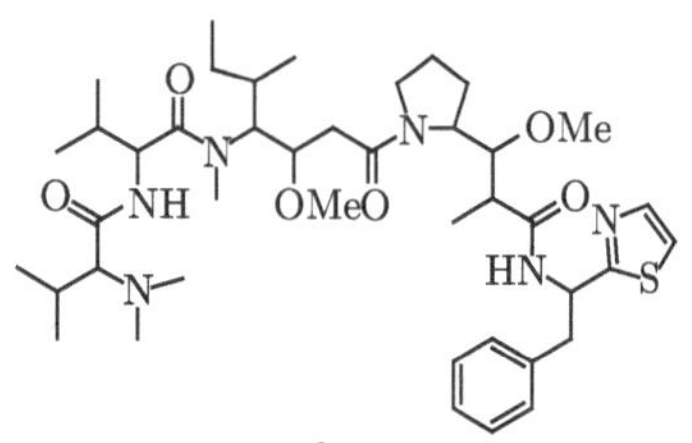

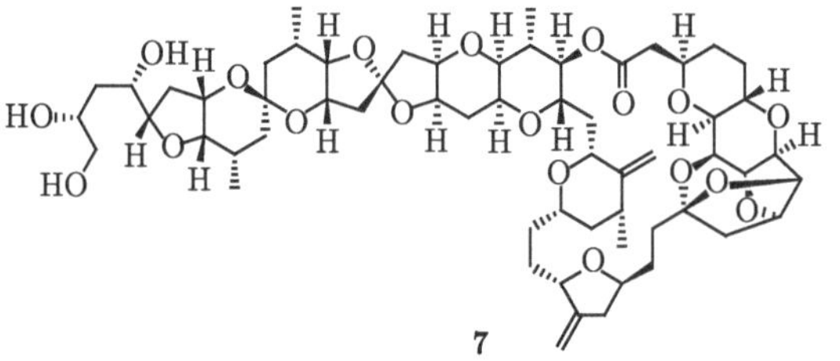

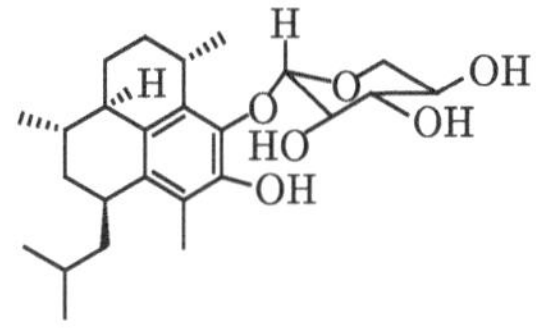

Research objects of marine natural product chemistry mainly include three large groups of marine organisms, namely marine plants, invertebrates, and micro-organisms. Secondary metabolites from marine sources differ greatly from those of terrestrial sources in both structure and bio-activity. The structural variations of marine natural products mainly rest on the rearrangement and high oxidation of the skeleton, bulkiness and complexity of molecules and numerous chiral centers in the structures. These metabolites can be classified into the following several large groups: terpenoids, steroids, alkaloids, polypeptides, macrolides, and lipids, etc. It is common to observe marine secondary metabolites with unusual functional groups, e.g., polyhalogenated metabolites, thio-methylamine containing compounds, sesqui- and diterpenoids with nitrile groups, isonitrile groups, and isothiocyanic functions. Up to now, biosynthetic processes of many compounds, e.g., manzamine-related tetrahydro-β-carboline alkaloid máeganedin A (**1**),[1] remain unknown. Some well-known compounds, such as brevetoxin B (**2**),[2] palytoxin (**3**),[3] and bryostatin 1(**4**)[4], represent typical examples of those which have been successfully structurally elucidated on the basis of the chemical method in combination with spectroscopic analysis (including x-ray diffraction).

Many marine metabolites display various biological activities, of which antiinflammation and cytotoxic activity are the most commonly encountered. America is the earliest country to develop drug research from marine sources. NIH and NCI annually dedicate more than half of their funds for natural product research into marine drug research. This enormous investment was richly repaid. By now, at least six new marine anticancer agents from NCI (for example, ecteinascidin 743(**5**),[5] dolastatin 10(**6**),[6] halichondrin B (**7**),[7] etc.) have been studied in clinical trials. Moreover, some additional promising new drug candidates have already been subjected to pre-clinical trials.

Marine bioactive substances are used as antitumor reagents and also potential and promising treatments of many other diseases. Pseudopterosin A (**8**),[8] a bioactive substance isolated from a Caribbean sea whip *Pseudopterogorgia elisabethae*, for example, showed significant anti-inflammatory and analgesic activities and use as a curative to dermatosis. The above examples were cases reported by NIH. In fact, there are many other marine drugs in different stages of pre-clinical studies.

Marine organisms seem to be inexhaustible sources of new compounds. In recent years, development in analytical techniques and bioassay models has made the isolation of new compounds from marine sources increase at an incredible speed. Novel compounds with unprecedented skeletons have been repeatedly reported[9]. These research achievements greatly enrich the content of organic chemistry, and promote their development in the related research fields of medicinal chemistry and molecular biology. On the other hand, it's worth noting that natural products from marine sources are only 10% of those from land-based organisms, although they have increased rapidly. Furthermore, from the view of organisms to be studied, only several thousand species of marine organisms are involved. It can be seen that there is still tremendous potential and development space in connection with the basic study and exploitation of marine organisms.

Due to limited length, this chapter will concentrate on the chemical constituents of marine organisms and their bio-active properties in terms of biological taxonomy, giving the reader an overview of research activities on marine natural products.

Section 2   Marine Plants

2.1   Mangrove Plants

Marine plants, mainly marine vascular plants and seaweeds, are the earliest to be studied as marine organisms. Marine vascular plants cover three large populations known as Pteridophyta, Gymnospermae, and Angiospermae. Almost all of the chemically studied marine vascular plants belong to Angiospermae. Recent studies of dicotyledon mangrove plants have been undertaken.

Mangrove plants are defined as the xylophyta which can merely survive in the region of an intertidal zone. A total of 84 species of mangrove plants have now been found, covering 38 genera in 24 families. These plants are distributed in the sea coasts of tropical and subtropical zones, ranging from north latitude 32° to south latitude 32° [10]. The constituents of mangrove plants can be sorted into sesquiterpenoids, diterpenoids, triterpenoids, steroids, alkaloids, tannins, flavonoids, lignanoids, cyclic polysulfides, etc.[11].

Diterpenoids from mangrove plants possess carbon skeletons commonly encountered in land-based plants, including labdane, secolabdane, daphnane, kaurane, etc. In the course of chemical investigation on mangrove plants, we isolated not only kaurane diterpenoids, but also diterpenoids with isopimarane (9) and secoatisane skeletons (10) from a Guanxi species of *Excoecaria agallocha* (Euphorbiaceae)[12]. More recently, two limonoid-type triterpenoid derivatives, named xyloccensin O (11) and P (12), were isolated from a Meliaceae plant *Xylocarpus granatum*, which enriched the wealth of triterpene carbon skeletons. The structure of xyloccensin O (11) was further confirmed by x-ray diffraction analysis[13].

Cyclic polysulfides are relatively uncommon constituents in mangrove plants, and only discovered in the genus *Bruguiera* (family Rhizophoraceae). From Guanxi *B. gymnorhiza*, we isolated a novel 15-membered macrocyclic polydisulfide with an unprecedented skeleton, known as gymnorrhizol (13). The molecule was highly symmetric, which was a challenge for the structural elucidation. In combination with a detailed analysis of the spectroscopic data, the structure was finally determined by x-ray diffraction[14]. Gymnorrhizol (13) showed an obviously close biogenetic relationship with several simple dithiolanes, such as brugierol (14), isobrugierol (15), and brugine (16), previously isolated from *B. conjugan*[15]. In an *in vitro* bioassay, gymnorrhizol (13) demonstrated inhibitory activity toward PTP1B with an IC$_{50}$ value of 7.95 μg/mL.

2.2   Seaweeds (Algae)

Seaweeds represent the earliest marine organisms known to human beings. Seaweeds contain more than 8000 species of flora, of which more than 100 species are edible. All seaweeds can be sorted into three groups known as red algae (Rhodophyta), brown algae (Phaeophyta), and green algae (Chlorophyta). In general, scientific reports on the chemical investigation of seaweeds are remarkably fewer in recent years, although all three groups were involved in studies.

Chemical substances from algae include polysaccharides, lipids, steroids, terpenoids, carotenoids, polypeptides, nitrogen-containing compounds, etc. Besides polysaccharides, the chemistry of algae is dominated by terpenes, particularly sesqui- and diterpenes that are superior to those from other marine organisms in both structural types and numbers of compounds. Terpenes from algae are frequently observed with substituted halogenated functional groups. Meanwhile, due to the co-existence of two parallel biosynthetic systems inside algae, namely isopentadiene and polypropylene approaches, it is common to find terpenes with a polypropylene moiety as substituents or fused subunits.

2.2.1   Rhodophyta (red algae)

Rhodophyta represents the largest group of seaweed species. Nearly 4000 species of red algae are classified into 558 genera. Secondary metabolites from red algae are mainly comprised of lipids, benzene derivatives, alkaloids, steroids, terpenes, etc. Compounds from red algae are heavily characterized by the halogen or polyhalogen substituents inside the molecule. Terpenes, especially sesquiterpenes, represent the most commonly encountered structural type.

Studies on the sesquiterpenoids from red algae of genus *Laurencia* (Ceramiales) were very active in recent years. Based on incomplete statistics, sesquiterpenes from genus *Laurencia* involved at least 25 kinds of carbon skeletons (Figure 14-1, calenzane through cyclococane)[16]. Many of the compounds have been reported to possess potential antibiotic or cytotoxic activities.

Figure 14-1   Common sesquiterpene skeletons from marine sources

Chamigrane sesquiterpenes are the most abundant and most diverse sesquiterpenes from red algae of genus *Laurencia*. The earlier discovered analogues, elatol (**17**), [1(15)$\underline{Z}$, 2$\underline{Z}$,

4<u>R</u>, 8<u>S</u>, 9<u>R</u>]-8, 15-dibromochamigr-1(15), 2, 11(12)-trien-9-ol (**18**), [1 (15) <u>E</u>, 2<u>Z</u>, 4<u>R</u>, 8<u>S</u>, 9<u>R</u>]-8, 15-dibromochamigr-1(15), 2, 11(12)-trien-9-ol (**19**) and isoobtusadiene (**20**), demonstrated significant cytotoxic activity against cell lines of P-388, A-549, HT-29 and MEL-28, with IC_{50} values all less than 1.0 μg/mL (some even as low as 0.025 μg/mL)[17]. Cuparene sesquiterpenes represent one of the earliest types of sesquiterpenes to be discovered. This kind of compound (**21** to **24**) was also found by our group to exist in *Laurencia okamurai*, from the East China Sea[18]. Laurinterol (**25**) and *iso*-laurinterol (**26**), two cuparene-derived sesquiterpenes isolated from Japanese *Laurencia*, exhibited strong inhibitory activity toward the marine bacteria *Alteromoas sp.*, *Azomonas agilis*, *Erwinia amyiovora*, and *Escherichia coli*.[19].

17

18

19

20

21 R_1=OOH, R_2=Br
22 R_1=R_2=Br

23

24

25

26

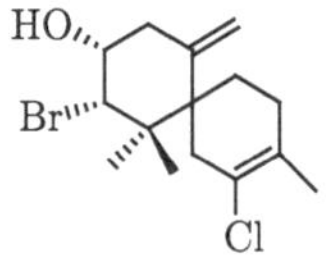

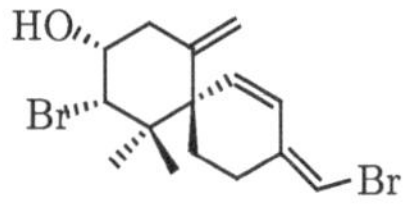

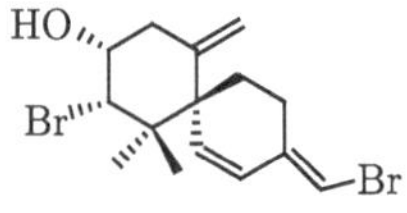

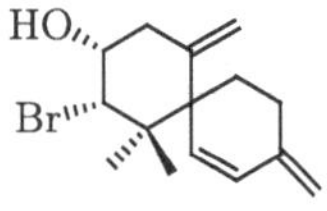

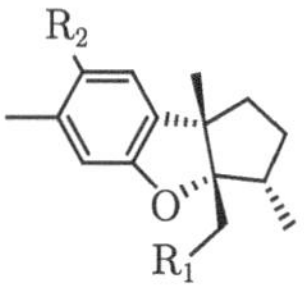

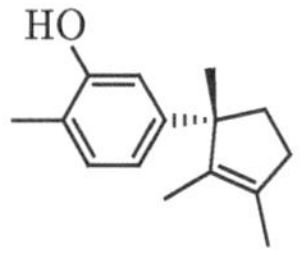

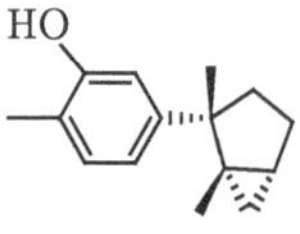

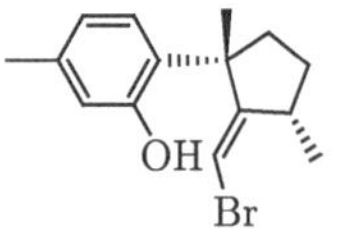

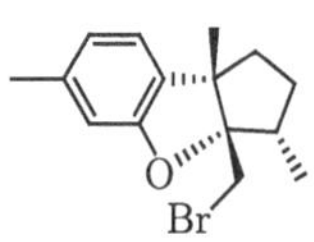

2.2.2 Phaeophyta (brown algae)

Brown algae are the smallest population of seaweeds, totaling 1500 species worldwide. Metabolites from brown algae involve all structural types of algae chemistry, of which diterpenes are the richest in amount and diversity. Brown algae diterpenes mainly display antioxidation, antibiotic, and cytotoxic bioactivities. Their primary structural feature consists of a polypropylene moiety acting as a substituent or a fused subunit. This may owe to the fact that two mixed biosynthetic approaches, isopentadiene and polypropylene, co-exist inside the organism.

27

28

29

30

32

31

33

34

35

36

37

38

39

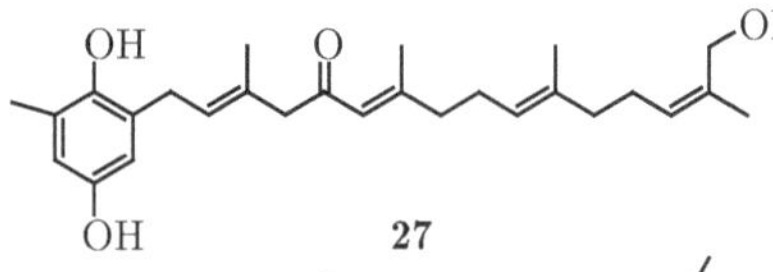

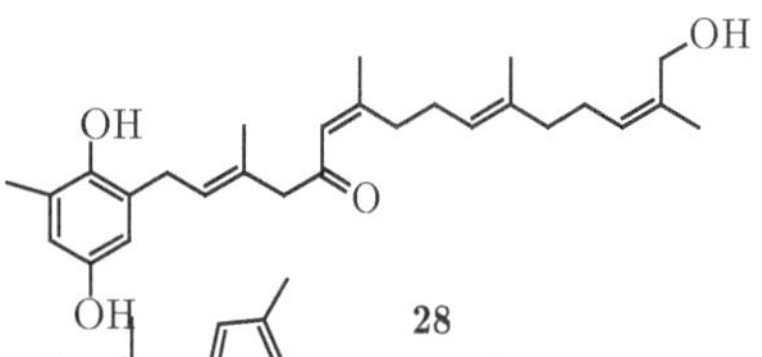

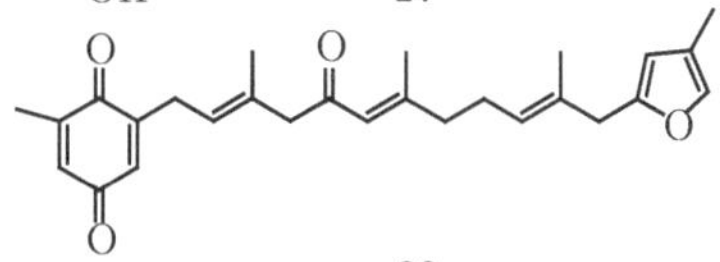

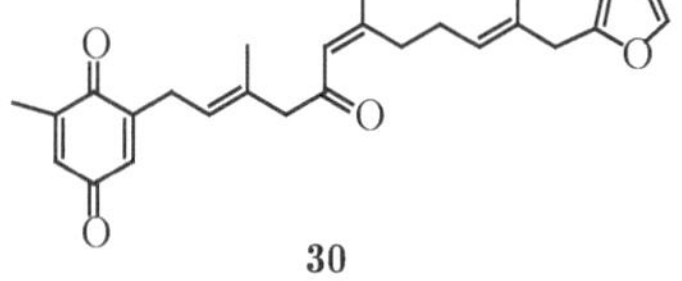

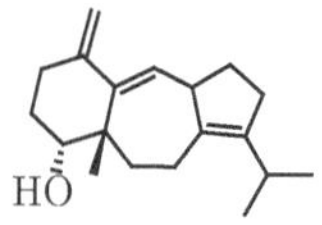

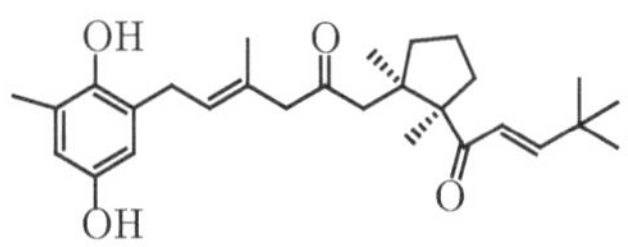

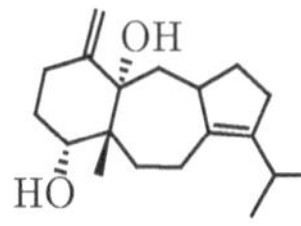

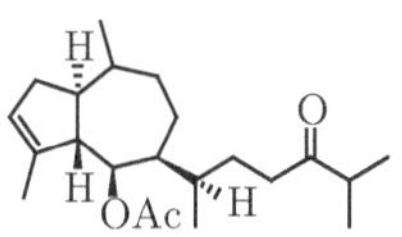

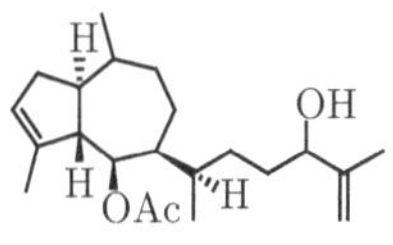

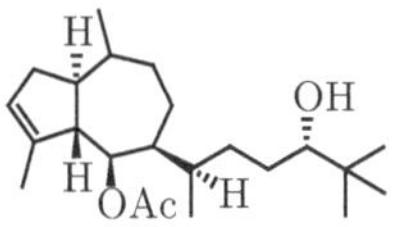

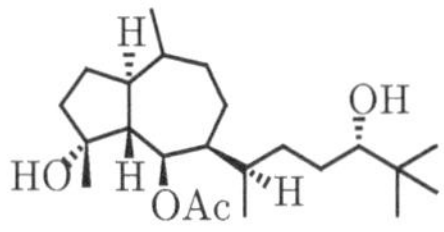

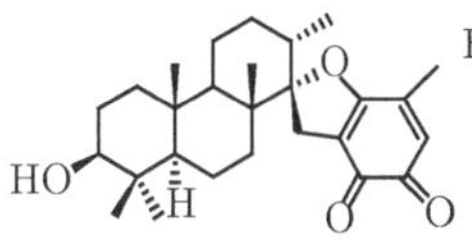

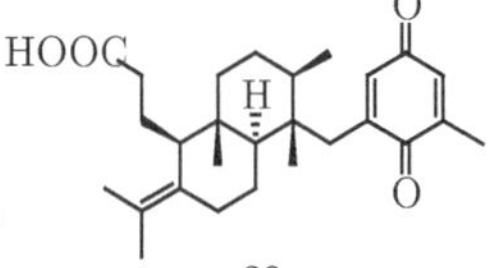

Linear diterpenes from brown algae are most likely to be present as unsaturated structures. Most linear diterpenes incorporate a polypropylene substructure to form complexes somewhat similar to those of vitamin E. The polypropylene subunit can be either phenol or benzoquinone, while the diterpene substructure demonstrates structural variety by isomerization of a double bond, polyoxidation, cyclization, etc. Examples include a series of analogues from *Cystoseira crinite* (**27** to **30**). Due to its unique structure, this kind of compound generally displays a potent antioxidative biological effect similar to that of vitamin E[20]. An analogue of the 7,11-cyclic condensation product, methoxybifurcarenone (**31**), exhibited a broad spectrum of antibiotic activities[21].

Cyclic diterpenes involve more than 20 carbon skeletons. The most commonly encountered structural types can be sorted into four groups known as xenicane (Figure 14-2), dolabellane (Figure 14-3), prenylgermacrane (Figure 14-4), and isocopalane and related types (Figure 14-5).

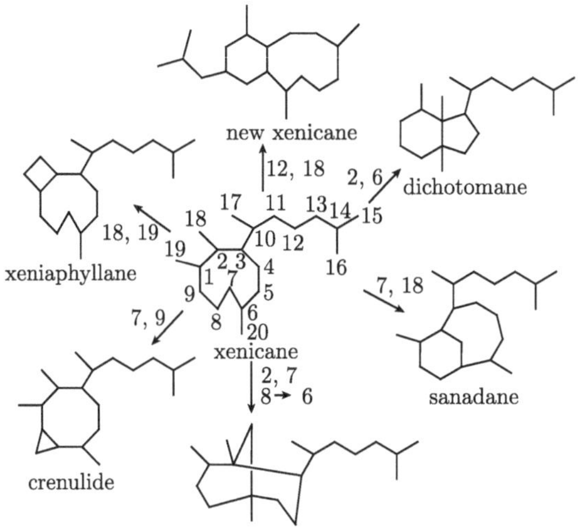

Figure 14-2　Common diterpene skeletons from marine sources: xenicanegroup

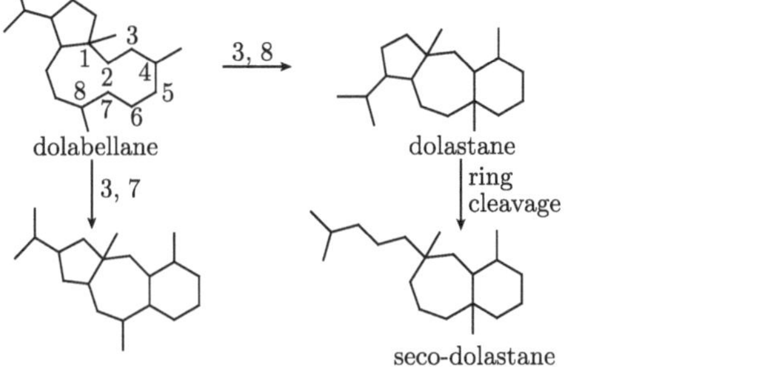

Figure 14-3　Common diterpene skeletons from marine sources: dolabellane group

Among the numerous cyclic diterpenes, those with hydroazulane and dolastane skeletons represent the most common structures. Amijitrienol (**32**) and 4-deoxyisoamijiol (**33**), two dolastane analogues isolated from brown algae *Dictyota linearia*, demonstrated moderate inhibitory activity toward *Staphylococcus aureus* and *Mucor mucedo*[22].Several hydroazulane derivatives (**34** to **37**) were obtained from *Cystoseira myrica*, and showed moderate cytotoxic activity against tumor cell KA3IT[23].

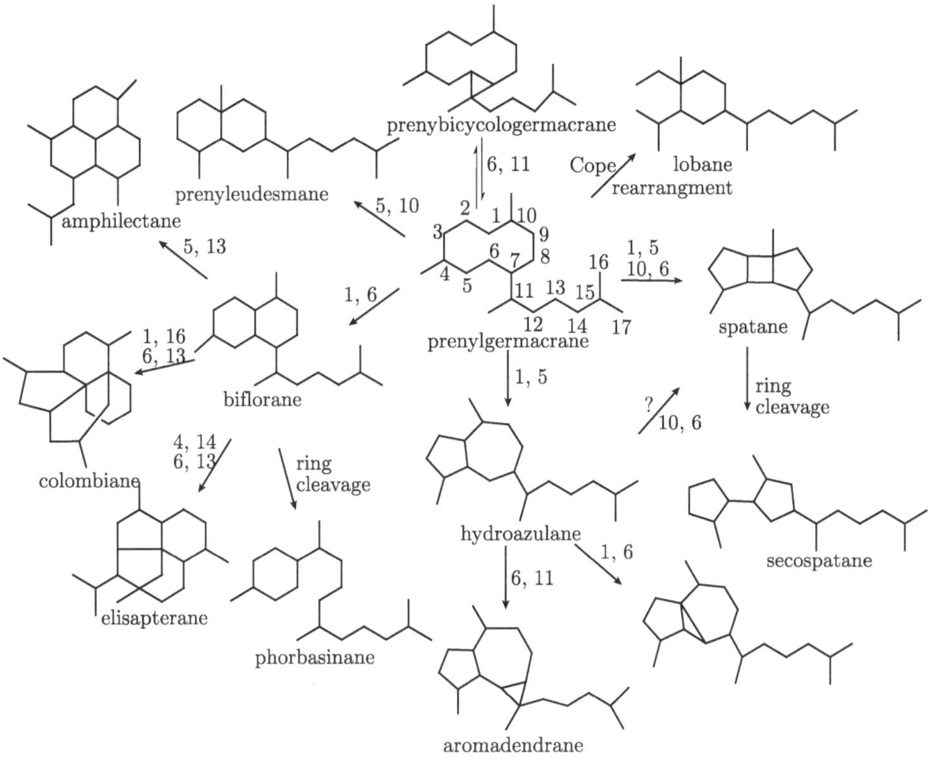

Figure 14-4   Common diterpene skeletons from marine sources: prenylgermacrane group

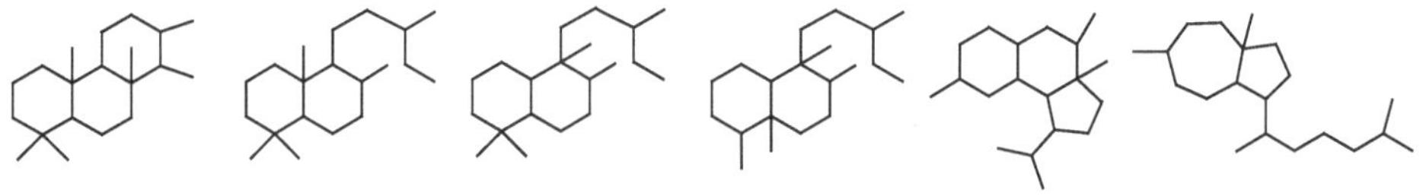

Figure 14-5   Other common diterpene skeletons from marine sources

Isocoplane diterpenes and related structures usually incorporate a polypropylene subunit, showing structural similarity to vitamin E. An example of such a structure is found in stypoldione (**38**), a tubulin polymerization inhibitor isolated from brown algae *Stypopodium zonale*. Stypoldione showed a strong narcotic effect on the reef-dwelling fish *Eupomacentrus leucostictus,* and was found to be an extremely potent inhibitor of synchronous cell division in a fertilized sea urchin egg assay[24]. A related derivative (**39**) from the same species of brown algae showed inhibition of tyrosine kinase[25].

2.2.3   Chlorophyta (green algae)

Total amount worldwide of green algae is between those of red and brown algae, but are the least reported in literature. Green algae chemistry involves structures of lipids, terpenes, steroids, polypeptides, alkaloids, carotnoids, flavonoids, benzene-like compounds, etc.

Terpenes from green algae are dominated by linear or monocyclic sesqui- and diterpenes, with the occasional occurrence of monoterpenes incorporating phenylpropene, which frequently co-appear with halogen substituents. These metabolites show major potential cytotoxic activity. Examples of such structures are depicted as **40** through **42** and were isolated respectively from green algae *Undota argenta*, *Tydemania expeditionis* and *Chilorodesmis fastigiata*. All three diterpene analogues possessed antimicrobial activities toward the human pathogenic bacteria *Staphylococcus aureus* and *Bacillus subtilis*, and completely inhibited cell division of fertilized sea urchin eggs at a concentration of 8 μg/mL[26].

The bis-indole alkaloid caulerpin (**43**) and analogues are red pigments regarded as a characteristic structural features of metabolites from the genus *Caulerpa*. The distribution of caulerpin is estimated at over 80% in various species of *Caulerpa*. Caulerpin and analogues were shown to promote root growth in lettuce seedlings with a corresponding activity to that of indoleacetic acid[27].

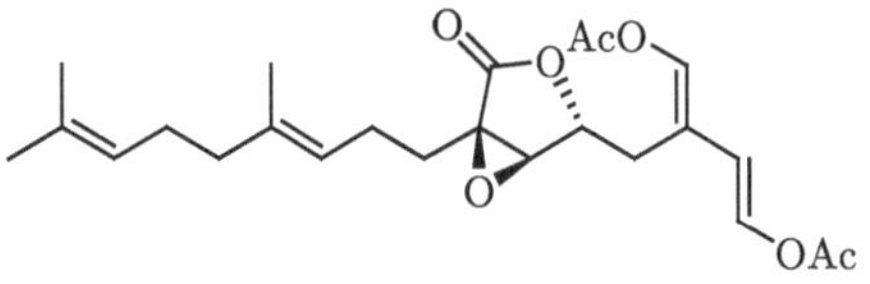

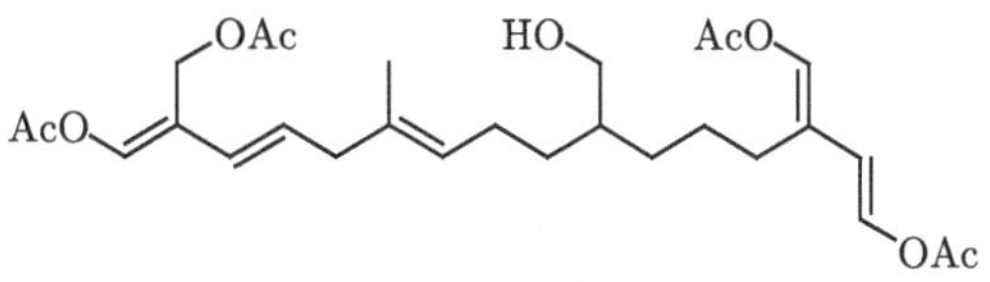

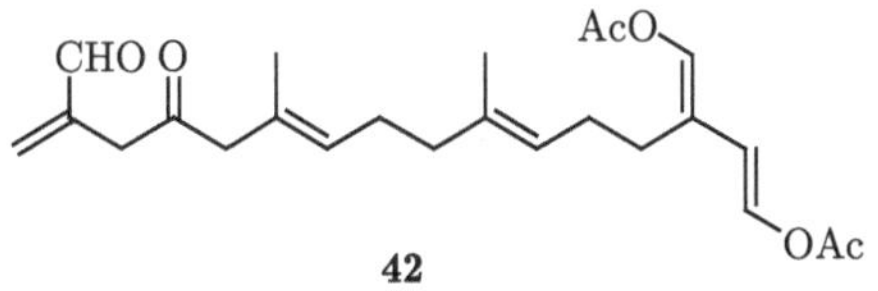

Polypeptide compounds of green algae come only from genus *Bryopsis*. The well-known compound kahalalide F (**44**) showed potential antitumor, antivirus, and antifungal bioactivities with minimum effective concentrations attaining parts-per-million levels[28]. A recent study revealed that the stereochemistry of Val-3 and Val-4 is crucial for the antitumor activity of kahalalide F[29]. It is currently in Phase I and II clinical trials as a drug candidate for prostate carcinoma.

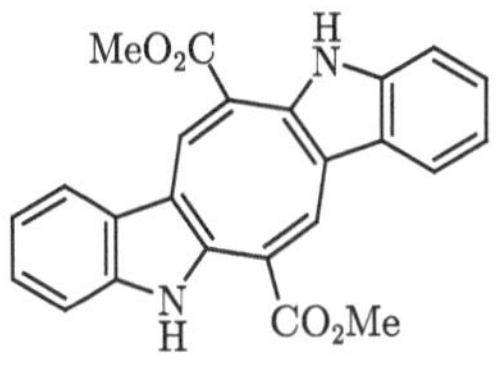

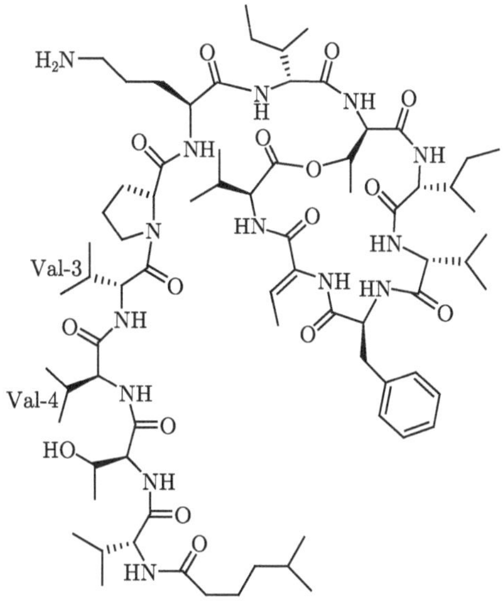

Section 3   Marine Invertebrates

Marine invertebrates represent the bulk of marine organisms, as well as a crucial field of marine natural product chemistry. The most active research fields currently involve coelenterates (Coelenterata), sponges (Porifera, Spongia), and molluscs (Mollucsca). This will introduce an overview regarding the above three phylla.

3.1   Coelenterata   (Coelenterates)

The chemically studied species of this phylum are mainly animals of subclass Octocorallia, of which most are gorgonians and soft corals with other types found in other populations.

3.1.1   Gorgonacea (gorgonians)

Gorgonians are commonly called sea fans or sea whips. They create their habitats from intertidal zones to a depth of 4000 m in sea water and are widely distributed from subtropical zones to the two poles. Research was focused on genera *Pseudopterogorgia, Briareum, Eunicea, Erythropodium, Eunicella, Isis,* and*Subergorgia*, along with some other species. In general, secondary metabolites from gorgonians are mainly composed of lipids, terpenes, and steroids[30].

Typical lipids from gorgonian sources consist of butenolide and prostaglandin derivatives. In 1966, the isolation of a novel bisbutenolide, ancepsenolide, from a Bimini gorgonian *Pterogorgia anceps*, greatly highlighted marine natural product chemistry[31]. Later in 1969, prostaglandin 15-*epi*-PGA$_2$ (**45**) was isolated from Caribbean gorgonian *Plexaura homomalla*,[32] and the quest for "drugs from the sea" was on in earnest. This event marked the rising of modern marine natural product chemistry.

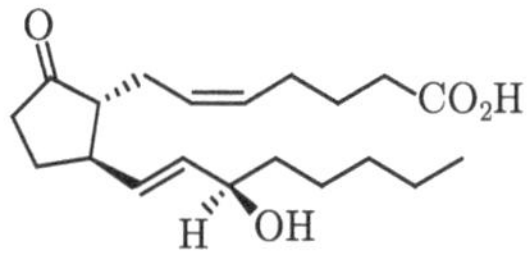

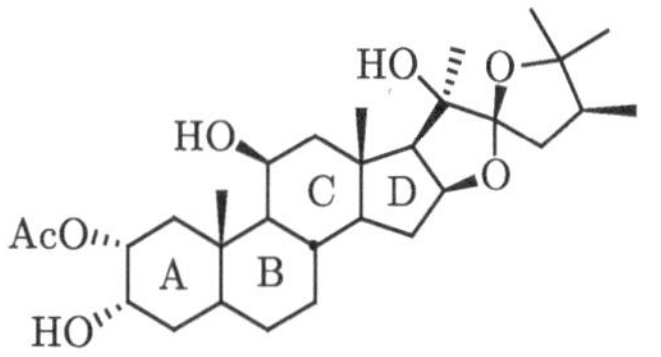

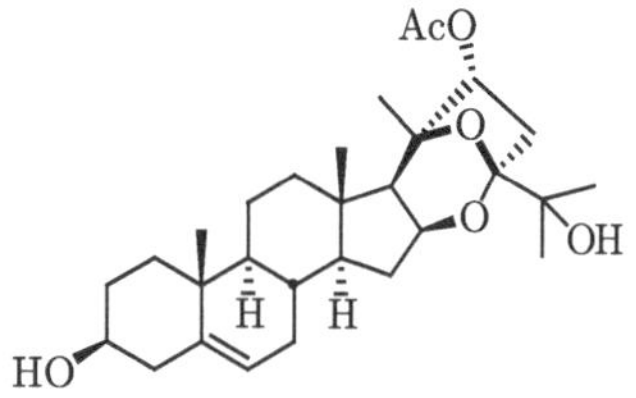

Steroids play an important role in gorgonian secondary metabolites. The structural diversity of steroids rests on the arylation of ring A, cleavage or oxidation of rings B, C, or D, the variety of side chain, etc. The 22-spiroketal steroid hippurin-1 (**46**) is a characteristic secondary metabolite from *Isis hippuris*[33]. Interestingly, an uncommon polyoxgenated steroid 24-ketal suberoretisteroid A (**47**) and analogues were isolated from a South China Sea gorgonian *Subergogia reticulata*. The structures of these compounds were elucidated on the basis of extensive spectral analysis and further confirmed by x-ray diffraction[34]. From another gorgonian (*Acanthogorgia vagae*) collected at the same region as that of *S. reticulata*, we unexpectedly obtained six 19-hydroxylated steroids[35]. This is the first report of 19-hydroxy steroids from a gorgonian and suggests a possible biosynthetic relationship between 19-nor and 19-hydroxy steroids.

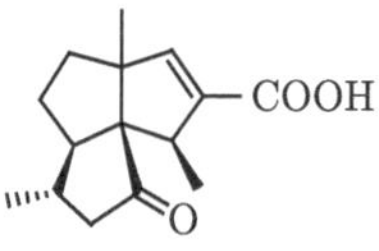

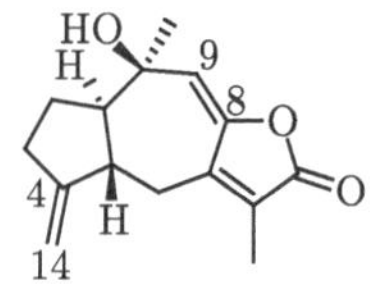

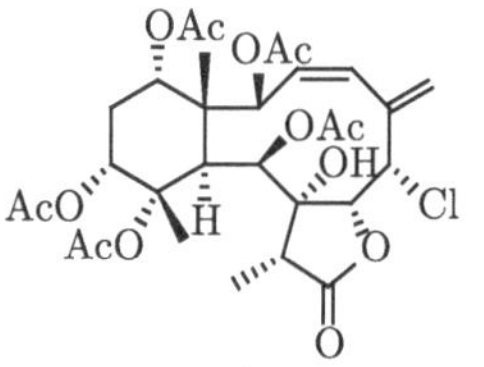

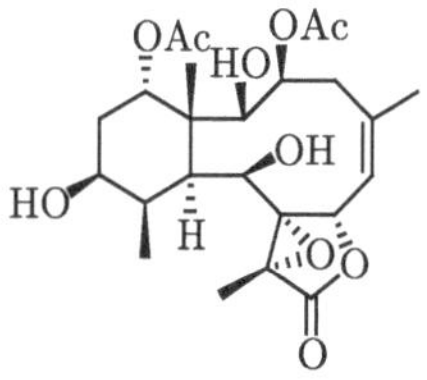

Among the secondary metabolites from gorgonians, terpene structures such as sesquiterpenes and diterpenes seem to be the richest in diversity. These types of sesquiterpenes can be sorted into six kinds of carbon skeletons known as guaiane, lindenane, bisabolane, subergane, suberosane and caryophyllane (Figure 14-1). The well-known compound subergorgic acid (**48**) is a typical structure of a subergane sesquiterpene. The pharmacological properties of subergorgic acid (**48**) are very interesting since it acts as a neurotoxin at high doses and as an antidote to soman at low doses. Further toxicological studies revealed that subergorgic acid may act on cholinesterase as a reversible inhibitor[36]. Sesquiterpenes with guaiane skeletons are commonly encountered in land-based plants, but fewer exist in marine organisms. We isolated meneverin A (**49**) and analogues from the South China Sea gorgonian *Menella verrucosa.* All the compounds contain the unusual $\Delta^{4,14}$exocyclic double bond, and a $\Delta^{8,9}$ double bond conjugated with a γ-lactone. The high instability of the sesquiterpenoids greatly challenged the full characterization of their stereochemistry and prevented a further bio-screening of their bioactivity[37].

Considering their biosynthetic connections, diterpenes from gorgonians can be classified into two groups, prenylgermacrane diterpenes and derivatives (Figure 14-3) and cembrane diterpenes and derivatives (Figure 14-6). Cembrane and related derivative skeletons such as briarane, eunicellane, secoeunicellane, cyclobutenbriarane, erythroane, aquariane, pseudopterane, and gersolane, constitute the stems of those gorgonian diterpenes possessing a large ring (nine to fourteen members). Of all the above structural types, briarane is the most characteristic of gorgonian diterpenes, and generally exhibits cytotoxic activity. The first example of briarane diterpene, briarein A (**50**), was obtained from *Briareum asbestinum*[38]. Recently, the Taiwanese gorgonian *B. excavatum* was found to produce a bioactive derivative excavatolide M (**51**) which displayed a very strong cytotoxic activity against tumor cell lines P-388, KB, A-549, and HT-29, with ED_{50} value toward P-388 as low as 0.001 μg/mL[39]. Other important sources of briarane diterpenes includes those from the *Junceella* genus. We have found that the secondary metabolite patterns of the South China Sea gorgonian*J. fragilis* were mainly briarane derivatives[40]. The new diterpene skeletons such as cyclobutenbriarane, erythroane, and aquariane are obviously rearranged derivatives of briarane (Figure 14-6)[41]. Gersolane diterpenes can couple to each other to form a dimer[42].

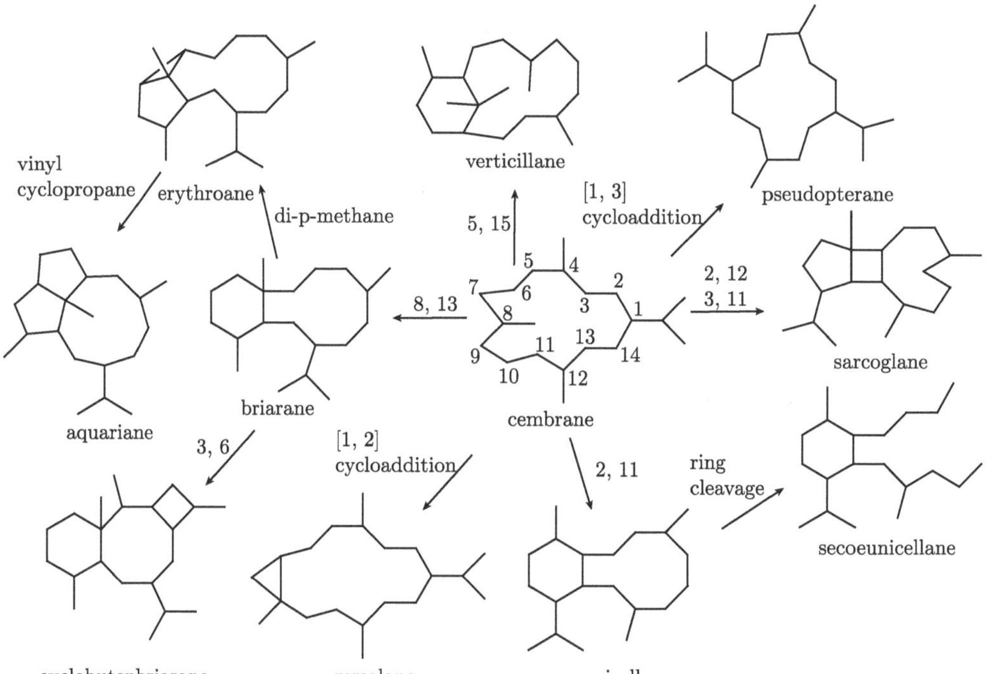

Figure 14-6   Common diterpene skeletons from marine sources:cembrane group

Carbon skeletons originating from prenylgermacrane are most likely to possess fused small rings (four to seven members), e.g., amphilectane, elisapterane, colombiane, biflorane, etc. (Figure 14-4). Up to now, all the so called "small-ring" diterpenes were isolated from animals of genus *Pseudopterogorgia*, which hold a close biogenetic relationship. An example of this class is pseudopterosin A (**8**), a diterpene pentoside isolated from the Caribbean sea whip *Pseudopterogorgia elisabethae*. Pseudopterosin A (**8**) showed anti-inflammatory and analgesic activities equivalent in potency to the industrial standard indomethacin[8] and has progressed to Phase I clinical trials as an anti-inflammatory under the aegis of Nereus Pharmaceuticals, Inc. Meanwhile, the international cosmetic giant Esteé Lauder has successfully developed it as a new kind of additive in its commercially available skin-care products.

3.1.2   Alcyonacea (soft corals)

As one of the most important members of Coelenterates, soft coral animals can be classified into six families. Chemical and biological research activities mainly involved *Simularia, Lobophytum, Nephthea, Sarcophyton*, and *Xenia* genera[30b,43].

Soft corals are a rich source of terpenes, polyoxygenated steroids, and prostaglandin derivatives. Prostaglandins mainly exist in animals of genus *Clavularia*, although they were first obtained from *L. depressum*[44]. Okinawan soft coral *C. viridis* produced three prostanoid-related analogues (**52-54**), all of which were substituted by halogens. Compound **52** showed cytotoxic activity against human T lymphocyte leukemia MOLT-4, human colorectal adenocarcinoma DLD-1, and human diploid lung fibroblast IMR-90[45]. Tricycloclavulone (**55**) and clavubicyclone (**56**) are examples of those with newly found carbon skeletons. Clavubicyclone (**56**) showed moderate growth inhibition activity toward cultured tumor cells of breast carcinoma MCF-7 and ovarian carcinoma OVCAR-3[46].

Similar to gorgonians, soft coral steroids exhibit their structural diversity as a polyoxygenated ketone or alcohol, cleavage on a core structure, alkylation of a side chain, etc. From the South China Sea soft coral *N. bayeri*, new polyhydroxylated steroids (nanjiol A (**57**) and analogues) were isolated, all of which possess a 20-hydroxy group and conjugated 4-en-3-one as structural features[47]. Soft coral metabolite patterns of genera *Nephthea* and *Litophyton* are characterized by 19-hydroxy steroids. In fact, the first marine 19-hydroxy steroid was isolated from genus *Litophyton*[48]. From a Chinese species of *Simularia*, we obtained two new 19-hydroxy sterol derivatives **58** and **59**. This is the first report of such compounds from *Simularia* sp.[49].

In addition to the common guaiane and furan types of structures, sesquiterpenes contain at least three other structural skeletons such as africanane, capnellane and illudalane (Figure 14-1). Capnellane sesquiterpene is a typical metabolite found from marine sources which mainly exists in the soft corals of *Capnella*. Compound **60** displayed significant cytotoxic activity against K562 leukemia[50].

Diterpene compounds include eleven types of carbon skeletons such as cembrane, amphilectane, xenicane, new xenicane, biflorane, lobane, aromadendrane, dolabellane, eunicellin, verticillane and sarcoglane. Considering their biosynthetic connections, these carbon skeletons can be categorized into three relatively independent groups, namely xenicane (Figure 14-2), prenylgermacrane (Figure 14-4), and cembrane (Figure 14-6). Diterpenes with cembrane and xenicane skeletons are relatively abundant in soft corals. Cembrane diterpenes have been found to exist in genera *Simularia, Lobophytum, Nephthea, Sarcophyton*, and *Scleronephthya*. Cembranes from both soft corals *Scleronephthya*[51] and *Sarcophyton*[52] have already been isolated and collected from the South China Sea. *Sarcophyton* sp. produced cembrane sarcophytonolide C (**61**), which possesses a rare C-2-C-18 γ-butyrolactone[52]. The soft coral *Simularia gardineri* yielded the cembrane dimer **62** coupled by an ester group. It showed cytotoxicity to cells P-388, A-549, HT-29, and MEL-28[53]. All the biscembranoids formed by C-C coupling were isolated from genus *Sarcophyton*. This group of bis-diterpenoids is likely derived by Diels-Alder reaction of two cembranes. A typical example is methyl tortuoate A (**63**), which was isolated from the South China Sea soft coral *S. tortuosum*. Methyl tortuoate A (**63**) showed in vitro cytotoxic activity toward human nasopharyngeal carcinoma CNE-2[54]. Xenicane diterpenes mainly exist in genus *Xenia*, family

Xeniidae. These kinds of compounds can also be found in soft corals *Nephthea* and *Eleutherobia*, as well as in animals with distant relationships, such as gorgonians, blue corals, and brown algae. Generally, these compounds display anti-inflammatory, antibiotic, and antitumor bioactivities. New xenicane diterpenes (Figure 14-2) with a new structural type were isolated from genus *Xenia*, whereas verticilane and sarcoglane were obviously derived from cembrane (Figure 14-6).

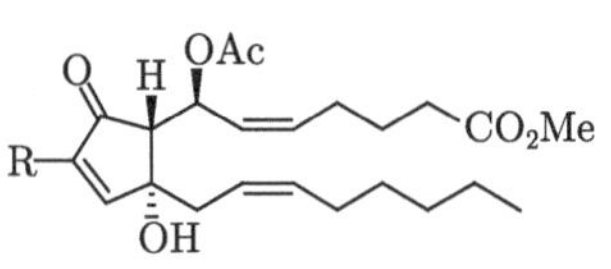

52 R=I
53 R=Br
54 R=Cl

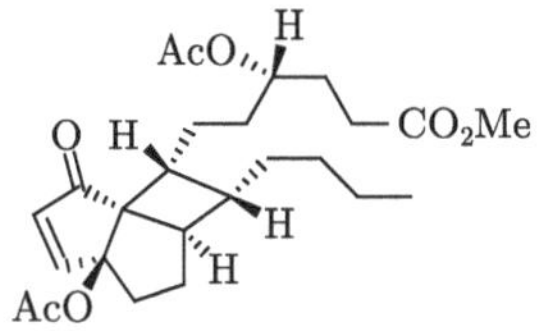

55

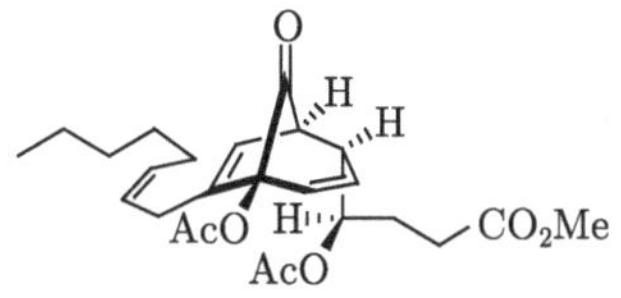

56

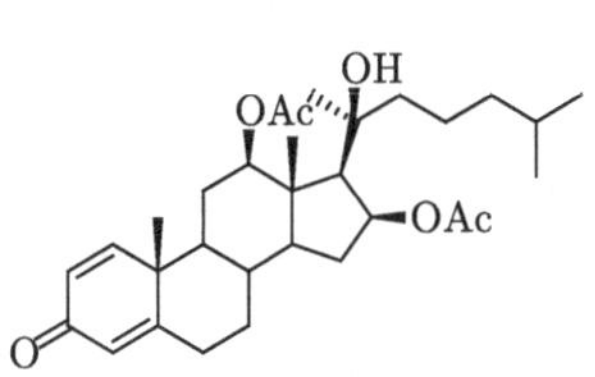

57

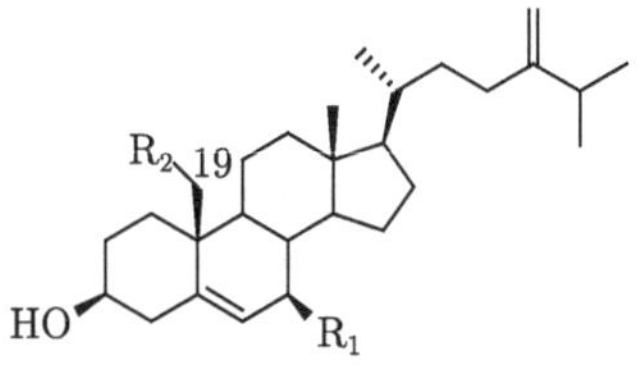

58 R$_1$=OH, R$_2$=OAc
59 R$_1$=OAc, R$_2$=OH

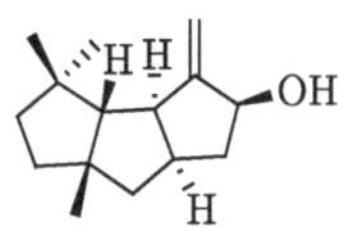

60

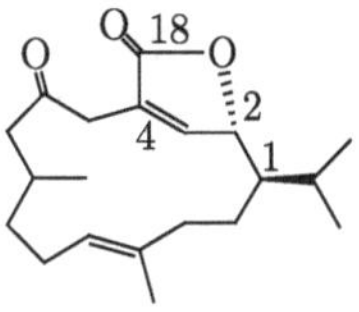

61

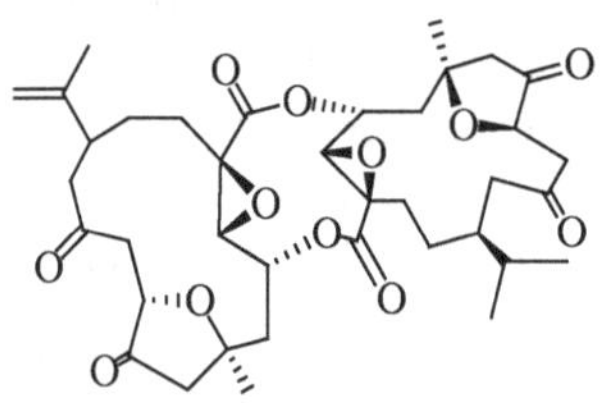

62

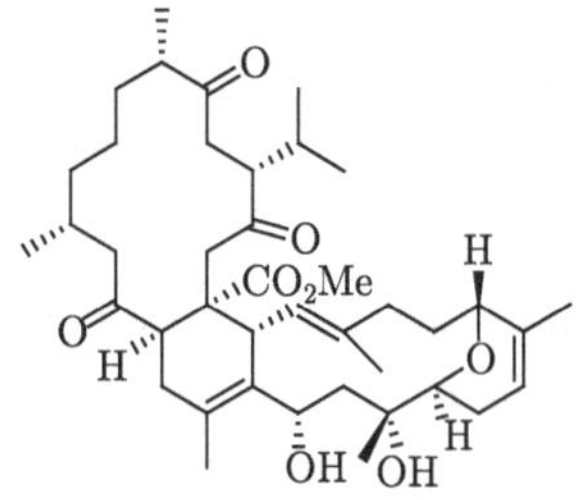

63

3.1.3   Hexacorallia (hexacorals)

Chemistry data on subclass Hexacorallia are quite limited, and only some steroids, lipids, and nitrogenous pigments are involved. Palytoxin (**3**), the noted marine toxin, makes such kinds of marine organisms well known. Palytoxin is an incredibly complex water-soluble substance isolated from *Palythoa toxica* (Zoanthideae, Zoanthidea) in 1971 with a molecular formula of $C_{129}H_{221}O_{54}N_3$. This polyhydroxylated molecule contains 64 stereocenters and 7 double bonds, totaling 71 stereochemical elements, and may have 271 isomers. Continuous efforts for the structural determination of palytoxin lasted for more than 10 years and came to an end in 1981,[3] and the effort is regarded as a great achievement in the field of natural product chemistry and exemplified as the apotheosis of structural elucidation for complex natural products by successfully using spectroscopic analysis in combination with the chemical method. The success of its total synthesis is still considered today by many to be the greatest synthetic accomplishment ever[55] and brings a continuous influence on the development of marine natural product chemistry and related fields for almost 30 years.

Palytoxin is considered to be one of the most toxic non-peptide substances known, being around ten times as toxic as tetrodotoxin. Meanwhile, palytoxin showed significant cytotoxic activity and completely inhibited the growth of Ehrlich ascites tumors in 100% of mice treated with a low dose of 84 μg/kg. Palytoxin is also found to be one of the strongest coronary artery constriction agents at over 100 times more potentcy than angiotensin II. An interesting fact is that the mechanism of palytoxin acting on the ion channel seems reverse to that of tetrodotoxin, keeping the Na^+ channel open. Research studies on palytoxin remain highly active in chemistry, pharmacology, and various research fields related to life sciences.

3.2   Porifera (Spongia)

Phylum Porifera, commonly known as Spongia, encompass about 1000 to 5000 species which can be classified into Calcispongea, Hexactinellida, and Demospongiae. Class Demospongiae is the largest group with the widest distribution. It is usual to observe the co-occurrence of symbiotic microbes in sponges[56].

Secondary metabolites from sponges include lipid, terpene, nitrogenous compound (alkaloid, amino acid, nucleoside, ceramide, cyclopeptide, etc.), polyether, macrolide peralcohol, polyacetylenic derivative, steroid, etc., structural types. These secondary metabolites not only demonstrate amazing structural complexity and novelty, but also display a series of biological properties such as antitumor, antivirus, and antimicrobial, as well as inhibitory activities toward various enzymes. Sponges are considered to be the crucial resources of potential lead compounds for clinical trials[57].

Lipids from sponges include glocerolipids, phospholipids, glycolipids, etc, and are mainly responsible for antifungal and antitumor bioactivities. Seven linear N-acyl-2-methylene-β-alanine methyl esters, hurghamides A to G, were successfully isolated from Red Sea sponge *Hipposongia* sp. This is the first report of such kinds of lipids isolated from marine sources[58].

Terpenes from sponges include sesquiterpenes, diterpenes, and triterpenes. Sesquiterpenes mainly involve carbon skeletons such as cadinane, eudesmane, maaliane, aristolane, bisabolane, pallescensinane, nakafurane, spiroseaquiterpnane, aromadendrane, axane, isodaucane, furodysinane, furodysininane, dysetherinane, reticulidinane, farnesane, bolinaquiane, friedodrimane, drimane, etc. (Figure 14-1). As for the diterpenes, over ten structural types have been discovered, which include cembrane, labdane, phorbasinane, biflorane, amphilectane, sphenolobane, isocopalane, clerodane, hamigerane, etc. (Figure 14-5, Figure 14-6). Sequi- and diterpenoids from sponges are frequently found to be substituted by isonitrile groups, isothiocyanic functions, form amino groups, or sometimes form derivatives by incorporating with benzoquinone or phenolic groups. These structural features are rarely observed in other kinds of marine organisms[9,59]. Similarly, naturally occurring sesquiterpene carbonimidic dichlorides are reported from sponges and related nudibranch predators[9,60]. Triterpenes are infrequently encountered in marine sources, and in most cases, their structures differ from those from terrestrial sources. Sponge triterpenes are most likely the dimers of sesquiterpenes or derivatives originated from the sesterterpene core, and more often contain a polypropylene substructure in the molecule[61].

Sesterpenes are relatively abundant in sponges and can be roughly classified into several structural types by the nature of carbon cyclozation (linear, monocyclic, bicyclic, tricyclic, tetracyclic, and pentacyclic sesterterpenes). The monocyclic sesterterpne manoalide (**64**) was earlier isolated from a California sponge *Luffariella variabilis*, and represents the first marine anti-inflammatory compound via selective inhibition of phospholipase A (PLA2)[62]. A structurally related compound, hurghaperoxide (**65**), was isolated from an unidentified sponge collected from the Red Sea, possessing an unusual hexacyclic peroxyl ether moiety in the molecule[63]. From the South China Sea sponge *Hyrtios erecta*, the scalarane sesterterpene hyrtiosin A (**66**) and analogues were obtained and structures confirmed by x-ray diffraction analysis[64]. Scalarane sesterpene is the most typical tetracyclic sesterpene from sponge sources, mainly existing in animals belonging to the genera *Hyrtios*, *Cacospongia*, *Phyllospongia*, *Strepsichordaia*, as well as molluscs preyed on by related sponges. Dysidiolide (**67**), a novel bicyclic sesterpene most likely derived from scalarane class by ring cleavage, was isolated from the Caribbean sponge *Dysidea etheria*. Dysidiolide proved to be the first example of a cdc25 protein phosphatase inhibitor, which might explain its potency in anti-cancer cells of A-549 and P-388[65]. The Indonesian sponge *Strepsichordaia aliena* produced the pentacyclic sesterpene honulactone A (**68**) and derivatives, all of which possess spiro-cyclopropane moieties. These metabolites showed significant cytotoxic activity toward cell lines P-388, A-549, HT-29 and MEL-28, with IC_{50} values as low as 1 μg/mL[66].

Structures of alkaloids from marine sponges demonstrate rich diversity and are sorted by structural features of simple alkaloid subunits such as pyrrole, indole, pyridine, isoquino-line, bromotyrosin, guanidine, quinolisidine, *bis-3-alkylpyridine* macrocyclic alkaloid, etc. A bis-indole alkaloid dragmacidin E (**69**) was isolated from a southern Australian sponge *Spongosorites* sp. as a potent inhibitor of serine-threonine protein phosphatases (PP_1 and PP_2A)[67]. Máeganedin A (**1**), a derivative of *bis-3-alkylpyridine* macrocyclic alkaloid as men-tioned above, was isolated from an Okinawan marine sponge *Amphimedon* sp., and showed significant antibacterial activity and cytotoxicity[1]. It represents one of most fascinating structural types of sponge alkaloids. In the course of our continuing efforts of chemical in-vestigation on Mediterranean sponges, a series of *bis-3-alkylpyridine* macrocyclic alkaloids such as saraines A-C, saraine-1 and -2, and isosaraine-1 to 3 were isolated from *Reniera sarai* and further investigation of the absolute stereochemistry of the related metabolites was done by modified Mosher's method[68]. More interestingly, an unidentified species of the same genus was found to produce an intriguing novel derivative of this class. Known as misenine (**70** and **71**), it possesses an unprecedented tetracyclic cage-like core with two macrocyclic rings. Misenine is very sensitive to the pH environment and could be present in two forms (**70** and **71**) in solutions of different acidity. The formation of the former (**70**) is favored in neutral or weakly basic conditions, since the proton attached the oxygen atom of the carbonyl group would enhance the C-N linkage and thus making C-11 display somewhat the nature of a secondary alcohol. On the other hand, increasing acidity would weaken the N/C–O-interaction, which lets the molecule prefer to adapt to the latter form (**71**) with C-11 mainly demonstrating the carbonyl property. The spatial proximity between the carbonyl (C-11) and the tertiary amine moiety (N-9) induce an interaction, known as a "proximity effect". This is the first observation of the proximity effect in a molecule from natural sources[69].

Cyclopeptides, polyethers, and macrolides represent secondary metabolites with great structural complexity and various bio-properties[57]. Okadaic acid (**72**) is a polyether-like toxin obtained from the sponge *Halichondria okadai* and was reported as the first-described potent substance with high specificity of reversible inhibitory effect against type 2A and type 1 phosphatases. It has already served as a unique "biochemical probe" tool in pharmaceutical studies related to the mechanism of inflammatory diseases[70]. A highly cytotoxic macrocyclic lactone polyether has been isolated from aMaldivian sponge *Spongia* sp., and is known as spongistatin 1 (**73**). Spongistatin 1 strongly inhibits mitosis by binding in the Vinca alkaloid domain of tubulin, and demonstrates an amazing cytotoxic activity toward L1210 murine leukemia cells (as high as 12 times that of dolastatin 10 (**6**))[71]. The total synthesis of spongistatin 1 has been completed[72]. The cyclic peptide cycloithistide A (**74**) has proved to be a potent antifungal agent,[73] while dihydrocyclotheonamide A (**75**) and papaumide A (**76**) showed inhibitory activities against serine protease and HIV-1$_{RF}$, respectively[74].

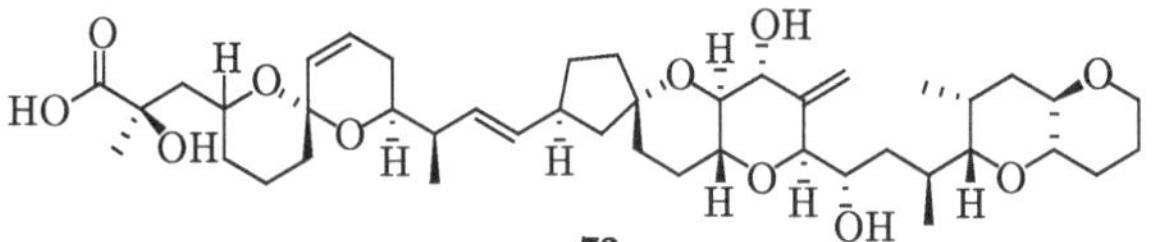

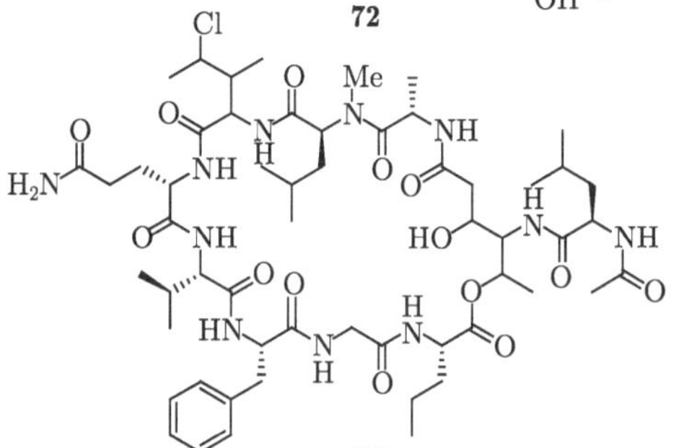

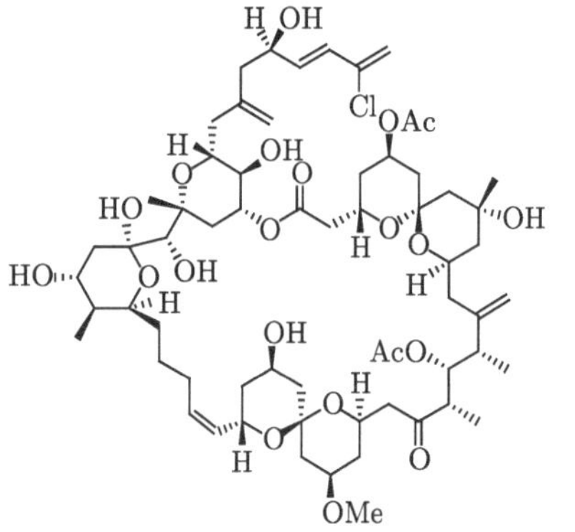

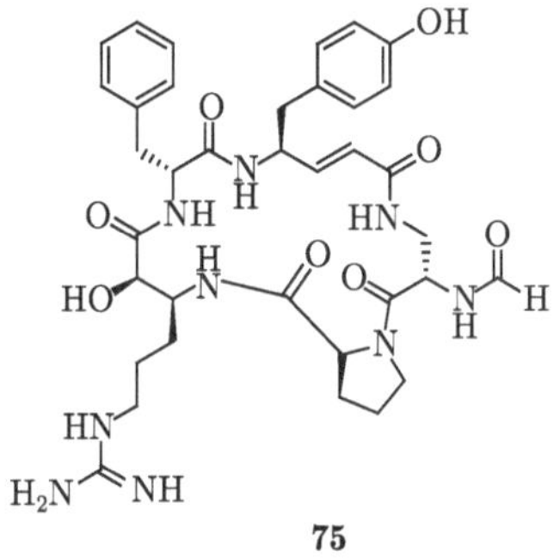

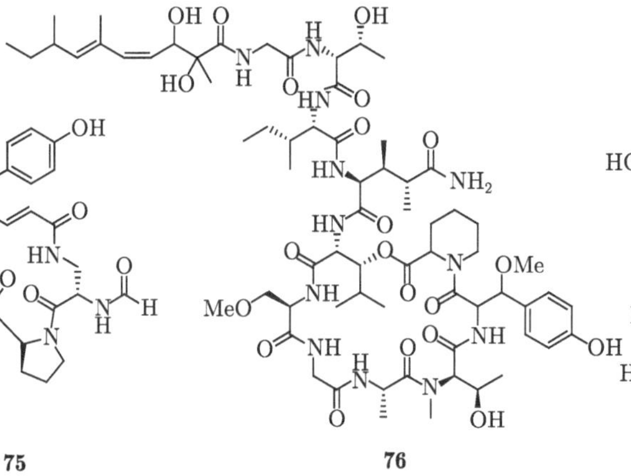

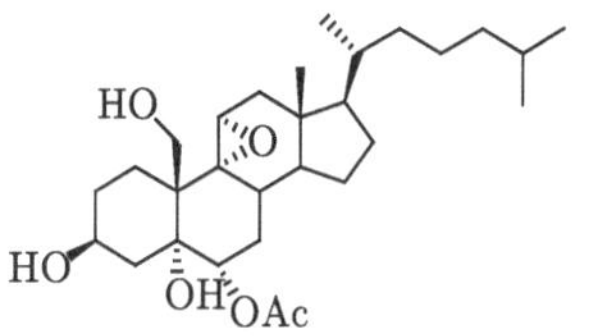

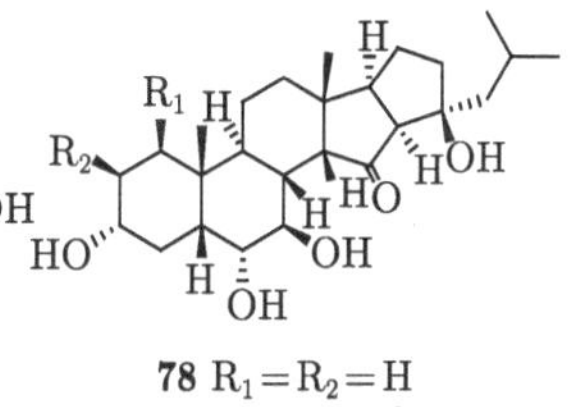

All the polyacetylenic derivatives from sponges exhibited linear structures, usually containing 30 or 46 carbons with some exceptions with miscellaneous carbons. These metabolites are responsible for antimicrobial, anticancer, antivirus, and enzyme inhibition bioproperties. Polyacetylenes are characteristic metabolites of the Mediterranean sponge *Petrosia ficiformis*, and over 20 analogues of C-46 class have been successfully isolated. These metabolites displayed the structural feature of a terminal 1-yn-3-ol-4-ene moiety. The absolute stereochemistry of some structures was elucidated by applying the modified Mosher's method[75].

Steroids from sponges showed a greater structural diversity than those from other marine organisms. The molecules varied not only by high oxygenation, ring cleavage, and arylation, but also dimerization in a symmetrical or unsymmetrical way, which made a contribution to the frequent occurrence of new structural skeletons. Polyhydroxylated steroids dysideasterol A (**77**) and analogues were obtained from the South China Sea sponge *Dysidea* sp., all of which possess $9\alpha,11\alpha$-epoxyl and 19-hydroxyl groups. The structures and relative configurations were confirmed by x-ray diffraction analysis, while the absolute stereochemistry was determined by the modified Mosher's method[76]. Generally, these steroids possessed antivirus, antitumor, antianaphylaxis and other biological activities. Antiallergic steroids were exemplified by xestobergsterol A and B (**78**, **79**), isolated from the Okinawan sponge *Xestospongia bergquistia* in 1992. Both xestobergsterol A and B (**78**, **79**) are unique pentacyclic steroids with both the C16/C23 bond and *cis* C/D ring junction. These compounds strongly inhibited histamine release from rat mast cells induced by anti-IgE. Their activity was found about 5200 times more potent than that of disodium cromoglycate (DSCG), a well-known antiallergic drug ($IC_m = 262$ pM). The effective inhibition of xestobergsterol A and B suggests that both compounds are hopeful as antiallergic drugs[77]. Total synthesis of xestobergsterol A was already achieved in 1999[78].

3.3   Mollucsca (Molluscs)

Research work concerning molluscs has focused on the subclass Opisthobranchia (Gastropoda). Lacking a hard protective shell, opisthobranch molluscs have to employ a series of chemical strategies to protect themselves from predators. All the related chemicals possess antifeedant, repelling, or strong cytotoxic activity, which provides a new approach, distinguished from traditional ones, to search for potential bioactive molecules from natural sources[79]. The earlier reported dolastatin 10 (**6**) was proven to be a potential antineoplastic pentapeptide isolated from sea hare *Dolabella auricularia*. Dolastatin 10 (**6**) demonstrated in vitro and in vivo efficacy in the DU-145 human prostate cancer model by microtubule-inhibitory and apoptotic effects, and was subjected to Phase II clinical trials[6].

Besides polypeptides, marine molluscs also produce numerous other compounds such as terpenes, steroids, carotenoids, alkaloids, macrolids, and polypropene derivatives. In the course of the chemical ecological study on marine molluscs, the chemistry of a Red Sea nudibranceh *Hexabranchus sanguineus*, and two South China Sea nudibranches, *Glossodoris ruformarginata* and *Phyllidiella pustulosa*, were investigated, resulting in the isolation of isocyanosesquiterpenes, isocyanoditerpenes, scalarane sesterterpenes, and a carotenoid containing an α,β-unsaturated cyclopentanone moiety. These compounds demonstrated the typical structural carbon skeletons from the related dietary sponges, which revealed a prey-predator relationship between them. Meanwhile, most of the compounds displayed ichthyotoxicity and cytotoxicity, indicating a defensive ecological role[80].

Marine molluscs seem to produce their secondary metabolites mainly by bio-accumulation, bio-transformation, or bio-modification of dietary chemicals. Some animals can bio-synthesize the necessary metabolites by their own enzyme systems. Therefore, the secondary metabolites of marine molluscs are closely related to those of their dietary organisms. It is then no surprise to observe the typical algae metabolites in the herbivorous molluscs belonging to orders of Sacoglossa and Anaspidea, whereas from carnivorous animals of orders Notaspidea and Nudibranchia, we can always find the characteristic metabolites of corals, sponges, tunicates, bryozoons, etc.[9,59,79,81]. One example is the antitumor cyclopeptide kahalalide F (**44**)[28], which was isolated from both green algae *Bryopsis* sp. and its predator *Elysia rubefescens* as well[82]. Obviously, the homology of metabolites from molluscs and related dietary organisms provides chemical evidence as a basis for the ecological study of marine organisms.

Section 4   Marine Microorganisms and Phytoplankton

Marine microorganisms and phytoplankton include marine bacteria, actinomycetes, fungi, archimycetes, cyanobacteria, prochlorophyta, eukaryotic microalgae, etc. Most of the currently studied marine microorganisms show high homology with terrestrial sources, and are virtually the varieties of land-based ones. Only a few populations, e.g., marine microalgae and actinomycetes, are real marine species.

It is not easy to summarize the structural types of secondary metabolites from marine microorganisms, due to the amazing complexity and diversity of their structures. In general, compounds related to cyclopeptides, alkaloids, halogen analogues, microlides, polyethers, and polyacetylenic derivatives are commonly encountered. These compounds are responsible for antibiotic, anti-inflammation, antineoplastic, antivirus, anticoagulant, immunoregulation, and blood pressure regulation bio-properties. In most cases, these metabolites are inclined to exhibit as marine toxins. An earlier example of naturally occurring toxin from marine sources is brevetoxin B (**2**), a sodium channel modulator isolated from the dinoflagellate *Gymnodinium breve* in 1981. Brevetoxin B represents an entirely new group of natural products made up of a single carbon chain locked into a rigid ladder-like structure consisting of 11 contiguous trans-fused octa-/hexacyclic ethers. The extraordinary structure of brevetoxin B (**2**) was finally established with the aid of x-ray diffraction[2]. Recently, three potential antitumor agents, lingshuiol (**80**), and lingshuiol A and B were isolated from the South China Sea dinoflagellate *Amphidinium* sp. They showed powerful cytotoxic activity against A-549 and L1210 cell lines in vitro with an IC_{50} of 0.21, 0.87, 0.045 $\mu g/mL$, and 0.23, 0.82, 0.49 $\mu g/mL$, respectively[83]. Indeed, microalgae also produce some structurally simple bioactive substances like curacin A (**81**). Curacin A is reported as a potent antimitotic agent by competitively inhibiting the binding of colchicine to tubulin, and is therefore, a promising candidate for new drug development[84].

The most stirring discovery regarding marine microorganisms in the past few years was the successful isolation and cultivation of the obligate marine actinomycete bacterium *Salinospora* sp. *Salinospora* sp., with previously undescribed 16S rRNA gene sequences, displays an obligate requirement of ionic sodium for growth, thus indicating a high level of marine adaptation. Its metabolite, salinosporamide A (**82**), displayed potent in vitro cytotoxicity against HCT-116 human colon carcinoma with an IC_{50} value of 11 ng/mL. This compound also displayed potent and highly selective activity in the NCIcs 60-cell-line panel, and thus, is a bright prospect as a new drug candidate[85]. The successful isolation and cultivation of *Salinospora* strains provides human beings an entirely new source of bioactive compounds, which is crucial for the increasing demand for new lead compounds in the pharmaceutical industry.

Section 5   Miscellaneous

Other marine organisms, such as tunicates and bryozoans, have also been the subjects of chemical studies. Quite a few reports involved tunicates that produced secondary metabolites such as lipids, terpenes, microlides, and nitrogenous compounds. Tunicate metabolites are characterized by their variety of nitrogenous compounds, including cyclopeptides, alkaloids, and nitrogen-containing dimeric steroids. These compounds mainly demonstrated antitumor cytotoxity. Among the discovered molecules and as two examples of nitrogenous compounds, tetrahydroisoquinoline alkaloid ecteinascidin 743 (**5**)[5] and cyclopeptide didemnin B (**83**)[86] were already tested in clinical trials.

A potent anticancer agent isolated from bryozoan *Bugula neritina* is known as bryostatin 1 (**4**). Research work began in 1968 and culminated in the structural elucidation as a 26-membered macrocyclic lactone in 1982, aided by crystallographic techniques[4]. Bryostatin

1 showed antitumor effects at extremely low dose levels, and Phase II clinical trials have already begun in patients with relapsed low grade non-Hodgkin's lymphoma, chronic lymphocytic leukemia, multiple myeloma, advanced renal cancer, as well as non-small cell lung cancer. Commercially available products of bryostatin 1 may be expected in the near future.

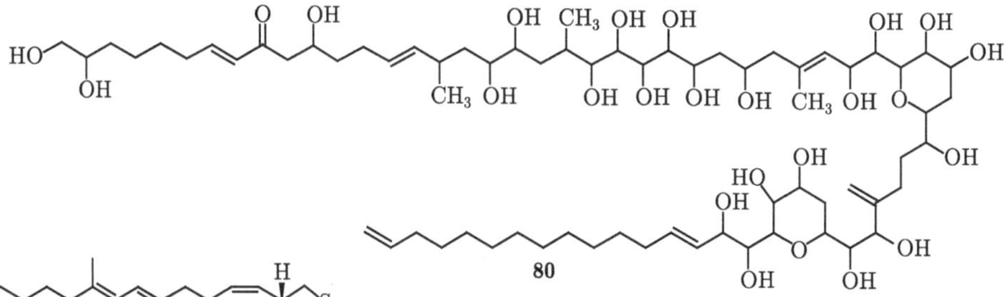

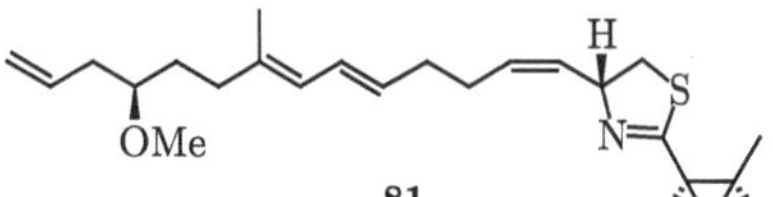

Moreover, isolations of marine toxins, such as tetrodotoxin (**84**)[87] and saxitoxin (**85**)[88], all play significant roles in marine natural product chemistry. The related discoveries greatly deepen and widen our understanding of the nature of the ocean and even nature itself, and furthermore, push the development of related scientific fields such as pharmacology, molecular biology, taxonomy, and ecology.

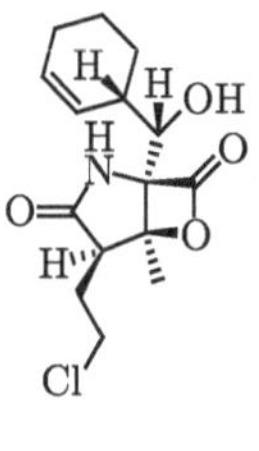

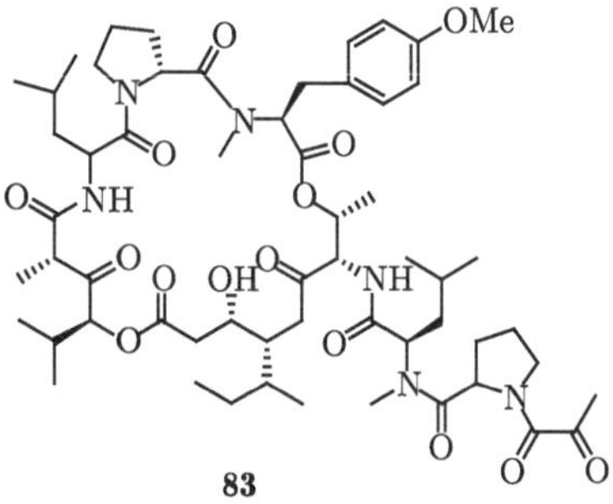

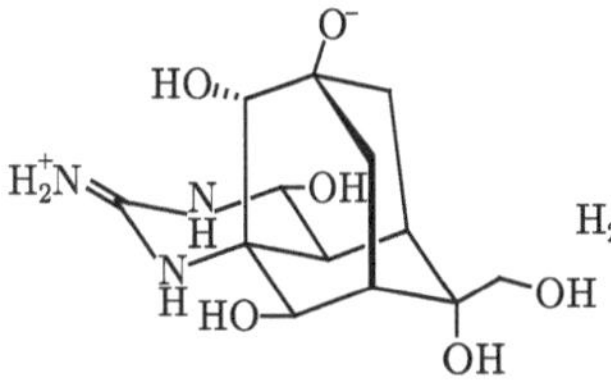

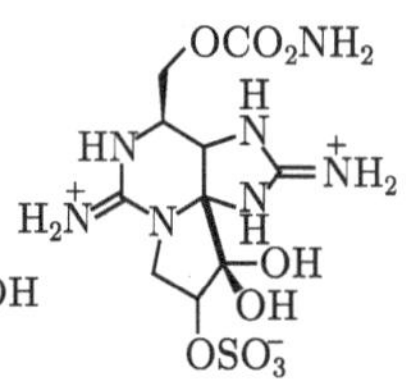

Recent developments in molecular biology and related scientific fields motivated the biological studies related to in vivo bio-transformation and bio-synthesis of marine secondary metabolites. Related research works were reported by Profs. William Gerwick, Guido Cimino, Raymond J. Andersen, and Mary Garson and co-workers who made primary contributions to the field.

Section 6   Examples: Isolation and Structural Elucidation of Uncommon Polyoxygenated 24-Ketal Steroids from the South China Sea Gorgonian *Subergorgia reticulata*

Animals of *Subergorgia reticulata* belong to the family Subergorgiidae (order Gorgonacea). Chinese species of *Subergorgia reticulata* are mainly distributed in the sea areas of Hainan, Guangdong, and Hong Kong[89].

Chemistry of genus *Subergorgia* comprises a variety of structurally novel compounds, including sesquiterpenes which possess different carbon skeletons such as subergane,[90−92] quadrone,[93] β-caryophyllene,[93,94] and 9,11-secosteroids as well[95,96]. During the process of chemical investigation on marine organisms from the South China Sea, an opportunity was presented to analyze the chemistry of *Subergorgia reticulate*, never studied so far. This research work resulted in the isolation of five new uncommon polyoxygenated steroids, namely suberoretisteroids A-E (**47, 86** to **89**), all displaying the rare 24-ketal functional group[34].

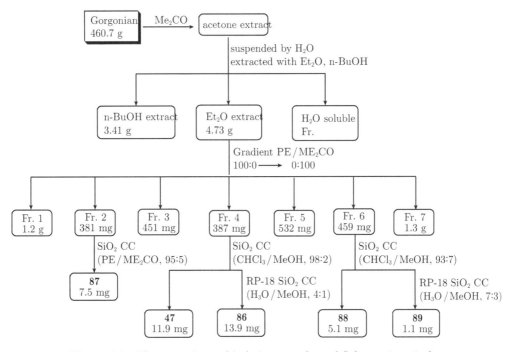

86 R$_1$=H, R$_2$=Ac   **88**   **89**
87 R$_1$=R$_2$=Ac

The extraction and isolation procedure of *Subergorgia reticulata* is shown in Figure 14-7.

Figure 14-7   The extraction and isolation procedure of *Subergorgia reticulata*

The structure and relative stereochemistry of suberoretisteroids A-E (**47, 86** to **89**) were elucidated on the basis of extensive spectral analysis in combination with chemical correlation and comparison with reported data of steroids obtained from Indian gorgonian *Gorgonella umbraculum*,[97,98] and further confirmed by x-ray diffraction analysis on a single crystal of **47**[99]. Suberoretisteroid A (**47**) displayed inhibitory activities in vitro against the growth of human lung adenocarcinoma cell A-549 and cervical carcinoma cell HeLa with inhibitions of 43.3% and 31.1% at a concentration of 10^{-6} mol/L, respectively.

Section 7   Epilogue (Outlook)

The twenty-first century is a new epoch of ocean exploration. Human beings have to rise against the increasing challenges of explosion in population, shortage in resources, and worsening of environment. The continuous decrease of land-based resources makes increasingly urgent and significant the quest for food and drugs from the sea. On the other hand, owing to the global change in climate and the day-by-day heavy environmental pollution, many marine organisms are never recognized, named, or numbered when they die out soundlessly.

It is, therefore, even more necessary to reinforce and accelerate the research work related to marine organisms. The thorough and systematic studies in biology and chemistry of marine organisms will help human beings to acquire some valuable, intrinsic information (e.g., pharmaceutical properties), and in turn, make it serve people in survival and development as soon as possible. America, Japan, and countries in Europe are leaders in the research fields of marine biology, marine natural product chemistry, marine chemical ecology, etc. This research trend was embodied by the increasing reports related to marine natural product chemistry that are published annually in the *Journal of Natural Products*. A large amount of patents protect their significant productions.

To cope with the great challenge of intellectual property rights and technical innovation after China was enrolled in the WTO, China recently put more attention and financial support on the basic study and exploitation of its marine sources. Under the guidance of state regulation, marine drug research was first enrolled as an independent topic in the state marine "863" program during the Tenth Five-year Plan period. Marine natural product programs focusing on the discovery of new drug lead compounds were initiated simultaneously.

Marine organisms are already, and will continue to be precious source of structurally novel bioactive compounds for new drug discovery. Following marine natural product research, numerous compounds with tremendous structural diversity are reported in increasing numbers, which greatly enriches organic chemistry, and continuously broadens knowledge regarding the nature of the ocean and even nature itself. The isolation of similar metabolites from different kinds of organisms not only provides us multiple sources for precious compounds which hold promise for subsequent artificial culture and industrialized manufacture, but also play an ecological role in elucidating the symbiosis, commensalisms, and prey-predator relationships among the organisms.

Many scientific disciplines study the marine environment today. In addition to chemistry, oceanic studies are undertaken in the fields of pharmacology, ecology, taxonomy, molecular biology, geology, physics, and material sciences. Interaction among these fields is advancing the development of cross-disciplinary techniques that will aid research work in natural products chemistry as well as other fields.

Studies of the ocean environment will not only yield an improved scientific understanding of marine ecology, but will likely also benefit mankind in many ways. It's important that any exploitation of the seas be carefully carried out so as to avoid damage to their sensitive ecology.

References

[1] Tsuda M, et al. Tetrahedron Lett, 1998, 39, 1207.

[2] Lin Y-Y, et al. J Am Chem Soc, 1981, 103, 6773.

[3] (a) Moore RE, et al. Science, 1971, 172, 495 (b) Moore RE, et al. J Am Chem Soc, 1981, 103, 2491.

[4] Pettit GR, et al. J Am Chem Soc, 1982, 104, 6846.

[5] Wright AE, et al. J Org Chem, 1990, 55, 4508.

[6] Pettit GR, et al. J Am Chem Soc, 1987, 109, 6883.

[7] Hirata Y, et al. Pure Appl Chem, 1986, 58, 701.

[8] Look SA, et al. J Org Chem, 1986, 51, 5140.

[9] Blunt JW, et al. Nat Prod Rep, 2005, 22, 15-61 and earlier articles in this series.

[10] Chen P-S. Acta Phytoecol Geobot Sinica, 1985, 9, 59.

[11] Zhao Y, et al. Chin J Nat Med, 2004, 2, 135.

[12] (a) Wang J-D, et al. Helv Chim Acta, 2004, 87, 2829 (b) Wang J-D, et al. Helv Chim Acta, 2005, 88.

[13] Wu J, et al. Org Lett, 2004, 6, 1841.

[14] (a) Sun Y-Q, et al. Tetrahedron Lett, 2004, 45, 5533 (b) Sun Y-Q, et al. Z Kristallogr NCS, 2004, 219, 121.

[15] (a) Kato, A. Tetrahedron Lett, 1972, (3), 203 (b) Kato, A. Phytochemistry, 1975, 14, 1458 (c) Kato, A. Phytochemistry, 1976, 15, 220.

[16] Mao S-C, et al. Chin Trad Herb Drugs, 2004, 35, 8.

[17] Juagdan EG, et al. Tetrahedron, 1997, 53(2), 521.

[18] Mao S-C, et al. Helv Chim Acta, 2005, in press.

[19] Vairappan CS, et al. Phytochemistry, 2001, 58, 517.

[20] Fisch KM, et al. J Nat Prod, 2003, 66, 968.

[21] Bennamara A, et al. Phytochemistry, 1999, 52, 37.

[22] Ochi M, et al. Bull Chem Soc Jpn, 1986, 59, 661.

[23] Ayyad SEN, et al. Z. Naturforsch, C: Biosci, 2003, 58, 33.

[24] (a) Gerwick WH, et al. Tetrahedron Lett, 1979, 145 (b) Gerwick WH, et al. J Org Chem, 1981, 46, 22.

[25] Wessels M, et al. J Nat Prod 1999, 62, 927.

[26] Paul VJ, et al. Phytochemistry, 1985, 24, 2239.

[27] (a) Maiti BC, et al. J Chem Res, (S), 1978, 126 (b) Schewede JG, et al. Phytochemistry, 1987, 26, 155.

[28] Hamann MT, et al. J Am Chem Soc, 1993, 115, 5825.

[29] Bonnard I, et al. J Nat Prod, 2003, 66, 1466.

[30] (a) Zhang W, et al. Chin J Nat Med, 2003, 1, 69 (b) Coll JC. Chem Rev, 1992, 92, 613.

[31] Schmitz FJ, et al. Tetrahedron Lett, 1966, 97.

[32] Weinheimer AJ, et al. Tetrahedron Lett, 1969, 5186.

[33] Kazlauskas R, et al. Tetrahedron Lett, 1977, 4439.

[34] Zhang W, et al. Helv Chim Acta, 2005, 88, 87.

[35] Zhang W, et al. J Nat Prod, 2004, 67, 2083.

[36] (a) Wu Z-D, et al. Acta Sci Nat Univ Sunyatseni, 1982, (3), 69 (b) Peng YC, et al. Chin J Mar Drugs, 1996, (1), 1 (c) Ai XH, et al. J South Chin Normal Univ (Nat Sci), 1999, (2), 75.

[37] Zhang W, et al. Helv Chim Acta, 2004, 87, 2919.

[38] Burks JE, et al. Acta Crystallogr, 1977, B33 (3), 704.

[39] Sung P-J, et al. J Nat Prod, 1999, 62, 457.

[40] Zhang W, et al. Helv Chim Acta, 2004, 87, 2341.

[41] (a) González N, et al. J Org Chem, 2002, 67, 5117 (b) Banjoo D, et al. J Nat Prod, 2002, 65, 314 (c) Taglialatela-Scafati O, et al. Org Lett, 2002, 4, 4085.

[42] Rodríguez AD, et al. Org Lett. 1999, 1, 337.

[43] Yan X-H, et al. Chin J Nat Med, 2005, 3, 65.

[44] Carmely s, et al. Tetrahedron Lett. 1980, 21, 875.

[45] Watanabe K, et al. J Nat Prod, 2001, 64, 1421.

[46] Iwashima M, et al. J Org Chem, 2002, 67, 2977.

[47] (a) Shao Z-Y, et al. J Nat Prod, 2002, 65, 1675 (b) Yan X-H, et al. Chin Chem Lett, 2005, 3, 356.

[48] Bortolotto M, et al. Steroids, 1976, 28, 461.

[49] Jia R, et al. Nat Prod Res, 2005, 19, 789.

[50] Morris LA, et al. Tetrahedron, 1998, 54, 12953.

[51] Yan X-H. Chin J Org Chem, 2004, 24, 1233.

[52] Jia R, et al. Helv Chim Acta, 2005, 88, 1028.

[53] El Sayed KA, et al. J Nat Prod, 1996, 59, 687.

[54] Zeng LM, et al. J Nat Prod, 2004, 67, 1915.

[55] (a) Armstrong RW, et al. J Am Chem Soc, 1989, 111, 7525 (b) Suh EM, et al. J Am Chem Soc, 1994, 116, 11205.

[56] (a) Long K-H, et al. Chin J Org Chem, 1985, 5, 360 (b) Fusetani N, et al. Chem Rev, 1993, 93, 1793.

[57] Huang, X-C, et al. Chin J Nat Med, 2005, 3, 1.

[58] (a) Guo Y-W, et al. Nat Prod Lett, 1997, 10, 143 (b) Guo Y-W, et al. J Asia Nat Prod Res, 2000, 2, 251.

[59] Garson MJ, et al. Nat Prod Rep, 2004, 21, 164.

[60] (a) Hirota H, et al. Tetrahedron, 1998, 54, 13971 (b) Wratten S, et al. Tetrahedron Lett, 1978, 1391 (c) Wratten S, et al. Tetrahedron Lett,1978, 1395 (d) Simpson JS, et al. Tetrahedron, 1997, 38, 7947 (e) Tanaka J, et al. J Nat Prod, 1999, 62, 1339.

[61] (a) Zhang WH, et al. J Nat Prod, 2001, 64, 1489 (b) Blackburn CL, et al. J Org Chem, 1999, 64, 5565 (c) Crews P, et al. Tetrahedron, 2000, 56, 9039.

[62] Potts BCM, et al. J Am Chem Soc, 1992,114, 5093.

[63] Guo Y-W, et al. Nat Prod Lett, 1996, 9, 105.

[64] (a) Yu Z-G, et al. 2005, 88 (b) Yu Z-G, et al. Z Kristallogr NCS, 2004, 219, 415.

[65] Gunasekera SP, et al. J Am Chem Soc, 1996, 118, 8759.

[66] Jiménez JI, et al. J Org Chem, 2000, 65, 6837.

[67] Capon RJ, et al. J Nat Prod, 1998, 61, 660.

[68] (a) Guo Y-W, et al. Tetrahedron, 1996, 52, 8341 (b) Guo Y-W, et al. Tetrahedron, 1996, 52, 14961 (c) Guo Y-W, et al. Tetrahedron Lett, 1998, 39, 463.

[69] Guo Y-W, et al. Tetrahedron, 1998, 54, 541.

[70] Bialojian C, et al. Biochem J, 1988, 256, 283.

[71] Bai R, et al. Mol Pharmacol, 1993, 44, 757.

[72] (a) Paterson I, et al. Angew Chem Int Ed Engl, 2001, 40, 4055 (b) Crimmins MT, et al. J Am Chem Soc, 2002, 124, 5661.

[73] Clark DP, et al. J Org Chem, 1998, 63, 875.

[74] (a) Nakao Y, et al. J Am Chem Soc, 1999, 121, 2425 (b) Ford PW, et al. J Am Chem Soc, 1999, 121, 5899.

[75] (a) Guo Y-W, et al. Tetrahedron, 1994, 50, 13261 (b) Guo Y-W, et al. J Nat Prod, 1995, 58, 712 (c) Guo Y-W, et al. J Nat Prod, 1998, 61, 333.

[76] Huang X-C, et al. Helv Chim Acta, 2005, 88, 281.

[77] Shoji N, et al. J Org Chem, 1992, 57, 2996.

[78] (a) Jung WE, et al. Org Lett, 1999, 1 (10), 1671 (b) Takei M, et al. Experientia, 1993, 49, 145

[79] (a) Karuso, P. In Bioorganic Marine Chemistry, P. J. Scheuer, Ed., Springer-Verlag: Berlin, 1987, Vol. 1, pp. 31-60 (b) Cimino G., et al. Curr Org Chem, 1999, 3, 327.

[80] (a) Guo Y-W, et al. Tetrahedron Lett, 1998, 39, 2635 (b) Gavagnin M, et al. J Nat Prod, 2004, 67, 2104 (c) Manzo E, et al. J Nat Prod, 2004, 67, 1701.

[81] Cimino G, et al. In *Bioactive Compounds from Natural Sources: Isolation, Structure Characterization and Biological Properties*; Tringali C, Ed; Taylor & Francis Ltd: London, 2001, pp. 578-637.

[82] Hamman MT, et al. J Org Chem, 1996, 61, 6594.

[83] (a) Huang X-C, et al. Bioorg Med Chem Lett, 2004, 14, 3117 (b) Huang X-C, et al. Tetrahedron Lett, 2004, 45, 5501.

[84] (a) Gerwick WH, et al. J Org Chem, 1994, 59, 1243 (b) Blokhin AV, et al. Mol Pharmacol, 1995, 48, 523.

[85] Feling RH, et al. Angew Chem (Int Ed), 2003, 42, 355.

[86] Sakai R, et al. J Med Chem, 1996, 39, 2819.

[87] McConnell OJ, et al. In The Discovery of Natural Products with Therapeutic Potential, Gullo VP, Ed, Butterworth-Heinemann, Boston, MA, 1994, 109.

[88] Kao CY. Fed Proc, 1973, 31, 1117.

[89] Zuo RL et al. In Marine Species and Their Distributions in China's Sea, Huang Z-G, Ed; China Ocean Press: Beijing, 1994. 295-297.

[90] Groweiss, A et al. Tetrahedron Lett, 1985, 26: 2379.

[91] Parameswaran, P S et al. J Nat Prod, 1998, 61: 832.

[92] Wang, GH et al. J Nat Prod, 2002, 65: 1033.

[93] Bokesch, HR et al. Tetrahedron Lett, 1996: 37, 3259.

[94] Wang, GH et al. J Nat Prod, 2002, 65: 887.

[95] Anjaneyulu, AS R. et al. Indian J Chem, 1997, 36B: 418.

[96] Aknin, M et al. Steroids, 1998, 63: 575.

[97] Subrahmanyam, C et al. Chem Res (S), 2000: 182.

[98] Anjaneyulu, ASR et al. Nat Prod Res, 2003, 17: 149.

[99] Zhang W et al. Acta Cryst, 2005, E61: 2884.

Wen ZHANG and Yue-Wei GUO

CHAPTER 15

Structural Modification of Active Principles from TCM

Section 1   Introduction

For thousands years of human history, natural products (NPs) derived from plants, animals, microbes, and marine organisms have played an important role for treating and preventing diseases. WHO has estimated that herbal medicines serve the health needs of about 80% of the world's population[1]. An analysis of the origin of the drugs developed between 1981 and 2002 showed that NPs or NP-derived drugs comprised 28% of all new chemical entities (NCEs) launched onto the market[2]. In addition, 24% of these NCEs were synthetic or natural mimic compounds, based on the study of pharmacophores related to NPs[3]. This combined percentage (52% of all NCEs) suggests that NPs are important sources for new drugs and are also precursors for further modification as drugs. The large proportion of NPs in drug discovery was attributed to the diverse structures and intricate carbon skeletons of NPs. In fact, several important drugs of modern medicine have come from medicinal plant, e.g., taxol, vinblastine, vincristine, topotecan, irinotecan, etoposide, teniposide, etc.[4]. Study found that NPs or related substances accounted for 40%, 24%, and 26% of the top 35 worldwide ethical drug sales in 2000, 2001, and 2002, respectively[5].

Traditional Chinese medicine (TCM) has a long history of serving people and tends to raise the natural defenses of the organism instead of trying to restore its natural functions. The earliest recorded TCM known as "Prescriptions for fifty-two diseases" was found in the *Changsha Mawangdui Han Tomb*; it was compiled around 1100 BC and contained 280 TCM prescriptions for the treatment of 52 diseases. The Chinese people have accumulated rich empirical knowledge of TCMs in treating diseases and have an enormous body of human clinical data on efficacy and toxicity. It is commonly understood in China that chemists and pharmacologists prefer to exploit Chinese medicinal plants for medical use, and discovering and developing lead compounds from TCMs is regarded as more effective and economic than other approaches[6].

Structural modifications to NPs to enhance activity or selectivity and reduce side effects or toxicity developed as organic chemistry grew in the late 19th century[7]. The main objective of this chapter is to review the evolution of several recent modern drugs which originated from TCMs or NPs, in an attempt to provide useful references for the designs and approaches for the modification of NPs or TCMs.

Section 2   Anti-malarial Drug Artemether

Artemether (**1**) might be the most famous case for research and development of a modern drug in China. It was developed independently by Chinese scientists from the discovery of the parent structure artemisinin (**2**), to the structural modification, preclinical studies, and clinical trials. It is also the first strictly TCM-based modern drug developed by Chinese scientists from a TCM formula to a single active ingredient and approved to enter the mainstream medicinal markets of Europe and the United States.

Malaria is the most common killer of all the parasitic diseases that affect humans. About 300 million people are infected every year and most of them are children. In sub-Saharan Africa, over a million children die annually from the disease (WHO 1998). Malaria is caused by the genus *Plasmodium*, of which four species infect humans: *P. falciparum* (malignant tertian), *P. vivax* (benign tertian malaria), *P. ovale* (ovale tertian) and *P. malariae* (quartan malaria). All are transmitted by the bite of an infected female Anopheles mosquito. Chloroquine and other quinine alkaloid derivatives have long been the drugs of choice in most regions where malaria is endemic. Parasite resistance to common drugs has become a widespread problem, so there is major interest in finding new compounds without drug resistance.

Chinese scientists gained inspiration from TCM books and records, and chose several TCM herbs used for the treatment of malaria and/or hemorrhoids as research targets. In 1972, they isolated artemisinin (2), a novel sesquiterpene lactone with peroxy bridge, from *Artemisia annua* L. (Qinghao, composites), which showed potent suppressive effects on chloroquine-sensitive strains and chloroquine-resistant *vivax* and *falciparum* malaria strains. This discovery broke the restriction that the molecule of antimalarial agents must contain an *N*-atom.

Artemisinin is a rapid acting, efficient, low toxicity, non-quinine agent against both drug-resistant and drug-sensitive malaria. Even if a patient suffers from severe cerebral malaria, the patient could rapidly recover after nasal use of artemisinin. But its two obvious deficiencies restricted clinical applications: poor bioavailability, because of its insolubility in water or oil, and the high recrudescent rate of parasites (45.8% in 28 days). In order to find a better anti-malaria agent, studies on the anti-malaria mechanism and structural modification of **2** were conducted.

The metabolic studies of **2** in the human body found that the urinary and fecal excretion of the unchanged drug were major routes of elimination. The other minor metabolites (Figure 15-1), characterizing the loss of the peroxy bridge functional group, showed negative reaction against *P. berghei* in mice, suggesting that the peroxy bridge of **2** was the pharmacophore. Another clue is that the Raney nickel hydrogenated product of **2** also lost bioactivity. Simultaneously, dihydroartemisinin (**3**) was found to be more active than **2**. Given that the introduction of a hydroxyl group into the molecular nucleus could not improve its water solubility, a plan of preparing its derivatives to search for more active and soluble compounds emerged. Over 50 derivatives, comprising three types (ethers (**4**), esters (**5**) and carbonates (**6**)) were semisynthesized and evaluated in 1977. Most showed greater activity than **2** and **3** (SD_{50} 6.5 and 3.65 mg/kg for **2** and **3**, respectively), and the top compounds are listed in Table 15-1[8].

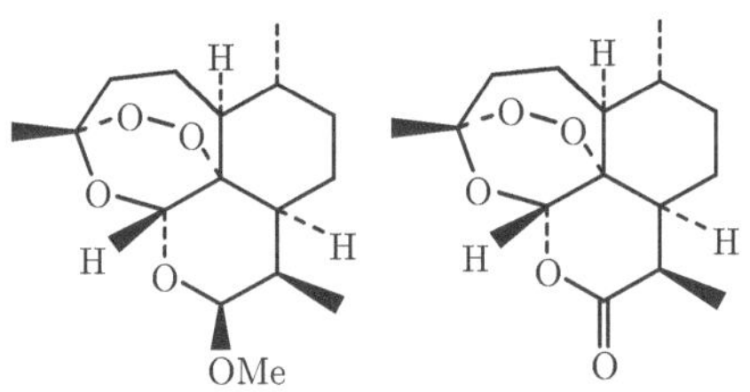

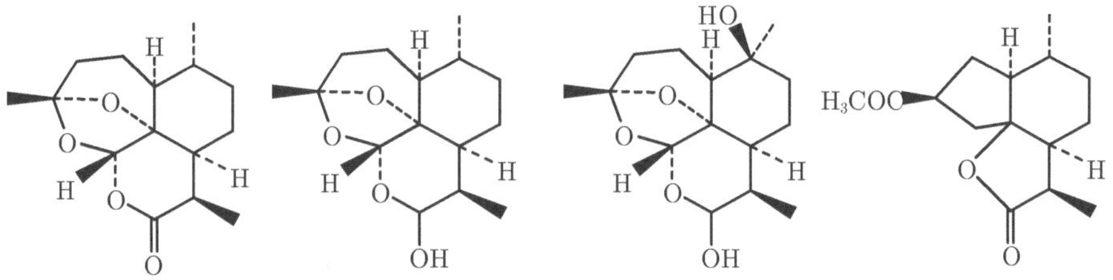

Figure 15-1   Structures of **1** and **2** and the metabolites of **2**

Table 15-1   The Efficacies of Derivatives of **2** on Choloroquinine-Resistant *P.berghei* Mice

Ethers	SD$_{50}$(mg/kg)	Esters	SD$_{50}$(mg/kg)	Carbonates	SD$_{50}$(mg/kg)
β-Me (**1**)	1.02	α-COEt	0.66	α-OCOC$_3$H$_7$	0.50
α-Me	1.16	α-COCH=CH-Ph	0.74	β-OCOC$_3$H$_7$	1.32
β-Et	1.95	α-COCH$_2$Ph	0.95	α-OCOC$_2$H$_5$	0.63
β-CH$_2$Ph	3.42	α-COCH$_3$	1.20	OCOPhg($\alpha + \beta$)	0.63
β-(CH$_2$)$_2$O(CH$_2$)$_2$CH$_3$	3.71	COPh-CH$_3$	1.73		

After comprehensive consideration of the bioactivity, chemical stability, toxicity, and product cost, artemether (**1**) and artesunate (**7**) were selected as drug candidates for further development. The former is highly soluble and easily prepared in oil as an injectable. Its recrudescent rate was 6.7% in 28 days. The latter is also easily dissolved in water and may be made in powder form for injection. The recrudescent rate was 50% in 28 days. When the treatment period was extended and doses were increased, the recrudescent rate was less than 10%.

The synthetic routes of the above derivatives are illustrated in Scheme 15-2. Dihydroartemisinin (**3**) was treated with alcohol to produce artemisinin ethers under catalysis by Lewis acid, e.g., BF$_3$·Et$_2$O or hydrochloric acid. The products were a pair of bioactive isomers (β-isomer and α-isomer). **3** was treated with acetic anhydride (or acyl chlorides) or chloroformate in pyridine to produce artemisinin esters (**5**) or carbonates (**6**).

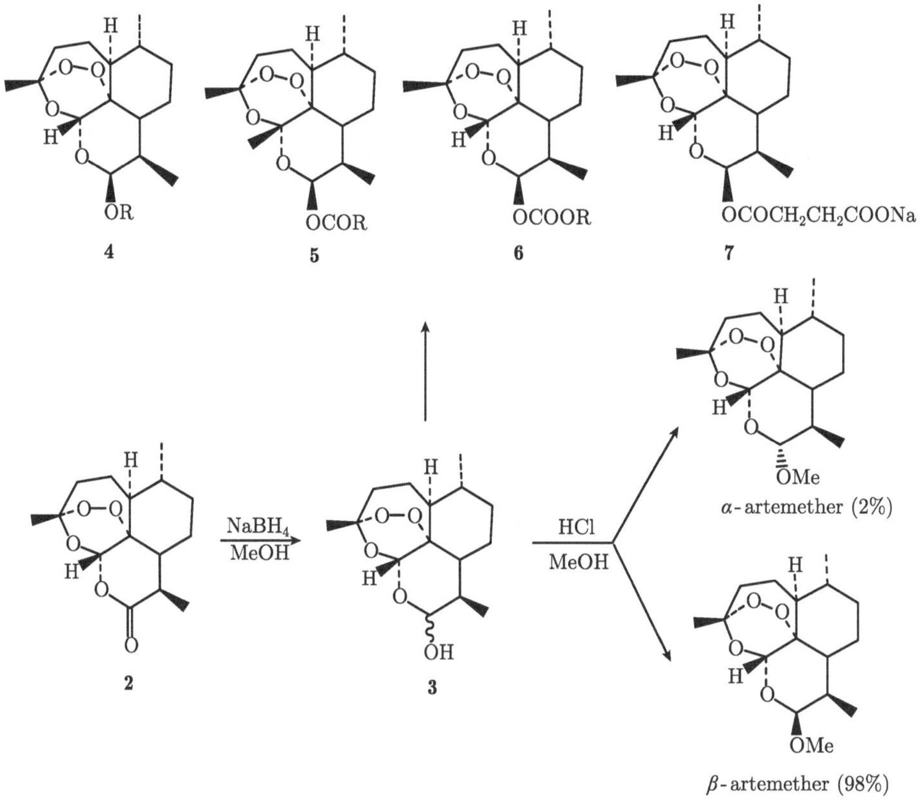

Scheme 15-2   Synthetic routes for derivatives of **2**

The antimalarial experiment of **1** was performed on mice infected with chloroquinine-resistant *P. berghei* by administration of intramuscular injection. The suppressive dose of **1** for 90% parasitemia (SD$_{90}$) was 1 mg·kg^{-1} and the minimal dose for clearing parasitemia

(MDCP) was 5.4 mg·kg^{-1}. In contrast, those of **2** were 6.2 and 75 mg·kg^{-1}, respectively. In monkey experiments, the oil injection of **1** had MDCP of 3 mg·kg^{-1}, and the minimal dose for no recrudescence was 12 mg·kg^{-1}, correspondingly, the oil injections of **2** were 30 and 60 mg·kg^{-1}, respectively. Toxicological and pharmacokinetic studies on **1** exhibited that no obvious side effects nor mutagenesis, carcinogenesis or teratogenesis were observed. Also, 1088 patients, including 829 *vivax malaria* and 259 *falciparum malaria*, were 100% cured without any liver, kidney, or heart side effects, and without any influence to pregnant women and their infants. Artemisinin, artemether, and artesunate were approved as the new antimalarial drugs by Chinese SFDA in 1986 and 1987. The clinical application of **1** in endemic regions, e.g., Thailand, Cambodia, Brazil, Togo, and Tanzania, also showed high efficacy on malaria patients.

Section 3   Anti-hepatitis Drug Bifendate

Schisandra chinensis (Turcz.) Baill is a commonly used TCM for the treatment of hepatitis. Pharmacological studies showed it could reduce the serum transaminase levels of hepatitis patients. Schizandrin C (**8**), a lignin isolated from this plant and having a potent liver protecting effect from damage induced by CCl$_4$ in mice, was selected as the lead compound for structural modification.

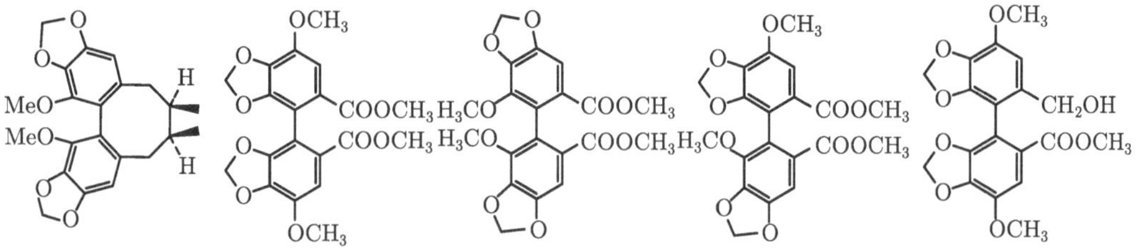

Figure 15-3   Structures of schizandrin C (**8**) and its derivatives **9** through **12**

Bifendate (biphenyldicarboxylate (**9**), an intermediate for synthesis of **8**, was found to have better anti-hepatitis virus effects than target compound **8**. Pharmacological studies indicated that α-bifendate could enhance liver detoxification and protect the liver from chemical damage in several animal models. Clinical applications of **9** on hepatitis patients have shown obvious efficacy on the reduction of serum glutamic-pyruvic transaminase (SGPT) levels and improving symptoms.

Bifendate (**9**) could not only lowered the CCl$_4$-induced high SGPT level in mice, but also protected the liver from damage. It also lowered the liver tissue GPT level of normal mice, CCl$_4$-treated or prednisone-treated mice, with no obvious influence on glutamic oxaloacetic transaminase (GOT) and aldolase activities. When **9** lowered the normal mouse GPT level, the GPT heart activity was not obviously changed. Furthermore, researchers found **9** to inhibit α-fetoprotein (AFP), a liver cancer marker. Further studies found some asymmetric biphenyl compounds to have strong inhibitory effects on cancer cells. The first total synthesis of **9** was conducted by Xie et al. via a five-step reaction starting from gallic acid (Scheme 15-2)[9]. The side products of this reaction were β-bifendate (**10**) and γ-bifendate (**11**).

Based on the bifendate structure, Zhang and Liu et al. designed and synthesized 4,4-dimethoxy-5,6,5′, 6′-dimethylenfrdioxy-2-hydroxymethyl-2′-carbonyl biphenyl, named "bicyclol (**12**)", containing a hydroxymethyl instead of the 2-methoxycarbonyl group of **9**. Oral administration of **12** at 50, 100, or 200 mg·kg^{-1} could protect mice livers from damage induced by CCl$_4$, D-galactosamine, paracetamol and B.C.G. plus lipopolysaccharides expressed in decreases of alanine transaminase (ALT) and aspartate transferase (AST), as well as morphological damage of liver tissues. **12** used at 5×10^{-4} mol/L showed obvious inhibitory effect on HBV-DNA replication, as well as secretion of HBeAg and HBsAg of a transgenic human hepato-carcinoma cell line HepG2.2.15. The results indicated that **12** is not an inhibitor of transaminase, but a cell membrane protector through elimination of free radicals. Moreover, **12** could protect hepatocyte nuclear DNA from damage and reduce the apoptosis induced by the immuno-stimulating compound concanavalin A[10]. The oral LD_{50} of **12** was over 5 g/kg in mice and rats. Furthermore, no noticeable alterations in subacute and chronic toxicity were demonstrated in rats and dogs. Bicyclol (**12**) has shown no mutagenicity or teratogenicity in animals[11].

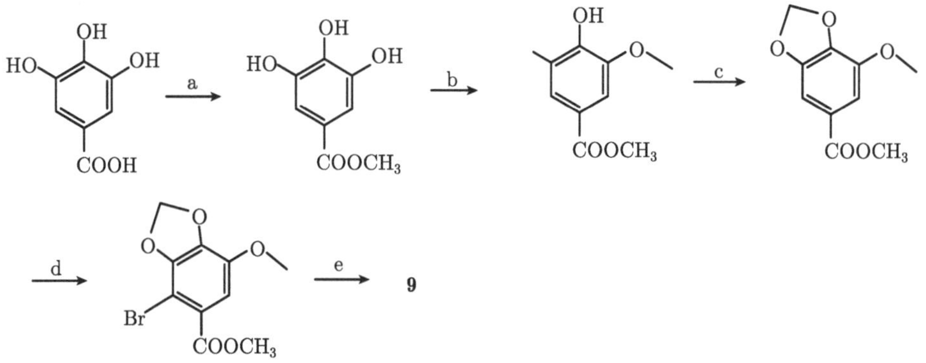

a) MeOH, H_2SO_4, 82%; b) $NaB_4O_7 \cdot 10H_2O$, Me_2SO_4, 79%; c) CH_2I_2, K_2CO_3, acetone, reflux, 69%; d) Br_2, $CHCl_3$, 64%; e) Ullmann reaction, 87%.

Scheme 15-2   The synthetic approach of bifendate (9)

A clinical trial of 2,200 patients with chronic hepatitis B (CHB) or chronic hepatitis C (CHC) was conducted with an oral dose of **12** at 75 mg daily for 24 weeks, then stopped and monitored for 12 weeks. Results showed no serious adverse events during the trial. The adverse event rate for the duration was 10.87%. The normalization rates of ALT for CHB and CHC patients were 60.37 % and 54.00 %, respectively, after 24 weeks of treatment. The sustained normalization rates of ALT were 71.88 % and 65.00 %, respectively, for the 12 weeks after treatment[12].

Section 4   Anti-AD Drug Huperzine A[13][14]

Huperzine A (HA, **13**) is a naturally occurring alkaloid isolated from the Chinese medicinal herb *Qian Ceng Ta* (*Huperzia serrata* = *Lycopodium serratum*) and related genera by Chinese scientists in the early 1980s[15]. Animal models and *in vitro* pharmacological studies demonstrated that **13** was a potent, selective, and reversible acetylcholinesterase (AChE) inhibitor, which crosses the blood-brain barrier smoothly and shows high specificity for AChE with a prolonged biological half-life[16]. It has been approved as a drug for the treatment of Alzheimer's disease (AD) in China. Because the preparation and usage of HA were permitted without patent protection, it is sold in the United States as a dietary supplement.

AD is a progressive, degenerative disease of the brain. The disease is the most common form of dementia affecting elderly people, with a mean duration of around 8.5 years between the onset of clinical symptoms and death. The incidence of AD increases with age, even in the oldest age groups: from 0.5% at 65, it rises to nearly 8% at 85 years old. Some 12 million people have AD, and by 2025, that number is expected to increase to 22 million.

Structure-Activity Relationship (SAR) of HA The powerful bioactivities of HA attracted scientists' interest in its structure-activity relationship. Over 100 HA analogues were designed and synthesized, which structurally changed on adding or removing the modifiable functional groups, or replaced the structural elements of HA, including the quinolinone ring, exocyclic ethylidene, primary amino, or three-carbon bridge ring, using similar or equipotential functions. Most HA analogues were far less active than HA, except for **14a** through **c**, which were C-10 axial methyl substituted analogues of HA. Although **14** has been found to have comparable or somewhat more potent anti-AChE activities than (±)-HA, the preparation of these analogues is quite laborious, and far more expensive than natural HA.

The structural modifications starting from natural HA were also performed. Due to the rigid conformation of HA, the alteration was focused on the pyridone ring and the primary amino group. Results showed that the reduction of two pairs of double bonds and the addition of substituents on the pyridone ring would result in a loss of activity.

This research indicated that the structural elements of HA were essential for the inhibitory activity on AChE. Based on this information, Zhu Da-Yuan designed a series of HA derivatives substituted on the prime amine group, comprising HA amides (**15a**) and HA Schiff's bases (**15b**) (Figure 15-3). Several derivatives of both types showed better activity than HA. After taking into account the bioactivity, chemical stability, and toxicity, ZT-1 (**16**) was selected as a drug candidate, named "schiperine". Structurally, ZT-1 is a condensation product of HA and 5-Cl-isovanillin (Scheme 15-3). ZT-1 has a longer duration time, lower toxicity, and better bioavailability compared with HA.

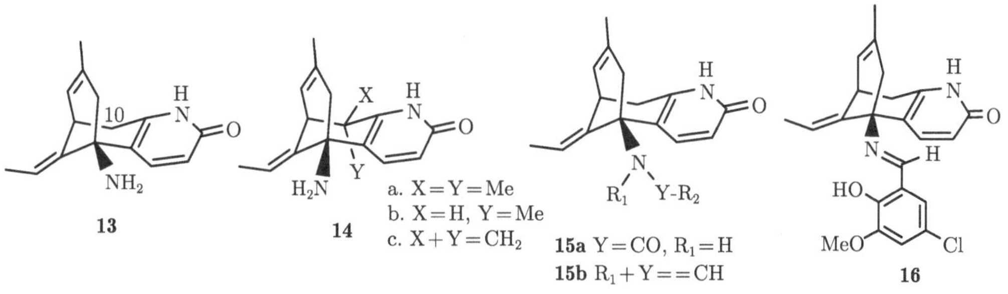

Figure 15-3   Structures of 13-16

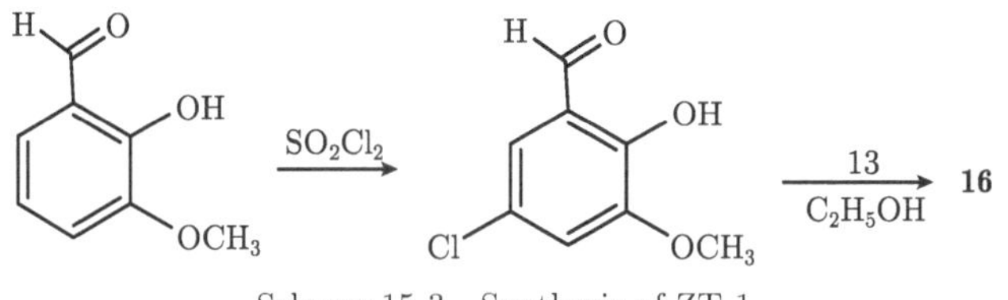

Scheme 15-3   Synthesis of ZT-1

The efficacy of ZT-1 on AChE inhibition and amelioration of memory deterioration were evaluated and compared with huperzine A, tacrine, and donepezil (Figure 15-4).

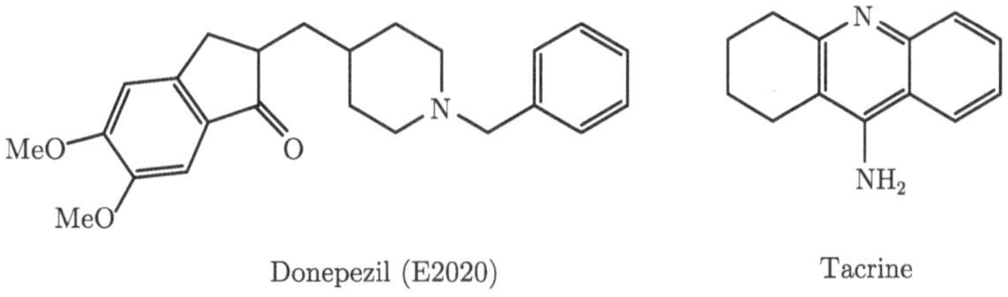

Donepezil (E2020)   Tacrine

Figure 15-4   Structures of donepezil and tacrine

In vitro experiments tested the inhibitory effect of AChE inhibitors on AChE of rat cortex and rat, bovine, and human cellular membranes. The results showed that the IC_{50} values of **16** were 83, 90, 99 and 78 nmol/L, respectively, comparable with those of huperzine A, which were weaker than those of donepezil. In contrast, ZT-1 inhibited butyrylcholinesterase (BuChE) at a much higher concentration than needed for inhibition of AChE compared with the others, demonstrating that ZT-1 has the highest specificity for AChE. These results indicted that ZT-1 had better selective inhibitory efficacy on AChE than the others. ZT-1's inhibitory effect on AChE is a reversible, mixed competitive type inhibition with a K_i value of 32 nM.

AChE inhibitory experiments of AChEIs were conducted in mice and rats. Oral administration of ZT-1 at doses of 0.5, 1.0, and 2.0 μmol·kg^{-1} showed dose-dependent AChE inhibitory effects. The inhibitory strength of ZT-1 is similar to HA, 16-fold of donepezil and 120-fold of tacrine. ZT-1 might inhibit persistant brain AChE in 6 hours, and 3 hours for BuChE. The latter was markedly shorter than that of huperzine A, denepezil and tacrine.

Animal behavior experiments for the improvement and amelioration of experimental memory dysfunction were performed on adult mice, rats and monkeys, as well as older monkeys. ZT-1 showed remarkable effects on improving the memory time on water maze and eight-arm maze mice models, with an effective dose of 0.05 to 0.1 mg/kg, equal to huperzine A, 1/6 of donepezil, and 1/15 of tacrine. Intramuscular injection of 1 to 15 μg ZT-1 showed obvious improved memory effects on older monkeys. The experimental rats did not display the resistant phenomenon after 7 to 8 days of continuous oral administration at 0.2 to 0.42 μmol·kg^{-1} ZT-1. Mice oral administration of 0.35 μmol·kg^{-1} ZT-1 could markedly postpone the 71% survival time, superior to huperzine A (53.6%) and donepezil (11.5%).

The acute and long-term toxicity experiments of ZT-1 on mice, rats, and monkeys showed that no obvious side effects were observed, and no mutagenic, carcinogenic and teratogenic activities were detectable. currently, ZT-1 has entered Phase II clinical trials in Europe.

Section 5   Anti-arrhythmia Drug Changrolin

Dichroa febrifuga Lour. has a more than 2000 years of history for the treatment of malaria. From this plant, β-dichroine (**17**) was isolated and proven to be the main active component against malaria. However, its strong emetic action limited clinical applications. Given that the antimalaria mechanism of **17** is that the carbonyl group, and the heteroatoms of the molecule form chelates with the trace metals of plasmodium, the modification of **17** was focused on enhancing the chelating capacity of the molecule. Among various analogues, changrolin (**18**, pyrrozolinum) was found to be effective on plasmodia of chickens with low toxicity. During the clinical trial of **18** on malaria patients, the numbers of ectopia cordis pulsatility were observed to be significantly reduced. **18** was, therefore, studied for the treatment of heart arrhythmia. Further research on animal models indicated **18** could prevent ventricular arrhythmia induced by aconitine, ouabain, $BaCl_2$ or ischemia/reperfusion (I/R), and improve stimulation of the ventricular fibrillation threshold. A multi-center clinical trial of 489 patients demonstrated that **18** had marked efficacy on ventricular premature beat and paroxysmal tachycardia, especially, on refractory arrhythmia patients. The total effective rate was 80%, and the main side effects were skin pigmentation (17.2%), a slight increase of SGPT, and mild gastrointestinal discomfort, dizziness, impotence, etc.[17].

A concise four step synthetic route of **18** is shown in Scheme 15-4,[17] which started from 2-aminobenzoic acid, obtained quinazoline through a condensation reaction with formamide, introduced a *p*-hydroxyphenamine substituent on the quinazoline ring via a Cl-atom, and finally yielded **18** by means of a Mannich reaction.

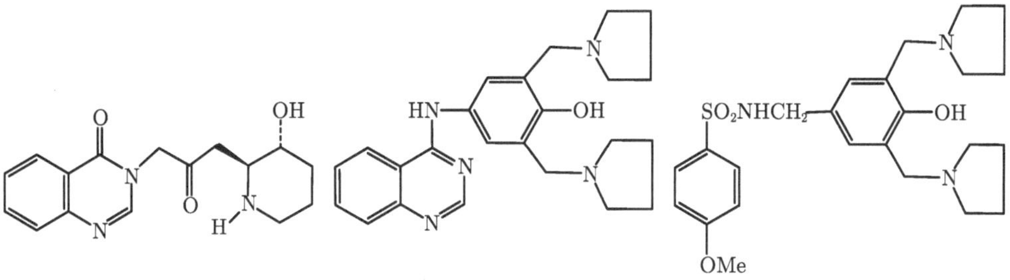

17   **18**   **19**

Figure 15-5   Structures of **13** through **16**

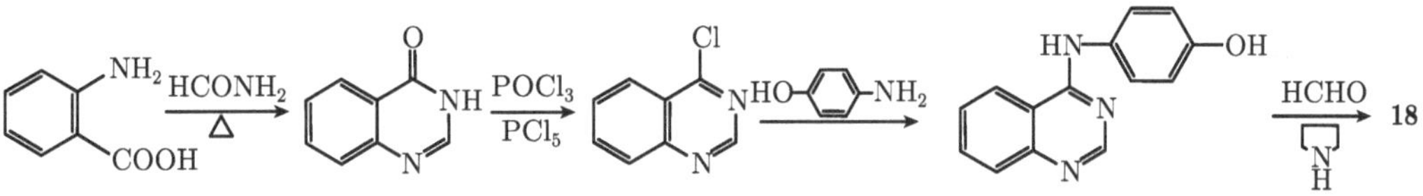

Scheme 15-4   Synthetic route of 18

Further structural optimization of **18** resulted in the finding of sulcadine (**19**, B87823)[18]. **19** could markedly reduce ventricular premature beat in mice induced by I/R with a 77% effective rate, as compared to 50% for **18**. The ED_{50} of **19** on preventing mice from aconitine-induced arrhythmia was 2.8 mg/kg, superior to 3.0 mg/kg of **18**. At the concentration of 30 μmol/L, **19** could reduce action potential amplitude (APA) and maximal rate of phase 0 depolarization (V_{max}), prolong 90% repolarization of action potential duration (APD_{90}), and effective refractory period (ERP) on the fast response action potential of papillary muscle in I/R guinea pig. **19** used at 100 μmol/L was also able to lower APA, V_{max} and spontaneous repolarization rate of phase 4 of the slow response action potential of dominant pacemaker cells of sinoatrial node in I/R rabbits. These results indicate that **19** might be a mixed blocker of Na^+, K^+, and Ca^{2+} ion channels.

In the molecule of **19**, the bis(pyrrolidinomethyl)-4-hydroxyphenyl fragment was reserved, with a 4-methoxy-N-methylbenzenesulfonamide group instead of the 4-aminoquinazoline of **18**. A concise, suitable for large-scale production synthetic approach is shown in Scheme 15-5.

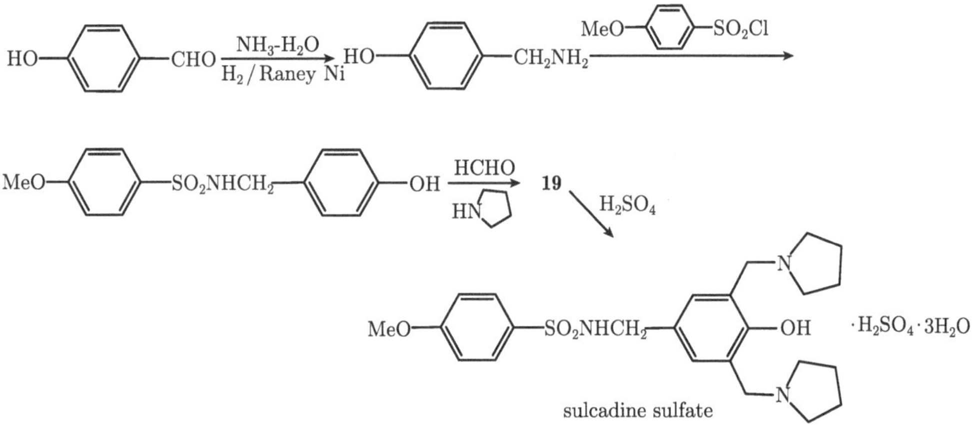

sulcadine sulfate

Scheme 15-5   Synthesis route of sulcadine sulfate

Section 6   Anti-neoplastic Drugs

NPs have been the mainstay of cancer chemotherapy for the past 30 years. NP and NP- derived drugs constitute a vast majority of the chemotherapeutic agents currently used for all types of cancers. Among the 92 drugs commercially available prior to 1983 in the United States, or approved worldwide between 1983 and 1994, approximately 62% can be related to a natural product origin, not counting those of biological origin such as interferon or recombinantly produced cytokines[19]. A common problem with many of the natural product-derived anticancer drugs now in clinical use is that the original natural product was too toxic. Chemists, therefore, had to make semisynthetic compounds based on the original natural product structures by selective modification.

6.1   Podophyllotoxin, Etoposide, and Teniposide

Podophyllotoxin (**20**)[20] is a well-known anti-cancer natural product, and is present in over 10 plants of the genus of *Dysosma*, *Podophyllum* or *Diphylleia*, of family Berberidaceae. The potent antitumor activity of **20** interested pharmacologists, but its strong gastrointestinal toxicity limited clinical applications.

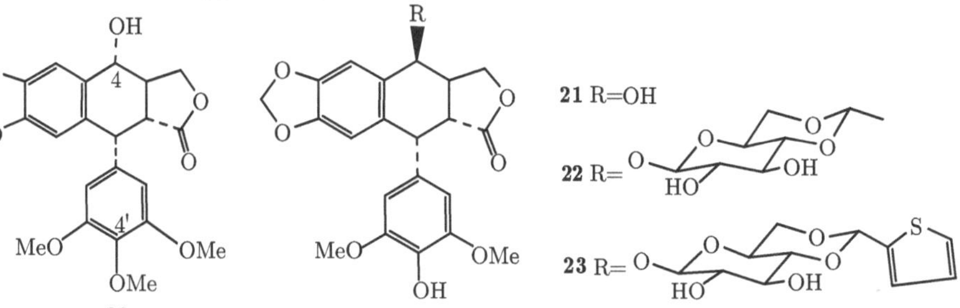

Figure 15-6   Structures of **20** through **23**

Podophyllotoxin (**20**) inhibits the formation of cell microtubules. In vitro, it binds to tubulin dimers giving podophyllotoxin-tubulin complexes. This stops the further formation of the microtubules at one end but does not stop the disassembly at the other end, leading to the degradation of the microtubules. Cells treated with **20** are arrested in the metaphase of the mitosis. This causes the cells to enter the mitosis, but the duplicated chromosomes will not be separated. In this way, the cells cannot duplicate and growth stops[21].

SAR studies of **20** showed that the 4α-OH is relative to the cytotoxicity, and when it was transformed to 4β-configuration to yield epipodophyllotoxin (4'-demethyl of **21**), the toxicity was obviously reduced. The *trans*-lactone D ring with a 2α, 3β-configuration is the core structure of **20** responsible for its anticancer activity. Any changes in this moiety will lower or eliminate the activity. Furthermore, when a hydroxyl replaces the methoxyl on C-4' of **20**, the activity will be enhanced. Such SAR unambiguously demonstrates that C-4 is the only molecular area tolerable to significant structural diversification. Therefore, research was carried out on the structural modification of **20** on the basis of keeping the units of 4-epimerization, *trans*-lactone D, C4'-demethyl, and free rotation of ring E. Glucosidation at C-4 of **21** displayed lower toxicity than the aglycone, but cytostatic activity was reduced to the same degree. Serendipitously, a radical change in mechanism of action and a quantum step in therapeutic utility were effected by acetalization of the 4- and 6-hydroxy groups of the glucopyrancose moiety using aldehydes, eventually leading to the discovery of the clinically important anticancer drugs etoposide (VP-16, **22**) and teniposide (VM-26, **23**)[22].

Etoposide (**22**), a clinically efficacious antineoplastic drug used for the treatment of testicular cancer and one of the most active single agents against small cell lung cancer, is the least toxic of all chemotherapeutic agents[23]. The synthetic route of **22** starting from **20** is illustrated in Scheme 15-6.

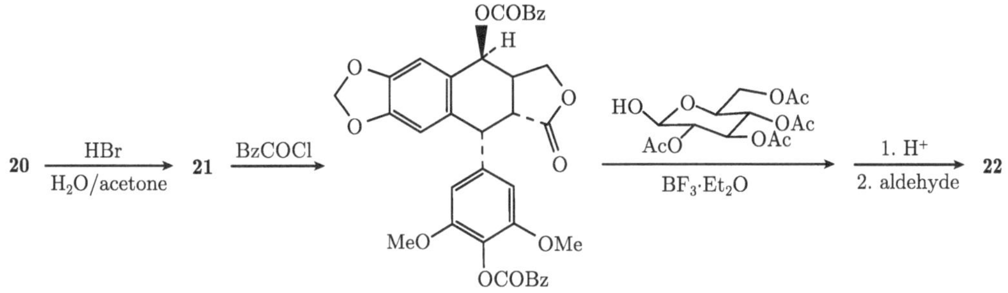

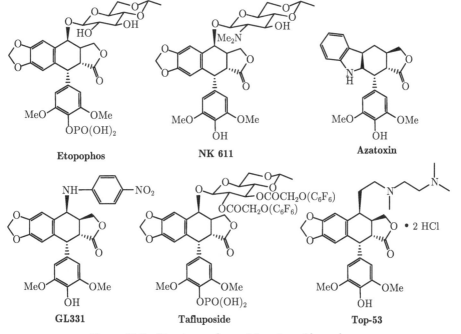

Scheme 15-6   The synthetic approach of 22 from 20

Etoposide (**22**) and teniposide (**23**) have completely different modes of action compared with their mother molecule. These compounds are topoisomerase II (topo II) inhibitors. Topo II is an enzyme that cleaves double-stranded DNA and seals it again after unwinding. It is crucial in the processes of DNA replication and repair. Functionally, topo II binds covalently to the broken DNA. **22** and **23** stabilize the DNA-topo II complex in such a way that resealing of the DNA strands becomes impossible. Cells that duplicate their DNA for mitosis are very sensitive to this mechanism. The overall effect of these anticancer drugs is cell arrest in late S or early G2 phase of the cell cycle[24].

Although **22** is widely used in cancer therapy, it still presents the deficiency of activity on a narrow spectrum of tumor types, low bioavailability, myelosuppression, and gastrointestinal toxicity. Therefore, extensive modifications on **22** have been carried out in order to obtain better therapeutic agents. Etopophos, launched in 1996 by Bristol-Myers Squibb, is a water-soluble phosphate ester prodrug of **22**. This prodrug is readily converted in vivo by endogenous phosphatase to the active drug **22**, and exhibited similar pharmacological and pharmacokinetic profiles to those of **22**. The in vivo bioavailability increased from 0.04% to over 50% through this prodrug approach[25]. Furthermore, several etoposide analogues (Figure 15-7), such as NK611, GL-331, azatoxin, TOP53, and tafluposide[26], which

Figure 15-7   Structures of promising etoposide analogues

exhibited either more potent activity or higher bioavailability, were subjected to preclinical or clinical studies. NK611, a 2″-dimethylamino analogue of **22**, improved bioavailability and potency greater than **22** in topo II inhibition and cytotoxicity assays against a variety of human cancer cell lines[27]. GL-331 was a 4-nitroanilino substituted analogue of **22**, which was shown to be active in many multidrug-resistant cancer cell lines[28]. Azatoxin was designed according to molecular modeling of the pharmacophore defined by topoisomerase II inhibitors. It was almost as potent as etoposide in inhibiting topoisomerase II and showed significant promise[29]. TOP-53 was a 4β-alkyl amino derivative of **22**, which displayed twice the activity of **22** against topoisomerase II and exhibits superior in vivo antitumor activity against several types of cancer compared to **22**[30]. Tafluposide (F 11782), a 2″, 3″-bis pentafluorophenoxyacetyl substituted analogue of **22** with a unique mechanism of interaction with both topoisomerase I and II, has shown outstanding antitumor activity in vivo against a panel of experimental human tumor xenografts[31].

6.2   Camptothecin

Camptothecin (CPT, **24**) is a potent antitumor agent first isolated by Monroe E. Wall and Mansukh C. Wani in 1958 from the extracts of *Camptotheca acuminata*, a deciduous tree native to China. CPT was also found in *Nothapodytes foetida*, *Pyrenacantha klaineana*, *Merrilliodendron megacarpum* (Icacinaceae), *Ophiorrhiza pumila* (Rubiaceae), *Ervatamia heyneana* (Apocynaceae) and *Mostuea brunonis* (Gelsemiaceae), species which belong to unrelated orders of angiosperms.

The success of CPT in preclinical studies in tumors of both colonic and gastric origin led to clinical investigations. Due to the negligible water solubility of CPT, these trials were initiated using the water-soluble sodium salt. The lesser efficacy of the open-ring salt form, accompanied by unpredictable and severe levels of toxicity associated with treatment, including life-threatening diarrhea, considerable hemorrhagic cystitis, and myelosuppression, resulted in suspension of the trials[32]. The discovery that the primary cellular target of CPT is type I DNA topoisomerase (topo I) was the breakthrough that revived interest in the drug in the mid 1980s[33]. Advances in the medicinal chemistry of CPT during the late 1980s and early 1990s resulted in semi-synthetic, water-soluble analogues, which are used clinically, and over a dozen new derivatives are currently under various stages of clinical development.

SAR of CPT

Early assessments identified the importance of the 20S chiral carbon for activity, and also noted a dynamic equilibrium between the closed-ring lactone and open-ring carboxylic acid forms (Figure 15-8)[34]. Furthermore, the pyridone moiety of the D-ring, the lactone moiety of the E-ring, and the planarity of the five-membered ring system were essential for anticancer activity. Hence, the C–D–E rings of CPT cannot be altered without severely affecting its efficacy. In contrast, modifications to the 9, 10, and 11 positions of the A-ring and the 7 position of the B-ring are generally well tolerated, and in many cases, enhance the potency of the CPT analogue both in vivo and in vitro[35]. Recently, a new family of CPT derivatives with an expanded seven-membered lactone E-ring has been developed. This class

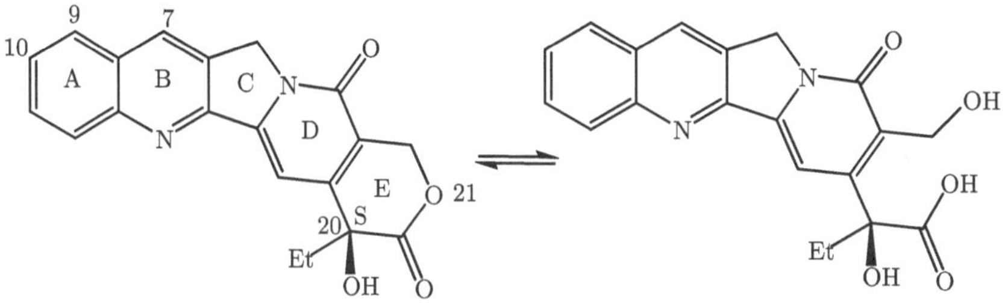

Figure 15-8   The dynamic equilibrium of camptothecin (24)

of analogues, known as homocamptothecins (hCPTs), has enhanced the plasma stability and reinforced the inhibition of topo I compared with conventional six-membered CPT[36].

Topotecan (TPT, 25) and irinotecan (CPT-11, 26) are two semisynthetic, water-soluble CPT derivatives, synthesized by modification of 10-hydroxycamptothecin, which demonstrated a broad-spectrum of activity in preclinical tumor models[37].

25 stabilizes the DNA-topoisomerase I covalent complex, albeit at concentrations four-fold higher than CPT. **25** is about two-fold less potent than CPT as a cytotoxic agent against L1210 cells in tissue culture. However, the therapeutic dosage range for **25** is virtually identical with that of CPT. Most importantly, **25** is reproducibly and significantly superior to CPT as an antitumor agent in mice bearing L1210 leukemia at the respective maximally tolerated doses ($173 \pm 16\%$ ILS for 25 as opposed to $118 \pm 6\%$ ILS for **24**).

The preparation of **25** is outlined in Scheme 15-7. **24** was converted to 10-hydroxycamptothecin by a reduction-oxidation sequence. Controlled catalytic reduction of **24** produced 1,2,6,7-tetrahydrocamptothecin, which, because of air sensitivity, was treated with lead tetraacetate immediately to afford a mixture of 10-hydroxycamptothecin, 10-acetoxycamptothecin, and camptothecin in an approximate ratio of 2:1:1. Treatment of this mixture with acetic acid and water resulted in the hydrolysis of 10-acetoxycamptothecin and provided a mixture of 10-hydroxycamptothecin and camptothecin. Due to the extremely poor solubility of 10-hydroxycamptothecin, only partial purification was effected for conversion to the desired **25** by treatment with an appropriately substituted amine, aqueous formaldehyde, and acetic acid.

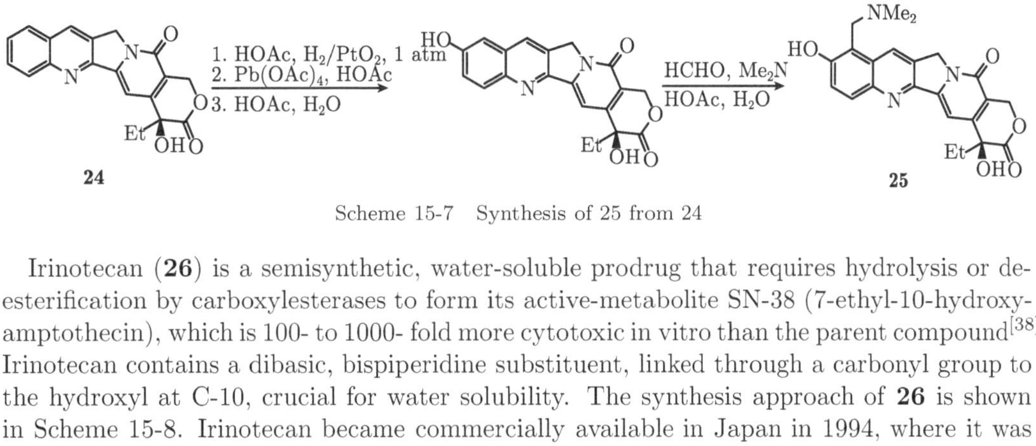

Scheme 15-7   Synthesis of 25 from 24

Irinotecan (**26**) is a semisynthetic, water-soluble prodrug that requires hydrolysis or de-esterification by carboxylesterases to form its active-metabolite SN-38 (7-ethyl-10-hydroxy-amptothecin), which is 100- to 1000- fold more cytotoxic in vitro than the parent compound[38]. Irinotecan contains a dibasic, bispiperidine substituent, linked through a carbonyl group to the hydroxyl at C-10, crucial for water solubility. The synthesis approach of **26** is shown in Scheme 15-8. Irinotecan became commercially available in Japan in 1994, where it was approved for the treatment of lung (small-cell and non-small-cell), cervix, and ovary cancer. Irinotecan was approved in Europe in 1995 as a second-line agent for colon cancer, one year before European approval of topotecan. Irinotecan was approved in the USA in 1996 for treatment of advanced colorectal cancer refractory to fluorouracil.

9-Aminocamptothecin (9-AC) is also a semisynthetic CPT derivative which showed outstanding preclinical activity against a wide spectrum of tumors, including breast, colon, lung, and prostate cancers, and melanoma[39]. Because of the poor aqueous solubility and poor clinical antitumor activity of 9-AC, some major efforts have been put into the design and synthesis of more derivatives that could be alternatives to the parent drug. To date, only one approach in prodrug design has yielded an agent that has progressed to clinical evaluation, i.e., 9-nitrocamptothecin (9-NC, rubitecan)], which acts as a partial prodrug of 9-AC. 9-NC has a nitro group in the C-9 position, is highly insoluble in water, and was initially identified as a precursor in the semisynthetic production of 9-AC. Since nearly all human cells are able to convert 9-NC to 9-AC, including tumor cells, it has proven difficult to identify whether its antitumor activity is directly associated with the parent drug alone, with 9-AC alone, or the combination of both[40].

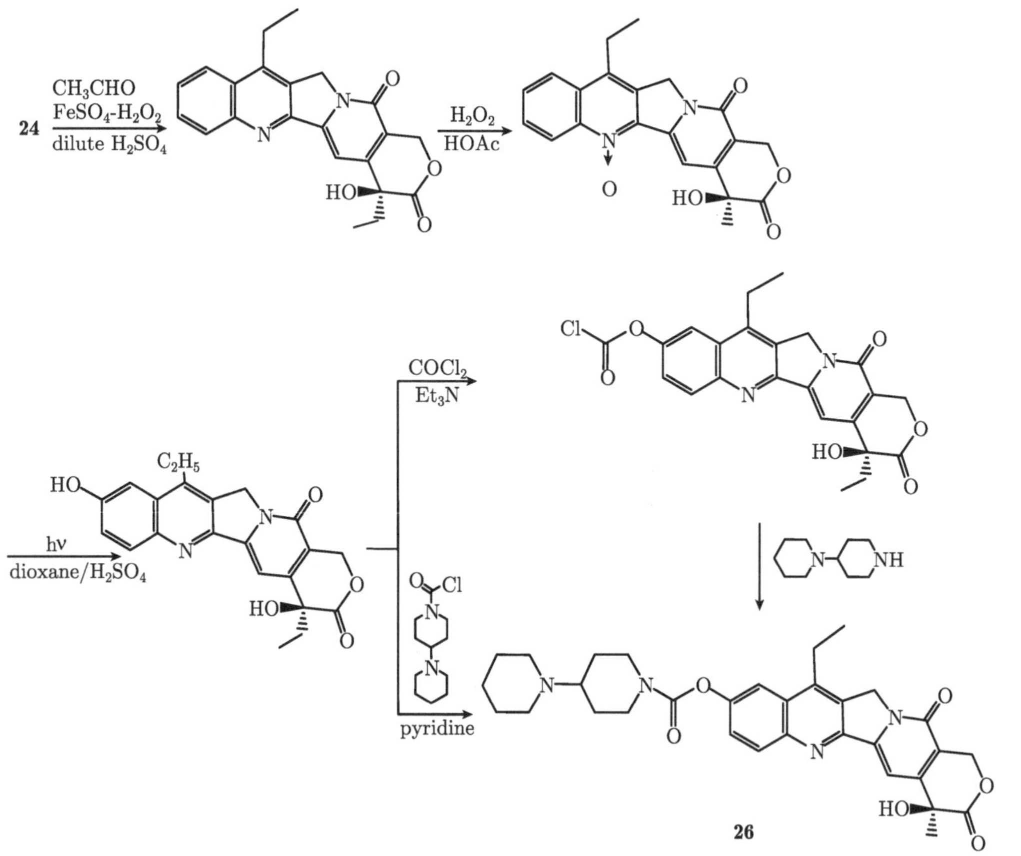

Scheme 15-8 Synthesis of 26 from 24

Chinese scientists made many efforts to study CPTs. They found that 10-hydroxycampto-thecin (HCPT) isolated from the fruits of *Camptotheca acuminate* had remarkable activity against a wide spectrum of tumors, including those of Ehrlich ascites tumor, Yoshida sarcoma cells and mice ascites hepatoma cells, with lower toxicity than CPT. HCPT has been approved in China for the treatment of hepatocarcinoma and head and neck cancer, e.g., salivary gland carcinoma[41]. For expanding the source of HCPT from the unitary plant, Zhu et al.[42] screened T-36 from 150 fungus strains, which could biotransform CPT into HCPT with 10% yield.

6.3 Vinca alkaloids

Vinca alkaloids, including the natural products vinblastine (**28**, VLB) and vincristine (**29**, VCR) isolated from *Catharanthus roseus* (Apocynaceae), and the semisynthetic derivatives vindesine (**30**, VDS) and vinorelbine (**31**, VRB), are antimitotic drugs that are widely and successfully used in the treatment of cancer[43]. *C. roseus*, known as "Chang Chun Hua" in Chinese, was used as an adjunct or in TCM for lowering blood pressure, tranquilizing effect, detoxicity, clearing heat, and calming the liver .

VLB and VCR were introduced into clinical practice in the early 1960s. Both of these *Vinca* alkaloids showed promising activity against Hodgkin's disease. Since then, VLB has demonstrated effectiveness against other lymphomas, neuroblastomas, and histiocytosis X. VCR has been used in acute lymphocytic leukemia and has been incorporated into several combinational chemotherapy regimens for the treatment of leukemias and solid tumors[44].

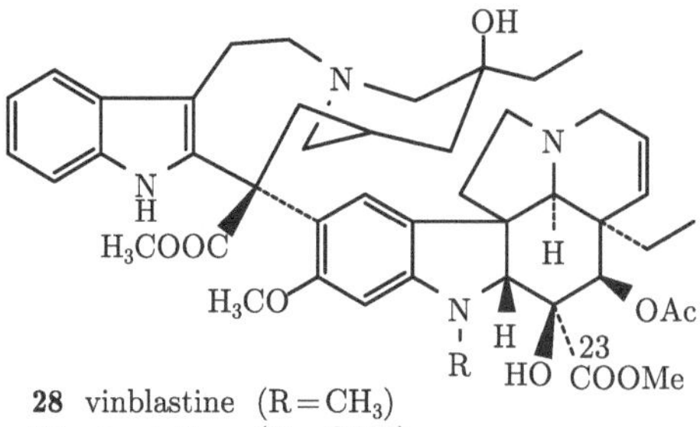

28 vinblastine (R=CH₃)
29 vincristine (R=CHO)

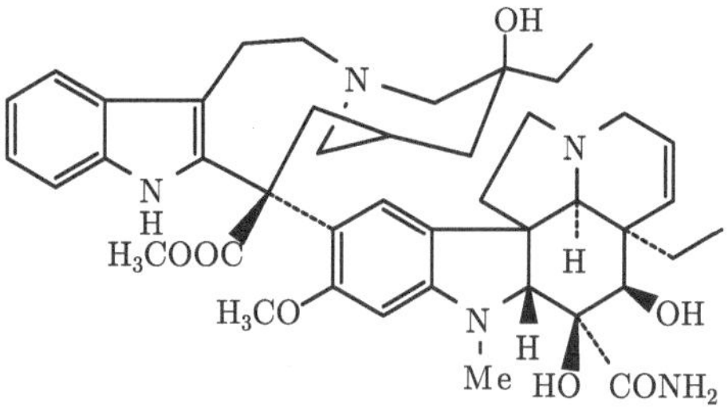

30 vindesine

Figure 15-9   Structures of *Vinca* alkaloids **28** through **30**

VLB and VCR were also found to have different side effects; the dose-limiting effect of VLB is leukopenia, and for VCR it is neurotoxicity. The dimeric *Vinca* alkaloid molecule presents several sites for chemical modifications that could further alter the spectrum of tumor response, change the therapeutic index, or reduce toxicity. VDS is a second generation *Vinca* alkaloid which was semisynthesized from VLB by either the preferential ammonolysis of the C-23 carbomethoxy group or the preferential hydrazinolysis and subsequent Raney nickel hydrogenolysis. These modifying procedures also effect the loss of the 4-*O*-acetyl group[45]. VDS has been clinically used for treatment of leukemia, lymphoma, melanoma, breast, and lung cancer.

Antitumor activities of VDS, VLB and VCR[44] VDS given i.p. daily completely inhibited the growth of both the Ridgeway osteogenic sarcoma at 0.4 mg/kg and the Gardner lymphosarcoma at 0.25 mg/kg. In contrast, VLB at 0.5 mg/kg was inactive against these two tumors. VDS caused inhibition of Mecca lymphosarcoma growth comparable to the inhibition by VCR (60 versus 68%), while VLB caused 41% inhibition. The Ca115 Shionogi mammary carcinoma did not respond to VDS, but both VLB and VCR caused inhibitions of 62 and 67%, respectively. Ca755 mammary adenocarcinoma showed a moderate response to VDS (46%) and VCR (39%), but no response to VLB. Sarcoma 180 ascites tumor (solid, s.c.), X5563 myeloma, and C3H mammary carcinoma were not responsive to VDS, VLB, or VCR.

In tissue cultures of Chinese hamster ovary cells, the minimum effective concentration of VDS that caused 10 to 15% accumulation of cell mitosis was 2.3×10^{-9} M. For VLB it was 2.2×10^{-8}M, and for VCR it was 7.3×10^{-9} M. The concentration of VDS needed for 40 to 50% accumulation of cell mitosis was 2.8×10^{-8} M; for VLB it was 8.3×10^{-8} M; and for VCR it was 2.4×10^{-8} M.

Acute toxicity studies showed that the LD₅₀ in rats for VDS given i.v. was 6.3 mg/kg, which is between VLB (10.0 mg/kg) and VCR (2.0 mg/kg).

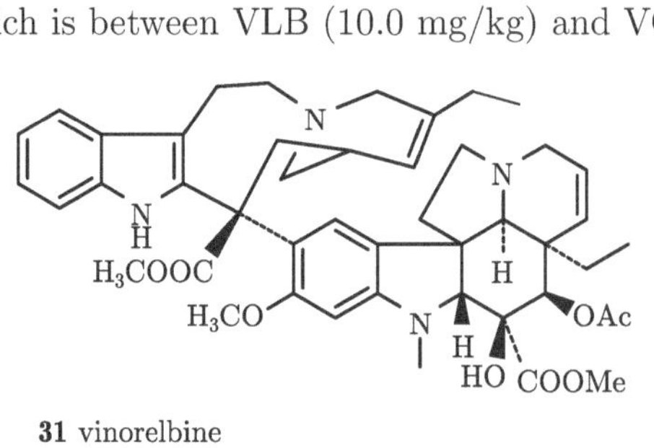

31 vinorelbine

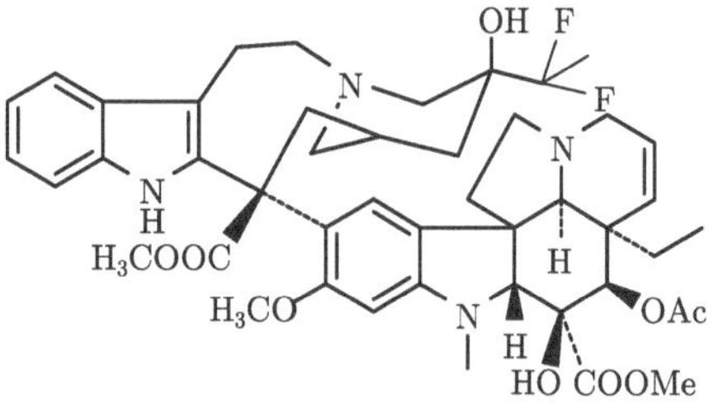

32 vinflunine

Figure 15-10   Structures of 31 and 32

Vinorelbine (VRB, **31**), the latest clinically approved *Vinca* alkaloid, has shown improved efficacy and reduced toxicity. It is effective in the treatment of non-small cell lung cancer (NSCLC), metastatic breast and ovarian cancer, and it shows promise in the management of lymphomas, esophageal cancer, and prostatic carcinomas. VRB demonstrated a broad spectrum of antitumor activity in preclinical studies and produced dose-limiting neutropenia in Phase I clinical trials. In Phase II studies, an overall response rate of approximately 30% was reported with single-agent VRB. Furthermore, in large, multicenter, randomized Phase III trials, treatment with VRB alone and in combination with cisplatin resulted in improved survival compared with controls. The drug was well tolerated, with granulocytopenia being the most commonly reported adverse effect. However, the incidence of fever and hospitalization associated with this granulocytopenia was exceptionally low. The recommended dose is 30 mg/m^2 administered weekly by intravenous injection or infusion[46].

A fifth new *Vinca* alkaloid, vinflunine (**32**, VFL), is a semisynthetic bifluorinated compound. VFL has demonstrated superior activity over VRB, VLB, and VCR in a range of murine tumors and human tumor xenografts[47]. For example, vinflunine showed high activity against RXF944LX kidney cell and NCI-H69 small cell lung xenografts, moderate activity against PAXF546 pancreatic cell, PC-3 prostate cell, and TC37 colon cell human tumor xenografts, achieving an overall response of 64% (i.e., activity against 7 of 11 xenografts tested). In contrast, only a 27% (3 of 11) response was observed for VRB tested concurrently with only moderate activity against RXF944LX kidney cell and TC37 colon cell xenografts. However, differences in terms of the inhibitory effects of VFL on microtubules dynamics and its tubulin binding affinities have been identified and appear to distinguish it from the other *Vinca* alkaloids. VFL induced smaller spirals with a shorter relaxation time, effects which might be associated with reduced neurotoxicity. Studies investigating the in vitro cytotoxicity of vinflunine in combinational therapy have revealed a high level of synergy when vinflunine was combined with either cisplatin, mitomycin C, doxorubicin or 5-fluorouracil. Furthermore, although VFL appears to participate in P-glycoprotein-mediated drug resistance mechanisms, it has proved only a weak substrate for this protein and a far less potent inducer of resistance than VRB. Interestingly, an in vivo study using a well vascularized adenocarcinoma of the colon has suggested that VFL mediates its antitumor activity at least in part via an antivascular mechanism, even at sub-cytotoxic doses[48,49].

6.4 Taxol

Taxol (paclitaxel, **33**), a novel diterpenoid chemotherapy drug, is the first unmodified natural product approved by the U.S. FDA in 30 years. The unique mode of action for **33** was found to be the stabilization of microtubule assembly. In 1992, **33** was approved for the treatment of refractory ovarian cancer, and subsequently, for the treatment of metastatic breast and lung cancers and Kaposi's sarcoma. The unique action of taxol spurred the development of a second-generation semisynthetic taxane, taxotere (docetaxel, **34**), approved in 1996 for treatment of anthracycline-refractory advanced breast cancer and now also used in lung cancer regimens.

Taxol was first isolated in 1971 by Wall and his collaborators. It showed significant activity against various leukemias, the Walker 256 carcinosarcoma, Sarcoma 180, and the Lewis lung tumor[50]. In spite of this promising spectrum of activities, progress on developing taxol as a drug was slow, largely because of the difficulty of isolating it from the bark of the western yew, *Taxus brevifolia*, and because of its lipophilicity which made formulation difficult. Interest in taxol was further heightened by the discovery that it has the unique property of acting as a promoter of microtubule assembly[51] and a determined effort resulted in adequate supplies of taxol for formulation studies and eventual clinical trials[52].

At the end of 1970s, the unique mechanism of action of taxol was discovered. Different from other anticancer drugs, such as VBL, VCS, and podophyllotoxin, taxol is an antimitotic agent which acts to promote the irreversible assembly of tubulin into microtubules[51,53].

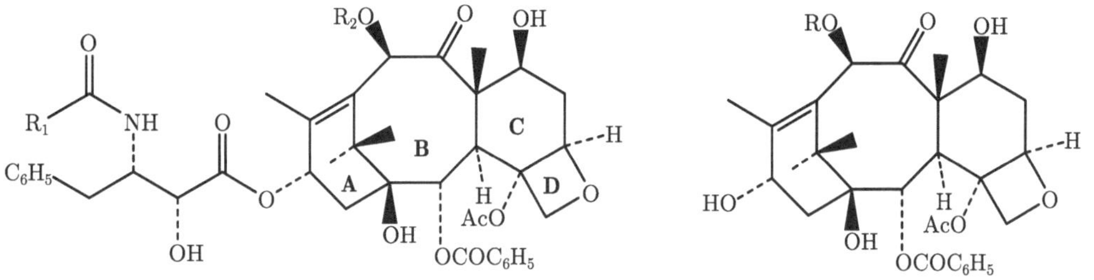

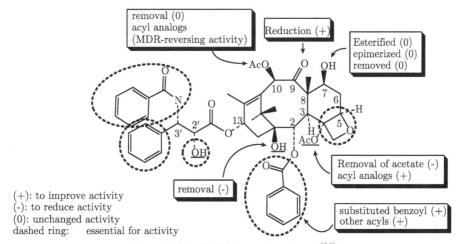

33 (taxol) $R_1 = C_6H_5$, $R_2 = Ac$
34 (taxotere) $R_1 = (CH_3)_3CO$, $R_2 = H$

R = H, 10-deacetylbaccatin III (**35**)
R = Ac, baccatin

Figure 15-11　Structures of **33** through **35**

The supply problem was elegantly resolved by a semisynthetic process from the available 10-deacetylbaccatin III (**35**), (Scheme 15-9), (Denis JACS, 1988). **35** can be readily extracted in high yield from the leaves of *Taxus baccata* L. (ca. 1 g/kg of fresh leaves)[54]. It is important to recognize that the yew leaves are quickly regenerated and, through prudent harvesting, large amounts of **35** can be continually produced with negligible effect on the yew population.

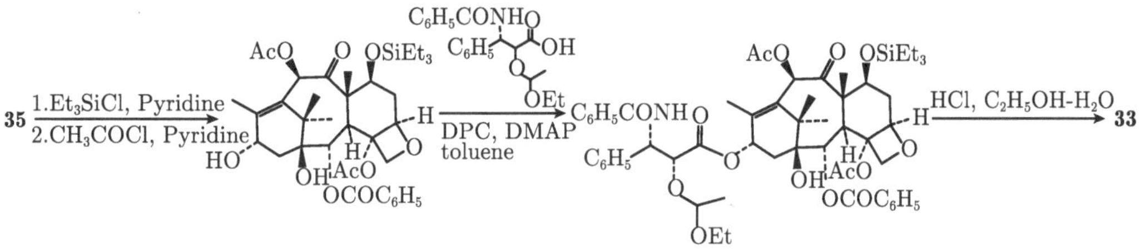

Scheme 15-9　Semisynthesis of taxol (33) from 10-deacetylbaccatin III (35)

Taxol SAR Summary Numbers of studies have led to a good general picture of the structure-activity relationships (SARs) of taxol, and they are summarized in Figure 15-12. In general, changes to the "southern hemisphere", comprising the C-14 and C-1 to C-5 positions, exert a major effect on taxol's activity. This is the case for changes at C-1 (although here the effects are not large), at C-2, and at C-4. Most changes to the C- and D-rings, including opening of the oxetane ring, almost always lead to loss of activity, although Appendino has found an intriguing example of a C-ring-opened analogue that retains significant activity[55].

Figure 15-12　SAR illustration of taxol[53]

Changes in the "northern hemisphere", comprising the C-6 to C-12 positions, appear to have less impact on taxol's activity, although these changes can still be important therapeutically. Ojima has developed some interesting taxol analogues modified at C-10 and at C-2 with usefully improved activities[56].

Changes in the side chain have also been made by several investigators, and it is expected that future "second generation" taxol analogues will have modifications both on the side chain and on the taxane ring system.

Taxotere (**34**) is a taxol analogue synthesized from 10-deacetylbaccatin III (**35**) (Scheme 15-10). It shows potent anticancer activity better than taxol[57]. It also has better pharmacological properties, such as improved water solubility, and acts at the microtubules. It enhances polymerization of tubulin into stable microtubule bundles leading to cell death. It is used for the treatment of patients with locally advanced metastatic breast cancer and nonsmall cell lung cancer[58]. The use of this drug is also associated with several side effects, such as bone marrow suppression, hypersensitivity reactions, vomiting, alopecia, etc.

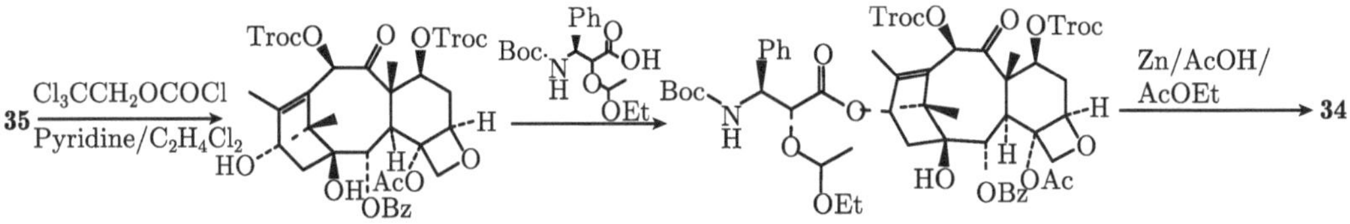

Scheme 15-10   Synthetic approach of taxotere (34) from 35

6.5   Cantharidin

Mylabris phalerata Pallas is an insect belonging to the family Meloidae, which has long been used as a medicinal agent in TCM for its aphrodisiac, phlogistic, vesicant, and toxic effects. Dried, powdered beetle, mylabris, has been used to treat furuncles and piles, deep ulcers, venomous worms and fistulae of tuberculous lymphadenitis[59].

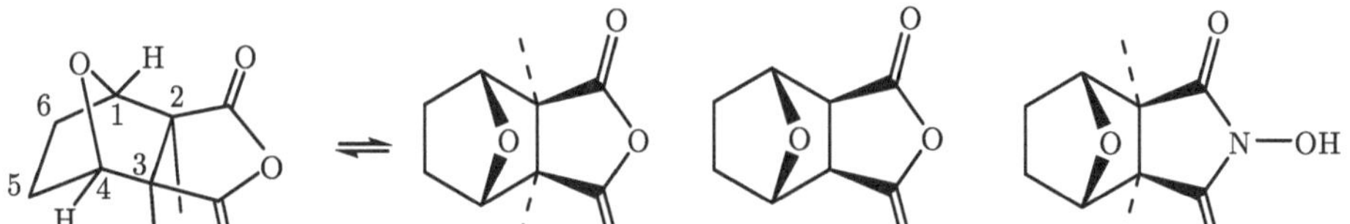

Figure 15-13   Structures of cantharidin (36) and its analogues

Cantharidin (**36**), the active substance of mylabris, inhibits some tumor cells (HeLa cells, murine ascites hepatoma or reticulocell sarcoma) but not others (murine erythroid leukemia cells in vitro, S180 in vivo and Walker tumour in rats)[59]. Clinical trials indicated that **36** had effects on patients with primary hepatoma, but the application was limited by its severe toxicity to mucous membranes, mainly in the gastrointestinal tract, ureter and kidney[59]. Norcantharidin (**37**), the demethylated analogue of **36**, appeared to cause the fewest nephrotoxic and inflammatory side effects. The acute LD_{50} of **37** is one fold higher than cantharidin (12.5 mg/kg). **37** inhibits the proliferation of several tumor cell lines (HeLa, CHO, CaEs-17, BEL-7402, SMMC7721 human hepatoma, HEP-2 and human epidermoid aryngocarcinoma) and transplanted tumors (embryonal adenocarcinoma or hepatoma 22 in mice)[60]. **37** has been approved for use in China for cancer therapy since 1984.

Pharmacological studies showed that both **36** and **37** are inhibitors of protein phosphatase 1 (PP1) and protein phosphatase 2A (PP2A)[61]. This inhibitor activity is necessary for growth inhibition and cytotoxicity provoked by these compounds. Protein phosphatases are involved in the regulation of multiple cellular processes, including apoptosis, signal transduction pathways, cell cycle progression, glucose metabolism and calcium transport, among others. Cantharidin (**36**) is considered to bind in the active sites of both PP1 and PP2A containing two metal ions in the ring-opened dicarboxylic acid form[62]. SAR studies of **36**[63] on the inhibition of PP1, PP2A$_1$, and PP2B showed that removal of the methyl substituents at C-2 and C-3 improved the inhibition of PP2B, but slightly decreased activity against PP1 and PP2A$_1$. In contrast, introduction of a C-1/C-4 position substituent drastically decreased the inhibition of all phosphatases. PP2B was found to be quite tolerant to modifications at the C-5 position.

The SAR results of **37** against PPs are summarized in Figure 15-14[64]. Norcantharidin dicarboxylate was determined to be a pharmacophore for PP inhibitors (Figure 15-14). The anhydride/dicarboxylate structure at the C-2/C-3 positions and the bridging ether oxygen are critical for the inhibition of all PPs.

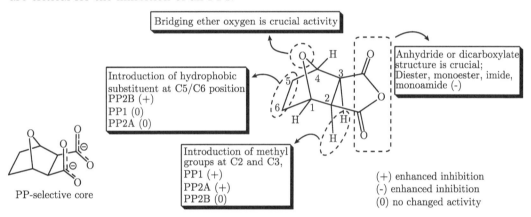

Figure 15-14   SAR of 37

Clinical trials with **37** as a monotherapeutic agent indicated that it was effective against primary hepatoma and against esophageal, gastric and cardiac carcinomas[59]. **37** increased the mean survival times of 285 reported cases with primary hepatoma from 4.7 to 11.1 months, and the l-year survival rate from 17 to 30%, as compared to 102 patients treated with conventional chemotherapy (5-FU, hydroxycamptotherine, VCS, thiophosphoramide and mitomycin)[59].

McCluskey et al. synthesized a series of cantharidin imides, and found that cantharimides (cantharidinimides) possessing a D- or L-histidine, are more potent inhibitors of PP1 and PP2A than norcantharidin and essentially equipotent with cantharidin[65]. The authors suggested to re-examine the original hypothesis that the ring opening dicarboxylate was crucial for the inhibition of PPs[65]. Prior to that, Chinese scientists designed the cantharidinimides for lowering toxicity through sustained-release **36** by hydrolyzation. N-Hydroxycantharidimide (**38**) and N-methycantharidinimide, two successful cantharidin amides, have also been applied for the clinical treatment of primary liver cancer. The LD$_{50}$ of **38** for mice i.v. is approximately 100-fold of **37**, and 500-fold of **36**.

6.6   Indirubin

Indirubin (**39**), a bisindole alkaloid isolated from *Indigofera tinctoria* L., is the active ingredient of the TCM formula *Danggui Longhui Wan*, which has been used to treat chronic diseases such as myelocytic leukemia as elsewhere[66]. This substance and its substituted derivatives are potent inhibitors of several kinases, such as glycogen synthase kinase-3β (GSK-3β, IC$_{50}$: 5 to 50 nM) and cyclin dependent kinases (CDKs, IC$_{50}$: 50 to 100 nM), competing with ATP in the ATP binding pocket[67]. Indirubin derivatives effectively inhibit the growth of various human tumor cell lines in the low micromolar range. Furthermore, indirubins cause a concentration-dependent cell cycle arrest at G1/S and/or G2/M phase in a wide spectrum of human tumor cell lines, resulting in induction of apoptosis[68]. **39** exhibited good antitumor activity and only minor toxicity in preclinical studies[69]. Clinical trials showed that indirubin has a definite efficiency against chronic myelocytic leukemia[70].

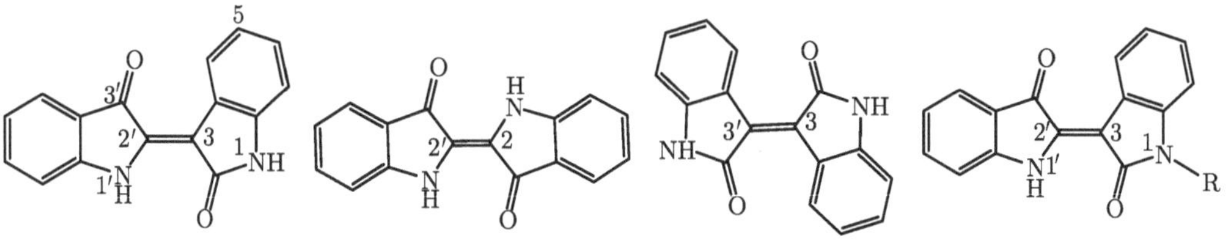

Indirubin (**39**)   Indigo (**40**)   isoindigo (**41**)   **42** R = Me, Et, Ac, CH$_2$COOEt

Scheme 15-15   Structures of **39** through **42**

Two natural isomers, indigo (**40**) and isoindigo (**41**), are inactive on tumor cells. Indirubin is hardly soluble in water or lipids, which makes formulation difficult. The main adverse clinical effects were gastrointestinal stimulated reactions, e.g., bellyache, diarrhea, and hematochezia. SAR studies on **39** against tumor cell lines showed that introduction of substituents (alkyl $\leqslant$ 4C, **42**) on N$_1$ improve the activity, possibly because the loss of a hydrogen bond between N-1 and the carbonyl of C-2 in **39** increases water solubility. Substituents on N-1 will lose activity with the loss of the planar structures. Analysis of the crystal structures of specific indirubin derivatives in complex with CDK2 and CDK2/cyclin A revealed that the ATP binding pocket within positions 5 and 3′ of the indirubin template should tolerate introduction of appropriate substituents to sustain or even tighten the binding to the enzyme[71]. Recently, modification on **39** was focused on the 3′-oxime analogues. Furthermore, indirubins bearing hydrophilic substituents in the 3′-oxime ether position were considered for increasing water solubility[68,72].

Section 7   Computer-aided Drug Design and Structural Modification

In the foregoing sections, we described the traditional development process of modern medicines from NP. In general, it consisted of 1) isolating an active NP as a lead compound; 2) synthesizing numbers of structural analogues for SAR studies; 3) designing the candidate according to the SAR; 4) preclinical studies, and 5) clinical trials. This process is very expensive and takes a long time. The average total cost per drug development varies from US $897 million to US $1.9 billion. The typical development time is 10 to 15 years.

With the development of molecular biology, we learned that drugs show bioactivity by blocking or activating the receptor or "target" (e.g., protein) related to the disease. Metaphorically, the active site of the target (protein) can be viewed as a lock, and the ligand (drug) can be thought of as a key. "Structural biology" and "computer-aided drug design (CADD)" were thus two new rising divisions to study the "lock" and "key", respectively. The former integrated chemistry, physiology, biology, and informatics, whose objects were the architecture and shape of biological macromolecules, by means of crystallography, NMR, ultra-fast laser spectroscopy, electron microscopy, electron cryomicroscopy (cryo-EM), dual polarization interferometry, circular dichroism, and so on. CADD uses computers to analyze the chemical structures of potential drugs and pinpoint the most promising candidates to match the structures of macromolecule targets.

CADD originated in the 1960s and remained localized at the level of quantitative structure-activity relationship (QSAR) until the end of 1980s. QSAR formalizes what is experimentally known about how a given protein interacts with some tested compounds. As an example, it may be known from previous experiments that the protein under investigation shows signs of activity against one group of compounds, but not against another group. Before the structure of the target is available, we do not know what the lock (target) looks like, but we do know which keys (ligands) work, and which do not. In order to build a QSAR model for determining why some compounds show signs of activity and others do not, a set of descriptors is chosen. These are assumed to influence whether a given compound will succeed or fail in binding to a given target. Typical descriptors are parameters such as molecular weight, molecular volume, and electrical and thermodynamical properties. QSAR models are used for virtual screening of compounds to investigate an appropriate drug candidate's descriptors for the target.

When target structure was made available in the 1990s, CADD entered a new phase of structure-based drug design. The most commonly used virtual screening method is molecular docking. Molecular docking programs predict how a drug candidate binds to a protein target. This software consists of two core components: a search algorithm (optimization algorithm), and an evaluation function (score function). The search algorithm is responsible for finding the best conformations of the ligand and protein system. In flexible docking, the conformation also contains information about the internal flexible structure of the ligand and, in some cases, about the internal flexible structure of the protein. Since the number of possible conformations is extremely large, it is not possible to test all of them. Therefore, sophisticated search techniques have to be applied. Examples of some commonly used methods are genetic algorithms and Monte Carlo simulations. The evaluation function provides a measure of how strongly a given ligand will interact with a particular protein. Energy force fields are often used as evaluation functions. These force fields calculate the energy contribution from different terms such as the known electrostatic forces between the atoms in the ligand and in the protein, forces arising from deformation of the ligand, pure electron-shell repulsion between atoms, and effects from the solvent in which the interaction takes place.

It is not possible to guarantee that the search algorithm will find the same solution as the true natural process, but more efficient search algorithms will be more likely to find the true solution if the evaluation functions properly reflect the natural processes. It is estimated that the application of CADD could reduce the cost of drug discovery up to 50%[73].

CADD was introduced into China in the 1970s. Chinese scientists achieved some progress in the methodology and practice of CADD. In searching the active conformation of a drug's interaction with a receptor, Chen and Jiang et al. developed a binding conformer search program for ligands (BCSPL) based on the 3D-structure of a receptor. According to this method, they studied the binding conformers of phosphoropeptidyl thrombin inhibitors. The results showed that both of the total binding energies and steric binding energies have good correlations with the inhibitory activities of these thrombin inhibitors, which confirmed the reliability of BCSPL and provided a different angle of view for the search of new thrombin inhibitors[74]. Furthermore, Chen et al. utilized comparative molecular field analysis (CoMFA), a three-dimensional quantitative structure-activity relationship (3D-QSAR) paradigm, to study the correlation between the physicochemical properties and the in vitro bioactivities of ginkgolide analogues. The correlation derived from CoMFA analysis has a good predictive capability. Based on the result of CoMFA analysis, they designed and synthesized a series of structure-optimized compounds. Among them, two compounds exhibited 2- and 4-fold activity of ginkgolides[75].

Donepezil (Figure 15-3) is an anti-Alzheimer's disease- (AD) drug which was designed based on the QSAR of piperidines and actylcholinesterase (AChE)[76]. Cholinesterase (ChE) inhibitors have been investigated for potential treatments of AD as it was hypothesized that cholinergic function was selectively and irreversibly affected in the dementia state. Initially, extensive chemical SAR and molecular modeling studies were performed on a series of piperidines. These studies were undertaken to determine the physicochemical properties of this series of compounds, including binding, docking and affinity characteristics. Such studies have provided insights to design the structure of new piperidine-type AChE inhibitors such as donepezil. X-ray analysis of the E2020-AChE complex disclosed the need for the high inhibition on AChE by donepezil. For example, donepezil is a narrow cylindrical ligand with a large dipole, it can fit into and slide along the inner surface of the narrow aromatic gorge of the AChE active site. The active site possesses a positive charge at one end and a negative charge at the other. Therefore, by electrostatic interactions, donepezil's dipole is in a complementary direction to that of AChE. In general, the benzyl group of donepezil binds to the bottom of the active site cavity and the indanone ring is located at the entrance to the cavity. Hence, both the indanone and benzyl rings are essential for optimal AChE inhibitory activity. Thus, such studies ensure that new inhibitors possess the optimum required characteristics. These results, in addition to the findings obtained from animal and in vitro models, warranted further investigation of this compound in humans.

Chen and Jiang studied the 3D-structure of huperzine A-AChE complex and that of donepezil-AChE, and found that both had different active sites binding to AChE. They designed a compound which combined the effective structural moieties of both of donepezil and huperzine A. However, bioactive tests showed donepezil to have only 1/850 the activity of huperzine A[77].

In summary, CADD is playing a more important role in drug discovery. It simulates the interaction between drugs and receptor, and indicates the action mechanism for further structural modification or designing a novel molecule. However, CADD could not resolve all problems in the drug discovery, because it is in a developing stage and will evolve further with the development of structural biology. Just like the metaphor of a lock and key, which is slightly oversimplified due to the fact that neither the ligand nor the proteins are completely rigid structures, their shapes are somewhat flexible and may adapt to each other. Therefore, CADD, in the current phase, is only one important part of drug discovery. It must be used closely with the other areas such as structural biology, chemical synthesis, and pharmacological evaluation to discover promising drug candidates.

References

[1] N. R. Farnsworth, O. Akerele, A. S. Bingel, D. D. Soejarto and Z. Guo. *Bull. WHO*, 1985, 63, 965.

[2] Newman DJ, Cragg GM, Snader KM. Natural products as sources of new drugs over the period 1981-2002. *J Nat Prod* 2003, 66: 1022-1037.

[3] Newman DJ, Cragg GM, Snader KM. The influence of natural products upon drug discovery. *Nat Prod Rep.* 2000, 17: 215-234.

[4] Kong JM, Goh NK, Chia LS, Chia TF. Recent advances in traditional plant drugs and orchids. *Acta Pharmacol Sin* 2003, 24 (1): 7-21.

[5] Butler MS. The role of natural product chemistry in drug discovery. *J Nat Prod* 2004, 67:

2141-2153.

[6]   Qin GW, Xu RS. Recent advances on bioactive natural products from Chinese medicinal plants. *Med Res Rev* 1998, 18(6): 375-382.

[7]   Sneader W. *Drug Discovery: The Evolution of Modern Medicines.* New York: Wiley, 1985

[8]   Li Y, Wu YL. An over four millennium story behind Qinghaosu (artemisinin): a fanstastic antimalarial drug from a traditional Chinese herb. *Curr Med Chem* 2003, 10: 2197-30.

[9]   Xie JX et al. Synthesis of schizandrin C analogs II. synthesis of dimethyl-4,4'-dimethoxy-5,6,5',6'-dimethylenedioxybiphenyl -2,2'-dicarboxylate and its isomers. *Acta Pharm Sin* 1982, 17: 23-27.

[10] LIU GT. The anti-virus and hepatoprotective effect of bicyclol and its mechanism of action. *Chinese J New Drugs* 2001, 10: 325-7.

[11] Liu GT, Li Y, Wei HL, Lu H, Zhang H, Gao YG, Wang LZ. Toxicity of novel anti-hepatitis drug bicyclol: A preclinical study. *World J Gastro* 2005, 11(5), 665-671.

[12] Yao GB, Xu DZ, Lan P, Xu CB, Wang C, Luo J, Shen YM, Li Q. Efficacy and safety of bicyclol in treatment of 2,200 chronic viral hepatitis patients. *Chinese J New Drugs Clin Rem* 2005, 24 (6): 421-426.

[13] Jiang HL, Bai DL, Luo XM. Progress in clinic, pharmacological, chemical and structural biological studies of huperzine A: a drug of traditional Chinese medicine origin for the treatment of Alzheimer's disease. *Curr Med Chem* 2003, 10: 2231-2252.

[14] Zhu DY, Tan CH, Li, YM. In *Medical Chemistry of Bioactive Natural Products*, 2006, Eds.: Liang, Xiao-Tian, Fang, Wei-Shuo. New York: Wiley. p143-182.

[15] Liu JS, Zhu YL, Yu CM, Zhou YZ, Han YY, Wu FW, Qi BF. The structures of huperzines A and B, two new alkaloids exhibiting marked anticholinesterase activity. *Can J Chem* 1986, 64: 837-839.

[16] Tang XC, Han YF. Pharmacological profile of huperzine A: a novel acetylcholinesterase inhibitor from Chinese herb. *CNS Drug Rev* 1999, 5(3): 281-300.

[17] Li LQ, Qu ZX, Wang ZM, Ceng YL, Ding GS, Hu GJ, Yang XY. Studies on changrolin: an antiarrhythmia drug. *Science in China, Ser.A* 1979, (07): 723-729.

[18] Bai DL, Chen WZ, Bo YX, Wang YP, Dong YL, Kang AL. The preparation and usage of N-substituted benzyl or phenyl aromatic sulfonamine compounds. *China patent* CN99124236.

[19] Cragg GM, Newman DJ, Snader KM. Natural products in drug discovery and development. *J Nat Prod* 1997, 60: 52-60.

[20] Keller-Juslen C, Kuhn M, Von-Wartburg A. Synthesis and antimitotic activity of glycosidic lignan derivatives related to podophyllotoxin. *J Med Chem* 1971, 14: 936-940.

[21] Stähelin HF, Von Wartburg A. The chemical and biological route from podophyllotoxin glucoside to etoposide. *Cancer Res* 1991, 51: 5-15.

[22] Canel C, Moraes RM, Dayan FE, Ferreira D. Podophyllotoxin. *Phytochemistry* 2000, 54: 115-120.

[23] Issell BF, Rudolph AR, Louie AC (1984) Etoposide (VP-16-213): an overview. In: *Etoposide (VP-16): Current Status and New Developments*. Eds. Issell BF, Muggia FM, Carter SK. Academic Press, New York, pp 1-13

[24] Stähelin HF. Activity of a new glycosidic lignan derivative (VP 16-213) related to podophyllotoxin in experimental tumors. *Eur J Cancer* 1973, 9: 215-21.

[25] Greco FA., Hainsworth JD. Clinical studies of etoposide phosphate. *Semin Oncol* 1996, 23 (suppl): 45-50.

[26] Liu YQ, Liu Y, Xuan T. Podophyllotoxin: Current Perspectives. *Curr Bioactive Comp* 2007, 3: 37-66.

[27] Pagani O, Zucchetti M, Sessa C, De Jong J, D' Incalci M, De Fusco M, Kaeser-Froehlich A, Hanauske A, Cavalla F. Clinical and pharmacokinetic study of oral NK611, a new podophyllotoxin derivatives. *Cancer Chemother Pharmacol* 1996, 38, 541-547.

[28] Lee KH. Novel Antitumour agents from higher plants. *Med Res Rev* 1999, 19: 569-596.

[29] Leteurtre F, Madalengoitia J, Orr A, Cuzi TJ, Lehnert E, Macdonald T, Prommier Y. Rational design and molecular effects of a new topoisomerase II inhibitor, azatoxin. *Cancer Res* 1992, 52, 4478-4483.

[30] Utsugi T, Shibata J, Sugimoto Y, Aoyagi K, Wierzba K, Kobunai T, Terada T, Oh-hara T, Tsuruo T, Yamada Y. Antitumor activity of a novel podophyllotoxin derivative (TOP-53) against lung cancer and lung metastatic cancer. *Cancer Res* 1996, 56(12): 2809-2814.

[31] Sargent JM, Elgie AW, Williamson CJ, Hill BT. Ex vivo effects of the dual topoisomerase inhibitor tafluposide (F 11782) on cells isolated from fresh tumor samples taken from patients with cancer. *Anticancer Drugs* 2003, 14(6):467-473.

[32] Pizzolato JF, Saltz LB. The camptothecins. *Lancet* 2003, 361: 2235-2242.

[33] Hsiang YH., Hertzberg R, Hecht S, Liu, LF. Camptothecin induces protein-linked DNA breaks via ammalian DNA topoisomerase I. *J Biol Chem* 1985, 260: 14873-14878.

[34] Wang X, Zhou X, Hecht SM. Role of the 20-hydroxyl group in camptothecin binding by the topoisomerase I-DNA binary complex. *Biochemistry* 1999, 38: 4374-4381.

[35] Redinbo MR, Stewart L. Kuhn P, Champoux JJ, Hol WG. Crystal structures of human topoisomerase I in covalent and noncovalent complexes with DNA. *Science* 1998, 279: 1504-1513.

[36] Bailly C. Homocamptothecins: potent topoisomerase I inhibitors and promising anticancer drugs. *Crit Rev Oncol Hematol* 2003, 45: 91-108.

[37] Kingsbury WD, Boehm JC, Jakas DR, Holden KG, Hecht SM, Gallagher G, Caranfa MJ, McCabe FL, Faucette LF, Johnson RK. Synthesis of water-soluble (aminoalkyl) camptothecin analogues: inhibition of topoisomerase I and antitumor activity. *J Med Chem* 1991, 34: 98-107.

[38] R. H. Mathijssen, W. J. Loos, J. Verweij, and A. Sparreboom. Pharmacology of topoisomerase I inhibitors irinotecan (CPT-11) and topotecan. *Curr Cancer Drug Targets* 2002, 2, 103-123.

[39] Sinha BK. Topoisomerase inhibitors. A review of their therapeutic potential in cancer. *Drugs* 1995, 49(1): 11-19.

[40] Hinz HR, Harris NJ, Natelson EA, Giovanella BC. Pharmacokinetics of the in vivo and in vitro conversion of 9-nitro-20(S)-camptothecin to 9-amino-20(S)-camptothecin in humans, dogs, and mice. *Cancer Res* 1994, 54(12): 3096-3100.

[41] Research group of Shanghai Institute of Materia Medica, CAS. Studies on the anticancer activity of 10-hydroxycamptothecin. *Chinese Med J* 1978, 58: 598-600

[42] Zhu GP, Lin LZ, Pan WJ, Zhou ZZ, Huang YC, Xie RY, Liang SF. Studies on biotransformation from camptothecin to 10-hydroxycamptothecin. *Chinese Sci Bull* 1978, 23: 761-762.

[43] Donehower RC, Rowinsky EK. Anticancer drugs derived from plants. In: *Cancer: Principles and Practice of Oncology*. Eds. De Vita VT, Hellman S, Rosenberg SA. pp. 409-417. Philadelphia: J B Lippincott Company, 1993.

[44] Sweeney MJ, Boder GB, Cullinan GJ, Culp HW, Daniels WD, Dyke RW, Gerzon K, McMahon

RE, Nelson RL, Poore GA, Todd GC. Antitumor activity of deacetyl vinblastine amide sulfate (vindesine) in rodents and mitotic accumulation studies in culture. *Cancer Res* 1978, 38: 2886-2891.

[45] Barnett CJ, Cullinan GJ, Gerzon K, Hoying RC, Jones WE, Newlon WM, Poore GA, Robison RL, Sweeney MJ, Todd GC, Dyke RW, Nelson RL. Structure-activity relationships of dimeric catharanthus alkaloids. 1. Deacetylvinblastine amide (vindesine). *J Med Chem* 1978, 21(1): 88-96.

[46] Jones SF, Burris HA 3rd. Vinorelbine: a new antineoplastic drug for the treatment of non-small-cell lung cancer. *Ann Pharmacother* 1996, 30(5): 501-506.

[47] Hill BT, Fiebig HH, Waud WR, Poupon MF, Colpaert F, Kruczynski A. Superior *in vivo* experimental antitumour activity of vinflunine, relative to vinorelbine, in a panel of human tumour xenografts. *Eur J Cancer* 1999, 35: 512-520.

[48] Kruczynski A, Hill BT. Vinflunine, the latest Vinca alkaloid in clinical development: A review of its preclinical anticancer properties. *Crit Rev Oncol Hematol* 2001, 40(2): 159-173.

[49] Bristol-Myers Squibb Company and Pierre Fabre Medicament declared that they were terminating their license agreement for the development of VFL, a chemotherapy agent under investigation for the treatment of advanced or metastatic bladder cancer and other tumor types on November 23, 2007. (http://www.fiercebiotech. com/press-releases/press-release-bristol-myers-squibb-ends-cancer-drug-deal-pierre-fabre)

[50] Wani MC, Taylor HL, Wall ME, Coggon P, McPhail AT. Plant antitumor agents. VI. The isolation and structure of taxol, a novel antileukemic and antitumor agent from *Taxus brevifolia*. *J Am Chem Soc* 1971, 93(9): 2325-2327.

[51] Schiff PB, Fant J, Horwitz SB. Promotion of microtubule assembly in vitro by taxol. *Nature* 1979, 277(5698): 665-667.

[52] Denis JN, Greene AE, Guénard D, Guéritte-Voegelein F, Mangatal L, Potier P. A highly efficient, practical approach to natural taxol. *J Am Chem Soc* 1988, 110: 5917-5919.

[53] Kingston DG. Recent advances in the chemistry of taxol. *J Nat Prod* 2000, 63: 726-734.

[54] Chauvière, G.; Guénard, D.; Picot. F.; Sénilh, V.; Potier. P. *C. R. Seances Acad. Sci., Ser 2* 1981, 293, 501-503.

[55] Appendino G, Danieli B, Jakupovic J, Belloro E, Scambia G and Bombardelli E. Synthesis and evaluation of C-seco paclitaxel analogues. *Tetrahedron Lett* 1997, 38, 4273-4276.

[56] Ojima I, Wang T, Miller ML, Lin S, Borella CP, Geng X, Pera P, Bernacki RJ. Synthesis and structure-activity relationships of new second-generation taxoids. *Bioorg Med Chem Lett* 1999, 9: 3423-3428.

[57] Bissery MC, Gunard D, Guéritte-Voegelein F, Lavelle F. Experimental antitumor activity of taxotere (RP 56976, NSC 628503), a taxol analogue. *Cancer Res* 1991, 51: 4845-4852.

[58] Pazdur R, Kudelka AP, Kavanagh JJ, Cohen PR, Raber MN. The taxoids: paclitaxel (taxol) and docetaxel (Taxotere). *Cancer Treat Rev* 1993, 19: 351-386.

[59] Wang GS. Medical uses of mylabris in ancient China and recent studies. *J Ethnopharm* 1989, 26: 147-162.

[60] Liu XH, Blazsek I, Comisso M, Legras S, Marion S, Quittet P, Anjo A, Wang GS, Misset JL. Effects of norcantharidin, a protein phosphatase type-2A inhibitor, on the growth of normal and malignant haemopoietic cells. *Eur J Cancer* 1995, 31A(6): 953-963.

[61] Shan HB, Cai YC, Liu Y, Zeng WN, Chen HX, Fan BT, Liu XH, Xu ZL, Wang B, Xian

LJ. Cytotoxicity of cantharidin analogues targeting protein phosphatase 2A. *Anticancer Drugs* 2006, 17(8): 905-911.

[62] McCuskey A, Keane MA, Mudgee LM, Sim AT, Sakoff J, Quinn RJ. Anhydride modified cantharidin analogues. Is ring opening important in the inhibition of protein phosphatase 2A? *Eur J Med Chem* 2000, 35: 957-964.

[63] Sodeoka M, Baba Y, Kobayashi S, Hirukawa N. Structure-activity relationship of cantharidin derivatives to protein phosphatases 1, 2A$_1$, and 2B. *Bioorg Med Chem Lett* 1997, 7(14): 1833-1836.

[64] Baba Y, Hirukawa N, Sodeokaa M. Optically active cantharidin analogues possessing selective inhibitory activity on Ser/Thr protein phosphatase 2B (calcineurin): Implications for the binding mode. *Bioorg Med Chem* 2005, 13: 5164-5170.

[65] McCluskey A, Walkom C, Bowyer MC, Ackland SP, Gardinera E, Sakoff JA. Cantharimides: A new class of modified cantharidin analogues inhibiting protein phosphatases 1 and 2A. *Bioorg Med Chem Lett* 2001, 11: 2941-2946.

[66] Tang W, Eisenbrand G. (1992). *Chinese Drugs of Plant Origin: Chemistry, Pharmacology and Use in Traditional and Modern Medicine.* New York: Springer-Verlag.

[67] Leclerc S, Garnier M, Hoessel R, Marko D, Bibb JA, Snyder GL, Greengard P, Biernat J, Wu YZ, Mandelkow EM, Eisenbrand G, Meijer L. Indirubins inhibit glycogen synthase kinase-3β and CDK5/P25, two protein kinases involved in abnormal tau phosphorylation in Alzheimer's disease. *J Biol Chem* 2001, 276(1): 251-260

[68] Merz, KH, Eisenbrand G. Chemistry and structure-activity of indirubins. In: *Indirubin, the Red Shade of Indigo.* Eds. Meijer L, Guyard N, Skaltsounis L, Eisenbrand G Roscoff. France, pp. 135-145, 2006.

[69] Ji XJ, Zhang FR, Lei JL, Xu YT. Studies on the antineoplastic effect and toxicity of synthetic indirubin. *Acta Pharma Sin* 1981, 16: 146-148.

[70] Han, R. Traditional Chinese medicine and the search for new antineoplastic drugs. *J Ethnopharmacol* 1998, 24: 1-17.

[71] Hoessel R, Leclerc S, Endicott JA, Nobel ME, Lawrie A, Tunnah P, Leost M, Damiens E, Marie D, Marko D, Niederberger E, Tang W, Eisenbrand G, Meijer L. Indirubin, the active constituent of a Chinese antileukaemia medicine, inhibits cyclin-dependent kinases. *Nat Cell Bio* 1999, 1(1): 60-67.

[72] Lee MJ, Kim MY, Mo JS, Ann EJ, Seo MS, Hong JA, Kim YC, Park HS. Indirubin-30-monoxime, a derivative of a Chinese anti-leukemia medicine, inhibits Notch1 signaling. *Cancer Lett* 2008, 265: 215-225.

[73] Taft CA, Da Silva VB, Da Silva CH. Current topics in computer-aided drug design. *J Pharm Sci* 2008, 97: 1089-1098.

[74] Jiang HL, Chen KX, Tang Y, Chen JZ, Li Q, Wang QM, Ji RY. Binding conformers searching method for ligands according to the structures of their receptors and its application to thrombin inhibitors. *Acta Pharm Sin.* 1997, 18(1): 36-44.

[75] Chen JZ, Hu LH, Jiang HL, Gu JD, Zhu WL, Chen ZL, Chen KX, Ji RY. A 3D-QSAR Study on ginkgolides and their analogues with comparative molecular field analysis. *Bioorgan Med Chem Lett* 1998, 8: 1291-1296.

[76] Tiseo P. Rational design of anti-dementia therapy. *Eur Neuropsychopharmacol* 1999, 9 (Suppl): S61-S68.

[77] Zeng FX, Jiang HL, Zhai YF, Liu DX, Zhang HY, Chen KX, Ji RY. Synthesis and pharmaco-
      logical study of huperzine A-E2020 combined compound. *Acta Chim Sin* 2000, 58: 580-587.

Da-Yuan Zhu and Chan

CHAPTER 16

Chemical Synthesis of Natural Products

The total synthesis of complex natural products has been the center of organic chemistry research. Not only does nature continue to present amazing and challenging molecules to attract organic chemists to synthesize, chemical synthesis of complex compounds will bring substantial advancement in organic chemistry and the application of natural products.

Professor T. Goto of Nagoya University believes that the purposes of synthesis of natural products are: 1) determining the structure; 2) providing the sample (when the availability is limited); 3) synthesizing derivatives; 4) developing new reactions and theories.

Before the wide use of spectroscopy and x-ray crystallography, chemical synthesis was once the authoritative way of determining the molecular structures of natural products. Although its role in structure determination has been considerably weakened, when confronted with trace amounts of non-crystal natural products, organic chemists can only identify the general structure with spectroscopy. The details have to be determined by total synthesis.

The configurations of C-5 and C-6 of lipoxin A were determined by total synthesis and the comparison of the four isomers[1].

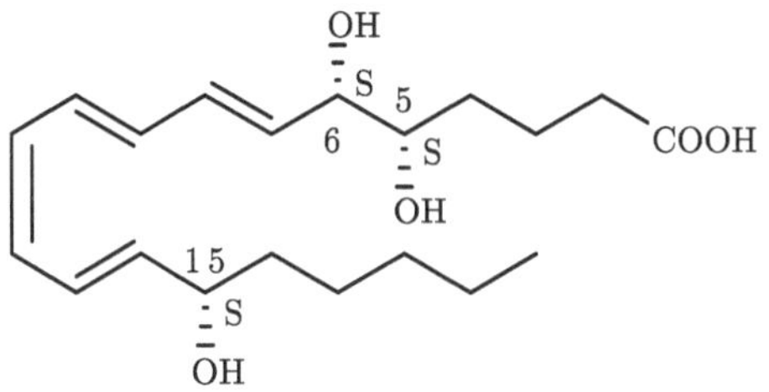

Another example is that the configuration of 12-OH of PUG, an anti-leukemia compound from a marine organism, was assigned α, not β, after its total synthesis[2].

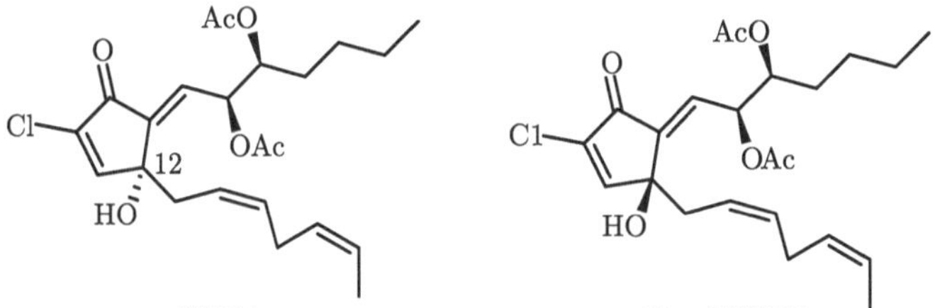

PUG4　　　　　　　　**12-epi-PUG3**

It is necessary to provide natural products by synthetic means when samples are too scarce to carry out further investigation on their properties, or when natural sources of some products cannot meet the market need. The ultimate solution is producing some natural products by chemical synthesis. Natural products are rarely produced by total chemical synthesis; a special exception is menthol. Takasago Company employed an asymmetric synthesis strategy to manufacture menthol with production of 1,500 tons per year[3].

Nevertheless, in most cases, these are not the objectives of synthesis of natural products. For some natural products with biological activities, the tendency is to synthesize not only the parent molecules, but also their derivatives to further explore the structure-bioactivity relationships to obtain more useful compounds. Many drugs are developed from derivatives of natural products.

In most cases, the synthesis of natural products represents the exploration of organic chemistry itself. The complexity of natural products presents a challenge to synthetic chemists and the synthesis of natural product is the reflection of synthetic art. To some extent, the history of chemical synthesis is the history of natural products synthesis. The tireless pursuits of natural products synthesis have confronted chemists with various difficulties that seemed too hard to overcome. In finding the solutions, they have developed various new approaches and continue to push chemical synthesis forward.

As natural products synthesis involves a broad range of compounds, it is impossible to cover the complete subject in one chapter and only limited examples will be given based on the classification of the natural products.

Section 1   Alkaloids

Alkaloids are the nitrogen-containing organic substances in plants (except proteins, peptides, amino acids and vitamins) and represent the largest class of natural products. Due to the diversity of their structures and the biological activities, many synthetic chemists are attracted to study them. Here are a few important examples of alkaloid synthesis.

1.1   Morphine

Morphine is an alkaloid discovered in the early 1800s. Since the first synthesis of morphine by Gate, its study has never stopped. There have been many total synthesis and relay syntheses published in the 1990s[4]. Among them, Overman's work is the most representative.

Overman's synthesis started from 2-allyl-cyclohexenone, which was elaborately converted to allylsilane by an asymmetric reduction, to finally give the two enantiomers of morphine. The allylsilane reacted with the A-ring precursor aldehyde and zinc iodide. The resulting imine cation intermediate was captured by the allyl group with high diastereoselectivity (more than 20:1). Heck reaction was used in the construction of the aryl-tertiary-carbon bond. After selective deprotection of the benzyl on phenolic oxygen and epoxidation of the cyclohexene, the final dihydrofuran ring formed (Scheme 16-1).

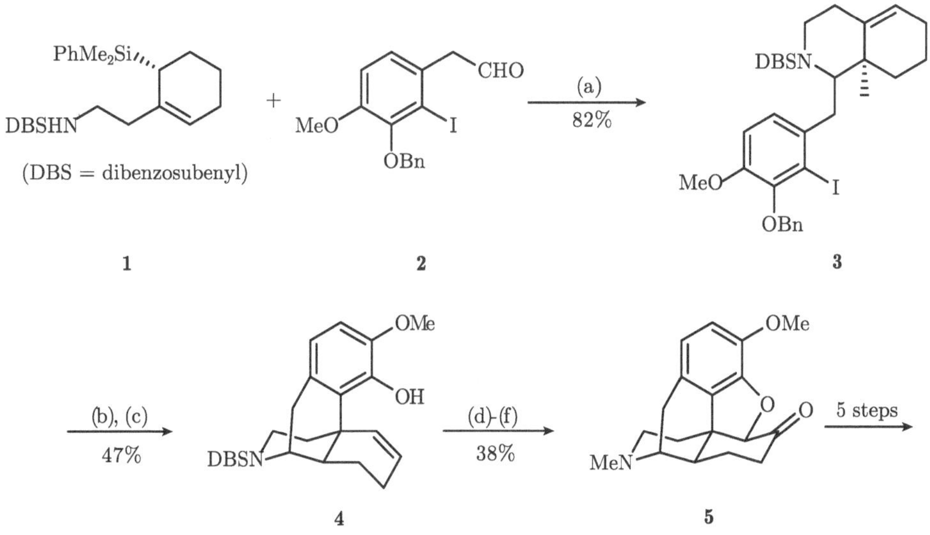

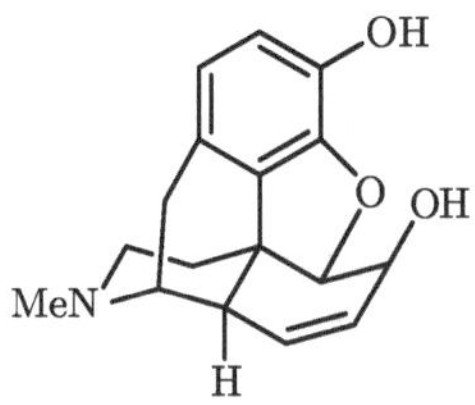

(-) Morphine
6

Scheme 16-1

a. ZnI_2; b. $Pd(O_2CCF_3)_2$, PMD; c. $BF_3 \cdot OEt_2$, EtSH; d. 3,5-dinitrobenzoic acid, CSA;

e. TPAP, NMO; f. H_2, $Pd(OH)_2$, HCHO

1.2 Strychnine

Strychnine, the first synthesized pure alkaloid, is a molecule with special significance in the history of organic chemistry and natural products. The synthesis study of strychnine reached its climax in the 1950s due to Woodward's[5] successful synthesis. This molecule, with seven rings and six chiral centers, was synthesized before the availability of powerful analytical instruments. It is regarded as a classic work.

After a 40-year break, scientists' interest on the synthesis of strychnine arose again in the 1990s. Six total syntheses were reported in the 1990s[6]. Of them, Rawal's concise racemic synthesis and Overman's enantioselective synthesis provide excellent examples. In Rawal's route, the successful combination of intramolecular Diels-Alder cycloaddition and intramolecular Heck reaction made the route fairly concise (Scheme 16-2).

Precursors **12** were subjected to intramolecular Diels-Alder cycloaddition in a sealed tube heating to obtain compound **13** in 99% yield, then formed lactam to construct C-ring after two intramolecular transformations. Compound **14** and **15** reacted to provide N-alkylation product compound **16** under mild conditions. The following intramolecular Heck reaction established F-ring non-enantioselectively. Double bonds in the D-ring transferred to the C-ring. Deprotection of hydroxyl and cyclization afforded isostrychnine in 10% yield, which is the best result at present.

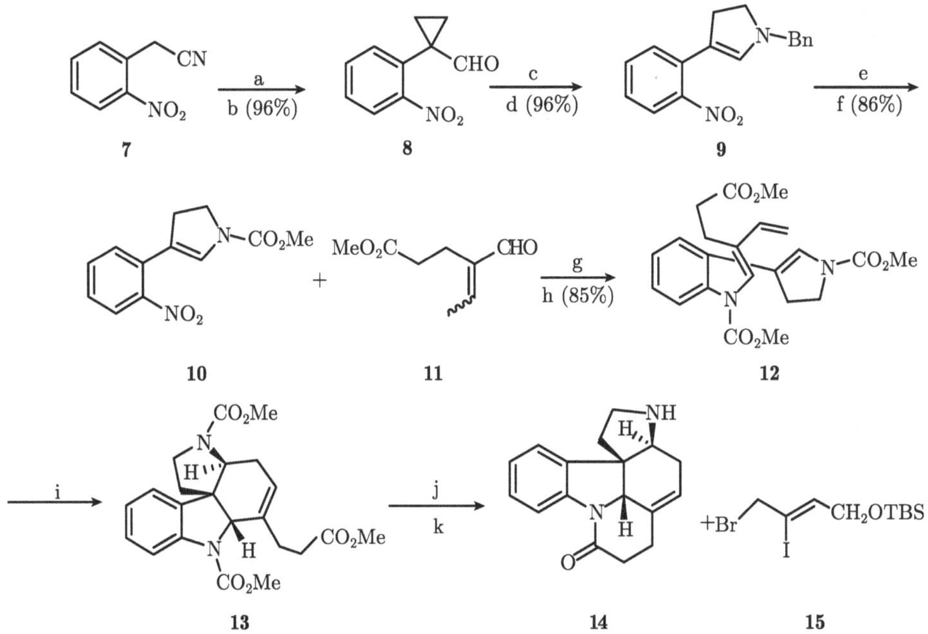

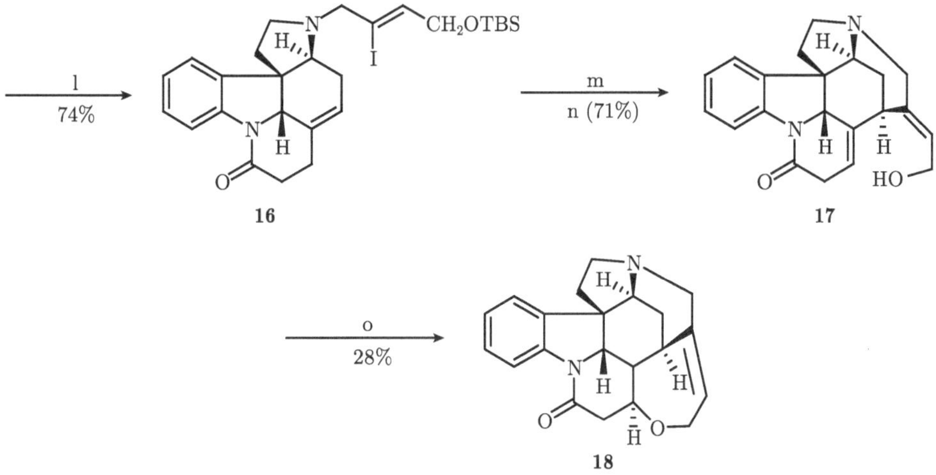

Scheme 16-2

a. Br(CH$_2$)$_2$Br, 50%NaOH; b. DIBAL-H; c. BnNH$_2$, Et$_2$O; d. TMSCl, NaI, DMF; e. ClCO$_2$Me, acetone;

f. 10%Pd-C, HCO$_2$NH$_4$; g. 23°C; h. ClCO$_2$Me, PhNEt$_2$; i. C$_6$H$_6$, 18.5^{o}C; j. TMSI, CHCl$_3$;

k. MeOH, 65°C; l. DMF, acetone, K$_2$CO$_3$; m. Pd(OAc)$_2$, Bu$_4$NCl, DMF, K$_2$CO$_3$, 70°C;

n. 2N HCl, THF; o. KOH, EtOH, 85°C

1.3   Camptothecin

Camptothecin possesses a good antitumor activity and there are nearly 20 records of its synthesis. Among them, Cai's[7] racemic approach in the 1970s and Comins's[8] enantioselective synthesis are two of the most representative examples. Cai's synthetic approach is illustrated in Scheme 16-3.

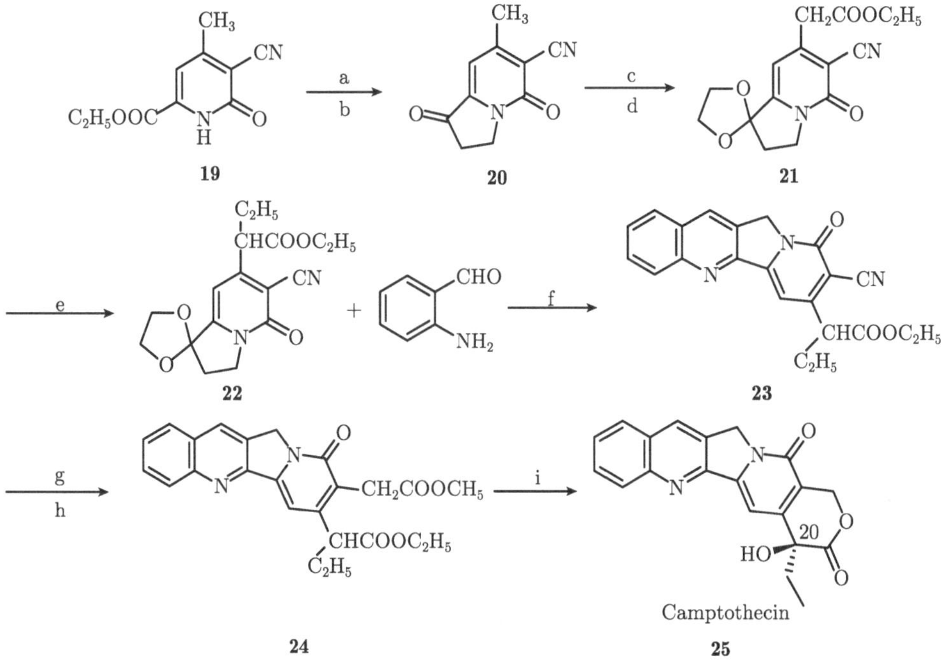

Scheme 16-3

a. K$_2$CO$_3$, DMF , methyl acrylate; b. HCl, HOAc; c. (CH$_2$OH)$_2$; d. (EtO)$_2$CO, NaH, PhCH$_3$;

e. C$_2$H$_5$I, NaH, DMF; f. HCl-HAc; g. Ac$_2$O-HAc, [H]; f. NaNO$_2$; i. CuCl$_2$, DMF, [O]

In Scheme 16-3, we can see that the formation of the BCD ring was quite concise. The key point of Cai's approach was the formation of compound **22** with a 20-aryl-tertiary-carbon for camptothecin, which paved the way for the subsequent steps. Another feature of this route is that all the reagents were very simple, and recrystallization and distillation are the major purification methods. No column chromatography was used, and the overall yield was as high as 18%.

In the early 1990s, Tagawa[9] modified Cai Juncha's approach to make it enantioselective (Scheme 16-4). The framework of the camptothecin derivatives was also constructed in this way.

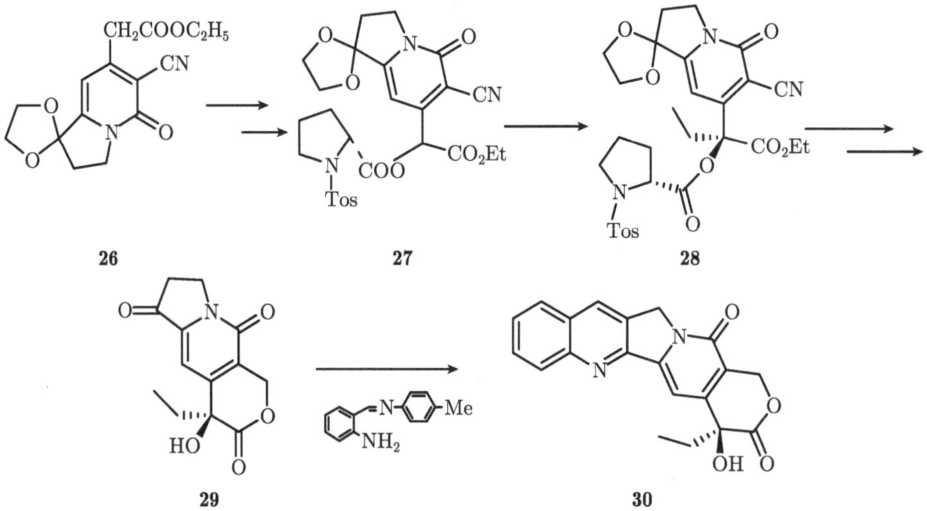

Scheme 16-4

Comins-constructed C-rings with an intramolecular Heck reaction provided a very concise synthesis (Scheme 16-5).

Comins's synthesis began with a two-step, one-pot reaction: 2-chloro-6-methoxypyridine was treated with t-butyl lithium, and the aromatic lithium intermediate was captured with formamide, then 1 eq t-butyl lithium was added directly to give the anion, which was quenched by iodine to give aldehyde **32**. Compound **32** was reduced with Et_3SH/TFA in methanol to give dimethyl ether **33**. Lithium-iodine exchange and addition of α-keto-8-phenylmethyl butyrate produced the intermediate enol-anion which was reacted with p-PhPhCOCl to give ester **34**. After recrystallization, enantiopure **34** was obtained. Compound **35** was obtained in three steps. Intramolecular Heck reaction was applied to give optically pure camptothecin after the coupling of compound **35** with **36**.

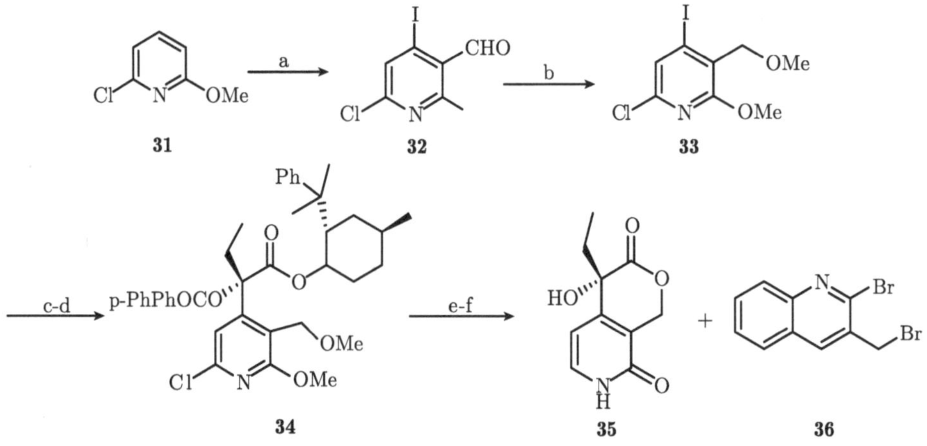

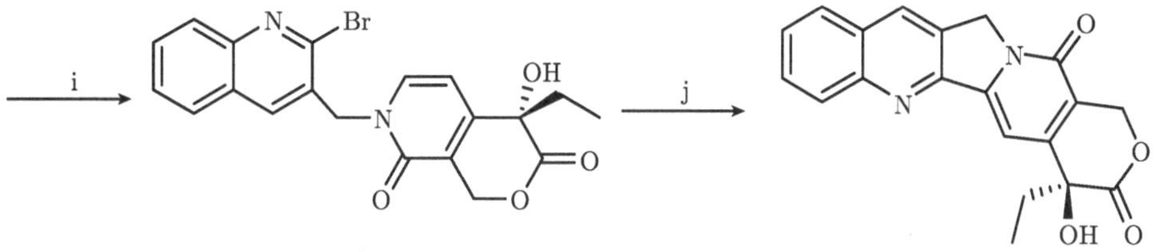

37   **38**

Scheme 16-5

a. i t-BuLi, THF, -78°C, 1h; ii (CH₃)₂NCH₂CH₂N(CH₃)CHO; iii. N-BuLi, -23°C, 2h; iv I₂(78%);
b. CH₃OH, EtSiH, TFA (98%); c. n-BuLi, THF, -78°C then α-keto-8-phenylmethyl butyrate;
d. p-PhPhCOCl (60% from **33**); e. 2N NaOH, EtOH (76%); f. TMSCl, NaI, CH₃CN,
Dabco, reflux; g. 6N HCl, 100°C(77%, two steps); h. H₂, Pd-C; i. tBuOK, DME,
reflux (87%); j. Pd(OAc)₂, Bu₄NBr, KOAc, DMF, 90°C, 3h (59%)

1.4   Reserpine

The synthesis of reserpine was one of Woodward's[10] representative works. His synthesis
route is illustrated in Scheme 16-6.

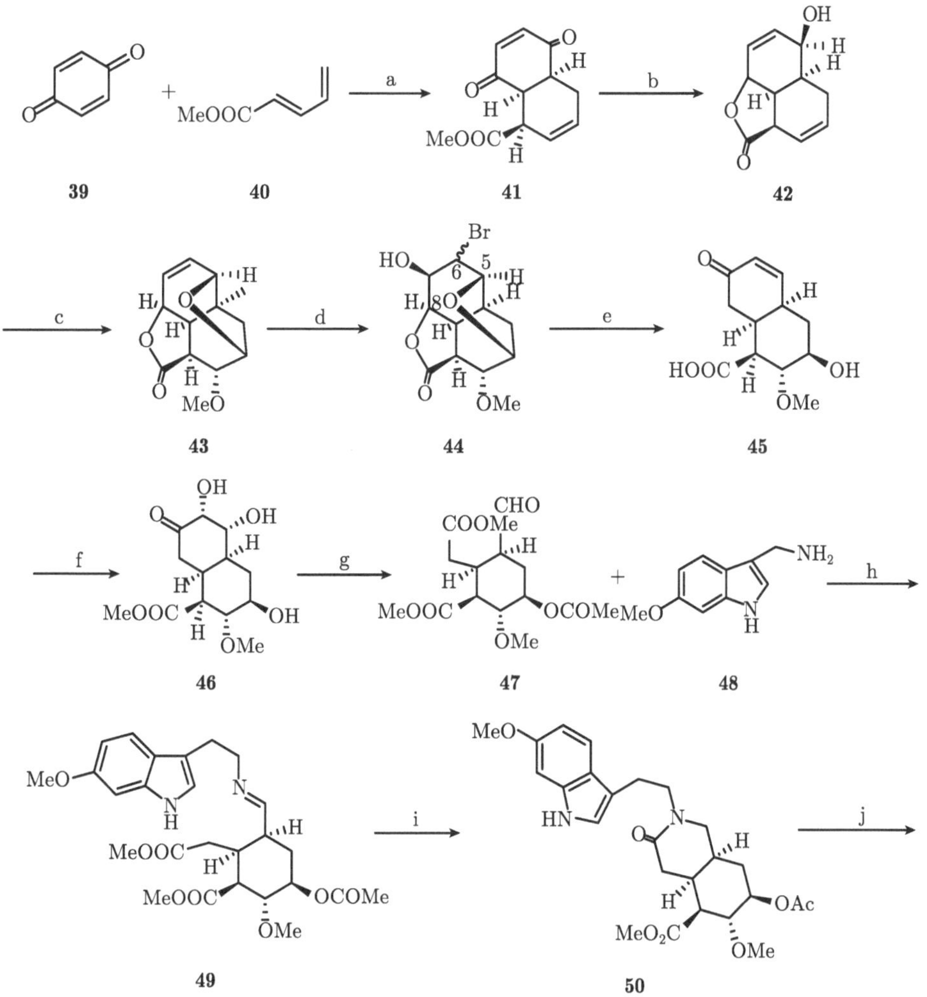

39   **40**   **41**   **42**

43   **44**   **45**

46   **47**   **48**

49   **50**

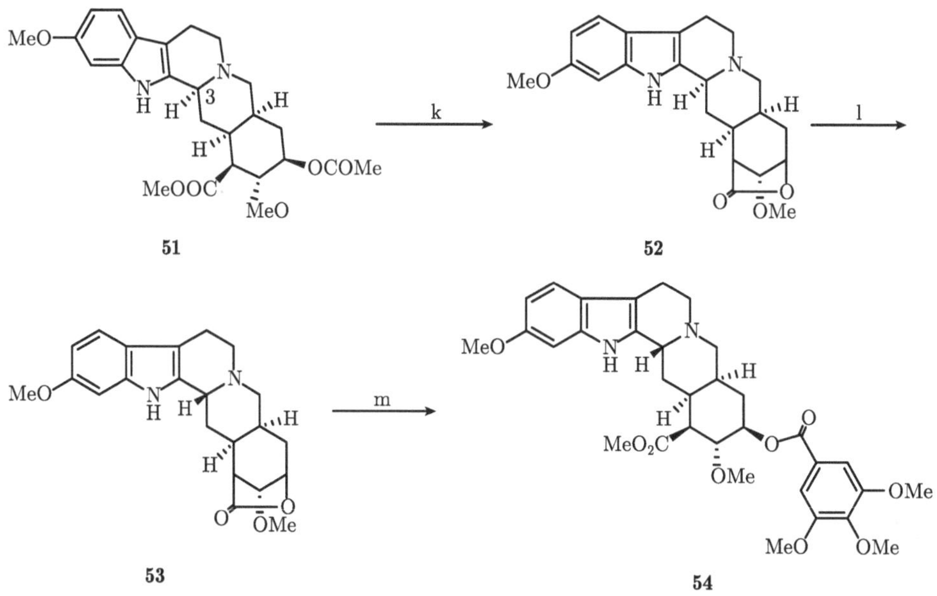

Scheme 16-5

a. PhH, reflux; b. Al(Oi-Pr)$_3$, i-PrOH; c. i.Br$_2$; ii. NaOH, MeOH; d. NBS/aq·H$_2$SO$_4$, 30°C;
e. CrO$_3$/HOA c; Zn/HOAc; f. i. CH$_2$N$_2$; ii Ac$_2$O; iii. OsO$_4$, NaClO$_3$, H$_2$O; g. i. HIO$_4$; ii. CH$_2$N$_2$;
h. PhH; i. NaBH$_4$/MeOH; j. i. POCl$_{3;ii.}$aq·NaBH$_4$; k. i. KOH/MeOH; ii. HCl; iii. DCC/Py;
l. t-BuCO$_2$H, xylenes; m. i. NaOMe, MeOH, reflux; ii. 3,4,5-(MeO)$_3$C$_6$H$_2$COCl, pyridine

The construction of the E-ring, which possesses five chiral centers, is the most difficult part of the synthesis of reserpine. Woodward used the Diels-Alder reaction to establish these chiral centers with a substrate-controlled strategy. Because of the steric hindrance, the double bond of **42** could be selectively brominated and simultaneously undergo an intramolecular etherification reaction. Treated with sodium methoxide, the bromo group was replaced by a methoxy group to give **43**. Thus, all the chiral centers in the E-ring were constructed. This work received great acclaim for its conciseness and ingenuity. The addition of double bond with NBS and water gave **44**. **44** was oxidized with chromic anhydride, and then treated with zinc and acetic acid, to give rise to the following transformation: 1) reduction of the C-O bond at C-8; 2) generation of the carbonyl group at C-1; and 3) formation of double bond by breaking up the 3,5 ester-bridge between C-Br bond at C-6 and C-5 to obtain **45**. Necessary protection and two oxidation reactions gave **47**, which was the key intermediary in the reserpine synthesis.

The coupling of **47** and **48**, and subsequent reduction, cyclization with POCl$_3$ and reduction again gave compound **51**. **51** was quite similar to reserpine, but there was a problem with the chiral configuration at C-3. After lactonization to give **52**, configuration inversion was realized by refluxing **52** with t-BuCO$_2$H in xylene to afford **53**. Racemic reserpine (**54**) was obtained after ring opening and esterification. Finally, chiral resolution provided the enantiopure natural product.

1.5   Maytansine

Maytansine is a macro ring lactam alkaloid with good anti-cancer activity. Since Corey[11] completed its total synthesis, there have been many publications on its synthesis. Herein, XQ Gu's approach[12] will be introduced.

This synthesis started with (-)-(2R,3R)-2-methyl-3-hydroxyl-butanedioic acid as the starting material. Esterification, reduction with LAH, and protection with acetone gave alcohol

55. **55** was converted to epoxide **57** through two steps, then **57** was reacted with thioacetal to generate **58** with one more carbon. As thioacetal was the dipole reversal form of carbonyl, it was ideal for the following segment connection.

Wittig reaction of **59** and **60** gave **61**. Because the target molecule has to construct multi-double bonds, Wittig reaction nicely meets this need to connect segments to obtain the opening chain precursor **61**. After reduction, protection and oxidation to aldehyde, and finally reaction with thioacetal **58, 61** was transformed to epimers **64** and **65**, which are separable. The S-isomer **64** was converted to R-isomer **65** by an oxidation and a subsequent reduction by using Noyori chemistry. **65** was methylated, deprotected and oxidized to give aldehyde. Again the Wittig-Horner reaction was used to connect **67** and **68** to generate compound **69**.

69 was reduced with NaBH$_4$ to give a pair of isomers and the desired isomer **71** was obtained after separation. Another isomer **70** was oxidized to give **69** which could go back to the reduction process. After protection of the hydroxyl group with TBS and hydrolysis of the ester group, Mukaiyama reaction[13] was found to be the ideal chemistry to generate **73**. Deprotection of MEM of **73** with 2-chloro-1, 3, 2-dithiaborolane gave **74**, which was treated with p-NO$_2$C$_6$H$_4$OCOCl to provide **75**. Then, maytansinol was synthesized according to Corey's procedure. Introduction of the side chain gave maytansine (Scheme 16-7).

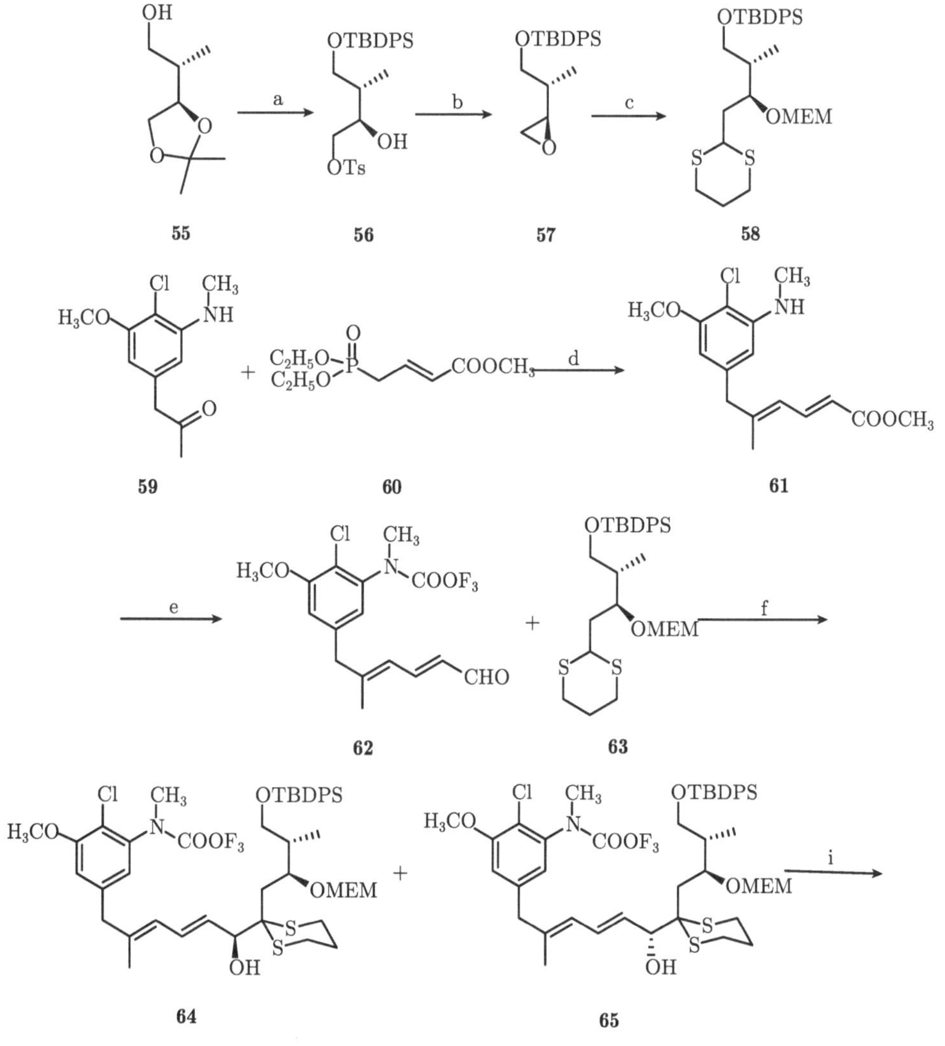

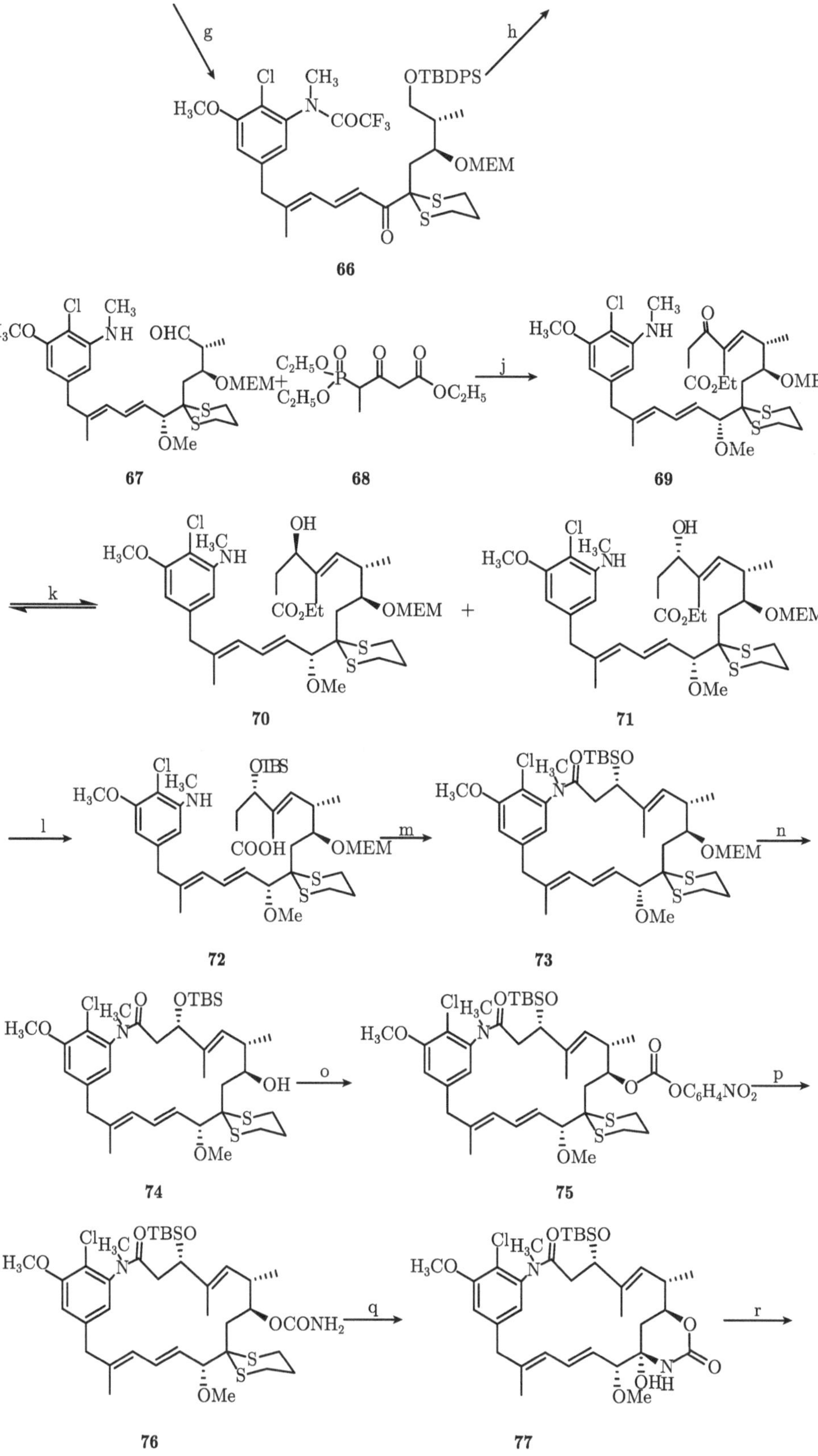
g
h
66
67
68
j
69
k
70
+
71
l
72
m
73
n
74
o
75
p
76
q
77
r

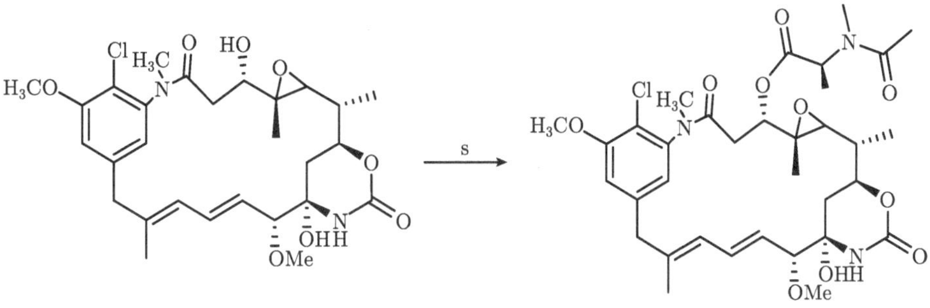

Scheme 16-7

a. i. TBDPSCl, triethylamine, DMAP, CH_2Cl_2 ii. 75%HAc ; iii. TsCl, pyridine; b. NaH, THF; c.i. LDA, dithiane , THF; ii. MEMCl, ethyl diisopropylamine; d.LDA, THF; e. i. LAH, $AlCl_3$; ii. TFAA, Py; iv. Ag_2CO_3, PhH. F. BuLi, THF; g. MnO_2 ; h. (R)-BANIL-LAH-EtOH; h. i. CH_2N_2; ii. TBAF, THF ; iii. [O]; j. LDA THF; k. $NaBH_4$; l. i. TBSCl, imidazol; ii. LiOH, DME; m.Mukaiyama's condition; n. 2-chloro-1,3,2-dithiaborolane, CH_2Cl_2 , -78^oC; o. $p\text{-}NO_2C_6H_4OCOCl$, Pyr.; p. NH_3/CH_3OH; q. $HgCl_2/CaCO_3$, ethyl diisopropylamine; r. i. HF/H_2O; ii. VO(acac)$_2$, t-BuOOH; s. N-acetyl-N-methyl-L-alanine, DCC, $ZnCl_2$

Section 2   Terpene

2.1   Monoterpene—Menthol

Monoterpene is defined as the molecule with a basic framework of head-to-tail combination of two isoprenes. Its framework is small, and the central problem in the synthesis is to control the stereochemistry.

Menthol is a very important monoterpene, and its world production reaches 3,500 tons per year, 30% of which is made by Takasago Company in Japan after it invented an industrial synthetic approach to produce optically pure menthols[14]. The pathway is shown in Scheme 16-8.

Using inexpensive β-pinene as the starting material, which was thermally cracked into myrcene; telomerization catalyzed by n-BuLi implemented the selective connection of the diethylamine group. The most important step was the isomerization reaction asymmetrically catalyzed by [Rh((S)-BINAP) (COD) ClO_4], generating the chiral methyl with an ee value more than 98%. This chiral methyl was applied to control the generation of the other two chiral centers in the following reaction. **83** was hydrolyzed to give R-citronellal and the ee value was up to 98%, which was higher than the value of its natural product. The ee value of natural citronellal currently is not greater than 80%. With treatment of zinc bromide, the substrate gave isopulegol **86** via a chair-like transition state. The 100% enantiopure product was obtained after low-temperature recrystallization. **86** was hydrogenated to give (-)-menthol.

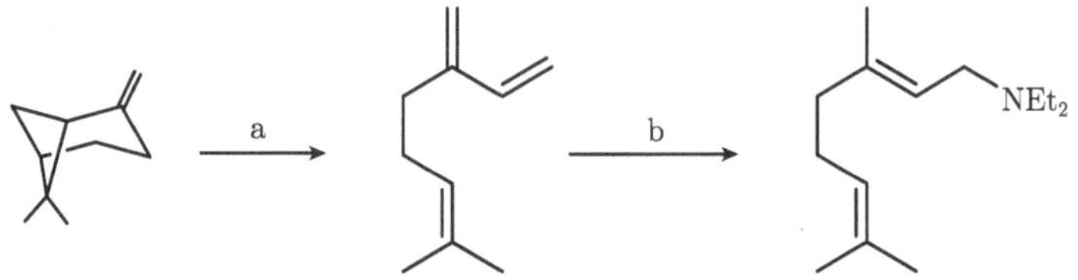

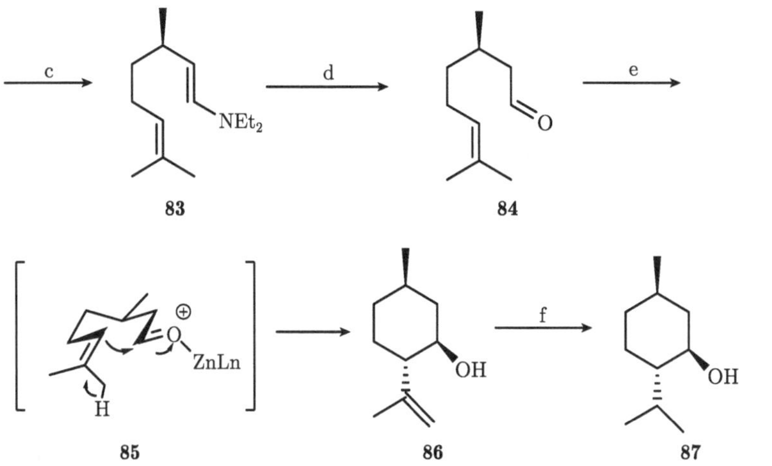

Scheme 16-8

1. a. thermal cleavage; b. n-BuLi (cat.), Et$_2$NH; c. [Rh((S)-BINAP)(COD)]ClO$_4$ (cat.),100°C;
d. aq.H$_2$SO$_4$; e. ZnBr$_2$; f. H$_2$, Ni

2.2   Sesquiterpene—Artemisinin

The basic framework of sesquiterpene can be recognized as a combination of three iso-prenes. Sesquiterpenes widely exist in plants, microbes, marine organisms, and certain insects, many of which have important biological functions and activities. Artemisinin is a very important sesquiterpene discovered by Chinese scientists as a new antimalarial drug with the characteristics of immediate effect and hypotoxicity. Since plasmodium already has resistance to the antimalarial drug quinine, artemisinin appeared to be more important. There have been many reports published on the synthesis of artemisinin. Herein XX Xu's work will be introduced[15] (Scheme 16-9).

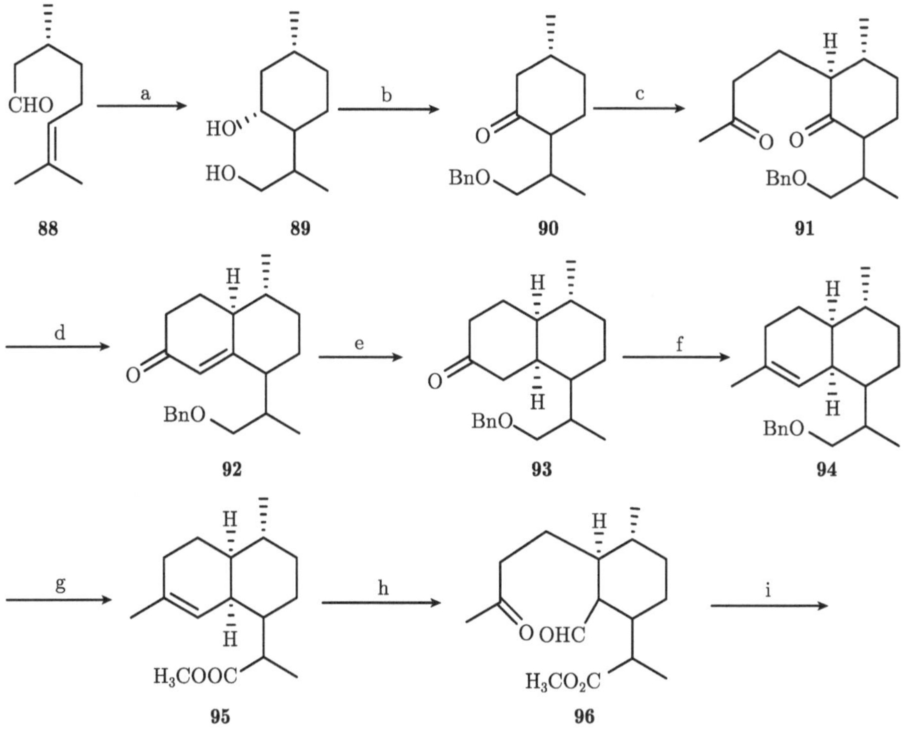

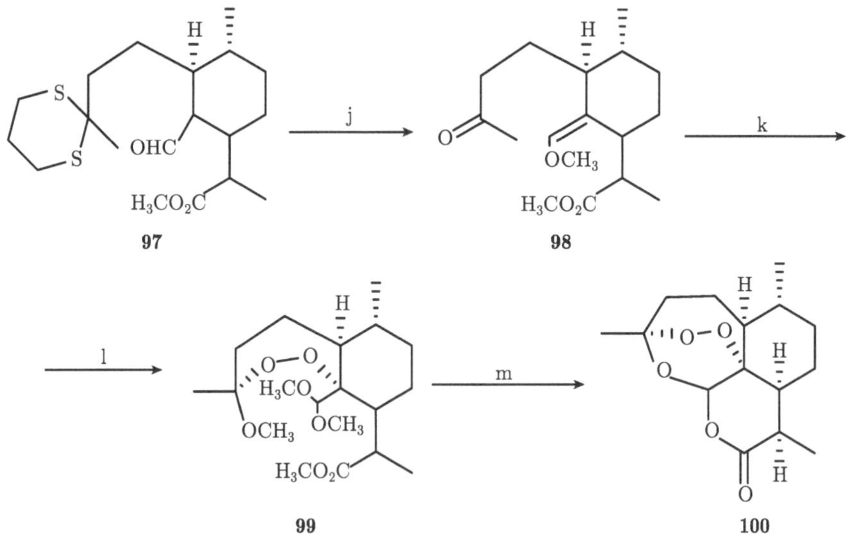

Scheme 16-9

1. a. i. ZnBr$_2$; ii. B$_2$H$_6$; H$_2$O$_2$, Na OH; b. i. PhCH$_2$Cl-NaH; ii. Jones oxidation; c. LDA, CH$_2$=C(Me$_3$Si)COCH$_3$ d. i. Ba(OH)$_2$·8H$_2$O; ii. (COOH)$_2$; e i. NaBH$_4$; ii. Jones oxidation; f i. MeMgI; ii. p-TsOH; g. i. Na- Liq.NH$_3$; ii. CH$_2$N$_2$; iii. NaBH$_4$·6H$_2$O;h O$_3$; MeS$_2$; i HS(CH$_2$)$_3$SH-BF$_3$·Et$_2$O; j i. HC(OMe)$_3$, p-TsOH; Xylene; ii. HgCl$_2$-CaCO$_3$-aq. CH$_3$CN; k. O$_2$, Rose Bengal, uv HClO$_4$

88 was first converted into the dihydroxy intermediate 89. Selective benzylation of the primary hydroxyl group followed by oxidation with Jones reagent gave the ketone 90. Kinetic deprotonation of ketone 90, the reaction of the resulting enolate with silylated vinyl ketone, provided 1, 5-diketone, and subsequent cleavage of the trimethyl silyl group gave 91. Ring closure of 91 with Ba(OH)$_2$ followed by acidification gave α,β-unsaturated ketone 92. Reduction of 92 with NaBH$_4$-Py followed by oxidation with Jones reagent afforded the ketone 93. When 93 reacted with MeMgI followed by dehydration with toluene-p-sulfonic acid, the mixture of 94 and its Δ^3-isomer was obtained in a 1:1 ratio. The pure 94 was separated by repeated flash chromatography.

94 was reduced, oxidized with Jones reagent, and finally esterified to give the desired product methyl dihydroarteannuinate 95. Ozonization of 95 afforded aldehyde-ketone 96. Selective protection of 96 with thioketal provided compound 97. 97 was reacted with trimethyl orthoformate, resulting in changing the acetal into the enol methyl ether. Removal of thioketal with HgCl$_2$-CaCO$_3$ gave the key intermediate 98. Oxidation of the methanolic solution of 98 in the presence of oxygen and Rose Bengal afforded compound 99. Treatment of 99 with 70% HClO$_4$ gave artemisinin 100.

2.3 Diterpene—Tanshinones, Triptolide

Tanshinones are diterpene compounds and have excellent biological activity. Because of the consistent structural framework feature of these types of compounds, they could be synthesized by common methods. One of the best approaches is using the Diels-Alder reaction to construct the framework. Lee[16] applied the ultrasonic-promoted reactions to the synthesis of tanshinone with satisfactory results. His synthetic route is illustrated in Scheme 16-10 with tanshinone IIA as the example.

The Diels-Alder reaction performed under heating conditions affected the yield. The stereoselectivity was low, which gave isomers in the proportion of close to 1:1. However, the reaction performed under the ultrasonic conditions could be conducted in mild conditions with higher yield, good selectivity and fewer isomers (< 5%).

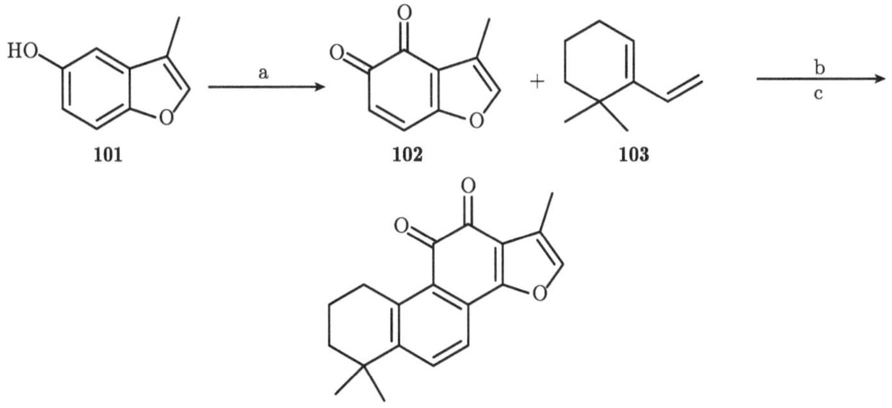

101 102 103 104

Scheme 16-10

a. Fremy's Salt; b. Ultrasound; c. DDQ, benzene, reflux; *O*-quinone **102** was unstable.

Triptolide exhibited many significant biological activities. Since Berchtold[17] completed its total synthesis, there have been many papers published on its total synthesis and derivatives. D Yang[18] adopted the radical cyclization reaction catalyzed by lanthanum trifluoroacetic acid to implement its synthesis stereoselectively, and is shown in Scheme 16-11.

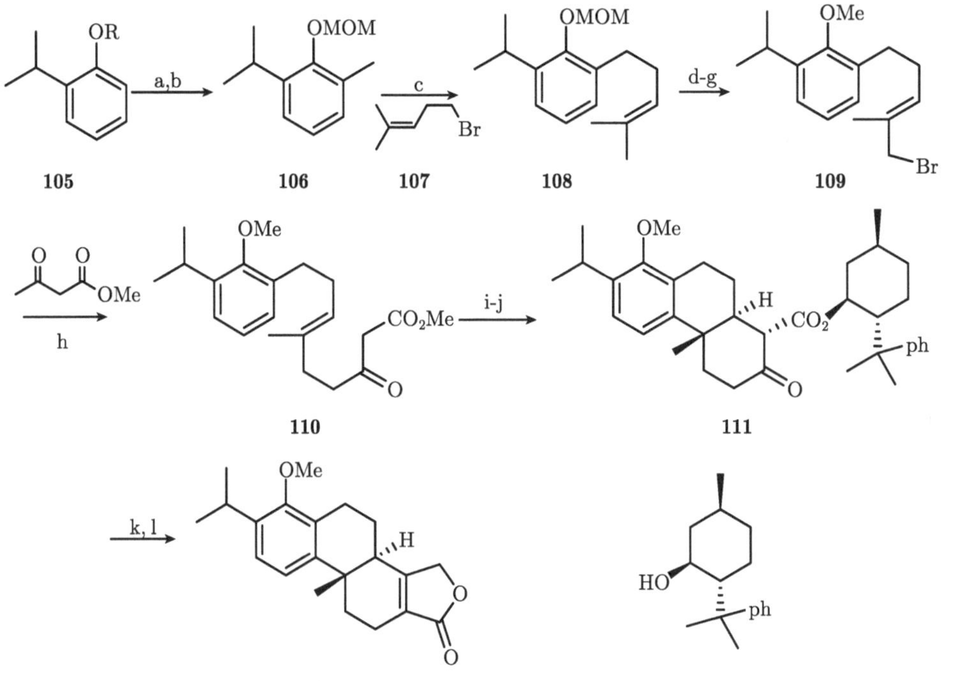

105 106 107 108 109 110 111 (+)112 113

Scheme 16-11

a. MOMCl, NaH, THF, 65°C, 2.5h, 99%; b. *n*-BuLi, THF, -40°C to 20°C, 2h, then -78°C, MeI, 1h, 98%; c. *s*-BuLi, THF, -78°C, **107**, 1h, 98%; d. TMSCl, $LiBF_4$, CH_3CN, -10°C to 25°C, 4h, 97%; e. Me_2SO_4, K_2CO_3, acetone, reflux, 2.5h, 100%; f. SeO_2, *t*-BuOOH, CH_2Cl_2, 0°C, 8h, then $NaBH_4$, MeOH, 73%; g. MsCl, Et_3N, CH_2Cl_2, -40°C to -20°C, 1h, then LiBr, THF, -20 to 25°C, 2h, 96%; h. $CH_3COCH_2COOCH_3$, NaH, *n*-BuLi, THF, 0°C, 1h, 85%; i. DMAP, toluene, 113; j. $Mn(OAc)_3{\cdot}2H_2O$, $Yb(OTf)_3$; k. KHMDS, $PhNTf_2$; l. i. DIBAL-H; ii. $Pd(PPh_3)_4$

In the synthesis of triptolide, D Yang, used the Tamelem methods[19] as a model reaction. She focused her main effort on asymmetric synthesis of the intermediate (+)-**112**. Protected with MOMCl, an anion of **105** was generated by treatment with *n*-BnLi and then with methyl iodide to form methylation intermediate **106**, which was converted to **108** by first treating with s-BuLi and subsequently with bromide **107**. Direct oxidation of **107** gave poor yield. When MOM of **107** was replaced by Me as a protective group, the oxidation furnished moderate yield and bromide **109** was prepared in three steps with satisfactory yield. Precursor **110** was prepared by a nucleophilic substitution of **109** with dianion of methyl acetoacetate. Transesterification of **110** with the chiral auxiliary **113** in DMAP and toluene, then catalytic reaction with $Mn(OAc)_3$ and $Yb(OTf)_3$ afforded a three-membered ring intermediate **111** in d.r. of 38:1.

Crisp's approach[20] was applied to construct α, β-unsaturated (+)-**112** with **111**. **111** underwent triflation, reduction and coupling reaction to give (+)-**112**.

With **112** at hand, (-)-**113** was first prepared by the following transformations: oxidation with CrO_3, demethylation with BBr_3, and reduction with $NaBH_4$. Then, oxidative epoxidaton of **113** with $NaIO_4$ afforded mono-epoxide intermediate (-)-**114**. Next, epoxidation of **114** with CF_3COCH_3 and ozone gave the di-epoxide **115**. Third, the epoxide moiety was introduced with $H_2O_2/NaOH$ system to give precursor **116**. Finally, reduction with $NaBH_4$ gave triptolide (Scheme 16-12).

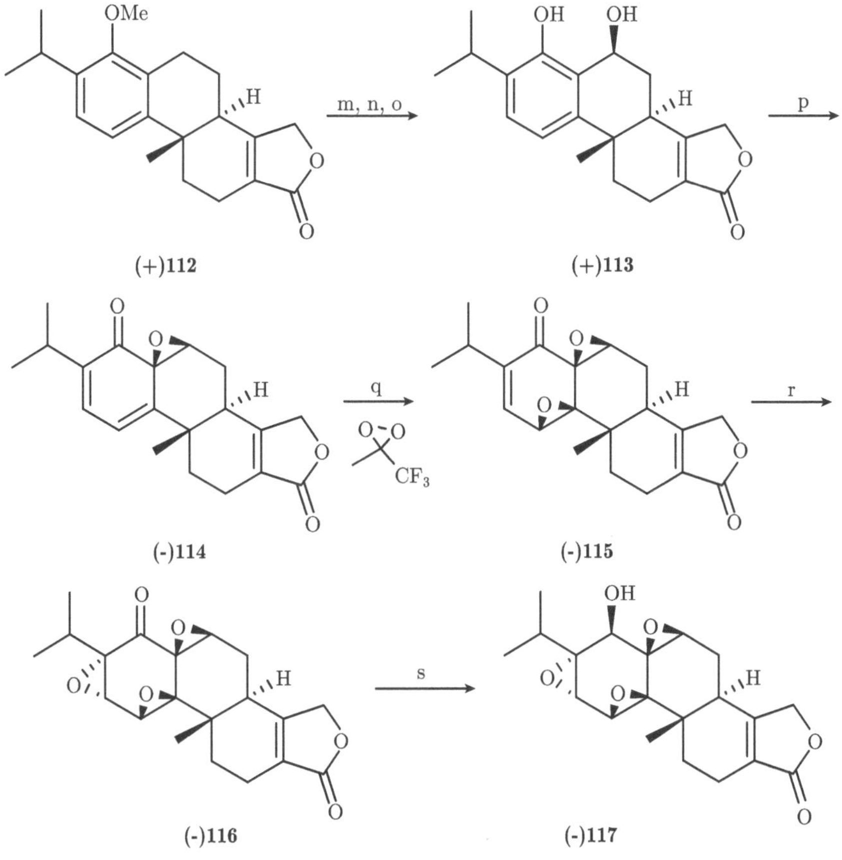

Scheme 16-12

a. CrO_3, Hac (aq), rt, 45% b. BBr_3, CH_2Cl_2, -78°C to rt, (99%); c. $NaBH_4$, CH_3OH, 0°C, 2h, (99%); d. $NaIO_4$, $MeOH/H_2O$, 0°C to 25°C, 1h, (96%); e. CF_3COCH_3, Oxone, $NaHCO_3$, CH_3CN/H_2O, Na_2(EDTA), 25°C, 4h, (70%), f. H_2O_2, NaOH, MeOh, 25°C, 3h, (96%); g.Eu(FOD)$_3$, CH_3OH, (49%).

Section 3   Flavonoids

The precursor of flavonoids can be synthesized through the following two strategies. The first, condensation of fragment C6C2 with C6C1, flavonoid precursor C6-C3-C6 **122** was obtained through condensation of 2-hydroxyacetophenone derivates **118** with benzaldehyde derivates **119** or through the intramolecular reaction of ester **121**, which was prepared with 2-hydroxyacetophenone derivates **118** and benzoyl chloride derivates **120**. The second, condensation of fragment C6C3 and C6, the flavonoid precursor could be prepared through acylation of phenol derivates **123** with cinnamoyl chloride derivates **124**. Substituents on ring A, ring B, and ring C may have an effect in the synthesis of specific compounds. Some modified approaches are also reported. General synthetic routes of flavonoids are shown in Scheme 16-13.

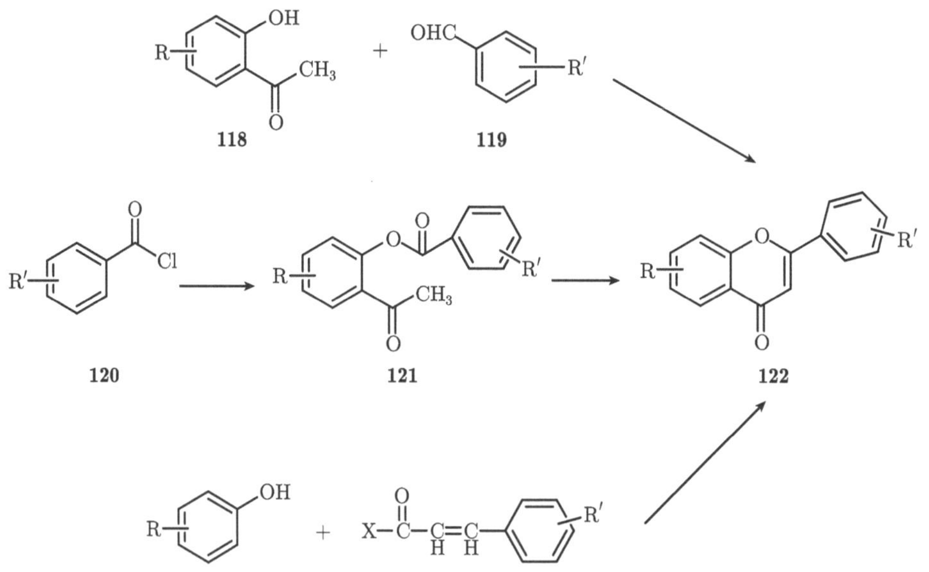

118   119

120   121   122

123   124

Scheme 16-13

3.1   Synthesis of chalcones and dihydroflavonoids

Chalcones are synthetic intermediates of many flavonoids; therefore, synthesis of these intermediates is an important procedure. The two approaches mentioned above can be used for this purpose. Generally, they can be prepared through aldol condensation of acetophenone derivates and benzaldehyde derivates with 50% to 60% KOH or NaOH as catalyst at 0 to 20°C. However, chalcones sometimes undergo ring closure to give dihydroflavonoids or a mixture of the two in basic or acidic conditions.

The main step in the total synthesis of licochalcone[21] was the preparation of the ring B fragment. Claisen rearrangement and esterification of intermediate **127** with butyric anhydride and N, N-dimethylaniline gave intermediate **128**, which was hydrolyzed to ring B fragment **129**. The yield of Claisen rearrangement would be low without esterification of the resulting phenolic hydroxyl group. Condensation of **129** with 4-hydroxyacetophenone gave chalcone **130** (Scheme 16-14). Due to lack of a 2'-hydroxyl group, **130** would not undergo ring closure to form a dihydroflavonoid. While those chalcones bearing a 2'-hydroxyl group would undergo ring closure to generate the corresponding dihydroflavonoid in 1% to 2% in base or acid overnight at room temperature, a 6'-hydroxyl group would accelerate the cyclization. Additionally, 3'-substituent or 4-hydroxyl groups decelerate the cyclization.

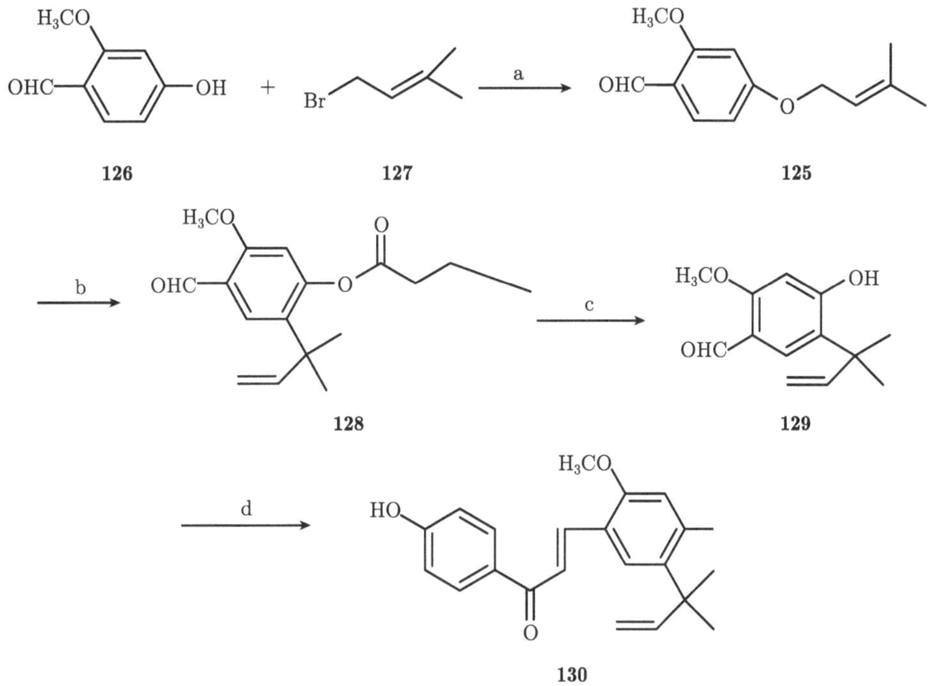

Scheme 16-14

a. K_2CO_3, acetone; b.N,N-dimethylaniline, butyric anhydride; 190-200°C;

c. NaOH, EtOH; d. 4-hydroxyacetophenone, HCl, EtOH

3.2   Synthesis of Flavonoids and Flavonols

Dehydrogenation of chalcones or dihydroflavonoids gave corresponding flavonoids. Dehydrogenation of dihydroflavonols gave corresponding flavonols. Additionally, they can also be synthesized through direct condensation. Some common approaches are presented below.

3.2.1   Baker-Venkatereman's method[22]

The key step for the synthesis of baicalein is intramolecular Claisen rearrangement. Treatment of intermediate **133** with KOH or NaOH in pyridine, benzene, or toluene gave β-diketone intermediate **134**. Cyclization of **134** with acid (H_2SO_4/EtOH) or weak base (NaOAc/AcOH) gave bacalecin **135**. Cyclization and simultaneous hydroxylation of **134** with peroxyformic acid gave 3-hydroxyl bacalecin **136** (Scheme 16-15). This approach can also be applied to the synthesis of flavonols.

3.2.2   Algar-Flynn-Oyamada's method[23]

Oxidation of 2'-hydroxychalcone **137** with H_2O_2 in basic condition gave corresponding flavonol **138**, which is an epoxide derivative. Then, an intramolecular nucleophilic substitution of **138** at the benzylic carbon and a concomitant oxidation afforded flavonol **139**. In contrast, nucleophilic substitution on the other side of the epoxide, i.e., α position of ketone, gave **142** as a byproduct (Scheme 16-16). Substituents and reaction temperature are the major factors for this reaction. First, for a substrate chalcone which contains 2- or 4-hydroxyl and does not contain 6'-methyl, the major product is flavonol. Second, for a substrate chalcone which contains 6'-methyl and does not contain 2- or 4-hydroxyl, the product will be a mixture of **142** and flavonol. Third, high temperature is beneficial to the formation of flavonol.

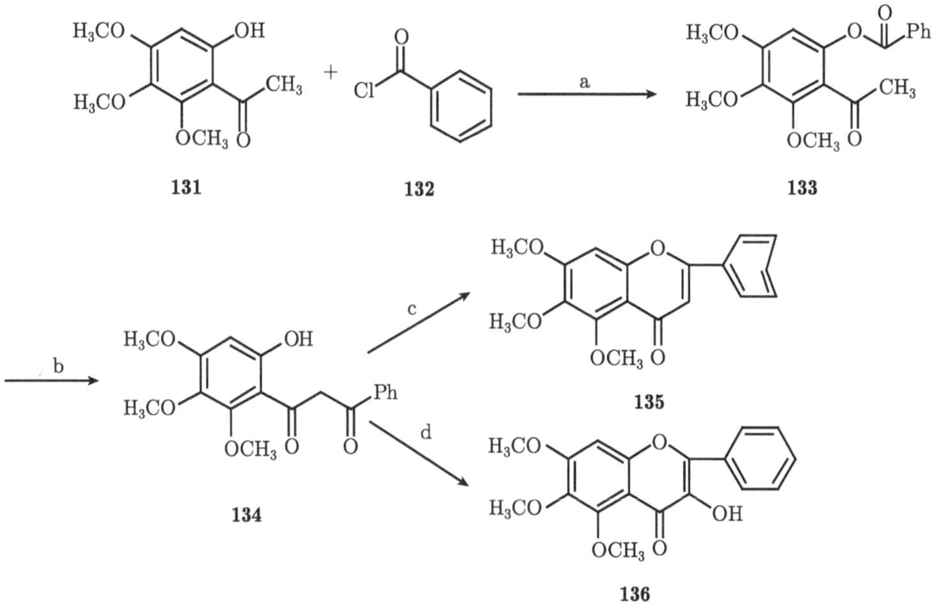

131   **132**   **133**

134   **135**

136

Scheme 16-15

a. K$_2$CO$_3$/Py; b.KOH/Py; c.H$_2$SO$_4$/EtOH; d. HCOOOH

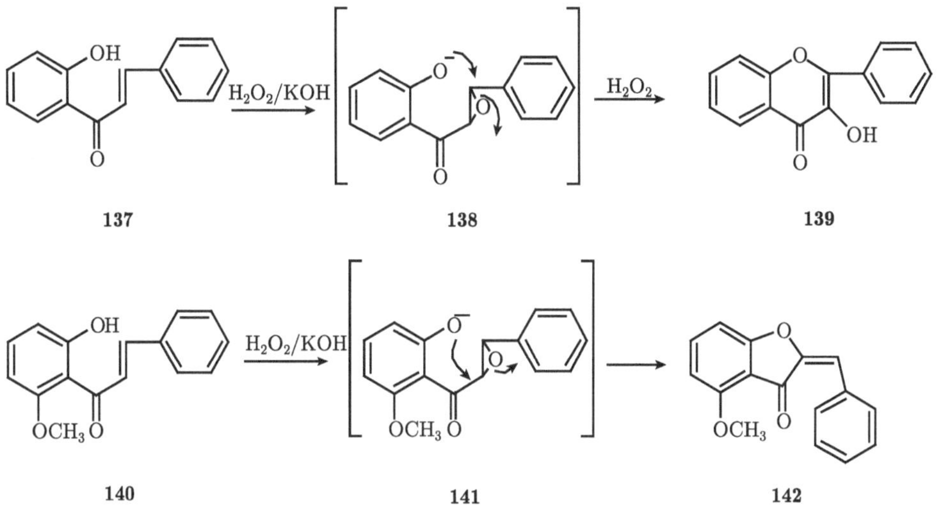

137   **138**   **139**

140   **141**   **142**

Scheme 16-16

3.2.3   Synthesis of isoflavone

There are many approaches to synthesize isoflavone. The classic one is cyclization of phenylbenzylketone derivates **145**. Many agents, such as ethyl orthoformate, Zn(CN)$_2$/HCl, DMF/POCl$_3$, and ethyl formate/Na, can be used in the cyclization step. However, these reagents have specific requirements for their structures, For example, ethyl orthoformate is only applied for the compounds which contain 2,3,6-trihydroxyphenyl substrate and DMF/POCl$_3$ is only applied to oxidative derivatives of resorcinol. The synthesis of formononetin **147** is one such example (Scheme 16-17).

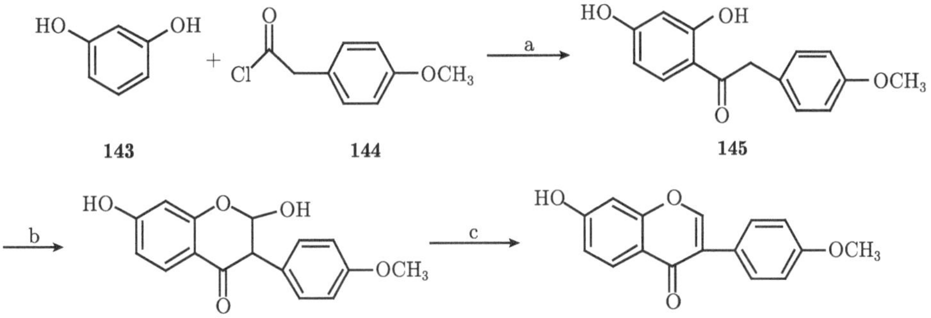

143   **144**   **145**

146   **147**

Scheme 16-17

a. BF$_3$; b. HCOOEt, Na; c. HCl

Section 4   Synthesis of Anthraquinones

Synthesis of anthraquinones was studied as early as in the mid-19th century. Biological functions of anthraquinones, especially the discovery of antitumor anthracycline antibiotics, serve as a major driving force for continued exploration of new synthetic approaches. Some major synthetic methods are introduced below.

4.1   Friedel-Crafts acylation

A classic approach is the double Friedel-Crafts condensation of phthalic anhydride derivatives and phenol derivatives. However, this method lacks the regioselectivity for asymmetric anthraquinones. Hayashi rearrangement is likely to occur in the intramolecular Friedel-Crafts condensation.

Alkylation with 3-bromophthalic anhydride derivative **148** catalyzed with SnCl$_2$ was regioselective (Scheme 16-18)[25].

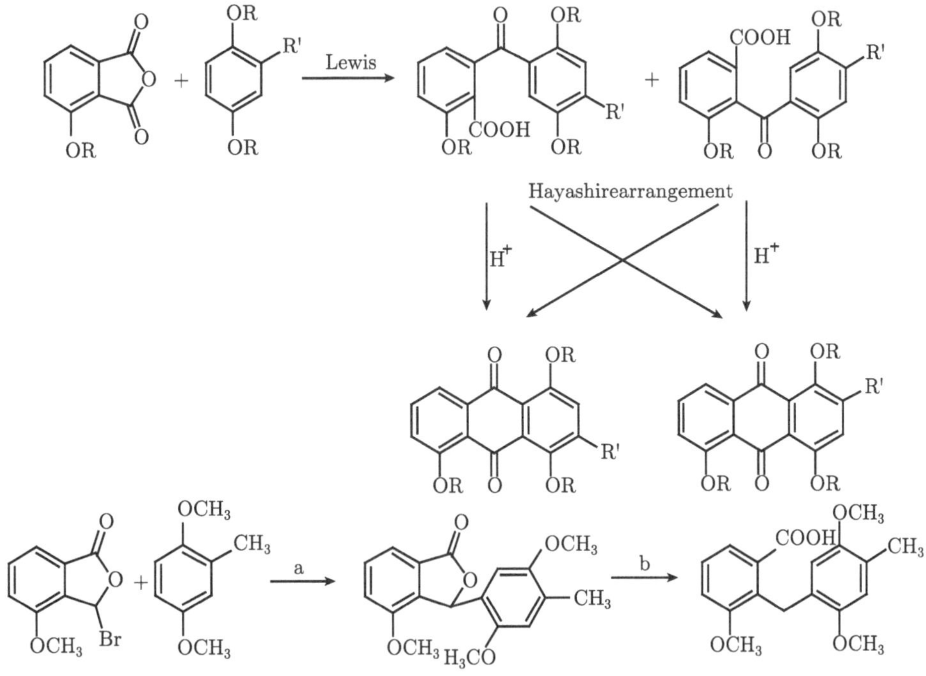

148   **149**   **150**   **151**

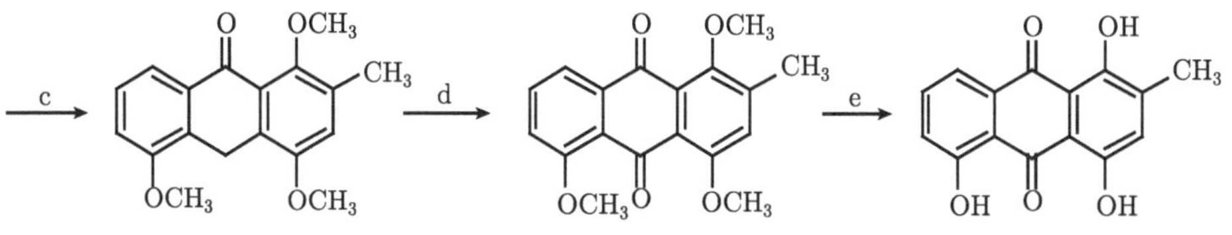

152   **153**   **154(Islandicin)**

Scheme 16-18

a. SnCl$_2$, CH$_2$Cl$_2$; b. Et$_3$SiH, TFA; c. TFAA, TFA; d. CrO$_3$, HOAc; e. AlCl$_3$, C$_6$H$_5$NO$_2$

4.2   Michael addition

The anion, provided by the treatment of 3-cyanophthalic anhydride derivative **155** with t-BuOK, underwent a Michael addition to cyclohex-2-enone derivative (Michael receptor) to afford anthraquinone (Scheme 16-19). The nitrile group in this reaction acted as both activating group and leaving group. Compared with Friedel-Crafts acylation, the advantage of this approach is its high regioselectivity[26].

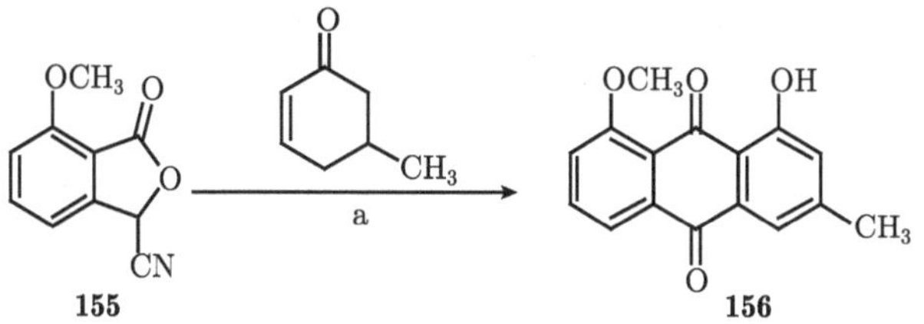

155   **156**

Scheme 16-19

a. t-BuOK, DMSO

4.3   Diels-Alder method

The Diels-Alder approach lacked the regioselectivity of the Friedel-Crafts method. Cano prepared **159** with 3-hydroxy-2H-pyran-2-one **157** and 6-chloronaphthalene-1,4-dione derivative **158** in the presence of lead oxide for 8 days at a 20% yield (Scheme 16-20)[27]. The 3-hydroxyl electron-donating group controls the regioselectivity of this reaction.

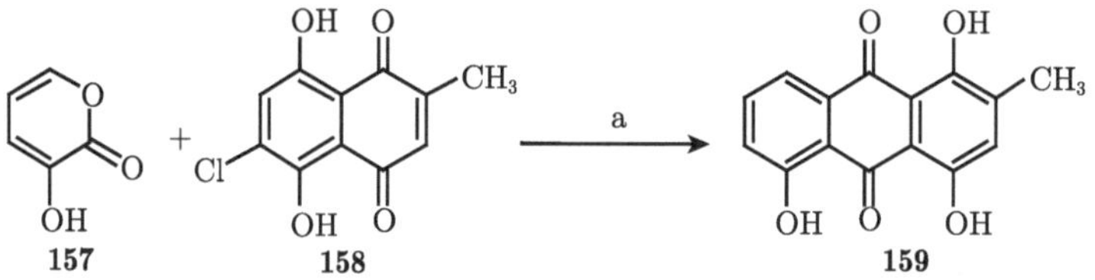

157   **158**   **159**

Scheme 16-20

a. PbO$_2$, CH$_3$CN

4.4   Ortho-metallization of *N,N*-diethylbenzamide[28]

The treatment of N, N-diethyl-3,5-dimethoxybenzamide **160** with t-butyllithium gave a carbanion ortho to amide group, which reacted with aldehyde **161** to give an alcohol intermediate. This intermediate reacted with tosic acid to yield an ester **162**, which was reduced to give **163**. Ring closure of **163** catalyzed by trifluoroacetic acid gave **164**, which was oxidized with CrO$_3$ to give anthraquinone **165**. Treatment of **165** with Py-HCl afforded catenarin **166**. However, treatment of **165** with BBr$_3$ gave a demethylation product erythroglancin **167**, as shown in Scheme 16-21.

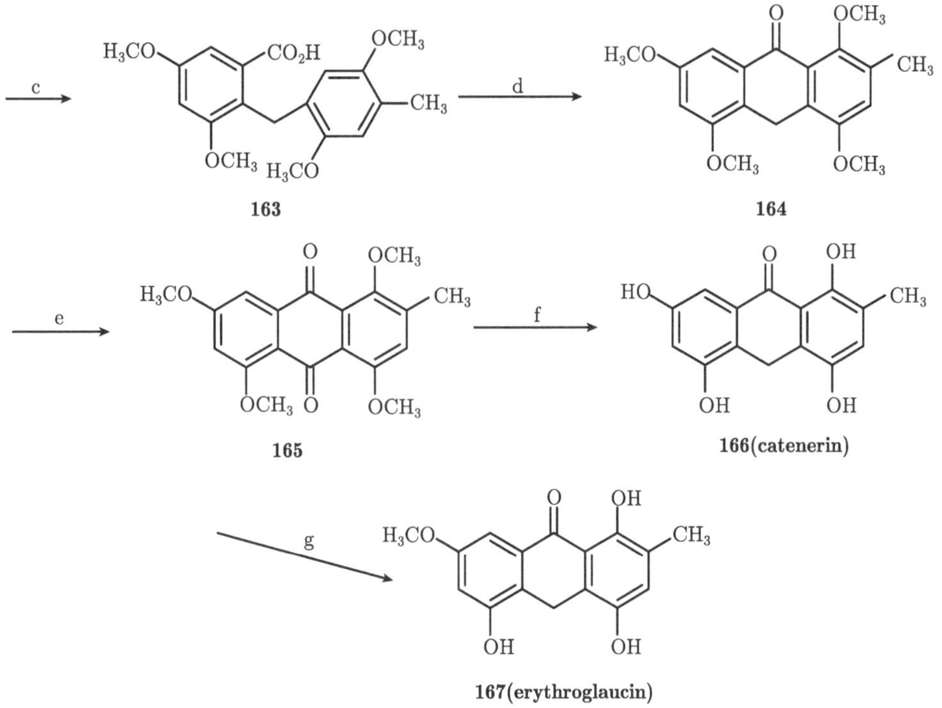

Scheme 16-21

a. BuLi, TMEDA-Et$_2$O; b. TsOH, PhMe; c. Pd/C, H$_2$, HAc; d. TFAA, CHCl$_3$; e. CrO$_3$, HOAc;
f. Py-HCl; g. BBr$_3$, CH$_2$Cl$_2$

Section 5   Lignans

Lignans are a class of natural products that are composed of oxidative condensation of C$_6$ and C$_3$ building blocks (allylphenol and its derivatives, propenylphenol and its derivatives). They are usually dimers. Podophyllotoxin and schizandrin-type lignans are representative compounds of this group.

5.1   Podophyllotoxin

The derivatives of podophyllotoxin, such as VP-16-123 and VM-26M, are anticancer drugs. Therefore, the synthesis of podophyllotoxin has attracted considerable attention. Two approaches are introduced here.

5.1.1   Cascade 1,4-1,2 addition approach[29]

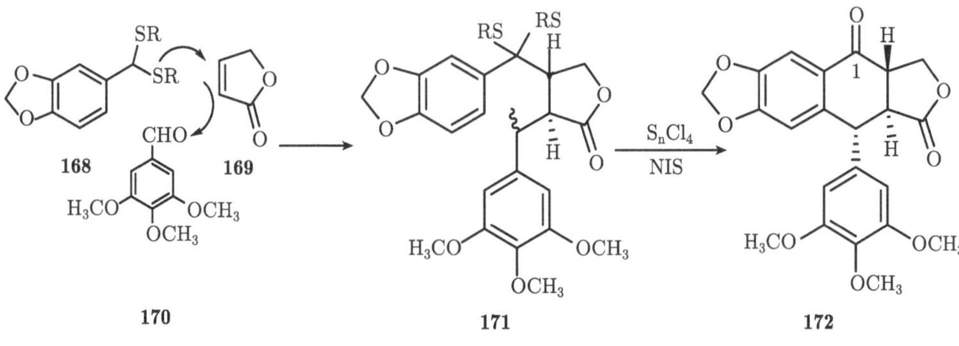

The anion intermediate provided by the Michael addition of the thioacetal anion **168** to the 2-butenolide **169** was trapped by an aryl aldehyde **170** to give **171**. Cyclization with $SnCl_4$ followed by hydrolization with NIS provided ($\pm$)-isopodophyllotoxone **172** in 60% yield or more. Reduction of **171** with Raney Ni followed by cyclization with $SnCl_4$ afforded a methylene on C-1 and gave ($\pm$)-deoxypodophyllotoxin in about 70% yield.

5.1.2   Diels-Alder reaction approach

There are a few reports on asymmetric synthesis of podophyllotoxin and its analogs. In 1996, Bush and Jones[30] finished an asymmetric synthesis of ()-podophyllotoxin in eight steps and in 15% overall yield. The key step is a Diels-Alder reaction of o-quinonoid pyrone **173** with the Feringa's dienophile **174** with high regio-, *endo*- and facial selectivity. The route is shown in Scheme 16-22.

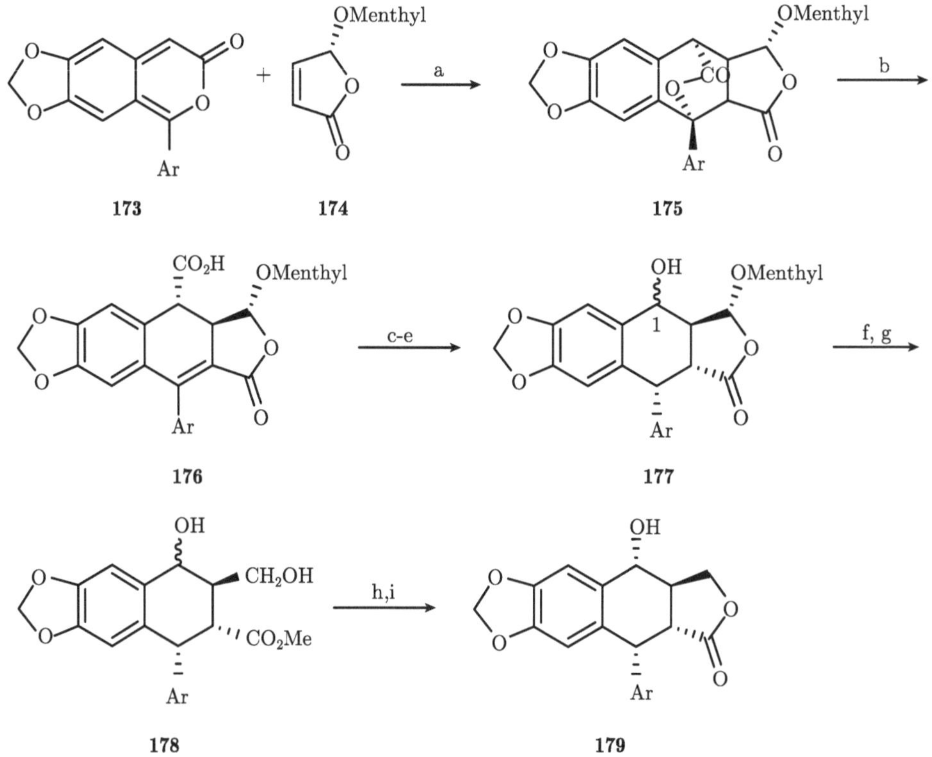

Scheme 16-22

a. 50°C, MeCN, base-washed glassware; b. AcOH, 49°C; c. H_2,10%Pd/C,EtAc; d. $Pd(OAc)_4$,
15 AcOH/THF; e.HCl, dioxane; f. CH_2N_2, Et_2O/MeOH; g.$LiEt_3BH$, THF, -78°C;
h. HCl, THF; i.$ZnCl_2$, THF, molecular sieves

The desired stereochemistries of all the substituents were established during the hydrogenation step of the unsaturated lactone. But two epimers at C-1 were formed after oxidation with $Pb(OAc)_4$. This pair of epimers could be separated and used in the next step. Finally, (-)-podophyllotoxin was obtained with 98% optical purity.

5.2   Schizandrin

Schizandrin-type lignans, which are characterized by their important biological activity and complex stereochemistry, have attracted many organic chemists for their sysnthesis.

Many synthetic routes have been reported. Herein, several representative routes will be introduced.

Takeya[31] synthesized schizandrin-type lignans with keto-condensation. This approach was easy, and the reagents used were inexpensive and commercially available. But this method could be only applied to compounds with symmetric substituents on the aromatic ring, such as wuweizisu C.

Aldehyde **180** and nitroethane were refluxed to form nitro-intermediate **181**, which was reduced with Fe-FeCl$_3$/conc. HCl to form phenylacetone **182**. Reductive coupling with TiCl$_4$-Zn in THF provided diastereoisomeric mixtures **183**. Treatment of the diol **183** with ethyl orthoformate in the presence of benzoic acid afforded a mixture of (Z)-**185** and (E)-**186**. The (E)-isomer could be converted to (Z)-iosomer by irradiation by UV (catalyzed with I$_2$). Catalytic reduction of the isomers with Pt black in AcOH furnished *syn-* and *anti-* dimethyl intermediates, respectively. Finally, oxidative coupling of the *syn-*dimethyl intermediate formed ($\pm$)-wuweizisu C **187**. The *anti-*intermediate could be transformed to the epimer of wuweizisu C with the same reaction.

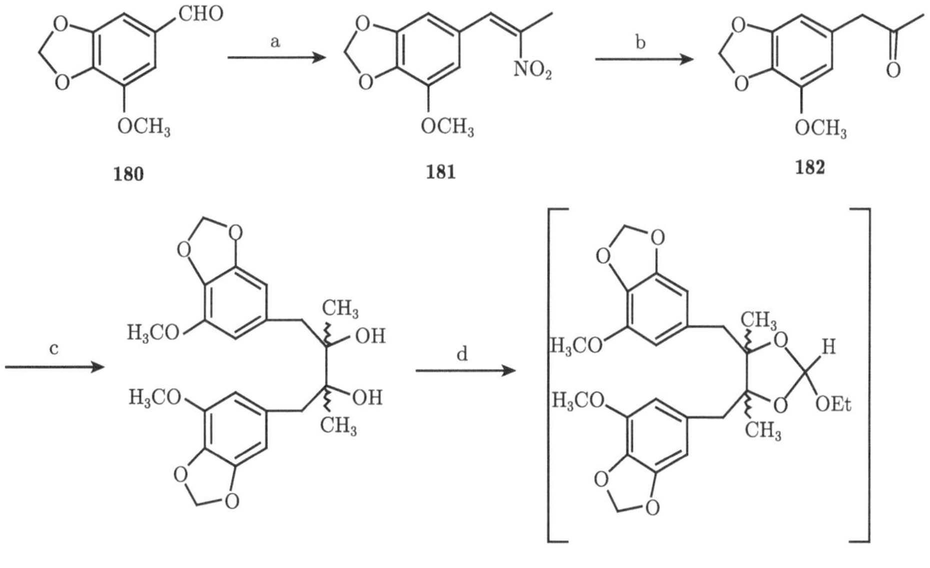

Scheme 16-23

a. EtNO$_2$, benzene, piperidine; b. Fe-FeCl$_3$, HCl; c. TiCl$_4$-Zn, THF;

d. triethyl orthoformate, benzoic acid

Mervic[31] et al. developed a method through which the compounds could be prepared from phenanthrene derivatives in high yield. Also, derivatives with different oxy- and aryl-substituents on the eight-membered ring could be synthesized through this approach. The synthesis of ($\pm$)-kadsurin is such an example.

The bromoaryl aldehyde **189** underwent a Witting reaction in the presence of LiCH$_3$ to afford a mixture of (Z)- and (E)-stilbenes. Irradiation in cyclohexane and THF in the presence of I$_2$ afforded phenanthrene derivative **191**, which was oxidized with OsO$_4$ and sulfur trioxide-pyridine complex to provide ketone **192**. A Grignard reaction with the magnesium derivative of ethyl bromide transformed **192** to a two-carbon-more **193**, which was found to lack the aromatic bromine substituent. Oxidation with lead tetraacetate afforded **194**, which was brominated and subsequently ring-closed with Zn-Ag or Zn-Cu to provide ($\pm$)-dibenzocycloocta-dienedione **195**. **195** was catalytically hydrogenated. The hydrogen only attacked the less steric hindrance face and the carbonyl of the ketone, which was coplanar with the aromatic ring resulting in a *cis*-isomer **196** in which the hydroxyl and two methyls were all α-placed. Methylsulfonylation of **196** with methylsulfonyl chloride in pyridine gave **197**. Reduction with LiAlH$_4$ followed by acetylation gave (±)-kadsurin **198**. (Scheme 16-24)

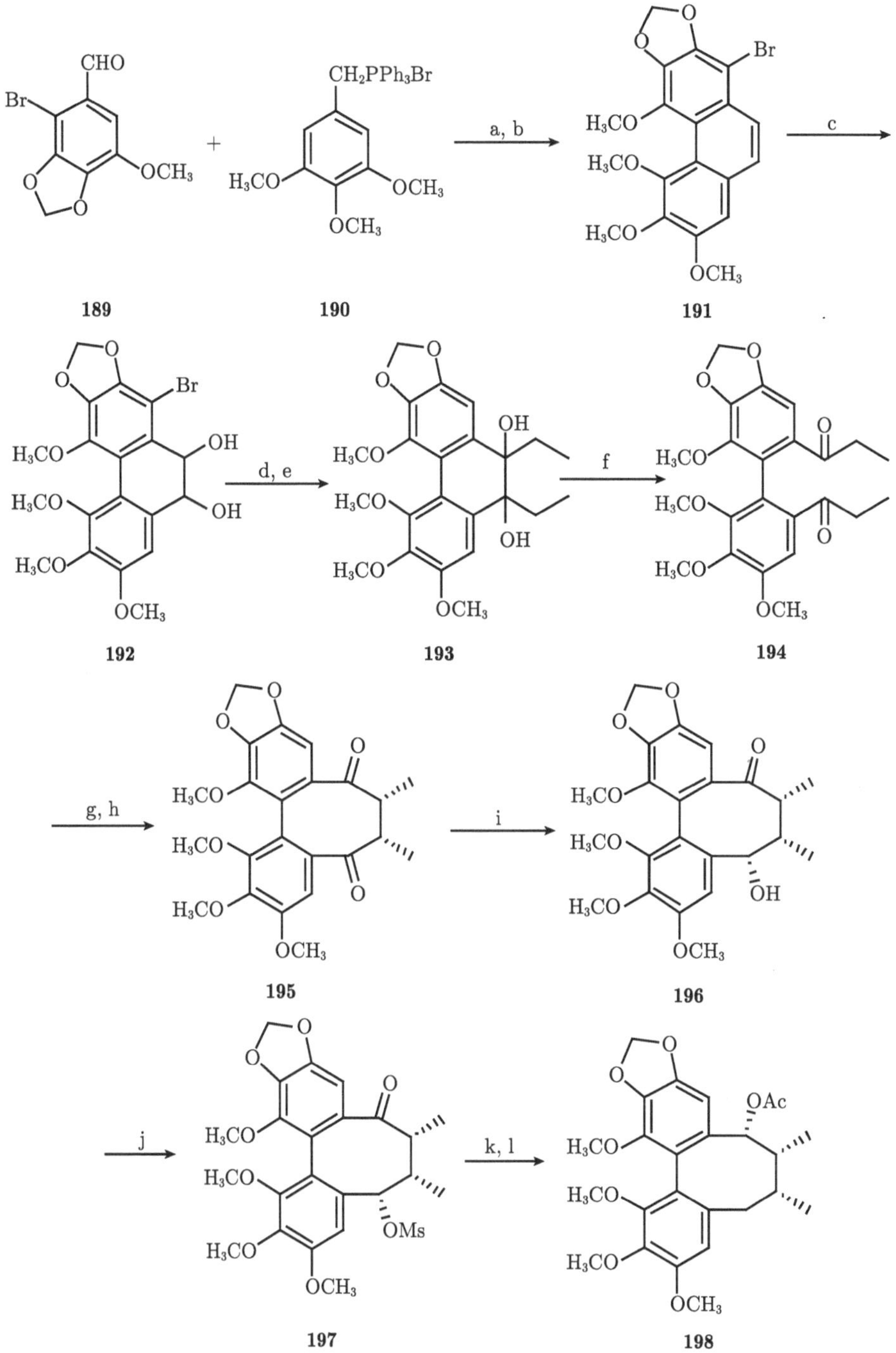

Scheme 16-24

a. LiMe; b.UV, I$_2$, THF, C$_6$H$_{12}$; c. OsO$_4$, Py; d. SO$_3$/Py; e. EtMgBr, PhH; f. Pb(OAc)$_4$, PhH, Py; g. Br$_2$, ether; h.DMSO/DMF, Zn/Agi.Pd/C, AcOH, H$_2$; j. MsCl, Py; k.LAH; l. Ac$_2$O

Tanaka[33] et al. used a enantioselective approach to establish the cyclyoctadiene ring and synthesized a variety of molecules, including wuweizisu C, gomisin J, gomisin N, and γ-schizandrin. The synthesis of wuweizisu C will be introduced as an example. The stereochemistry was achieved through an enantioselective catalytic hydrogenation of unsaturated ester **197** which could control the generation of the other chiral centers. The aromatic group was introduced to the lactone **200** through an aldol condensation followed by elimination. Finally, oxidative coupling with iron perchlorate followed by hydrogenation and subsequent reduction furnished wuweizisu C **203**. (Scheme 16-25).

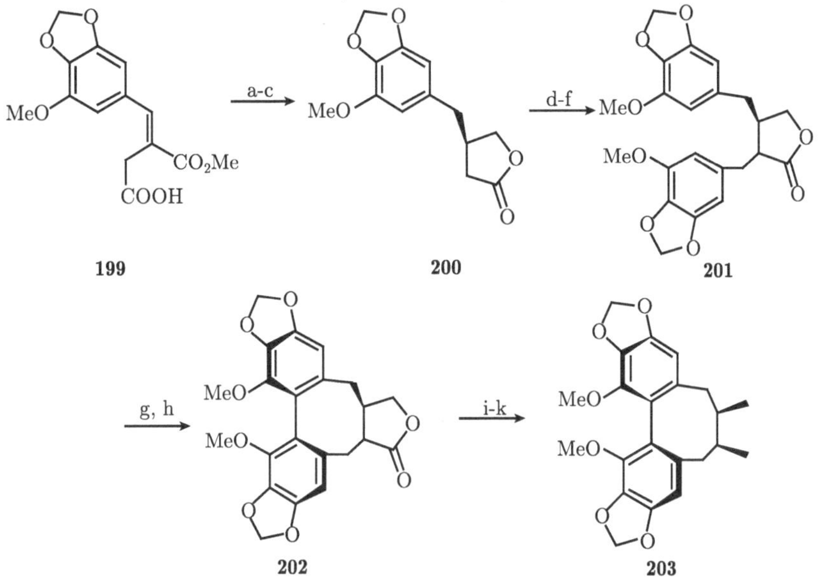

Scheme 16-25

a. (S, S)-MOD-DIOP, H_2, Rh(COD)$_2$BF$_4$; b. Ca(BH$_4$)$_2$, KOH; c. HCl; d. LiN(I-Pr)$_2$, 3-methoxy-4, 5-methylenedioxybenzaldehyde; e. Ac$_2$O, Net$_3$, DMAP; f. DBU; g. Fe(ClO$_4$)$_3$, CF$_3$COOH, CH$_2$Cl$_2$; h. H_2, Pd/C; i. i-BuAlH; j. MeSO$_2$Cl, Net$_3$; k. LiBHEt$_3$

Section 6   Synthesis of Macrolide Antibiotics

The structural and stereochemical complexity of macrolide antibiotics have drawn considerable attention from organic chemists. Synthesis of these compounds has accelerated the innovation of reaction methodology. As a result, the efficiency of stereoselective synthesis of this group of natural products has been altered dramatically. Here we can only show three representative macrolides, oleandomycin, fluvirucin B$_1$, and macrolactin A, as examples. Although all of these compounds belong to the family of macrolide antibiotics, the various strategies applied to construct the macrocycles are worthy of discussion.

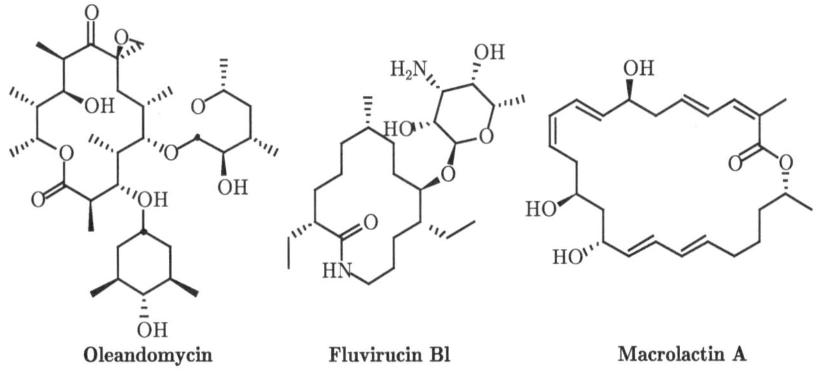

Oleandomycin   Fluvirucin Bl   Macrolactin A

6.1   Oleandomycin

Oleandomycin has similar structure and biological activity as erythromycin. There are many chiral centers in its structure. Since 1990, three syntheses were reported by Tatsuda[34], Paterson[35] and Evans[36], respectively, which gave excellent samples for the stereoselective construction of this class of molecules. Being aware of the contiguity of chiral centers, Evans accomplished the synthesis concisely via a substrate-controlled strategy. Evans was the first one who applied carboximide auxiliaries to asymmetrically synthesize polypropionate-derived adducts. The construction of dipropionate fragments via the aldol reaction of α,β-keto imide **204** with an aldehyde **205** along with the synthesis of the C(1)-C(8) fragment is a very efficient and elegant example. Treatment of **204** with Ti(i-PrO)Cl$_3$ and triethylamine generated chemoselective enolization of the ketone at C(4). This (Z)-enolate selectively reacted with **205** to form **206** (Scheme 16-26).

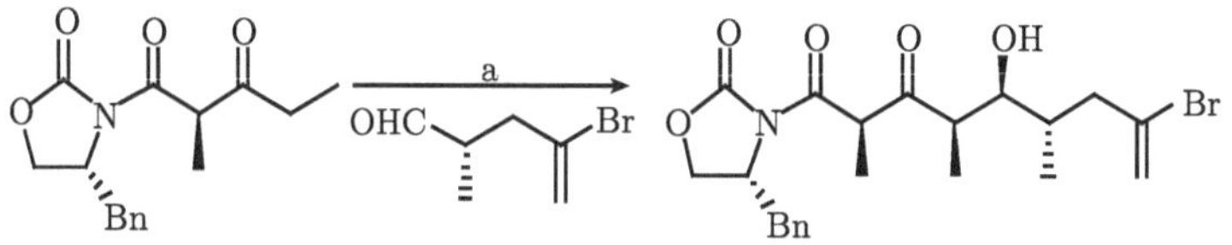

Scheme 16-26

a. Ti(i-OPr)Cl$_3$

For the synthesis of the C(9)-C(14) fragment, the corresponding stereochemical result was achieved through the use of stannous triflate in the enolization reaction. Product **207** is an 83:17 mixture of diastereomers. The adduct was proven to contain the *anti* stereochemical relationship, which was in contrast to the result of the reaction of Ti (IV) enolates. A reduction of **207** with Me$_4$NBH(Oac)$_3$ afforded 1,3-*anti* diol **208** (Scheme 16-27). This reagent is ideal for the diastereoselective directed reduction of β-hydroxy ketones to 1,3-*anti* diol.

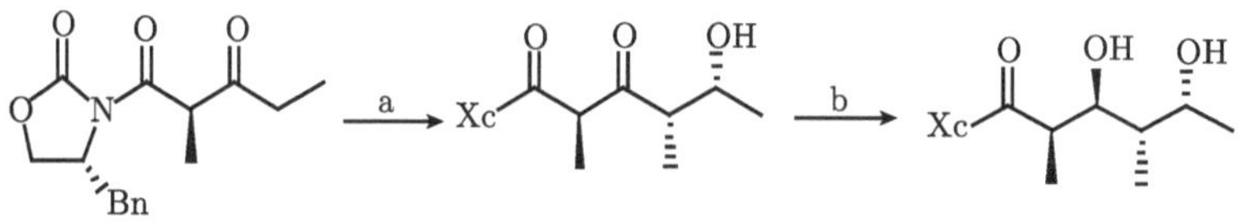

Scheme 16-27

a. Sn(OTf)$_2$, MeCHO; b. Me$_4$NBH(OAc)$_3$

209 was prepared from **206** through several steps. The convergent coupling of two highly functionalized fragments **209** and **210** is a key step in the Evans synthesis. Pd(0)-catalysis was innovatively used in this step. Reaction of vinylstannane **209** with acid chloride **210** [Pd$_2$(dba)$_3$, i-Pr$_2$NEt, C$_6$H$_6$] gave unsaturated ketone **211** in 85% yield. With the C(1)-C(14) fragment in hand, Evans chose the 9-(S) hydroxyl to stereochemically control the preparation of the C(8) epoxide through a Sharpless epoxidation. Thus far, the acyclic fragment **212**, which contains all of the requisite chiral centers, was obtained in only 11 steps from **204**.

Finally, Evans applied Yamaguchi's method[33]. Epoxy seco acid **213** underwent macrocyclization to provide **214** (Scheme 16-29), which subsequently underwent deprotonation and oxidation to afford synthetic oleandolide.

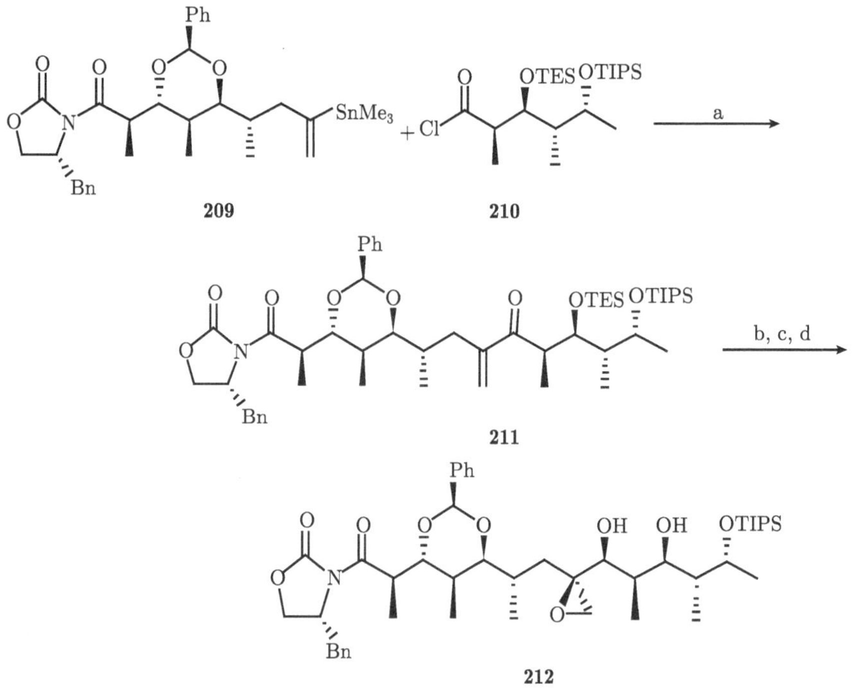

Scheme 16-28

a. Pd(dba)$_3$, i-Pr$_2$Net; b. HF.Py; c. Zn(BH$_4$)$_2$; d. VO(acac)$_2$, t−BuOOH

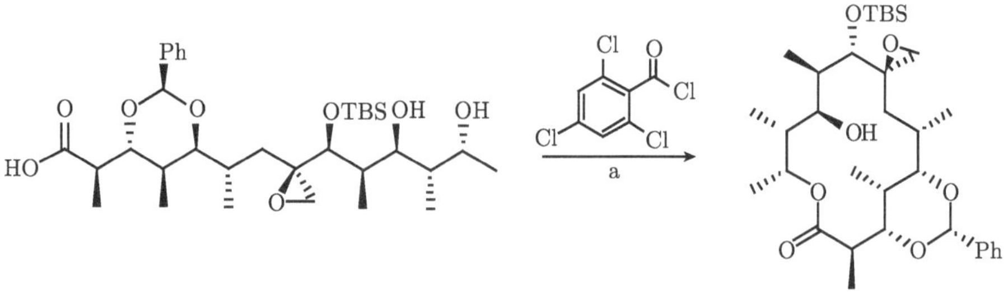

Scheme 16-29

a. DMAP, i−Pr$_2$NEt, CH$_2$Cl$_2$

6.2   Fluvirucin B$_1$

In contrast to oleandolide, fluviricin B$_1$ has fewer chiral centers which are remote from each other and thus it is difficult to use substrate-controlled methodologies. Moreover, the deceptive simplicity of this macrolide and the highly flexible ring make marcrocyclization difficult. Two research groups reported their synthesis, and both applied transition-metal catalysis to construct the remote chiral centers and form the 14-membered ring[38].

Hoveyda used **215** as the starting material, which was prepared through Sharpless kinetic resolution of the racemate catalyzed by Ti(IV). Diastereoselective ethylmagnesiation of **209** catalyzed by Cp$_2$ZrCl$_2$, as developed by Hoveyda, formed **216** (Scheme 16-30), which was subsequently reacted in situ with excess N-tosyl aziridine to afford fully functionalized C(6)-C(13) fragments in 97:3 diastereoselectivity.

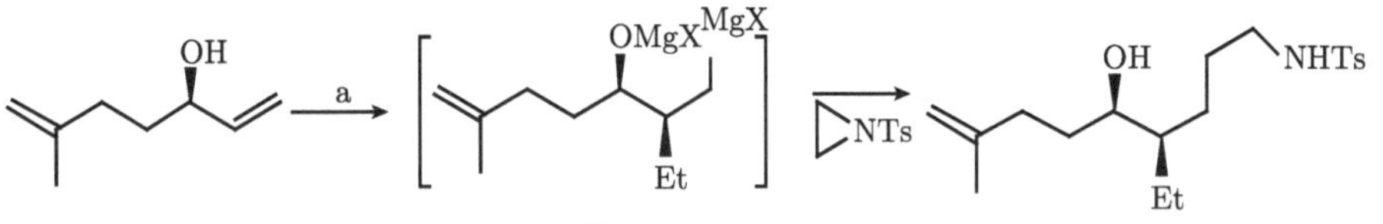

Scheme 16-30

a. EtMgBr, Cp$_2$ZrCl$_2$

The synthesis of the other fragments also uses the Haveyda catalytic, enantioselective ethylmagnesiation reaction. The substrate was 3,5-dihydrofuran. Treatment of ethylmagnesium bromide with excess of **218** and catalytic **219** formed **220** in 99% ee. Hydromagnesiation of alcohol **220** with n-PrMgBr and 3 mol% Cp$_2$ZrCl$_2$ provided **221**, which then underwent in situ coupling with vinyl bromide [3 mol% (Ph$_3$P)$_2$NiCl$_2$] to afford **222** in 72% overall yield (Scheme 16-31).

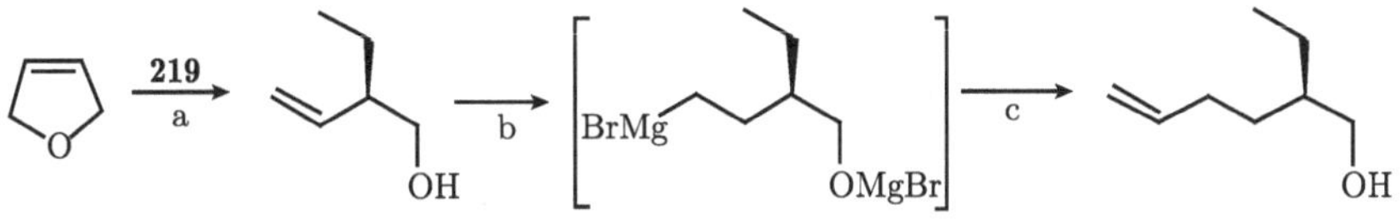

Scheme 16-31

a. EtMgBr; b. n−PrMgBr, 3mol% Cp$_2$TiCl$_2$; c.vinyl bromide, 3mol% (Ph$_3$P)$_2$NiCl$_2$; **219** = (R,R)thylene-1,
2-bis(η^5-4,5,6,7-tetra-hydro-1-indenyl)titanium® -1,1′-binaphth-2,2′diolete

Oxidation of **222** provided the corresponding acid, which was reacted with fragment **217** to give **223**. Glycosylation of **223** followed by ring-closing metathesis in the presence of Schrock catalyst provided **225** in 92% yield. Glycosylation after the ring-closing reaction of **223** did not produce the desired product. Catalytic hydrogenation of **225** furnished the desired product in 80% yield and 98% de, which was deprotonated to obtain fluvirucin B$_1$(Scheme 16-32).

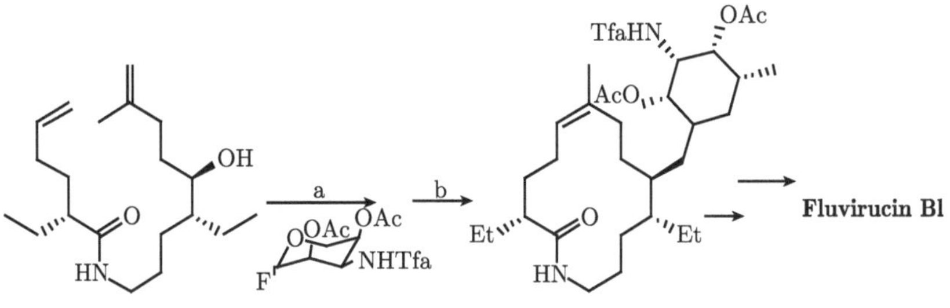

Scheme 16-32

a. SnCl$_2$, AgClO$_4$, b. **224**, PhH; **224**=Mo(CHCMe$_2$Ph)[N(2,6-(i−Pr)$_2$C$_6$H$_3$)][OCMe(CF$_3$)$_2$]$_2$

6.3   Macrolatin A

Macrolatin A is a polyene macrolide. Biological study showed that it had a prophylactic effect on T-lymphoblast cells against HIV replication. This novel macrolide is not readily available from natural sources, so its biological study relies on a laboratory synthesis. The unusual synthetic challenge has attracted a number of groups to develop efficient synthetic approaches. Three groups[39]have independently reported synthesis of the macrolatin A core structure. Two of them documented the total synthesis of the target molecule.

The first total synthesis of macrolatin A was reported by Smith. The researchers conducted a careful study on the construction of C(9)-C(10) bond through both possible per-
mutations involving vinyl stannane and iodide reaction in Pd(0)-catalyzed cross-coupling
reactions. For the synthesis of fragments containing C(7), C (13) and C (15) chiral centers,
enantioselective allylation of propynal with both (R)- and (S)-diisopinylcampheyl allyl bo-
ranes was used to give alkynes **226** and **232** in 90% ee. Protection of the secondary alcohol
and ozonolytic cleavage of the terminal alkene in **226** provided aldehyde **227**, which was
treated with $CrCl_2$, LiI, and $Bu_3SnCHBr_2$ to generate **229**. This vinyl stannane was used
as the starting material for a series of chemoselective catalytic transformations that afforded
231, including cross-coupling with (Z)-3-iodopropenoic acid **229** and **228**. The carboxyl
group was unprotected when the alkyne group underwent hydrostannylation (Scheme 16-33).

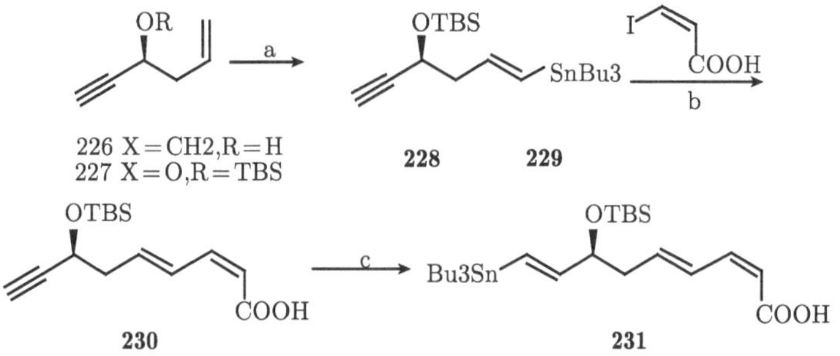

226 X=CH2,R=H
227 X=O,R=TBS   **228**   **229**

230   **231**

Scheme 16-33

a. $Bu_3SnCHBr_2$, $CrCl_2$, LiI, DMF; b. $PdCl_2(MeCN)$;c. $PdCl_2(PPh_3)_2$, Bu_3SnH

The enantiomers of **226** and **232** were used as the starting materials for the synthesis
of the other fragments. They were converted to alkyne **233**, which was chemoselectively
hydrostannylated to afford vinyl stannane **234**. **234** underwent tin-halogen exchange to
give **235**. Coupling of this vinylic iodide with vinyl stannane **236** provided the C(9)-C(24)
fragment **237**, which was oxidized to provide an aldehyde. The aldehyde was transformed
to vinyl iodide **238** by utilizing the Stork reagent (Scheme 16-34). This fragment contains
the correct (E)-double bond and chiral centers of the C(10)-C(24) region of macrolactin A.

In Smith's synthesis, the final cyclization reaction was applied between the C(9)-C(10)
bond. They used two precursors: the C(9) stannane/C(10) iodide and the C(10) stan-
nane/C(9) iodide. The protecting group was optimized (R=Et) in the ring closure reaction.
Protected macrolactin A was obtained in 56% yield, which was desilylated to generate synthetic macrolactin A.

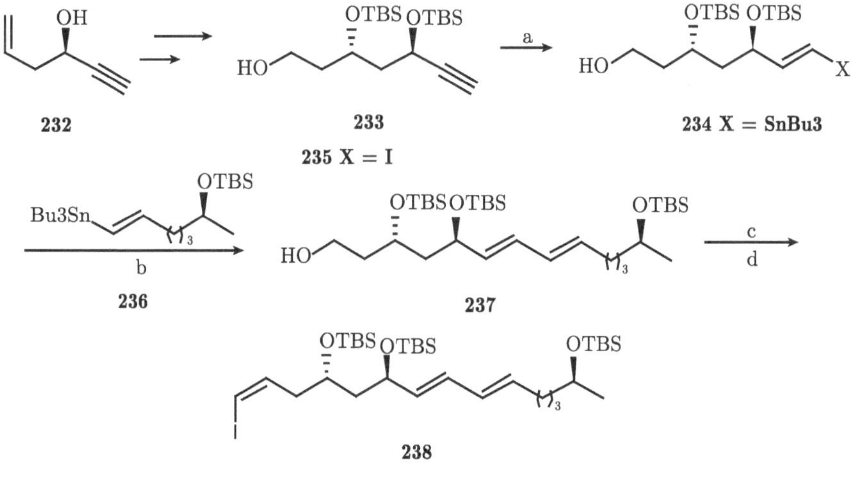

232   **233**   **234** X = SnBu3
                 235 X = I

236   **237**

238

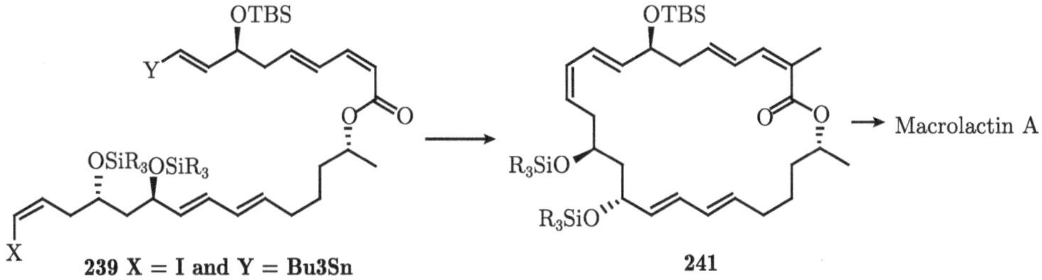

Scheme 16-34

a. PdCl$_2$(PPh$_3$)$_2$, n-Bu$_3$SnH; b.PdCl$_2$(MeCN); c. Dess-Martin; d. (Ph$_3$PCH$_2$I)I, NaHMDS

Bibliography

[1] Xu R S, *Natural Product Chemistry*, Beijing: Science Press, 1993.

[2] Wu Y L, Yao Z J, *Modern Synthetic Organic Chemistry*, Beijing: Science Press, 2001.

[3] K.C. Nicolaou and E. J. Sorensen, *Classics in Total Synthesis,* New York: VCH., 1996.

[4] Karl J. Hale, *The Chemical Synthesis of Natural Products*, Sheffield: Academic Press Ltd., 2000.

References

[1] E.J. Corey, W-G. Su, l985, *Tetrahedron Lett.*, 26, 28l. (b)J. Adams, B.J Fitzsimmons, Y Girand, Y.Leblanc, J.F. Evans, J. Rokach, 1985, *J Am. Chem. Soc.*, l07, 464.

[2] (a) H. Nagaoda, H. Miyaoka, T Miyaoshi, Y Yamada, 1986, *J Am. Chem. Soc.*, 108, 5019. (b) M.Suzuki, Y Morida, A. Yanagisawa, R. Noyori, 1986, *J Am. Chem. Soc.*, I08, 502l. (c) M. Suzaki, Y. Morida, A. Yanagisawa, B. Baker, P.J. Scheuer, R. Noyori, 1988, *J Org. Chem.*, 53, 286.

[3] (a) Kagan, H.B. *Bull. Soc. Chim. Fr.* l988, 846, (b) Scott. J. W *Top. Stereochem.* l989, 19, 209, (c) Crosby, J. *Tetrahedron* 1991, 47, 4789, (d) Akutagawa, S. In *Organic Synthesis in Japan: Past. Present. and Future*, Tokyo Kagaku Dozin, Tokyo. 1992, p. 75.

[4] (a) J. E. Toth, P R. Hamann, and P L. Fuchs, l988, *J Org Chem.*, 53, 4694. (b) K. A. Parker and D.Fokas, 1992, *J Am. Chem. Soc.*, ll4, 9688. (c) C. Y Hong, N. Kado. and L. E. Oveman 1993, *J Am.Chem. Soc.* 115, l l028. (d) D. Trauner, J. W. Bats, A. WenerY, and J. Mu1zer l998, *J Org Chem.*, 62, 5250.

[5] R. B. Woodward, M. P Cava, W D. Ollis, A. Hunger, H. U. Daeniker and K. Schenker, 1954, *J. Am. Chem. Soc.*, 76, 4749.

[6] (a) G. Stork, l992, reported at the Ischia Advanced School of Organic Chemita Ischia Porto, Italy, 2l September, (b) P Magnus, M. Giles, R. Bonnert, C.S. Kim, L. McQuire, A. Merritt, and N. Vicker, l992, *J Am. Chem. Soc.*, 114, 4403, (c) P. Magnus, M. Giles, R. Bonnert, G. Johnson, L. MeQuir, M.Deluca, A. Metritt, C.S. Kim, and N. Vicker, l993, *J Am. Chem. Soc.*, l 15, 8l l6, (d) M.E. Kuehne and F. Xu, l993, *J Org. Chem.*, 58, 7490, (e) S.D. knight, L.E. Overman, and G.ariaudeau, 1993, *J Am. Chem. Soc.*, l 15, 9293, (f) V.H. Rawal and S. Iwasa, 1994, *J. Org Chem.*, 59, 2685, (g) S.F. Martin, C. W. Clark, M. Ito, and M. Mortimore, 1996, *J Am. Chem. Soc.*, 118, 9804.

[7] Shanghai No. 5 Pharmaceutical Plant, et al., l978, *Scinitia Sinaca* 21, 87.

[8] (a) D.I. Comins, H. Hong and G. Jianhua, l994, *Tetrahedron Lett.*, 35, 533l. (b) D.L. Comins,

M.F. Baevsky and H. Hong, 1992, *J Am. Chem. Soc.*, 1 14, 10971.

[9]   Ejima, X., et a1., 1990, *J Chem. Soc. Perkin I* 27.

[10]   Woodward, R. B., et al., 1958, *Tetrahedron*, 2, 1.

[11]   Corey, E.J. et al., 1980, *J. Am. Chem. Soc*,, 102, 6603.

[12]   Xu R S. *Natural Product Chemistry*, Beijing: Science Press, 1993, p. 220.

[13]   Bald, E., et al., 1975 *Chem. Lett.*, 1, 163.

[14]   Noyori, R., *Asymmetric Catalysis in Organic Synthesis*, John Wiley & Sons Inc., 1994, Chapter
       3, p. 96.

[15]   X.X. Xu, J. Zhu, D.Z. Huang, W.S. Zhou, 1986 *Tetrahedron* 42, 819.

[16]   J. Lee and J.K. Synder 1989, *J Am. Chem. Soc.* 1 11, 1522.

[17]   Sher, F.T.; Berchtold, G. A., 1977, *J. Org. Chem.* 42, 256.

[18]   Yang, D., Ye X.-Y., Xu, M., 2000, *J. Org. Chem.* 65, 2208.

[19]   van Tamelen, E. E., Demers, J. P., Taylor, E. G, Koller, K., 1980, *J. Am. Chem. Soc.* 101,
       5424.

[20]   Crisp, G. T., Meyer A. G., 1992, *J. Org. Chem.* 57, 6972.

[21]   Xu R S et al. 1979, *Acta Chimica Sinica,* 37, 289.

[22]   (a) Baker, W., 1933, *J. Chem. Soc.*, 1381. (b) Mahal, H.S, , 1934, *J. Chem. Soc.*, 1767.

[23]   (a) Alger, J., 1934, *Proc. R.. Irish Acad.*, 42B, 1. (b)Oyamnada, T, 1935, *Bull. Chem.
       Soc.Japan*, 10, 182.

[24]   Joshi, P.C. et a1., 1934, *J. Chem. Soc.* 513.

[25]   K.S. Kim, 1979, *Tetrahedron Lett.* 4, 331.

[26]   G. A. Kraus et a1., 1983, *J. Org. Chem.* 48, 3439.

[27]   P Cano , 1983, *J Org Chem.* 48. 5373.

[28]   S. O. de Silva ot al., 1979, *J Org. Chem.* 44, 4802.

[29]   Zieger, F.E. , 1978, *J. Org. Chem.* 43, 985.

[30]   E. J. Bush and D.W Jones, 1996, *J. Chem. Soc. Perkin* 151.

[31]   Takeya, T. , 1985, *Chem. Pharm. Bull.* 33, 3599.

[32]   (a) Marvic, M. , 1977, *J. Am. Chem. Soc.* 99, 7673. (b) Ghera, E. , 1978, *J. Chem. Soc.
       Commun.* 480.

[33]   (a) M. Tanaka, H. Mitsubashi, M. Maruno, and T. Wakamatsu, 1994, *Tehahedron Lett*, 35,
       3733, (b) M. Tanaka, T. Ohshima, H. Mitsuhashi, M. Maruno, and T. Wakamatsu, 1995,
       Tetrahedron, 51, 11693.

[34]   K Tatsuda, T. Ishiyama, S. Takima. Y Koguchi. And H. Gunji, 1990, *Tetrahedron Lett*, 31,
       709.

[35]   I. Paterson, R.D. Norcross, R..A. War, P. Romea, and M.A. Lister, 1994, *J. Am. Chem. Soc.*
       116, 11287.

[36]   D.A. Evans, A.S. Kim, R. Metternich, and V.J. Novachk, 1998, *J. Am.Chem. Soc.* 120, 5921.

[37]   J. Inanaga, K. Hirata, H. Saeki, T Katsuki, and M. Yamagushi, 1979, *Bull Chem. Soc.* 52,
       1989.

[38]   (a) A.F. Houri, Z. Xu, D.A. Cogan, and A.H. Hoveyda, 1995, *J. Am. Chem. Soc.* 117, 2943,
       (b)Z. Xu, C.W. Johannes, S.S. Salman, and A.H. Hoveyda, 1996, *J. Am.Chem.Soc.* 1 18, 10926,
       (c)Z. Xu, Z.W Johannes, A.F. Houri, D.S. La, D.A. Cogan, G.E. Hofilena, and A.H. Hoveyda,
       1997, *J. Am. Chem. Soc.* 119, 1032, (d) B.M. Trost, M.A. Ceschi, and B. Konig, 1997, *Angew*

Chem., Int Ed Engl, 36, 1486.

[39] (a) G. Pattenden and R.J. Boyce, 1996, *Tetrahedron Lett*, 37, 3501. (b)Y. Kim, R.A. Singer and E.M. Carreira, 1998, *Angew Chem., Int Ed Engl.*, 37, 1261. (c) A.B. Smith, and G.R. Ott, 1998, *J. Am. Chem. Soc.*, 120, 3935.

Wen-Hu Duan

Index